CHEMISTRY: A UNIFIED APPROACH

CHEMISTRY: A UNIFIED APPROACH

J. W. BUTTLE, B.Sc., A.R.I.C.
University of London, Goldsmiths' College, London

D. J. DANIELS, B.Sc., Ph.D., F.R.I.C.
University of Bath

P. J. BECKETT, B.Sc., Ph.D., A.R.I.C.
North Kesteven Grammar School

LONDON
BUTTERWORTHS

ENGLAND:	BUTTERWORTH & CO. (PUBLISHERS) LTD. LONDON: 88 Kingsway, W.C.2
AUSTRALIA:	BUTTERWORTH & CO. (AUSTRALIA) LTD. SYDNEY: 20 Loftus Street MELBOURNE: 343 Little Collins Street BRISBANE: 240 Queen Street
CANADA:	BUTTERWORTH & CO. (CANADA) LTD. TORONTO: 14 Curity Avenue, 374
NEW ZEALAND:	BUTTERWORTH & CO. (NEW ZEALAND) LTD. WELLINGTON: 49/51 Ballance Street AUCKLAND: 35 High Street
SOUTH AFRICA:	BUTTERWORTH & CO. (SOUTH AFRICA) LTD DURBAN: 33/35 Beach Grove

First published 1966
Second Edition 1970

Suggested U.D.C. number: 54(075)

ISBN O 408 70012 2

Printed in Great Britain by Page Bros. (Norwich) Ltd.

CONTENTS

v

PREFACE TO THE SECOND EDITION

We have taken the opportunity in this new edition of incorporating the rationalized units based on the m.k.s. system and published as the Système International (SI), although we have retained references, where appropriate, to earlier units. We have also added a concluding chapter on the growth and structure of the chemical industry to provide a background to the industrial applications mentioned throughout the book.

Minor changes include an increase in the number of questions, to provide a wider range and variety, and an up-dating of industrial statistics.

P. J. Beckett
J. W. Buttle
D. J. Daniels

PREFACE TO THE FIRST EDITION

We believe that, for too long, the teaching of Chemistry has been handicapped by a forbidding mixture of fact, recipe and mysticism. This has been due, in large measure, to the content of the examination syllabus, with its emphasis on descriptive aspects and uncomprehending memorizing; the teacher, aware of his responsibilities to his students, has thus felt obliged to follow this pattern and provide his pupils with 'all the facts'. There is, however, no need for the textbook to emulate the teacher: rather, we think, it should extend, stimulate and provoke. Perhaps then the Science student will be encouraged to make fuller use of his critical faculties and be more prepared to participate in discussion.

This particular book is based upon our own A level Chemistry courses, but we think it should also prove of value to Open Scholarship candidates and for University students up to the level of a General degree. Much of the material is also relevant to the industrially-biased O.N.C. courses.

It has been our intention to provide the student with a book that is complementary to his laboratory notes. We have therefore omitted the practical detail which could obscure the main issues involved and render the book unwieldy, although a chapter has been devoted to experimental techniques, to indicate the type of evidence upon which chemical theories may be based and reaction mechanisms formulated. Principles rather than uncorrelated fact have been emphasized throughout the book, in the belief that they are not always so difficult to understand as their belated entry into present examination syllabuses might suggest. At the same time, we have tried to introduce modern knowledge and ideas in a manner compatible with present examination requirements without, we hope, being too precise in our approach and terminology. It would be unwise to suggest that a chemical formula always gives a wholly accurate impression or that a chemical equation conveys a full account of what actually takes place. We have endeavoured to give those representations that seem most relevant to the context. The teacher can, in fact, play a very useful role by discussing, from time to time, the limitations inherent in the symbolism of the subject. The student should not be encouraged to take a definition too precisely or to believe, for example, that chemical bonding can, at the present time, be completely elucidated in terms of electrons or electron clouds. (The electron itself illustrates perfectly the dilemma that awaits the chemist who insists on the 'cut and dried' approach.) In short, the student should be made fully aware that Chemistry, like the other sciences, is imperfectly understood and is in a constant state of flux, with theories being rendered obsolete, new techniques being devised and the frontiers of the subject being continually extended.

The book is really in three sections. In the first Part, the physical aspect of the subject is developed in order to make the rest of the book more meaningful. We make no apology for discussing free energy and entropy, nor for introducing such things as s, p, d and f orbitals. The second Part is a comparative review of the chemical elements from the point of view of the Periodic Table. Because so much chemistry is the chemistry of the non-metals and anions, we have progressed through the Periodic Classification from right to left. We have found it convenient to deal with the elements of Groups 1, 2 and 3 as a whole and to regard the Transitional Metals as members of the d block of elements rather than as members of undefined subgroups of the representative elements.

The last Part of the book deals with Organic Chemistry. Here the emphasis is on the functional group and the characteristic property. We have not found it necessary to perpetuate the division between aliphatic and aromatic compounds; instead, the aromatic nucleus has been regarded as a modifying influence upon the functional group. Because we feel that they are a rationalizing influence in Chemistry, we have used reaction mechanisms widely.

We have tried to provide a broad view of Chemistry, in the hope that the student will get things better in perspective. For instance, we have discussed topics such as the chemistry of inheritance, respiration and photosynthesis, the lanthanides and actinides and various geological aspects of the distribution of the elements, as well as covering the more important industrial processes. Throughout the book the student has been introduced to systematic nomenclature, and at the end of each Chapter are included selections of questions which we hope will prove helpful to him. An Appendix gives a collection of relevant questions from past examination papers, and we should like to acknowledge the help and co-operation of the various Examination Boards in permitting their questions to be reproduced here. We are also very grateful to the publishers for their advice and assistance.

P. J. Beckett
J. W. Buttle
D. J. Daniels

CHAPTER 1

THE DEVELOPMENT OF CHEMISTRY

Homo sapiens is believed to have appeared on this planet about a million years ago. He is distinguished from other animals by his ability to make tools for himself, and here there are signs of inventiveness and the beginnings of technology. Yet nine-tenths of all the scientists who ever lived are alive at the present time and the publication of scientific papers has been increasing at an exponential rate (*Figure 1.1*).

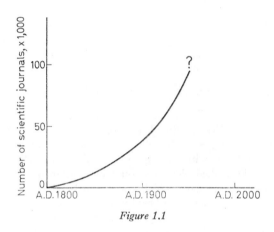

Figure 1.1

There are possibly several reasons for the relatively late development of Science. Early man was preoccupied with self-preservation, with hunting and with being hunted. Settlement of tribes in river valleys, stratification of occupation and emphasis on farming rather than hunting were essential to provide the climate in which ideas could germinate, take root and flourish. It is no accident that we find evidence of an impressive technology in pottery, metals and dyes by the River Nile, nor that the Babylonians, living between the Tigris and the Euphrates, carried out a remarkable number of astronomical observations and incorporated them into the study of astrology.

Much of the achievement of the Egyptians was the result of practical necessity. With the Greeks, particularly from 400 B.C. to about A.D. 100, contemplation and speculation blossomed at the expense of application. The slave system then in vogue could hardly be expected to commend the virtues of physical effort. Religion also

1

played its part. The works of the gods were believed to be beautiful and perfect—it was clear that the physical world was impure, and so mathematics was favoured rather than experimentation with material things. Whatever the reasons, there developed a period of intellectual activity that separated the Greeks from all previous civilizations.

It would be wrong to suggest that Greek science was entirely without experimental foundation. Aristotle was apparently a brilliant practical biologist; furthermore, he so systematized biology that his methods of classification survived until

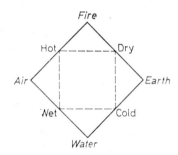

Figure 1.2. The 'four elements'

comparatively recent times. But theories often had their roots in mystical and spiritual notions. Empedocles had the idea of conflict between Love and Strife very much in mind when he suggested that all things were made up of Four Elements, Fire, Earth, Water and Air, with Fire opposed to Water but allied to Air, and so on (see *Figure 1.2*). The allied elements could be combined together to give the qualities Wet, Hot, Dry and Cold which could also be allied or opposed to each other: for example, Hot was allied to Dry but opposed to Cold.

This theory of Empedocles was not without merit, and one can sympathize with several ideas implicit in it. Thus, the three states of matter correspond to Earth, Water and Air, whilst Fire is the agency by which one can be converted into the other. Aristotle adopted this theory and added a fifth element, 'quintessence', which was supposed to have even greater powers of transmutation. In this form, the theory survived for two thousand years and served to bring some unity at least to the study of matter.

The Greeks also contemplated the physical as distinct from the chemical nature of matter: the Atomists, notable among whom was Democritus, argued that matter was made up of indivisible particles or *atoms*, whilst their opponents claimed that matter was continuous and could be subdivided indefinitely. The Continuists had the support of Aristotle, and so it was the latter view that largely prevailed, although there was nothing that we would regard as evidence to support either alternative.

Eventually the influence of Greece waned and the arena was dominated until the Fifth Century by Rome. The Roman temperament appeared more practical and militant than the Greek, and the ecclesiastics of the growing Christian Church were

2

not always inclined to look with favour upon those who challenged authority and dogma with original and imaginative thought. So intellectual stagnation set in and the power of the Church was such that, long after the Roman Empire had decayed, barely a ripple disturbed the calm of conformity.

The historian refers to the period from the Fifth to the Thirteenth Centuries as the Dark Ages. Parts of Southern Europe were occupied during this time by the Arabs, but it would be less than just to equate their arrival upon the scene with the onset of twilight. In fact, these invaders from the East brought with them much knowledge from the Orient and established in Europe the practice of Alchemy. Two incentives to alchemical investigation were the 'elixir of life', that potion which would be capable of conferring everlasting life upon the consumer, and the 'philosopher's stone', which would possess the ability to change all base metals into gold. Although the search for these proved abortive, various chemical skills were acquired in the process, but it must be admitted that they were often used to deceive the gullible and exploit the wealthy. Alchemy was not entirely without its philosophers: theories, many of them appearing very fanciful in retrospect, were advanced to explain certain phenomena and direct the course of investigation. For instance, it was supposed that all metals were compounded of sulphur and mercury —if these were not very pure, then a base metal such as lead would be formed, whereas gold or silver would be the product of much greater purity. If the mercury and sulphur were of exceptional purity, then it was believed that the philosopher's stone would emerge.

Throughout the Middle Ages, contact with Ancient Greece was slight and was confined to a few classical texts such as those of Aristotle on logic. In the Thirteenth Century, however, interest in the past awakened and a large number of Greek works were translated and made available to scholars. The climate at once became favourable to the artist and creator. It was a time of *Renaissance*, of rebirth of the glories of the past. Idealism was the currency and the cultured man of many parts the product. Inevitably the position of the Church was questioned but in no sense was the new movement atheistic. In Britain a Franciscan, Roger Bacon, preached the gospel of experimental method, not in opposition to the Bible but as a natural complement of religion.

It has been suggested that Renaissance painters, by closely observing nature, made man more aware of his environment. The Sixteenth Century certainly saw considerable activity in the field of astronomy, culminating in Kepler's laws, which brought some sort of order to astronomical observation, and in the revision by Copernicus of the Ptolemaic solar system. Copernicus suggested that the sun, and not the earth, was the centre of the solar system; this modification provided a simpler picture, without the need for planets to describe epicycles as they rotated around the earth (*Figure 1.3*).

The Seventeenth Century is one of great importance to the physical sciences. As the century opened, an Italian, Galileo Galilei, was challenging many of the theses of Aristotle by submitting them to experimental test. For instance, he showed (although he was not the first in this respect) that bodies of different weights fell to the ground at the same rate, whereas Aristotle had taught that the heavier the body, the faster its fall. Aristotle had, of course, been deceived by air resistance, and who, seeing a leaf falling gently to earth, would have been prepared to challenge him? The real measure of Galileo is that he enquired of nature openly and without prejudice; he can be regarded as the founder of modern science. In the course of

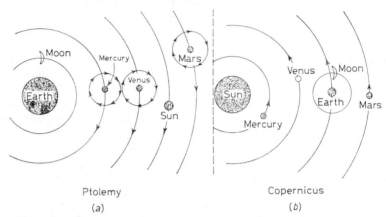

Ptolemy

(a)

Copernicus

(b)

Figure 1.3. Comparison of the solar systems of Ptolemy and Copernicus

his life, he carried out a remarkable number of experiments, and he laid the foundations for mechanics by making forces objects of measurement and revealing some of the consequences of their interaction.

Galileo died in 1642, and in that same year Isaac Newton was born. Newton, in a work of monumental synthesis, fused the mechanics of Galileo and the laws of Kepler into his own Laws of Motion and the Law of Gravity. Swept away was Aristotle's conception that a force was necessary to maintain a steady motion of a body; instead, the movement of the planets around the sun was seen to be the result of the force of gravity between the bodies deflecting the planets from a straight path.

Scientific investigation during this period was not always carried out without opposition; nor were the resulting theories always well received. Opposition to new ideas often comes from an Establishment which might have little to gain from change and has to overcome a certain complacence in accepting it. (It is easy to place all the blame for irrational resistance to experiment at the door of the Church. To do so, however, is quite unjust. There is evidence that, but for the encouragement that Copernicus received from certain members of the Church, he would never have dared publish his heliocentric theory which, at that time, had little to support it.) Galileo himself fell foul of the academics of the day on more than one occasion. Scientists do not have a monopoly of all those virtues which might ensure a sympathetic response, and Galileo certainly did not help his cause by his immodesty and self-deception. It is revealing, as a study of character, to compare two excerpts from letters written by Galileo and Newton:

4

Galileo: 'Your dear friend and servant, Galileo, has been for the last month perfectly blind so that this heaven, this earth, this universe, which I by my marvellous discoveries and clear demonstrations have enlarged a hundred thousand times beyond the belief of the wise men of bygone years, henceforward is for me shrunk into such a small space as is filled with my own bodily sensations.'

Newton: 'I know not what the world may think of my labours, but to myself it seems that I have been but as a child playing on the sea-shore: now finding some prettier pebble or more beautiful shell than my companions while the unbounded ocean of truth lay undiscovered before me.'

Never again, after Galileo and Newton, would sceptics prefer the pronouncements of Aristotle to the evidence of their own eyes. Before the Seventeenth Century was out, Robert Boyle had further undermined the position of Aristotle. Although his name is perpetuated in the law concerning the effect of pressure on gases, Boyle was also successful in the field of chemical investigation, and it is clear from his writings that he believed in the atomic nature of matter. He had something to say on the subject of Elements: 'I now mean by Elements . . . certain Primitive and simple . . . bodies which, not being made of any other bodies or of one another, are the Ingredients of which all those call'd perfectly mixt Bodies are immediately compounded and into which they are ultimately resolved.' There is no prejudice in this definition of Boyle's about how many elements there might be: so the chemist was now in a position to regard all substances as elements until he had shown otherwise.

Boyle also pointed the way to a new conception of combustion. He showed that when various substances were burnt in air, they increased in weight, although he mistakenly attributed the increase to the absorption of heat, which at that time was regarded as a material thing. It was believed about this time that when a metal was burnt it lost 'phlogiston' and that it was the presence of phlogiston in the first place that gave the metal its 'noble' quality. Whilst Boyle had discovered the facts that could destroy the phlogiston theory, he reprieved it by his faulty interpretation. A hundred years were to pass before Lavoisier proved that the increase in weight accompanying combustion was due to the absorption of a substance from the atmosphere and thus rendered the position of the Phlogistonists untenable.

This was a period of revolution. The French Revolutionaries executed Lavoisier as a representative of the old order, and yet it was the same Lavoisier who was instrumental in the establishing of a new order in Chemistry. This other revolution was ushered in softly and without violence; it doubtless went unnoticed by kings and statesmen, but such has been its impact that all our lives are in some way affected, and today the word Science has reached a stature such that advertisers and politicians alike seek to create the impression that they are most aware of its implications. This particular revolution concerned the application of scientific method to the elucidation of the structure of matter, and the rest of this Chapter will be devoted to the growth of the atomic model.

THE ATOMIC MODEL

Between 1799 and 1802, Proust carried out a series of experiments which led to the enunciation of the Law of Constant Composition, according to which the proportions in which elements combine are 'fixed by nature'. Berthollet instantly challenged the validity of this law by claiming that, unless one constituent crystallized out or distilled over, elements could combine in variable proportions. It is clear that Berthollet was confused about the concept of purity, a fact that was appreciated by Proust when he defended his position:

> The attraction which makes sugar dissolve in water may or may not be the same as that which makes a definite quantity of charcoal and hydrogen dissolve in another quantity of oxygen, to form the sugar of our plants; but we can see clearly that the two sorts of attractions are so different in their results that it is impossible to confound them.

Proust was at least aware of differences between mixtures and compounds and physical and chemical changes. Nevertheless, substances with variable composition do exist (p. 79), and these are often described as 'Berthollides', although it is open to question whether they are genuine compounds.

At about the same time, John Dalton was investigating the extent of the dissolution of gases in water and the effect of mixing gases upon the final pressure. He sought to explain both partial pressures and solubilities in terms of a particulate structure of matter. Four years later, in 1807, the ideas of Dalton were first set out in print by Thomas Thomson in his 'System of Chemistry'. With rare charm and generosity he writes:

> We have no direct means of ascertaining the density of the atoms of bodies; but Mr. Dalton, to whose uncommon ingenuity and sagacity the philosophic world is no stranger, has lately contrived an hypothesis which, if it prove correct, will furnish us with a very simple method of ascertaining that density with great precision. Though the author has not yet thought fit to publish his hypothesis, yet as the notions of which it consists are original and extremely interesting, and as they are intimately connected with some of the most intricate parts of the doctrine of affinity, I have ventured, with Mr. Dalton's permission, to enrich this work with a short sketch of it.
>
> The hypothesis upon which the whole of Mr. Dalton's notions respecting chemical elements is founded, is this:
>
> When two elements unite to form a third substance, it is to be presumed that one atom of one joins to one atom of the other, unless when some reason can be assigned for supposing the contrary. Thus oxygen and hydrogen unite together and form water. . . . If we represent an atom of oxygen, hydrogen . . . by the following symbols:

Oxygen O   Hydrogen ⊙

then an atom of water . . . will be represented by the following symbols:

water O⊙

This example reveals a major limitation of the Dalton theory: even if the existence of atoms is accepted, there is still the difficulty of establishing the precise number of atoms in a particular molecule. Although Dalton himself was aware of the possibility of a 'compound atom' containing more than one atom of a particular element, he believed, for some reason, that water contained one atom only of hydrogen and of oxygen. Consequently, when water was analysed on this basis, the atomic weight of oxygen (that is, the number of times one atom of the element is heavier than one atom of hydrogen) appeared to be eight. Yet the answer to this problem was at hand. . . .

On the last day of 1808, Gay-Lussac presented a paper to the Philomathic Society in which he dealt with investigations he had carried out on the combination of gases. As a result of this research, Gay-Lussac was able to present the law which still bears his name:

> Not only, however, do gases combine in very simple proportions . . . but the apparent contraction of volume which they experience on combination has also a simple relation to the volume of the gases, or at least to that of one of them.

Here was support indeed for an atomic theory. And yet Dalton so missed the point that he felt compelled to cast doubt upon the accuracy of Gay-Lussac's experiments!

> The truth is, I believe, that gases do not unite in equal or exact measures in any one instance; when they appear to do so, it is owing to the inaccuracy of our experiments.

Even Gay-Lussac himself felt bound to support the laws of both fixed and variable proportions:

> We must first of all admit, with M. Berthollet, that chemical action is exercised indefinitely in a continuous manner between the molecules of substances, whatever their number and ratio may be, and that in general we can obtain compounds with very variable proportions. But then we must admit at the same time that—apart from insolubility, cohesion and elasticity, which tend to produce compounds in fixed proportions—chemical action is exerted more powerfully when the elements are in simple ratios. . . . In this way we reconcile the two opinions and maintain the great chemical law. . . .

It was Avogadro who in 1811 brought reason to this situation. Discussing the significance of Gay-Lussac's law, he wrote:

> The first hypothesis to present itself . . . is the supposition that the number of integral molecules in any gases is always the same for equal volumes. . . .

Avogadro even realised that it was now possible to determine the relative masses of molecules and he calculated, for example, that the molecule of oxygen was about fifteen times as heavy as a molecule of hydrogen. One important aspect of Avogadro's contribution was that it focused attention upon the molecule as a group of atoms, instead of the vague 'compound atom' that other chemists were in the habit of talking about. But no one seemed ready to develop the ideas of Avogadro.

The growing acceptance of the atomic theory resulted in attention being directed towards the determination of atomic weights. But because there was at the time no accepted way of knowing the precise formula of a compound, the value determined was often the equivalent, that is, the weight of element combined with one

7

gramme of hydrogen. For example, eight grammes (g) of oxygen are found to combine with 1 g of hydrogen to form water; if one takes the formula of water to be HO, then one naturally assumes that the atomic weight of oxygen is eight.

The key to this problem lay with Gay-Lussac's law and Avogadro's hypothesis. Since one volume of hydrogen combines with one volume of chlorine to produce two volumes of hydrogen chloride, then, by Avogadro's hypothesis, one molecule of hydrogen combines with one molecule of chlorine to produce two molecules of hydrogen chloride. Therefore, half a molecule of hydrogen combines with half a molecule of chlorine to produce one molecule of hydrogen chloride. Now, because atoms, in accordance with Dalton's theory, cannot be split by chemical means, half a molecule of hydrogen or chlorine must contain at least one atom and the gases are possibly diatomic.

Cannizzaro in 1858 revived Avogadro's hypothesis for the purpose of determining atomicities and atomic weights. By measuring the weights of equal volumes of the gaseous element and gaseous compounds of it (and therefore equal numbers of molecules) and determining the amount of the element in each compound, he arrived at the following sort of information:

Name of Substance	Weight of one molecule	Weight of hydrogen present in one molecule
Hydrogen	$2x$	$2x$
Hydrogen chloride	$36 \cdot 5x$	$1x$
Methane	$16x$	$4x$
Ammonia	$17x$	$3x$
Hydrogen sulphide	$34x$	$2x$

Eventually, after a large number of compounds had been investigated and the smallest amount of hydrogen found was always half of the hydrogen molecule, the conviction that the hydrogen molecule is diatomic could no longer be repressed. In Cannizzaro's own words:

Compare the various quantities of the same element contained in the molecule of the free substance and in those of all its different compounds and you will not be able to escape the following law: The different quantities of the same element contained in different molecules are all whole multiples of one and the same quantity which, always being entire, has the right to be called an atom.

Similar experiments and reasoning permit the atomicity of the other gaseous elements and compounds to be determined, for instance, the formula for water:

2 volumes of hydrogen + 1 volume of oxygen → 2 volumes of steam

by Avogadro's hypothesis:

2 molecules of hydrogen + 1 molecule of oxygen → 2 molecules of steam

8

But hydrogen and oxygen are both diatomic. Therefore

1 molecule of steam contains 2 atoms of hydrogen and 1 atom of oxygen

That is, the formula of water is H_2O.

Also, since 8 g of oxygen combine with 1 g of hydrogen, and 16 g of oxygen with 2 g of hydrogen, the atomic weight of oxygen is 16·00.

An important relationship between vapour density and molecular weight also follows from the diatomicity of hydrogen:

$$\text{Vapour density} = \frac{\text{weight of gas}}{\text{weight of equal volume of hydrogen (at same temp. and pressure)}}$$

by Avogadro's hypothesis

$$= \frac{\text{weight of } n \text{ molecules of gas}}{\text{weight of } n \text{ molecules of hydrogen}}$$

$$= \frac{\text{weight of 1 molecule of gas}}{\text{weight of 1 molecule of hydrogen}}$$

$$= \frac{\text{weight of 1 molecule of gas}}{\text{weight of 2 atoms of hydrogen}}$$

But molecular weight $= \dfrac{\text{weight of 1 molecule of gas}}{\text{weight of 1 atom of hydrogen}}$

Clearly

molecular weight = 2 × vapour density

Since the vapour density of a volatile substance is readily determined, this relationship is often used to calculate the molecular weight.

Avogadro Constant

The number of molecules in a mole of a compound (or the number of atoms in a mole of an element) is constant for all compounds, since

Weight of 1 molecule of substance = molecular weight × weight of 1 atom of hydrogen

$$\text{Number of molecules in 1 mole} = \frac{\text{molecular weight, } M}{\text{wt. of 1 molecule}}$$

$$= \frac{M}{M \times \text{wt. of 1 atom hydrogen}}$$

$$= \frac{1}{\text{wt. of 1 atom hydrogen}}$$

$$= \text{a constant}$$

9

This constant is called the Avogadro constant. X-ray measurements suggest that it is about 6.023×10^{23}. The mole (Appendix 1) is a very important concept, because it involves this constant number of molecules, and it is of particular application where the phenomena involved are functions of the number of particles present.

It follows from Avogadro's hypothesis that the Avogadro constant of molecules will occupy a certain fixed volume under specified conditions of temperature and pressure. The volume occupied by 1 mole of any gas is called, naturally enough, the Gramme Molecular Volume and equals 22.4 dm^3 (litres) at s.t.p. [standard temperature, $0°C$ ($=273$ K), and pressure, 1 atm ($=101\ 325$ N m^{-2})].

Cannizzaro's method of determining atomic weights was only applicable to gaseous elements. In the case of solid substances, it was necessary to know the equivalent (the weight displacing 1 g of hydrogen). The relationship between the atomic weight, A, and the equivalent, E, of an element is readily derived:

If x is the weight of one atom of hydrogen, then the weight of one atom of the element is Ax, and since

$$E \text{ g of an element displaces 1 g of hydrogen}$$

(E/Ax) atoms of the element displace $(1/x)$ atoms of hydrogen, i.e.
1 atom of the element (dividing by E/Ax) displaces (A/E) atoms of hydrogen

But this is the valency, V. That is, $A = E \times V$

This means to say that the valency as well as the equivalent of the element must be known before its atomic weight can be calculated. It is because of this that the law put forward by Dulong and Petit in 1819 was of great utility to the nineteenth-century chemist:

The atoms of all simple substances have the same capacity for heat.

Because there is the same number of atoms in a mole of any element (the Avogadro constant), it follows that the 'atomic heat', i.e. the product of atomic weight and specific heat capacity, should be constant. With very few exceptions this is approximately true for solids at ordinary temperatures:

	J mol^{-1} K^{-1}*		J mol^{-1} K^{-1}
Magnesium	25·6	Gold	26·0
Zinc	25·6	Chromium	26·4
Silver	25·6	Bismuth	26·4
Phosphorus	26·0	Iron	26·8

A knowledge of the specific heat capacity of the element can therefore be instrumental in determining atomic weights.

For example, the specific heat capacity of nickel is 0.46 J g^{-1} K^{-1}. The atomic weight, therefore, is from Dulong and Petit's law $26.4/0.46$ or approximately 57. The equivalent is 29.35 and so the valency, $A/E \sim 57/29.35 = 2$ (the nearest integer).

The accurate atomic weight, $E \times V$, is therefore $29.35 \times 2 = 58.70$.

*See Appendix 1 for Units.

A further aid to the determination of the atomic weights of non-volatile elements came in 1819 with the enunciation by Mitscherlich of the law of isomorphism. This can be stated: 'Substances which have similar crystalline form (isomorphous substances) possess similar formulae.' This law has its limitations, but it permitted the valency of several elements to be elucidated by comparison with isomorphous compounds of known formulae.

CLASSIFICATION OF THE ELEMENTS

The abandonment in the eighteenth century of the Aristotelian concept of 'element' opened the way for the discovery of new elements. As the list of simple substances grew, it was natural that chemists should attempt to systematize the subject by classification. Lavoisier divided the elements then known into three broad groups which included metals and non-metals in separate categories, and this undoubtedly represented a step forward. But the availability of the newly-determined atomic weights provided a new basis for classification. Döbereiner (1829) noted the existence of families of three (called 'triads') in which the central element had an atomic weight approximately the average of the other two, for example

chlorine	bromine	iodine	and	calcium	strontium	barium
35·46	79·92	126·91		40·06	87·63	137·36

Whilst this correlation was partly fortuitous and was not always maintained as new families were discovered, it at least served to focus attention upon the possible connection between atomic weight and chemical properties.

The term 'triad' was replaced by another musical term in Newlands's Law of Octaves (1865):

The number of analogous elements generally differ either by seven or some multiple of seven; in other words, members of the same group stand to each other in the same relation as the extremities of one or more octaves in music.

To maintain this pattern, however, Newlands found it necessary on occasion to allot only one place to two elements:

H	Li	Be	B	C	N	O
1	2	3	4	5	6	7

F	Na	Mg	Al	Si	P	S
8	9	10	11	12	13	14

Cl	K	Ca	Ti	Cr	Mn	Fe
15	16	17	18	19	20	21

Co and Ni	Cu	Y	Zn	In	As	Se
22	23	24	25	26	27	28

Mendeleev, two years later, carried Newlands's work an important stage further (*Table 1.1*). He stressed the periodicity of the arrangement and, by arranging the elements in sub-groups, not only revealed a closer family relationship but obviated the need for ascribing the same number to more than one element. He also avoided the criticism that was levelled at Newlands 'on the score of it having been assumed that no elements remain to be discovered'. Not only did Mendeleev leave appropriate blanks in his Periodic Table, but he predicted with uncanny accuracy the properties of some undiscovered elements. Thus, for 'ekasilicon' which, when it was eventually found, was named germanium:

	'Ekasilicon' (predicted values)	Germanium (observed values)
Atomic weight	72	72·5
Density	5·5	5·47
Density of chloride	1·9	1·9
B.p. of chloride	<100°C	86°C

Mendeleev seemed to sense that there was a more fundamental quantity than atomic weight and was prepared to overlook the atomic weight if this meant keeping a family relationship. For instance, unlike Newlands, he inverted the positions of tellurium and iodine. He also emphasized the connection between the Group Number of the element and the valency exhibited in the formation of the hydride and oxide. (The valency in the case of the former reaches a maximum at Group 4 whilst with the oxides it rises progressively to Group 8.)

There can be little doubt of the calibre of Mendeleev's achievement. It should be added, however, that one of his contemporaries, Lothar Meyer, not only anticipated several aspects of Mendeleev's Periodic Table but, by plotting the *atomic volume* (that is, atomic weight/density) against the *atomic number* (the number of the element in the Periodic Table), revealed the periodicity in a singularly elegant manner (p. 34).

Table 1.1. Mendeleev's Periodic Table

(Dashes represent elements unknown to Mendeleev)

Series	Group I	Group II	Group III	Group IV	Group V	Group VI	Group VII	Group VIII
1	H							
2	Li	Be	B	C	N	O	F	
3	Na	Mg	Al	Si	P	S	Cl	
4	K	Ca	—	Ti	V	Cr	Mn	Fe Co Ni
5	Cu	Zn	—	—	As	Se	Br	
6	Rb	Sr	Y	Zr	Nb	Mo	—	Ru Ph Pd
7	Ag	Cd	In	Sn	Sb	Te	I	
8	Cs	Ba	Dy	Ce	—	—	—	—
9	—	—	—	—	—		—	
10	—	—	Er	La	Ta	W		Os Ir Pt
11	Au	Hg	Tl	Pb	Bi	—	—	
12	—	—	—	Th	—	U		

Ingenious though these classifications were, they were without theoretical foundation. Explanation was eventually forthcoming in terms of a more sophisticated atomic model, but at that time chemists regarded the atom as a solid homogeneous ball. Although Faraday's laws of electrolysis suggested that there was also an atomic form of electricity, fifty years later, in 1881, Helmholtz was saying:

> Now the most startling result of Faraday's laws is perhaps this. If we accept the hypothesis that the elementary substances are composed of atoms, we cannot avoid the conclusion that electricity also . . . is divided into definite elementary portions which behave like atoms of electricity.

Before the century was out, J. J. Thomson, investigating the rays emanating from the cathode of a discharge tube, had discovered these 'atoms of electricity', or electrons. He found that they were attracted towards positively-charged plates, that they all possessed the same negative charge, had negligible mass and travelled with approximately the same speed.

The discovery of radioactivity in 1897 by Becquerel also revealed the heterogeneous and electrical nature of the atom. Three main types of radioactivity exist:

α-rays, with the mass of the helium atom and 2 positive charges

β-rays, which consist of rapidly-moving electrons

γ-rays, made up of electromagnetic waves of very high frequency.

The atom was then taken to consist of positively charged masses (protons) embedded in a sea of electrons. In order to maintain the neutral structure of the atom, the number of protons was equal to the number of electrons.

Radioactive particles were invaluable weapons in the scientist's fight to discover more of the detailed structure of the atom. Rutherford, in 1911, bombarded thin aluminium foil with α-particles and found that only a very small proportion were reflected back. Evidently the latter had come under the influence of the positively-charged part of the atom and so this must form only a small part of the atom (*Figure 1.4*). Rutherford's atomic model then consisted of a very small nucleus containing the positive charge and most of the mass, around which the electrons rotated—rather like a solar system in miniature. This analogy, however, is not quite correct. A moving electron should emit radiation and, losing energy, gradually move into the nucleus. To overcome this objection, Bohr (1913) postulated the existence of 'closed' orbits; provided the electrons remained in the same orbit, no energy was emitted. This modification incorporated the relatively new quantum theory. When an electron moved from an inner to an outer orbit, a quantum of energy was absorbed, whilst if it moved from an outer to an inner orbit, a quantum of energy was emitted. Impressive support for this theory comes from the line spectra of the elements, each line being associated with a particular amount of energy.

A later modification, by Sommerfeld, associated differently-shaped orbits with characteristic energies to account for the fine structure of spectra. This Bohr–Sommerfeld model has held considerable appeal for the chemist because it has provided some theoretical justification for the Periodic Table (see Chapter 2) and also an interpretation of valency (see Chapter 3).

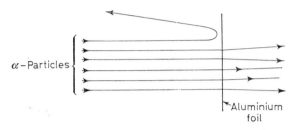

Figure 1.4. Rutherford's experiment

It was shown in the 1920's that electrons possess wave properties and so resemble light in having both wave and corpuscular qualities. According to Heisenberg's Uncertainty Principle, either the momentum or the position of an electron may be accurately known but not both simultaneously. The reason for this uncertainty can be roughly explained by saying that in measuring the momentum of an electron, its position will be disturbed by the observing mechanism. This idea has led to the present so-called wave-mechanical picture of the atom for which, however, simple models have had to be abandoned since an electron can no longer be represented by a point charge.

In 1932, the neutron, a particle having the mass of the proton but without electrical charge, was discovered. The nucleus of atoms could now be explained more satisfactorily. The atomic number, i.e. that of the element in the Periodic Table, had come to be identified with the number of electrons outside the nucleus. It was now seen to be equal to the number of protons in the nucleus, and the difference between the atomic number and the atomic weight was equal to the number of neutrons. For example, the atomic number of oxygen is eight; this means that it is the eighth element in the Periodic Table and that it has eight extranuclear electrons and eight protons in the nucleus. The atomic weight of sixteen is the result of there being eight neutrons as well as eight protons in the nucleus. Soddy, in 1913, had proposed the name 'isotope' for atoms of the same element having different atomic weights. We can now see that such atoms differ only in having different numbers of neutrons, but because they have the same number of electrons, they have the same chemistry and are justifiably regarded as being the same element. The atomic weight of the element as determined by the chemist is the arithmetical mean of the weights of the isotopes present. In the case of chlorine, for example, two isotopes of atomic weight 35 and 37 exist, in the proportions 75·4 and 24·6 per cent, respectively, giving, as the average atomic weight, 35·457. (For accurate physical determinations, the atomic weights of the elements are referred to the most abundant naturally-occurring isotope of carbon, with a relative mass of 12.)

The atomic weight has come to be less and less important to the chemist and the atomic number more and more. Mendeleev showed good intuitive sense

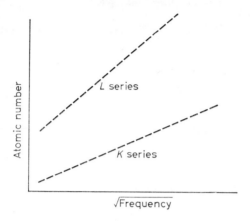

Figure 1.5

when he ignored the anomalous atomic weights of certain elements in the construction of his Periodic Table. We can now see that the reason for tellurium having a larger atomic weight than iodine is the excessive number of neutrons; these do not affect the chemical properties, and so the atomic weight is not relevant. The properties of the elements are really periodic functions of the atomic numbers. Moseley, by a very timely experiment in 1914, focused attention upon the atomic number; he found that, if the numbers of the elements were plotted against the square root of the frequency of the x-ray spectra, a straight line resulted (*Figure 1.5*). Because of this correlation, Moseley was able to determine the atomic numbers of several elements (notably transition elements and lanthanides).

The development of the concept of the element and the model of the atom has been the result of continued application of *Scientific Method*. Although this should not be taken to be a rigid procedure, it has come to be regarded as broadly involving the following sequence:

(*a*) gaining of information through experiment and observation
(*b*) correlation of this information
(*c*) development of hypotheses (guesses) to explain the observed facts
(*d*) designing of new experiments on the basis of these hypotheses, to provide more definite evidence on which theories and laws can be established.

It is clear that, working with limited information, theories eventually become outmoded and at some stage a replacement is needed. Always the theory becomes more satisfactory and the working model more sophisticated. But, it seems, perfection is never achieved. The completion of one successful experiment usually suggests the beginning of another; the story is never complete and the words of

15

Newton quoted above still seem appropriate. It is in this spirit, and with the awareness that the theories and principles quoted are unlikely to stand the test of time, that the reader should approach this book.

SUMMARY OF BASIC LAWS AND DEFINITIONS

The Atomic Theory
 (*a*) Matter is made up of atoms
 (*b*) Atoms of the same element are identical (but note the existence of isotopes)
 (*c*) Chemical reaction takes place between simple whole numbers of atoms.

Law of Conservation of Mass/Energy
 Mass/Energy can be neither created nor destroyed.

Law of Constant Composition
 All pure samples of the same compound contain the same elements combined in the same proportion by weight (but note the existence of the non-stoichiometric Berthollides).

The Law of Multiple Proportions
 If two elements, *A* and *B*, can combine together in more than one way, then the weights of *A* which separately combine with a fixed weight of *B* are in a simple ratio.

The Law of Reciprocal Proportions
 If two elements, *A* and *B*, can separately combine with a third element, *C*, then the weights of *A* and *B* which separately combine with a fixed weight of *C* are the weights by which *A* and *B* might combine with each other (or simple multiples of them).

 (This law may be restated as the Law of Equivalents: substances combine together in the ratio of their equivalents.)

 The *atomic weight* of an element is the number of times one atom of the element is heavier than one-twelfth of an atom of carbon 12.

 (The *mass number* of an isotope is the whole number nearest to the actual atomic weight on the same scale; for example, the oxygen isotope, with atomic weight 15·995, is given the mass number 16.)

 The *equivalent* of a substance is the weight of it which will combine with, or displace, 1 g of hydrogen, 8 g of oxygen or the equivalent of any other substance.

 The *valency* of an element is the number of hydrogen atoms which will combine with (or displace) one atom of the element.

QUESTIONS

1. Correlate the four laws of chemical combination with the atomic theory. Can Berthollides be reconciled with any aspects of these laws and theory?

2. Comment on the statement: 'The structure of an atom represents a miniature solar system'.

3. A metal chloride has a vapour density of 95·0 and the percentage by weight of chlorine in the compound is 74·6. Calculate the atomic weight of the metal.

4. An element has a specific heat capacity of 0·113 J g^{-1} K^{-1} and an equivalent of 39·67. Find the atomic weight of the element and the valency state in the compound considered.

5. The following values were found for five compounds each containing silicon:

| Vapour density | 16·1 | 31·1 | 46·2 | 85·0 | 267·8 |
| per cent Si | 87·5 | 90·5 | 91·4 | 16·6 | 10·4 |

Calculate the atomic weight of silicon from these figures.

6. Modify Dalton's Atomic Theory so that it is in accord with modern views.

7. Who, of the following, do you think has had the greatest impact on the course of chemical development, and why? Dalton, Mendeleev, Cannizzaro, Lavoisier, Boyle.

8. As far as Plato was concerned, all things with which we have contact through sense experience, approach, with more or less precision, the 'idea' of the thing. (You can see his mathematical leanings in this philosophy: for example, we know in our minds what the perfect circle is like, but we can never quite draw it.) How far do you think the scientists' predilection for models is an acknowledgement of the Platonic view?

CHAPTER 2

THE ATOM

An atom can be briefly described as consisting of a very small, dense nucleus, around which *electrons* circulate in a comparatively large volume.

The volume of the atomic nucleus is about 10^{-44} m^3 and it consists of positively-charged *protons* and electrically neutral particles known as *neutrons* which have a mass almost identical with that of the protons. These are collectively known as *nucleons* and are equal numerically to the *mass number*. The number of protons themselves represents the *atomic number* of the atom.

The electrons are sufficient in number to balance the positive charge on the nucleus, each electron having a negative charge numerically equal to the positive charge of a proton, although the mass of an electron is only about $\frac{1}{1840}$ of that of a nucleon. The *total* volume of an atom is very large compared with the volume occupied by the nucleons, being of the order of 10^{-30} m^3, so that the volume of the nucleus is only about 10^{-14} that of the atom.

Some of the properties of these fundamental particles are shown in *Tables 2.1(a)* and *2.1(b)* while *Table 2.1(c)* illustrates the particulate compositions of some representative elements.

Table 2.1(a). The atom

Radius of		m
an electron		2×10^{-15}
a nucleus	H	$1 \cdot 5 \times 10^{-15}$
	U	9×10^{-15}
an atom		$\sim 1 \times 10^{-10}$

Table 2.1(b). Fundamental particles

		Electron	Proton	Neutron
Rest mass	g	$9 \cdot 1 \times 10^{-28}$	$1 \cdot 67 \times 10^{-24}$	$1 \cdot 67 \times 10^{-24}$
Absolute charge	C	$-1 \cdot 6 \times 10^{-19}$	$+1 \cdot 6 \times 10^{-19}$	0
Mass relative to H $= 1$		0·000 54	1·007 6	1·008 9
Relative charge		-1	$+1$	0
Designation		$_{-1}^{0}e$	$_{1}^{1}$H or $_{1}^{1}p$	$_{0}^{1}n$

18

Table 2.1(c). Composition of some representative atoms
(the most common isotopes (p. 26) have been selected)

Element	Atomic number	Mass number	Number of		
			protons	neutrons	electrons
H	1	1	1	0	1
He	2	4	2	2	2
Na	11	23	11	12	11
U	92	238	92	146	92

THE ELECTRONIC STRUCTURE OF THE ATOM

Electrons orbiting around a nucleus are accelerated towards the centre and, since they are charged, they should be continually losing energy and therefore tend to collapse into the nucleus. This tendency must be balanced by an outwardly directed force in order that the structure should remain stable; this force was suggested by Bohr to be the centrifugal force, acting as a result of the rotation of the electrons in *shells* around the nucleus.

Planck had suggested in 1900 that energy is 'quantized', that is, energy changes only occur in 'packets' or *quanta* and not continuously. In 1913, Bohr utilized this quantum theory and proposed that an electron revolved in a 'closed' orbit, being associated with a definite fixed energy. Any energy changes then occurred as discrete quanta when an electron was transferred from one orbit to another of different energy, the energy change, ΔE, being an integral multiple, n, of the frequency, ν, of the spectral line associated with the transition, i.e.

$$\text{Quantum} = \Delta E = nh\nu$$

where h is Planck's constant ($6 \cdot 625 \times 10^{-34}$ J s).

As spectra are manifestations of the changes in energy of electrons, their observation can be made to yield much useful information. The *emission spectra* of all the elements consist of lines occurring in different parts of the electromagnetic spectrum (p. 523) and may be observed by passing an electric discharge through the elements in the gaseous phase, or by using the more complicated *arc spectra*, produced by striking an electric arc between electrodes of the element. In the spectra so obtained, the lines represent electrons of the element which, having been excited to higher *energy levels* (orbits), are returning to lower energy states. The return of an electron from different, higher-energy, levels or shells to a certain lower energy level gives rise to a series of spectral lines. The energy picture for different transitions of the single excited electron of the hydrogen atom is given in *Figure 2.1*.

In Bohr's original theory, successive closed shells were shown to be capable of accommodating two, eight, eighteen and thirty-two electrons, respectively. However, in 1923, Pauli propounded his *Exclusion Principle*, according to which each electron in an atom is unique, so that two electrons of equal energy must be spinning in opposite directions.

As a result of this principle and an examination of spectra, it became necessary to split the shells of Bohr's atom into sub-shells (orbits) of differing eccentricities. Because the energy of the electrons in a given orbit is constant, the maximum number per orbit is two, the electrons spinning in opposite directions on their own axes (see *Figure 2.2*).

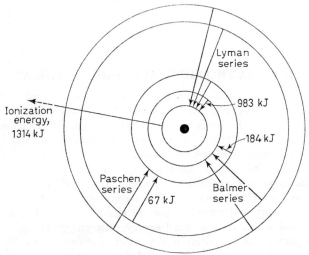

Figure 2.1. Energy picture of the hydrogen atom (diagrammatic)

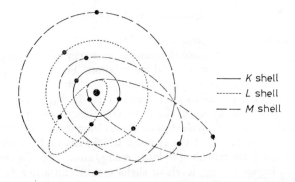

Figure 2.2. Bohr–Sommerfeld atomic model (diagrammatic)

So far in this Chapter electrons have been referred to as particles, but it was demonstrated by Davisson and Germer in 1927 that electrons can be diffracted by the atoms in a crystal and, therefore, some wave properties must be associated with them. One result of this is that the position of an electron cannot be precisely located at any particular instant, but only the region in which there is the greatest probability of finding it. The probability can be given physical reality as an *electron cloud*, the density of which increases with the probability. Such regions are called *orbitals*, each of which, when full, contains two electrons.

If this wave theory is used to calculate the probability of finding an electron at different points in space, and a continuous line is drawn around the nucleus such that there is about a 90 per cent chance of finding the negative charge associated with the electron between the nucleus and the line, then the boundary surface can be said to represent the orbital.

TYPES OF ORBITAL

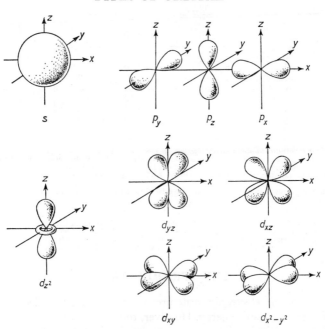

Figure 2.3. Atomic orbitals

The first, *K*, shell contains one orbital which is spherically symmetrical and is referred to as the *1s orbital*.

The second, *L*, shell contains one *s* orbital (the *2s*) and three others, known as *2p orbitals*, which each consist of two lobes and which are in three mutually perpendicular directions; the total complement of electrons in this second shell is thus eight.

The third, *M*, shell, as well as containing a *3s orbital* and three *3p orbitals* can accommodate five orbitals—not all equivalent in energy—known as *3d orbitals*, thus allowing a further ten electrons to enter this shell, giving eighteen in all.

21

The next N, shell, can hold a further fourteen electrons in a set of seven *4f orbitals*, giving a total of thirty-two electrons (*Table 2.2*).

The shapes of some of these orbitals are illustrated in *Figure 2.3*.

Table 2.2. Accommodation of Electrons

Principal quantum number, n	Angular quantum number	Electron type	Magnetic quantum number, m	Spin quantum number, s	Number of electrons accommodated
K 1	0	s	0	$\pm\frac{1}{2}$	2
L 2	0	s	0	$\pm\frac{1}{2}$	2 $\Big\}$ 8
	1	p	$+1, 0, -1$	$\pm\frac{1}{2}$	6
M 3	0	s	0	$\pm\frac{1}{2}$	2
	1	p	$+1, 0, -1$	$\pm\frac{1}{2}$	6 $\Big\}$ 18
	2	d	$+2, +1, 0, -1, -2$	$\pm\frac{1}{2}$	10
N 4	0	s	0	$\pm\frac{1}{2}$	2
	1	p	$+1, 0, -1$	$\pm\frac{1}{2}$	6
	2	d	$+2, +1, 0, -1, -2$	$\pm\frac{1}{2}$	10 $\Big\}$ 32
	3	f	$+3, +2, +1, 0, -1, -2, -3$	$\pm\frac{1}{2}$	14

There are four quantum numbers characterizing an electron. The principal quantum number, n, gives the number of the shell. The angular quantum number, l, is concerned with the eccentricity of the orbital while the magnetic quantum number, m, takes account of the behaviour of the electron in an external magnetic field. The fourth quantum number, s, which can be either $+\frac{1}{2}$ or $-\frac{1}{2}$, is concerned with the direction in which the electron spins.

The angular and magnetic quantum numbers are related to the principal quantum number as follows:

$$l = 0 \ldots (n-1) \qquad m = +l \text{ to } -l$$

If the Pauli exclusion principle (which postulates that no two electrons in the same atom have exactly the same set of quantum numbers) is now applied, we arrive at the result that the first shell can hold two s electrons, the second shell two s electrons and six p electrons, the third shell two s electrons, six p electrons and ten d electrons and the fourth shell two s, six p, ten d and fourteen f electrons.

THE PERIODIC TABLE

In 'building up' the electronic structure of an element, the electrons enter the orbitals of lowest available energy. However, this does not mean that the shells are always filled successively, as there is some overlap in energies between orbitals of different shells after the second. The order in which the various orbitals are occupied, that is the order of increasing energy, is shown in *Table 2.3(a)*; it is this selective filling of orbitals which accounts for the framework of the Periodic Table of the elements (*Table 2.3(b)*).

Period One

The first shell contains one s orbital and can thus only accommodate two electrons; the *electronic configuration* of the first electron entering the orbital is designated as $1s^1$ (the first figure represents the number of the shell and the superscript the number of electrons present in the orbital). This electronic configuration is that of the hydrogen atom in the *ground state*, i.e. that of lowest energy. With two electrons, the configuration is $1s^2$. It represents the electronic structure of the helium atom; the electrons must be of opposite spin and these two configurations are often shown as

1	2
H □	He ⊞

This $1s$ orbital, and hence the first shell, is now full and addition of a further electron must take place in the shell of next higher energy.

Period Two

The third element, lithium, thus has the configuration $1s^22s^1$ and belongs to Group 1 of the Periodic Table. Beryllium, of atomic number four, is represented $1s^22s^2$ or $K2s^2$, and belongs to Group 2.

Unlike the first shell, the second shell has three p orbitals also available and the next electron must enter one of these. Thus boron, of atomic number five, has the configuration $K2s^22p_x^1$, and is the first member of Group 3. The placing of the next electron is indicated by *Hund's rule* which states that, under normal circumstances, no available orbital of any one type will take up a second electron until all the available orbitals of that type have acquired one electron each, e.g. the p orbitals will each receive one electron before any pairing occurs. The next element, carbon (Group 4) will accordingly have the configuration $K2s^22p_x^12p_y^1$; nitrogen, atomic number seven, will be $K2s^22p_x^12p_y^12p_z^1$ (Group 5), oxygen (in Group 6) $K2s^22p_x^22p_y^12p_z^1$, fluorine (Group 7) $K2s^22p_x^22p_y^22p_z^1$ and neon (Group O) $K2s^22p_x^22p_y^22p_z^2$.

3 4	5	6	7	8	9	10
Li Be	B	C	N	O	F	Ne
2s 2s 2s 2p	2s 2p	2s 2p	2s 2p	2s 2p	2s 2p	2s 2p

For most purposes it is not necessary to distinguish the various p orbitals from each other; for example, nitrogen may be quoted as $K2s^22p^3$.

Period Three

The configurations of the next group of eight elements, sodium to argon, follow an analogous pattern, except that it is the third shell which is in the process of being filled (although not completely at this stage).

11 12	13	14	15	16	17	18
Na Mg	Al	Si	P	S	Cl	Ar
3s 3s 3s 3p	3s 3p	3s 3p	3s 3p	3s 3p	3s 3p	3s 3p

Period Four

The two elements following, potassium and calcium, have electrons in the $4s$

23

Table 2·3(a) Framework of the Periodic Table in Terms of Orbitals and Energy Levels

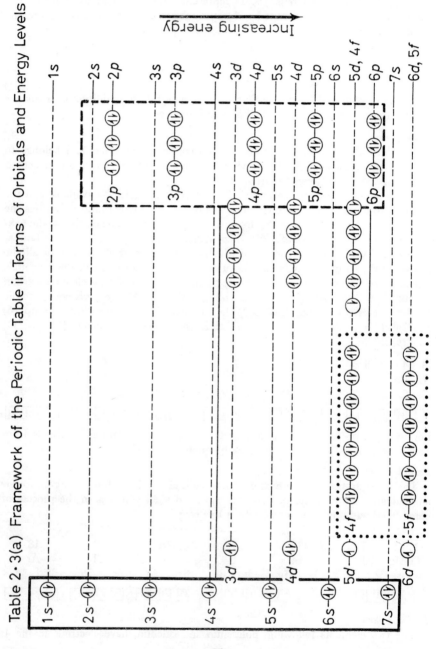

Table 2·3(b) The Periodic Table — Long Form

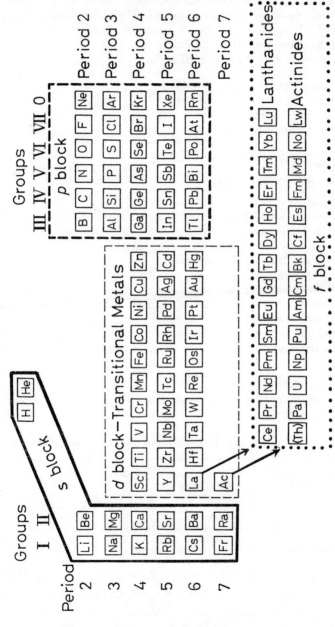

25

orbital, but reference to *Table 2.3(a)* shows that after this is full, the next available set of orbitals is not the $4p$ but the $3d$. These will therefore be the next to be occupied, giving rise in all to ten elements called 'transitional elements', with atomic numbers from 21 to 30. After this the $4p$ orbitals fill up, thus accounting for elements of atomic numbers 31 to 36.

19	20	21	......	25	......	30		31	
K	Ca	Sc		Mn		Zn		Ga	
$4s$	$4s$	$4s$	$3d$	$4s$	$3d$	$4s$	$3d$	$4s$	$4p$ etc.

Period Five

This s, d, p sequence is now repeated with elements 37 to 54.

Periods Six and Seven

For periods six and seven this sequence is modified to include the seven $4f$ orbitals in the former and the seven $5f$ orbitals in the latter but the basic principles remain the same, until ultimately the largest known atom, that of the element with an atomic number of 104 is reached.

It can be seen that all the elements of Group 1 have one electron in the outer shell, those of Group 2 have two electrons in the outer shell, and similarly for the other Groups. As the chemistry of elements is to a large extent dictated by the outer electronic configurations, the elements in any one Group have great chemical similarity and there is therefore a certain *periodicity* in traversing the table. Some meaning has now been given to the ideas of Mendeleev expressed in the previous Chapter.

Nuclear stability

Since the number of electrons in an atom is not directly dependent upon the number of neutrons present, it is possible for atoms to have the same number of

Table 2.4. Isotopic composition of some elements

Element	Atomic number	Stable isotopes (in order of abundance)	Element	Atomic number	Stable isotopes (in order of abundance)
H	1	1, 2	V	23	51
C	6	12, 13	Fe	26	56, 54, 57, 58
O	8	16, 18, 17	Sn	50	120, 118, 116, 119, 117, 124, 122, 112, 114, 115
Na	11	23	I	53	127
S	16	32, 34, 33	Hg	80	202, 200, 199, 201, 198, 204, 196

electrons and protons, and therefore the same chemical properties, but yet contain different numbers of neutrons and hence possess different atomic weights. These different atoms of the same element are called *isotopes*. The atomic weight of an element, as used chemically, is then an average value representative of the different isotopes present and their relative proportions. For example, the percentage abundance of the two isotopes of chlorine of mass numbers 35 and 37 are 75·4 and 24·6, respectively, giving the atomic weight of chlorine as 35·46.

The stable isotopes of some representative elements are given in *Table 2.4.*

	H₂	HD	D₂
Melting point	13·8K	16·5K	18·5K
Boiling point	20·3K		23·5K
Latent heat of fusion, J kg⁻¹	0·118	0·155	0·197
Latent heat of vaporization, J kg⁻¹	0·93	1·11	1·27

If the mass number of an element is small, then the relative differences in the isotopic masses can be appreciable, and the isotopes can show some marked divergencies in properties. This point is well illustrated by hydrogen (mass number 1) and its main isotope, deuterium, D (mass number 2).

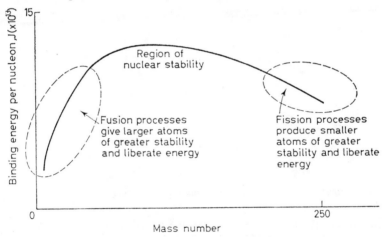

Figure 2.4.   Stability of atomic nuclei

The nature of the forces, millions of times more powerful than those between a proton and an extranuclear electron, which operate in the nucleus and bind together the protons and neutrons, is still not understood. It has been suggested that the interconversion

$$\text{neutron} + \text{proton} \rightarrow \text{proton} + \text{neutron}$$

provides the basis for the force of attraction; certainly a neutron can decay into a

27

proton and an electron, with the emission of energy. Furthermore, many stable atoms possess equal (often even) numbers of protons and neutrons, although the elements of high atomic weight possess more neutrons than protons.

Notwithstanding the nature of the nuclear force, the stability of the nucleus is revealed by the fact that the mass of the nucleus is not exactly equal to the total of the masses of the individual nucleons. Mass (equivalent to energy in terms of the Einstein equation, $E = mc^2$, where c is the velocity of light) is lost as the nucleons so interact as to produce a more stable situation. The energy required to separate

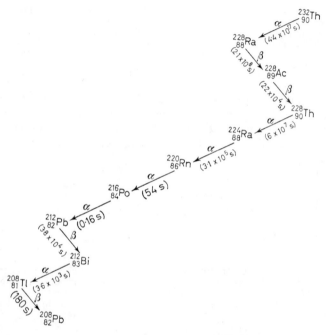

Figure 2.5. A representative, naturally occurring radioactive series

the nucleons of the nucleus is known as the *binding energy* of the nucleus. *Figure 2.4* shows the dependence of binding energy per nucleon upon the mass number, and it is clear from this that the most stable elements tend to be of intermediate mass number and that conversion of elements at the extremes into these more stable elements would result in the release of energy. This is the basis for the nuclear processes known as fusion and fission (p. 29).

Many of the elements of very high mass number (all the natural isotopes of elements 84 to 92) are spontaneously unstable and undergo *radioactive decay*, that is, they emit various particles of very high energy as they form a more stable nucleus (*Figure 2.5*).

For example, the loss of a helium nucleus (an α-particle) leaves a nuclear residue of mass number 4 units and atomic number 2 units less than the original atom:

$$^{226}_{88}\text{Ra} \rightarrow {}^{222}_{86}\text{Rn} + {}^{4}_{2}\text{He}$$

The loss of an electron (a β-particle) from the nucleus involves the conversion of a neutron into a proton, leaving the mass virtually unchanged but giving a new element with atomic number one unit more than the original:

$$^{228}_{88}\text{Ra} \rightarrow {}^{228}_{89}\text{Ac} + {}_{-1}^{0}e$$

The disintegration of an unstable isotope is not influenced by any external conditions, and so the rate of decay is proportional to the number of unstable atoms present; i.e. it corresponds to a first-order reaction (p. 128). Clearly, the rate will decline as the number of unstable atoms remaining becomes less, and so it will take an infinitely long time for decay to become complete. Hence we refer to the *half-life* of a radioactive element, or the time taken for half of the atoms to decay.

A knowledge of the half-lives of radioactive elements can be useful in calculating the effects of 'fall-out' from nuclear explosions and in dating rocks and fossils. For example, radioactive carbon-14 (^{14}C) is produced in the atmosphere by the absorption of a neutron (activated by cosmic radiation) into an atom of nitrogen:

$$^{14}_{7}\text{N} + {}^{1}_{0}n \rightarrow {}^{14}_{6}\text{C} + {}^{1}_{1}p$$

Plants, whilst alive, constantly absorb this radioactive carbon into their system by photosynthesis; death, however, puts a stop to this and so the proportion of this isotope decreases as it undergoes decay and is not replenished. A knowledge of the proportion and the half-life (5700 years) for this isotope permits calculations of the approximate time of death to be carried out—values that can be of great importance to geologists and anthropologists. Similar calculations, based on the proportion and half-lives of various uranium isotopes, suggest that the earth was formed about 5 000 million years ago.

Artificially radioactive isotopes can be made by bombardment of atoms with atomic particles, e.g. in the atomic pile. Because of the radiation which these emit on decay, they act as 'tracers' in the elucidation of reaction mechanisms; for example, carbon dioxide containing ^{14}C is used in photosynthesis—compounds subsequently found to contain this particular isotope can then be assumed to have been formed, directly or indirectly, from carbon dioxide. In this way, progressively more detail is obtained about the reactions taking place.

The best-known isotope which undergoes fission, that is, splits into two or more smaller atoms of roughly comparable mass, is uranium-235 (^{235}U). In each fission process, more neutrons are produced than are used, so that further fission is induced and the process, once initiated, can become a chain reaction with consequent explosion, provided that the material exceeds the critical size, i.e. retains sufficient neutrons and allows sufficient heat build-up, e.g.

$$^{235}_{92}U + ^{1}_{0}n \rightarrow ^{236}_{92}U \rightarrow ^{143}_{56}Ba + ^{90}_{36}Kr + 3\,^{1}_{0}n + 28pJ\ fission^{-1}$$

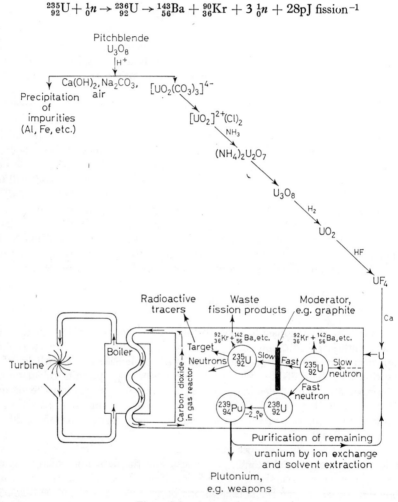

Figure 2.6. Atomic fission

This is the situation for pure ^{235}U; natural uranium contains about 140 times as much ^{238}U as ^{235}U. As the former is an excellent absorber of neutrons, there is little prospect of a chain reaction developing, and natural uranium is therefore far safer than it might have been thought at first to be the case. By using a 'moderator for slowing down neutrons, however, it is possible to ensure a favourable balance, as

^{238}U does not readily capture slow neutrons; ^{235}U therefore still undergoes appreciable fission. This is what happens in the slow reactor, where the heat generated is converted into electricity. At the same time, the capture of neutrons by ^{238}U leads to the 'breeding' of fissile material, in the shape of plutonium which, like ^{235}U, has military potential (*Figure 2.6*):

$$^{238}_{92}U + ^{1}_{0}n \rightarrow ^{239}_{92}U \xrightarrow{-\beta} ^{239}_{93}Np \xrightarrow{-\beta} ^{239}_{94}Pu$$

With the fast reactor, ^{235}U is enriched over the 238 isotope by taking advantage of the slight difference in their masses, commonly by the fractional diffusion of volatile compounds such as uranium hexafluoride.

At very high temperatures, such as those of stellar atmospheres and thermonuclear bombs, *fusion* of hydrogen atoms or its isotopes can take place, with the formation of helium. The binding energies (*Figure 2.4*) are such that considerable energy is evolved, e.g.

$$^{2}_{1}H + ^{3}_{1}H \rightarrow ^{4}_{2}He + ^{1}_{0}n + 2 \cdot 8 \text{ pJ}$$

relative mass

$$\underbrace{2 \cdot 015 \quad 3 \cdot 017}_{5 \cdot 032} \qquad \underbrace{4 \cdot 003 \quad 1 \cdot 009}_{5 \cdot 012}$$

i.e. $0 \cdot 020$ units of mass are converted into $2 \cdot 8$ pJ fission^{-1} of energy.

SUMMARY

The electronic arrangement of an atom is governed by the permissible values taken by the four quantum numbers, which represent the availability of suitable orbitals. According to the order in which these stable orbitals are filled, the following types of outer electronic configuration result, giving the characteristic structure of the Periodic Table:

$ns^{1 \text{ and } 2}$	s block of representative elements
$ns^2 \, np^{1 \ldots 6}$	p block of representative elements
$(n-1) \, d^{1 \ldots 10} ns^2$	d block of transitional elements
$(n-2) f^{1 \ldots 14} (n-1) \, d^1 ns^2$	f block of inner transitional elements

Some atomic nuclei are unstable and undergo decay, at a rate indicated by the half-life characteristic of the nucleus in question. Absorptions of radiations, e.g. α- and β-particles, can lead, by subsequent disintegration, to artificial radioactivity.

The conversion of less stable into more stable atoms results in the evolution of large quantities of energy; in fission processes, larger atoms rearrange as smaller ones, while in fusion, small atoms like hydrogen are converted into larger atoms such as helium.

QUESTIONS

1. Calculate the abundance of the two isotopes of bromine (of mass numbers 79 and 81) from its atomic weight.

2. Dr. Prout in 1815 suggested that all atomic weights were simple multiples of that of hydrogen. Discuss this hypothesis in the light of modern knowledge.

3. Discuss the reasons why some atoms are unstable.

4. The count rate (s^{-1}) of a radioactive isotope fell from 1 000 to 250 in an hour. Find the half-life of the isotope.

5. Given that the Avogadro constant is $6 \cdot 02 \times 10^{23}$ mol^{-1}, calculate the energy which would be released by the reaction

$$ {}^1_1p + {}^3_1H \rightarrow {}^4_2He $$

6. In the Periodic Table (p. 25), hydrogen and helium are placed apart from the main table. Comment on this and discuss alternative ways of accommodating these two elements.

7. The atomic weight of lead found in two different minerals is $206 \cdot 01$ and $207 \cdot 9$. Suggest a reason for this discrepancy.

8. Write an essay on the uses of radio-isotopes in industry and medicine.

9. Rewrite and complete in the form of equations, the radioactive disintegrations of the following naturally occurring, unstable nuclides:

[e.g. $^{227}Ac(\beta^-, a)$ is $^{227}Ac + {}_{-1}^{0}e \rightarrow {}^{223}_{86}Rn + {}^4_2He$]

$^{210}Bi(\beta^-, a)$, $^{14}C(\beta^-)$, $^3H(\beta^-)$, $^{223}Fr(\beta^-, \gamma, a)$,

$^{231}Pa(a, \gamma)$, $^{226}Ra(a, \gamma)$, $^{222}Rn(a)$, $^{228}Th(\beta^-)$,

$^{238}U(a, \gamma)$

10. Calculate the potential difference required to accelerate an electron from rest to a velocity of 10^6 m s^{-1}

11. 'It was inevitable that we should come to know more of matter than of the less tangible energy, and that we should seek to quantify matter much earlier than energy. Now, with our realisation of the equivalence of mass and energy, and with our discovery of an increasing number of atomic 'particles', it is clear that our ideas must go back into the melting-pot. In short, we have not advanced as far as we had hoped in the understanding of such basic terms as mass, matter and energy.' Discuss.

CHAPTER 3

THE CHEMICAL BOND

Attention must now be turned to the ways in which atoms can combine to form molecules.

As seen in Chapter 2, the noble gases have completely full inner shells and an outer octet of electrons (except helium, which has only two electrons in its electron shell), each electron in every orbital being paired with another of opposite spin. The noble gases are monatomic and form only a very limited number of compounds. A reasonable deduction is therefore that a fairly stable state for any atom to exist in is that in which its electronic configuration has been modified to correspond to that of the nearest noble gas. This is typical of all the elements except, to some extent, those of the transition series.

An atom can attain such a configuration in one of three ways: by the acceptance or removal of electrons or by sharing electrons with other atoms. These possibilities give rise to the two basic types of valence—*electrovalence* and *covalence*.

Although it is customary to refer to compounds as being of one type or the other, it must be realized that few compounds are ever completely covalent or electrovalent: the majority of compounds have bonding of an intermediate character.

THE ELECTROVALENT (IONIC) BOND

An electrovalent bond involves a transference of one or more electrons from one atom to another, with the resultant formation of charged particles or ions.

Simple electronic transfer to produce an **electro**valent bond is illustrated by sodium chloride:

$$\underset{1s^22s^22p^63s'}{\text{Na}} \quad \rightarrow \quad \underset{1s^22s^22p^6}{\text{Na}^+} + e^- \quad \text{(now with core of noble gas, neon)}$$

$$\underset{1s^22s^22p^63s^23p^5}{\text{Cl}} + e^- \rightarrow \quad \underset{1s^22s^22p^63s^23p^6}{\text{Cl}^-} \text{(now with core of noble gas, argon)}$$

Ionization energy—The removal of an electron from an element requires a certain amount of energy, expressed in terms of the *ionization energy* (*Table 3.1*). The main factors which influence the size of this energy are

(*i*) *the distance of the electron from the nucleus*: the larger the atom, the greater will be the distance between the positively charged nucleus and the negative electron which is to be removed, and hence, since electrostatic attraction varies inversely as the square of the distance separating opposite charges, the lower will be the ionization energy;

(*ii*) *the charge on the nucleus*: the greater the positive charge on the nucleus, the greater will be the attractive forces between it and the outer electrons, and hence the

Table 3.1. Some representative first ionization energies

Element	Ionization energy kJ mol⁻¹	Element	Ionization energy kJ mol⁻¹
H	1 314	Li	519
He	2 372	Na	494
Ne	2 080	K	418
Ar	1 522	Cu	745
Xe	1 163	Ag	732
Cl	1 255	Tl	590

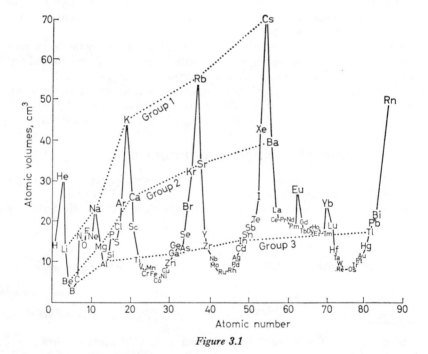

Figure 3.1

greater will be the ionization energy. The situation is, however, complicated by the varying shielding powers of different types of electrons separating the nucleus from the valency electrons; unpaired inner electrons of the transition elements do not screen the nucleus as effectively as do paired electrons.

Since the number of atoms in a mole of any element is a constant (Avogadro's constant, $6 \cdot 023 \times 10^{23}$), atomic weight/density will be a function of the volume of an atom. This ratio, called the *atomic volume*, is the volume (in cm³) occupied by 1 mole of the element. The periodic variation in atomic volume with atomic number reflects the modern classification of the elements (compare *Table 2.3* and *Figure 3.1*). The connection between the size of atoms and their ionization energies is well illustrated by the elements of the first and second Periods (*Figures 3.1* and *3.2*).

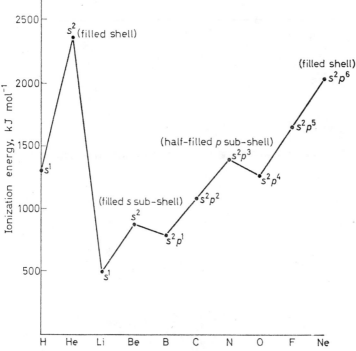

Figure 3.2. First ionization energies of Periods 1 and 2

As further electrons are removed from an atom, the ionization energy increases, since there is then a net positive charge on the atom. The removal of electrons from a completely filled shell generally requires far more energy than is normally associated with chemical changes, so that atoms or ions possessing only completely filled shells are generally stable (but see Chapter 17). *Figure 3.3* shows not only the increase of successive ionization energies but the excessive energy required to remove an electron from a complete shell.

Electron affinity—When electrons are added to an atom, there will be an absorption or release of energy. This energy is, somewhat misleadingly, referred to as the electron affinity. The only normal anions to be formed with release of energy are those of the halogens. The energy required to add a second electron to an ion already negatively charged results in the electron affinity having positive values see *Table 3.2*).

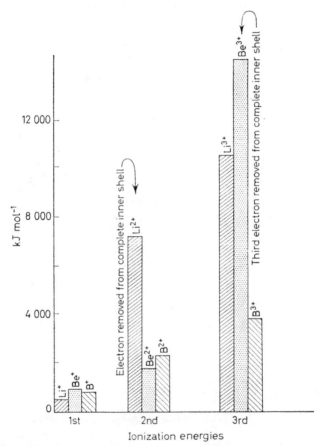

Figure 3.3. Comparison of ionization energies of lithium, beryllium and boron

Lattice and solvation energies—It would seem reasonable to suppose that an ionic compound would be formed when the ionization energy of the one element was balanced by the electron affinity of the other. As the halogens are the only elements to liberate energy when forming ions with a noble gas type configuration, this would suggest that the only ionic compounds possible would be the halides. In fact, many other ionic compounds exist; the previous discussion has omitted considerations such as the energy released in forming a crystal lattice (that required to break down one mole of a crystal lattice is called the *lattice energy*) or that evolved by an ion when it becomes solvated (the *solvation energy*). Thus the above considerations are

Table 3.2. Electron affinities ($kJ\ mol^{-1}$)

$F+e\rightarrow$ F^-	$Cl+e\rightarrow$ Cl^-	$Br+e\rightarrow$ Br^-	$I+e\rightarrow$ I^-	$O+e\rightarrow$ O^-	$O+2e\rightarrow$ O^{2-}	$S+2e\rightarrow$ S^{2-}	$Se+2e\rightarrow$ Se^{2-}
-335	-381	-331	-305	-142	$+640$	$+397$	$+423$

not by themselves sufficient to decide whether a compound is predominantly ionic or not but, when intelligently used, they can provide an indication of the likely type of bonding.

The total energy changes involved in the formation of a compound can be linked together by means of the Born–Haber cycle. This is illustrated for the case of sodium chloride in *Figure 3.4* (see also p. 121).

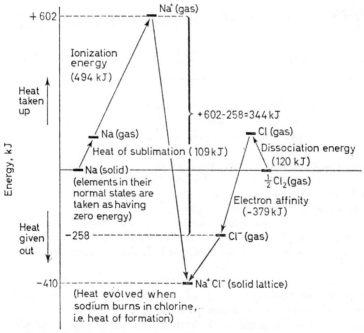

Figure 3.4. Energy changes accompanying the formation of sodium chloride
Energy required for formation of Na^+Cl^-(gas): $(109 + 494 + 120 - 379) = +344\ kJ$
therefore, lattice energy $= (+344 - (-410)) = 754\ kJ\ mol^{-1}$

Electronegativity—A measure of the tendency for a *combined* atom to pull electrons towards itself can be determined approximately from the average of the ionization energy and electron affinity of the uncombined atom, since both ionization energy and electron affinity are a measure of the electron-attracting capacity of an atom. Originally, electronegativity values were based on the difference between experimental and calculated bond energies (p. 119) because any difference between these two values was taken to represent the difference between a polarized and a pure covalent bond (*Table 3.4*). A direct theoretical assessment of electronegativity can now be made, using Coulomb's law (p. 39), but, nevertheless, the application of electronegativity is largely in qualitative terms. The elements on the right-hand side of the Periodic Table have the greatest electronegativities, and in any one Group the electronegativity falls as the atomic number increases. This general pattern is in accord with the values of atomic volumes.

Thus, of the naturally occurring elements, caesium has the smallest electronegativity, whilst fluorine is the most electronegative element (*Table 3.3*).

Table 3.3.   Some representative electronegativity values

(from E. Cartmell and G. W. A. Fowles, *Valency and Molecular Structure*, 3rd ed., p. 140, London, Butterworth, 1970)

H 2·1						
Li 1·0	Be 1·5	B 2·0	C 2·5	N 3·0	O 3·5	F 3·9
Na 0·9	Mg 1·2	Al 1·5	Si 1·8	P 2·1	S 2·5	Cl 3·0
K 0·8	Ca 1·0		Ge 1·8	As 2·0	Se 2·4	Br 2·8
Rb 0·8	Sr 1·0		Sn 1·8	Sb 1·9	Te 2·1	I 2·5
Cs 0·7	Ba 0·9					

The greater the difference in electronegativity between two combining atoms, the larger will be the ionic character of the bond; caesium fluoride is therefore a highly ionic compound, while carbon dioxide and silicon sulphide are typical covalent compounds.

To summarize, it can safely be said that the elements of Group 1 of the Periodic Table (i.e. those which only need to lose one electron to attain an s^2p^6 configuration) will be those which most readily form positive ions (*cations*), followed by the elements of Group 2, which require to lose two electrons. Further, as each Group is descended, the atomic radii become larger and loss of electrons becomes easier. In Groups 3 and 4 there will be a progressively smaller tendency to form ions with s^2p^6 configurations because this would entail the loss of three and four electrons, respectively.

In the same way it may be expected that the elements which would most readily form negative ions (*anions*) by a gain of electrons would be those of Group 7 (the halogens), followed by those of Groups 6 and 5, in that order.

It must be stressed that the conclusions drawn from this discussion are of a very general nature and exclude considerations of the transition elements. The changes in the Transition Periods are neither as great nor as regular as for the non-transitional elements, since inner quantum levels are being filled with electrons. This aspect will be dealt with in more detail in Chapter 16.

Properties of Electrovalent Compounds

The ionic bond confers very distinct properties, but it must be remembered that very few compounds are completely ionic, and so deviations from these properties must be expected.

An important feature of ionic solids is that the forces holding the ions together are of an electrostatic nature and have no preferential direction. They are said to be coulombic, as they obey Coulomb's law:

$$F \propto \frac{z_1 e \times z_2 e}{r^2}$$

where e = electronic charge

z_1, z_2 = the two valencies of the ions

r = distance between the charges

F = a force of attraction if the charges are unlike and of repulsion if like.

The solid therefore will consist of ions regularly disposed with regard to packing and charge considerations (see Chapter 4). No discrete molecules exist in the lattice and the structure is firmly held together in all directions; it will therefore be hard. Because of the large electrostatic forces which have to be overcome to separate the constituent particles, ionic compounds have a high melting point. For similar reasons the liquid produced has a high boiling point. The larger the distance between the ions, the smaller the force of attraction, in accordance with Coulomb's law, and generally the lower the m.p. and b.p. Thus, whereas lithium fluoride boils at 1 680°C, sodium chloride boils at 1 440°C and potassium iodide at 1 320°C. Similarly, ions of multiple charge experience correspondingly larger forces and have higher boiling points, e.g. magnesium oxide, $Mg^{2+}O^{2-}$, does not even melt until 2 800°C while calcium chloride, $Ca^{2+}Cl_2^-$, boils at 1 600°C.

A further feature is that, when fused, an ionic compound is capable of carrying an electric current. This is transported by the ions present in the liquid, the positively charged cations migrating to the cathode and the negatively charged anions to the anode.

Ionic compounds only dissolve in solvents which themselves are of a polar nature (i.e. compounds such as water and liquid ammonia, in which there is an uneven distribution of charge or dipole; see p. 43). Briefly, the explanation is that, for the solid to dissolve, the ions must somehow acquire a corresponding stability to counteract the potential energy of the crystal lattice which is destroyed. This they can do if the solvent molecules can solvate some or all of the ions, thus giving them a solvation energy. Polar solvents are able to interact in this way, and hence the solid will tend to dissolve. Differences in solubility between compounds can thus be partly attributed to differences in lattice and solvation energies. A further consideration is that high dielectric solvents (i.e. insulators) such as water can insulate any ion which has dissolved from the resultant oppositely charged solid,

39

thus hindering its return to the lattice. Such solvents will therefore assist the dissolution of ionic compounds.

When in solution, the reactions of ionic compounds are the reactions of their constituent ions; thus, all bromides will instantaneously react with silver nitrate solution to precipitate silver bromide:

$$Br^- + Ag^+ \rightarrow AgBr \downarrow$$

THE COVALENT BOND

(a) Simple Covalency

This involves the sharing of electrons of opposite spin from different atoms. The sharing normally continues until no unpaired electrons remain in the outer orbitals. It has been shown that in ionic compounds the atoms usually acquire a noble-gas-type configuration; by the process of sharing electrons this is often the case also with covalently bound atoms, although here the atom does not exert complete control over all the electrons. It is also possible for elements in Period 3 and above to acquire more than eight electrons in their outer shells by this process of sharing, because of the availability of d and sometimes f as well as the usual s and p orbitals; it is this fact that is responsible for many marked differences between the elements of Periods 2 and 3.

Covalent bond formation is best explained by some examples. Consider first the molecule of hydrogen. The hydrogen atom has the configuration $1s^1$, so that by sharing this electron with a second hydrogen atom, both atoms acquire the stable s^2 configuration. This process can be represented in two ways:

$$\underset{s^1}{H\cdot} + \underset{s^1}{H\cdot} \rightarrow \underset{s^2}{H{:}H} \text{ (or } H\text{—}H)$$

or, in terms of orbitals

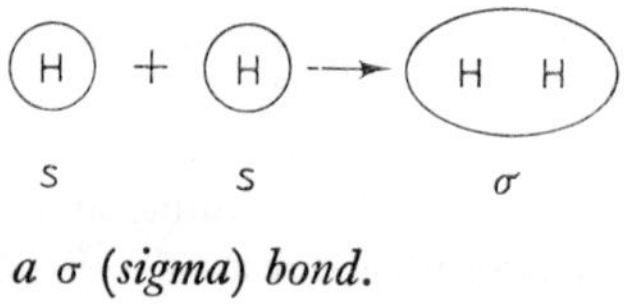

i.e. *the two s orbitals form a σ (sigma) bond.*

Similarly, for the case of chlorine, the electronic configuration of the atoms is $KL3s^23p^5$, which can be alternatively represented as

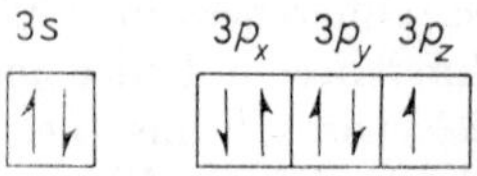

40

One of the p electrons is unpaired so that, by sharing another p electron from a second chlorine atom, a stable configuration is obtained:

$$\overset{\bullet\bullet}{\underset{\bullet\bullet}{:Cl}}\,\overset{\bullet}{} \quad + \quad \times\overset{\times\,\times}{\underset{\times\,\times}{Cl}}\,\overset{\times}{} \quad \longrightarrow \quad \overset{\bullet\bullet}{\underset{\bullet\bullet}{:Cl}}\,\overset{\bullet}{\times}\,\overset{\times\,\times}{\underset{\times\,\times}{Cl}}\,\overset{\times}{} \quad (\text{or } Cl-Cl)$$

$$s^2p^5 \qquad\qquad s^2p^5 \qquad\qquad s^2p^6$$

(Only the electrons in the valency shell are shown, since these are the ones mainly affected by bond formation. The electrons are represented as $\cdot$ and $\times$, although after chemical union the electrons of equal energy from the different atoms are indistinguishable from one another.)

In terms of orbitals, this bond formed by the p orbitals can be represented as

$$P_x \qquad\qquad\qquad P_x \qquad\qquad\qquad \sigma$$

i.e. *the p orbitals form a σ (sigma) bond.*

As a further example, consider carbon tetrachloride. Carbon initially has the configuration $K2s^22p^2$, and to complete its octet it requires a share in one electron from each of four chlorine atoms

$$\cdot C \cdot \quad + \quad 4\times\overset{\times\,\times}{\underset{\times\,\times}{Cl}}\,\overset{\times}{} \quad \longrightarrow \quad \begin{matrix} \overset{\times\times}{\times Cl\times} \\[-2pt] \times Cl\times C\times Cl\times \\[-2pt] \times Cl\times \\[-2pt] \times\times \end{matrix} \quad \text{or} \quad \begin{matrix} Cl \\ | \\ Cl-C-Cl \\ | \\ Cl \end{matrix}$$

two shared electrons being equivalent to a single chemical bond. In its state of lowest energy, the carbon atom has only two unpaired electrons (the p electrons) and hence, before it can form four chemical bonds, one of the s electrons must be promoted to the vacant p orbital, thus giving four unpaired electrons and allowing a valency of four:

C

$1s^2$	$2s^2$	$2p^2$			$\longrightarrow$	$1s^2$	$2s$	$2P_x$	$2P_y$	$2P_z$
↑↓	↑↓	↑	↑			↑↓	↑	↑	↑	↑

Ground state   Excited state

The description of the carbon tetrachloride molecule in terms of orbitals is more complicated than those so far discussed. The four orbitals required of the carbon

41

atom can be described as formed by a combination (*'hybridization'*) of the one s and the three p orbitals

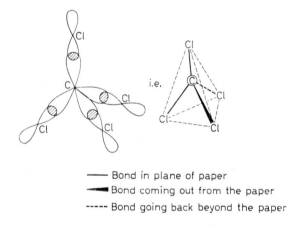

The bonds are then formed by the overlap of these orbitals with the appropriate p orbitals of the chlorine atoms

—— Bond in plane of paper

◄▬ Bond coming out from the paper

----- Bond going back beyond the paper

Covalency implies that there is an even distribution of charge between the atoms making up the molecule. As with ionic compounds, however, a completely covalent structure is rare. If only two atoms are linked together and are identical, as in the chlorine molecule, then their electronegativities must of necessity be equal and the electrons forming the bond will be symmetrically placed between the atoms, with the result that the molecule will be non-polar. However, if one of the atoms has a greater electronegativity than the other, then it will attract the electrons towards its end of the bond and the other atom will become relatively positive; there will thus be a tendency towards an ionic nature, although in this case the electrons have not been completely transferred from the one atom to the other, and the tendency will be greater the greater the difference in electronegativity. This polar nature is very common and is well represented by the hydrogen halides, $H^{\delta+} \longrightarrow X^{\delta-}$ (*Table 3.4*).

Table 3.4

HX	Electronegativity of X	Percentage ionic character
HF	3·9	43
HCl	3·0	17
HBr	2·8	11
HI	2·5	5

The displacement of charge gives rise to a *dipole moment* (by analogy with the magnetic moment of a magnet, the dipole moment of a bond is equal to the displaced charge multiplied by the length of the bond; for a molecule, the vector sum of the bond dipole moments gives the dipole moment of the molecule; see p. 511). The values for the hydrogen halides are shown in *Table 3.5*.

Table 3.5

HX	Dipole moment, $\times 10^{-30}$C m
HF	6·3
HCl	3·3
HBr	2·6
HI	1·3

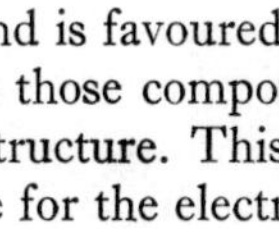

Dipole moment $= l \cdot \delta+$

The formation of a covalent bond is favoured when the elements concerned are of similar electronegativity and in those compounds where a small cation of high charge would be the alternative structure. This high charge would confer on the cation such a great attractive force for the electrons of the anion—an effect known as *polarizing power*—that the latter would be pulled towards the cation, resulting in a return to a covalent structure (as the anion is polarized). This is the basis of *Fajans' rules:*

1. The larger the charge on the cation or anion, and
2. the smaller the size of the cation and/or the larger that of the anion, the higher

43

will be the percentage of covalent bonding involved (*Figure 3.5*).

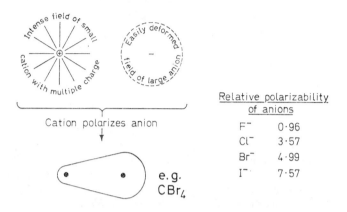

(*a*) Considerable polarization: structure largely covalent

Cation polarizes anion

e.g. CBr₄

Relative polarizability of anions	
F⁻	0·96
Cl⁻	3·57
Br⁻	4·99
I⁻	7·57

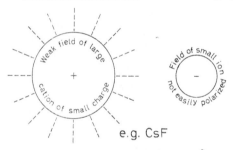

(*b*) Negligible polarization: structure largely ionic

e.g. CsF

Figure 3.5. Polarization of anion by cation

(b) Dative Covalency

Covalency has been defined as the sharing of electrons between two atoms, one electron being supplied from each atom. There is, however, a further possibility, namely that one of the atoms supplies both of the electrons for the bond. This type of bond is often referred to as a coordinate bond or as dative covalency, but there is no fundamental difference between this and ordinary covalency. The conditions necessary for the formation of such a bond are that the *donor atom* shall have a pair of electrons (a *lone pair*) not already involved in bond formation and that the *acceptor atom* requires two electrons to attain a stable configuration. As its name implies, this type of bond is particularly common in coordination compounds (p. 322), but a suitable illustration is provided by aluminium chloride. If we con-

44

sider the molecule $AlCl_3$, in which the constituent atoms are held together by covalent bonds, then the aluminium atom has only six electrons in its outer orbitals:

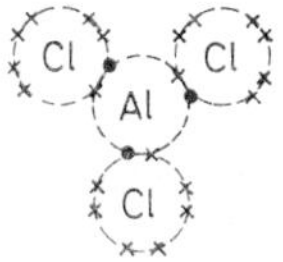

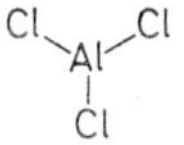

The chlorine atoms, however, each have three pairs of electrons not involved in bonding. If one of these atoms were to donate a pair of electrons to another aluminium atom in a second $AlCl_3$ unit, then this second aluminium atom would be surrounded by eight electrons. Similarly, one of the chlorine atoms attached to this aluminium atom can donate a pair of electrons to the first aluminium atom, completing its octet and conferring stability on the resulting dimeric molecule, Al_2Cl_6:

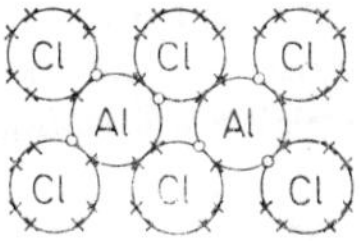

(Aluminium chloride exists mainly as this dimer at temperatures just above sublimation point but increasingly as the monomer at temperatures above 400°.)

A coordinate bond may be represented by an arrow pointing from the donor to the acceptor atom

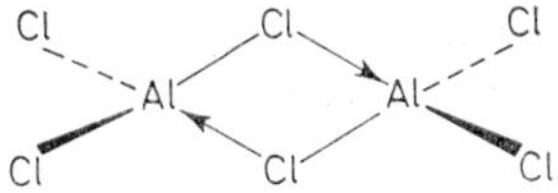

Properties of Covalent Compounds

Covalent molecules are usually discrete individuals held together by strong bonds in definite directions. As the atoms which form a bond do so by sharing electrons, it may be supposed that they will approach each other in such a way that the relevant electronic orbitals will overlap as much as possible. It was seen in Chapter 2 that p, d and f orbitals are directional and it is therefore to be expected that covalent molecules will have definite fixed shapes.

Consider methane, CH_4. There will be eight electrons in the outer valency orbitals of the carbon atom (four from the carbon and one from each of the four hydrogen atoms), divided into four pairs. As electrons are all negatively charged, these four pairs will arrange themselves as far apart as possible. This is equivalent to saying that they will be tetrahedrally arranged:

$H\hat{C}H = 109°28'$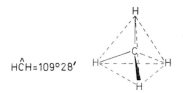

This is in agreement with the shape of the carbon tetrachloride molecule which has earlier (p. 41) been shown in terms of hybridization of the s and three p orbitals of the valency electrons of the carbon atom.

In the case of the ammonia molecule, NH_3, there will still be eight electrons in the outer orbitals of the central atom (five from the nitrogen and one from each of the hydrogen atoms), but of the four pairs only three will be involved in bond formation and the fourth will exist as a lone pair. As there are still four pairs altogether, they will still be orientated towards the corners of a tetrahedron; the lone pair, however, will reside closer to the nitrogen atom than the pairs in the N—H bonds and will exert a repulsive effect on these bonds. This will make the bond angles contract somewhat. Thus the ammonia molecule is pyramidal, with $H\hat{N}H = 108°$ (as compared with the tetrahedral angle of 109°28')

The lone pair on the nitrogen atom can be donated to a proton, thus forming the tetrahedral ammonium ion (it must be stressed that, in the ammonium ion, all four bonds are equivalent and no one bond can be identified as being the coordinate link):

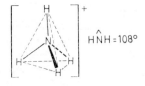 $H\hat{N}H = 108°$

Similarly, in Group 6, oxygen exerts a valency of two, but in its hydride, water, there are two lone pairs. These will exert a greater distorting effect on the bond pairs than the one lone pair does in the ammonia molecule, and although the overall structure is still based on a tetrahedron, the molecule will be V-shaped, with an even smaller bond angle:

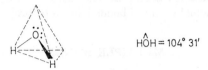

$H\hat{O}H = 104° \; 31'$

Although the bonds in covalent molecules are themselves very strong, there are no great forces between individual molecules. In the crystalline state, the various molecules are conveniently packed but the only forces holding the molecules together are weak forces (*van der Waals' forces*, p. 63) such as that produced when two adjacent molecules distort the electronic clouds of each other and induce a weak electrostatic attraction between them, that is, form an *induced dipole*

2 molecules

As a result, crystals of most covalent compounds are easily broken, soft, and have low melting and boiling points. There are, however, some structures which are bonded throughout by covalent links; these will be very hard and have high melting points. A notable example of such a structure is that of diamond, which may be thought of as a *giant molecule* (see p. 78). In some crystals the properties are also appreciably modified by the presence of the weak bond, known as the hydrogen bond (p. 52).

The contrast between ionic and covalent compounds is well illustrated by the change in melting points of the chlorides of the elements of the Second Period in proceeding from left to right, i.e. from a predominantly ionic to a predominantly covalent molecular compound:

	NaCl	MgCl$_2$	Al$_2$Cl$_6$	SiCl$_4$	PCl$_3$	SCl$_2$	Cl$_2$
m.p.°C	800°	712°	193°	−70°	−112°	−78°	−102°

(at $2 \cdot 26 \times 10^5 \text{N m}^{-2}$)

Unlike ionic compounds, fused covalent compounds will not conduct an electric current because of the absence of any charged particles to act as transferring agents (see Chapter 6):

	NaCl	MgCl$_2$	Al$_2$Cl$_6$	SiCl$_4$
Molar conductivity, $\Omega^{-1} \text{cm}^2 \text{mol}^{-1}$	133	29	$1 \cdot 5 \times 10^{-6}$	0

Also, in contrast to electrovalent compounds, covalent substances generally

47

dissolve in non-polar solvents but not in polar solvents, although in water hydrolysis may occur; e.g. silicon tetrachloride is hydrolysed to an indefinite hydrate of silica, $SiO_2 . nH_2O$. Their reactions with other substances in solution usually take place slowly because actual chemical bonds have to be broken and reformed.

MULTIPLE BONDS

Sometimes two atoms combine together by sharing not two but four or even six electrons, giving rise to multiple bonds. Nitrogen normally exists as a diatomic molecule; it can readily be seen that for each nitrogen atom to have an octet of electrons in its outer orbitals, each atom must share three electrons with the other:

$$\overset{xx}{\underset{x}{N}}\,\overset{}{\underset{}{x}} \;+\; \overset{oo}{\underset{}{\overset{}{o}N}} \longrightarrow \; {}^{x}_{x}N{}^{o}_{o}\,{}^{x}_{x}N{}^{o}_{o} \quad \text{or} \quad N\!\equiv\!N$$

From this it might be thought that all these bonds are equivalent and that the bond strength is three times that of a single bond, but this is not the case. The electronic configuration of the nitrogen atom is $K2s^2 2p^3$, i.e.

1s	2s	$2p_x$	$2p_y$	$2p_z$
↑↓	↑↓	↑	↑	↑

and the bonding electrons are the three p electrons, which give rise to two types of bond.

It has been shown above (p. 41) that when the p orbitals overlap in an end-on manner, they form an orbital which embraces the two atoms in the same way as when two s orbitals overlap; such bonds, whatever types of orbital they are formed from, are known as σ (**sigma**) **bonds**. A second possible mode of interaction between p orbitals is by sideways overlap; in this case the resultant orbital consists of two negative charge clouds—one above and one below the line joining the centres of the two atoms. The bond formed by this process is called a π (**pi**) **bond**.

Both types of bond are exhibited in the nitrogen molecule (*Figure 3.6*).

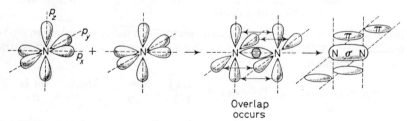

Overlap
occurs

Figure 3.6. The nitrogen molecule

In view of its importance in organic chemistry, a suitable element to consider more fully is carbon. It has been shown that the four bonds in methane are equivalent and pointing towards the corners of an imaginary regular tetrahedron, and that the s and three p orbitals do not retain their individual identities but instead they give rise to four equal orbitals ($[sp^3]$ hybrids) separated by the tetrahedral angle (p. 42).

This accounts very satisfactorily for compounds in which carbon is exerting a valency of four, but other carbon compounds exist in which, for example, each carbon atom is only joined to three other atoms, e.g. ethene, C_2H_4. In this case hybridization of the s orbital and of two of the p orbitals of carbon occurs, to produce three $[sp^2]$ hybrid bonds, all in one plane and directed towards the corners of an imaginary equilateral triangle. The p orbital remaining is at right angles to the plane of the triangle:

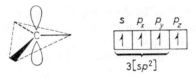

This lone p orbital can interact with one in the same plane on an adjacent atom to yield a π bond, as in the nitrogen molecule. The molecule of ethene can thus be pictured as shown in *Figure 3.7*.

i.e. $H_2C=CH_2$

Figure 3.7. The ethene molecule

It is also possible for carbon to use the s orbital and one p orbital to form two $[sp]$ hybrid orbitals.

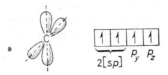

This will produce two linear σ bonds and there will be two p orbitals remaining which will be at right angles both mutually and also to the direction of the σ bonds.

These p orbitals can interact to give two π bonds, as for example in ethyne, C_2H_2 (*Figure 3.8*).

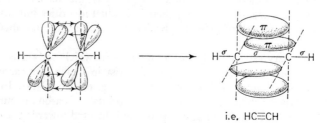

i.e. HC≡CH

Figure 3.8. The ethyne molecule

This molecule can therefore be described as being surrounded by a cylindrical cloud of negative charge, equal in magnitude to that of four electrons (cf. N≡N).

A further example is afforded by but-1,3-diene, $CH_2=CH—CH=CH_2$. The formation of the π bonds in this molecule can be represented as

but there is obviously a probability that the p orbital on the second carbon atom will interact with that on the third

This is equivalent to saying that the complete molecule will be enveloped in a *delocalized* π-type charge cloud and that all three of the C—C bonds will exhibit some double bond character (a phenomenon often called *resonance*). That this is so is shown by the intermediate lengths of the bonds in the molecule, as compared with normal single and double bonds between carbon atoms:

C—C
154 pm (1·54Å)

$H_2C=CH—CH=CH_2$
135 pm   144 pm   135 pm
(1·35Å)   (1·44Å)   (1·35Å)

C=C
133 pm (1·33Å)

50

One other compound is worthy of consideration, namely benzene, C_6H_6. This is often formulated as

H
|
C
H—C   C—H
H—C   C—H   or more simply as
C
|
H

Each carbon atom is joined to three other atoms; this is an example of sp^2 hybridization, and the benzene molecule is a regular, planar hexagon, each carbon atom having one spare p orbital perpendicular to the plane of the ring. These six orbitals can interact together in either of the two ways shown:

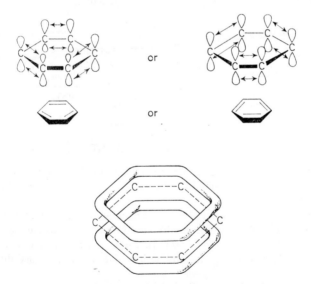

or

or

Figure 3.9. The benzene ring

As there is an equal probability of the two forms occurring, the actual structure is intermediate between the two, and the individual π orbitals are fused into one large delocalized π cloud (see above). Such a view accounts very satisfactorily both for the observed equality in length of all the C—C bonds in the molecule and for the chemical properties of this compound (p. 382).

51

THE HYDROGEN BOND

A hydrogen atom is so small that it very nearly represents a 'point' charge, as a result of which, although it has only a valency of one, another suitable atom can be held by an attraction of mainly electrostatic character. Because of the electrostatic nature of the force, only atoms which have a large electronegativity will form such hydrogen bonds; indeed this phenomenon appears to be confined to fluorine, oxygen, nitrogen and possibly sulphur. Chlorine, although highly electronegative, does not appear to form hydrogen bonds (but see chloral hydrate, p. 464), presumably because it is too large to approach sufficiently closely. A hydrogen bond may be represented by:

$$\overset{\delta-}{-X}-\overset{\delta+}{H}\cdots\overset{\delta-}{Y}-$$

$$\underbrace{\qquad\qquad}$$

H bond

Reference to *Table 3.5* will show that the energy associated with a hydrogen bond is very small in comparison with the bond energies of normal chemical bonds, but nevertheless the presence of hydrogen bonding between molecules can produce very marked effects, as examples in later Chapters will show.

Table 3.5. Bond energies (i.e. energy needed to *break* the bond)

Hydrogen bond	Bond energy, $kJ\ mol^{-1}$	Covalent bond	Bond energy, $kJ\ mol^{-1}$
$H\cdots F$	41·8	H—F	433
$H\cdots O$	29·4	O—H	461
$H\cdots N$	8·4	N—H	391

THE METALLIC BOND

About 70 per cent of all the elements are metals, and a discussion of valency would be incomplete if it did not include some mention of the way in which metallic atoms are held together.

Normally in metals there is simply a close packing of identical atoms, each atom having eight or twelve equidistant neighbours, i.e. it is said to have a *coordination number* of eight or twelve (see Chapter 4). Such a structure confers the typical properties of metals such as high melting point, high density, malleability, ductility and opacity.

Covalent bonding in the normal sense is precluded, not only because a large number of electrons would be required, but also because such a large number of suitable orbitals is not available. The only way in which a covalent structure could exist with such a large coordination number would be by having a *resonating* structure.

Resonance implies that the real structure has an intermediate position between the various possible structures which can be drawn conventionally. For example,

52

for a coordination number of six and a valency of three for all the atoms in a structure, the following possibilities can be drawn:

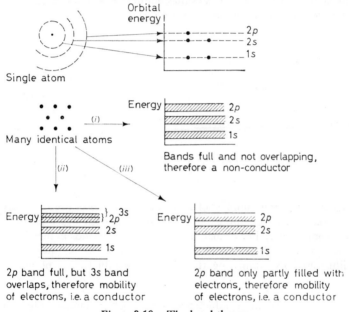

This would logically lead to a situation in which the valency electrons would be mobile. The picture thus emerges of identical cations in a 'sea' of mobile electrons; these electrons are the valency electrons of the original atoms, now no longer connected to any one atom but instead acting as the 'cement' binding the cations

Figure 3.10. The band theory

together. This idea of mobile electrons immediately explains the good electrical and thermal conductivities of metals generally. In ionic compounds, the ions carry the current whereas, in metals, electrons perform this service.

An extension of this model can be used to explain the different conducting powers of metals and also the phenomenon of *semi-conduction* in certain elements, e.g. silicon, germanium, grey tin and tellurium. The energy of each electron in the 'sea' of electrons is clearly not the same because the electrons may not all have come from the same type of orbital. Further, when many atoms come together to form a metal there are many electrons present which are in

53

identical quantum states. By Pauli's principle (p. 22) this situation is not permissible, so the energy levels 'expand' into energy bands (*Figure 3.10*). If (*i*) the bands do not overlap and (*ii*) each readily available band is completely filled with electrons, the element is a non-conductor at low temperatures. At high temperatures, enough energy may be supplied to promote some of the electrons to empty energy bands of greater energy, thus making conduction possible. The nearer an empty band is to the highest filled band, the lower will be the temperature at which conduction begins, and as the temperature is increased above this point, so the conductivity will increase. This behaviour characterizes a semi-conductor. The presence of impurities can alter the semi-conductor properties considerably, either by providing fresh empty energy bands or by altering the position of existing bands relative to each other (*Figure 3.11*). Semi-conductors find application in *transistors* and as *photoelectric devices*, e.g. in exposure meters.

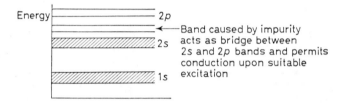

Figure 3.11. Effect of an impurity on a non-conductor

The higher the atomic number of an element, the closer are the energy levels; the energy bands also get progressively nearer to each other as the quantum numbers become larger. The availability of conduction bands in metals is increased by the overlapping of the energy levels for those with higher quantum numbers. This is illustrated by comparing the electrical conductivities of potassium and copper

$$\text{Atomic conductances } (\Omega^{-1} \text{ cm}^{-1} \div \text{ atomic volume in cm}^3)$$
$$\text{K} \qquad 0.004$$
$$\text{Cu} \qquad 0.084$$

The latter element is a better conductor than the former, even though they both possess only one unpaired electron: in copper, however, there is a greater overlapping of bands, as the unpaired electron is in a higher quantum shell.

OXIDATION STATE

Oxidation can be defined as a reaction involving partial or complete loss of electrons (and reduction, as a reaction involving a gain of electrons). For example, in the reaction

$$Fe^{2+} + \tfrac{1}{2}Cl_2 \rightarrow Fe^{3+} + Cl^-$$

iron is oxidized and chlorine reduced.

Many reactions involve oxidation and reduction, although their names do not suggest this:

Displacement such as $Cu^{2+} + Zn \rightarrow Cu + Zn^{2+}$ involves oxidation of the dissolving metal and reduction of the metal being displaced from solution.

Combination between metal and non-metal involves oxidation of the metal and reduction of the non-metal: $2\,Mg + O_2 \rightarrow 2\,Mg^{2+}O^{2-}$.

If a metal loses one electron per atom, it is said to have gained an oxidation number or state of $+1$; a non-metal gaining one electron per atom is similarly said to acquire an oxidation number of -1. In the case of the formation of potassium chloride from the elements

$$K + \tfrac{1}{2}Cl_2 \rightarrow K^+ Cl^-$$
Oxidation number $\quad 0 \qquad 0 \qquad +1\ -1$

Because the usual covalent bond involves elements of different electronegativity, with the less electronegative element acquiring a slight positive charge at the expense of the other, the above concept of oxidation state still holds, and the oxidation number can be regarded as the charge which each atom would carry in the hypothetically ionized state. Thus, in the compound $H^{\delta+}-Cl^{\delta-}$, hydrogen is in an oxidation state of $+1$ and chlorine of -1.

The application of simple rules permits the oxidation state of an atom in any given compound to be worked out:

(*i*) atoms in the elemental form are in zero oxidation state;

(*ii*) hydrogen in compounds other than the ionic hydrides is in the $+1$ oxidation state;

(*iii*) oxygen in compounds is in the -2 oxidation state, except in peroxides (-1) and oxygen difluoride ($+2$);

(*iv*) the halogens in halides are in the -1 oxidation state.

The oxidation state of a particular atom in an ion or compound can be determined by breaking the latter down into its component atoms. The algebraic sum of the oxidation states of the individual atoms is then equated to the charge on the complete entity. This can be made clear by some examples:

(*a*) Phosphoric acid has the formula H_3PO_4; each hydrogen atom is in oxidation state $+1$, each oxygen atom is in oxidation state -2, giving a total of -5 for the hydrogen and oxygen atoms together. As the charge on the molecule is zero, the atom of phosphorus must be in oxidation state $+5$.

(b) Potassium dichromate, $K_2Cr_2O_7$

$$+ 2 \ ? \ -14 = 0$$

The sum of the oxidation states of the two chromium atoms is 12 and the oxidation number of the chromium is therefore $+6$.

SUMMARY

Orbitals are, on the whole, more stable when either empty or containing their full complement of two electrons of opposite spin. Chemical reaction, leading to the formation of chemical bonds, involves the vacating or filling of orbitals.

Metals tend to lose electrons in the course of chemical reaction to give positive ions and are thus described as 'electropositive'. In the presence of non-metals, valency electrons are transferred to the non-metallic atoms and negative ions are formed, so that non-metals are said to be 'electronegative'. The chemical bond so formed is called *electrovalent* or *ionic*. In the absence of non-metals, the valency electrons form a 'cement' binding together the tightly-packed metallic ions. Both this *metallic bond* and the ionic bond give substances capable of conducting electricity: in the case of the metal itself, the valency electrons are free to move under an applied potential, even in the solid state. With ionic compounds, however, electrical conduction is the result of migration of ions and not electrons; these are free to move only in the liquid state, and so ionic compounds conduct electricity only after melting or dissolving has broken down the ionic lattice.

Reaction occurs between atoms of non-metals by sharing electrons. For single bonds, overlapping of orbitals takes place along the line joining the centres of the atoms to give σ- (sigma) bonds, e.g.

p-orbital   p-orbital σ-bond

Overlapping of any remaining p-orbitals at right angles to the σ-bond produces a π-(pi) bond with electron clouds above and below the single bond: this is the situation for the formation of multiple bonds

p-orbitals π-bond

Because there are no charged particles available in these covalent compounds, they are incapable of conducting electricity, even in the liquid state. If the covalent link takes place between atoms of differing electronegativity, the bonding electrons

56

will not be shared equally (i.e. the electron cloud will not be uniform), and there will be some separation of charge. The molecule will, in fact, be a dipole

○ more electronegative than •

When hydrogen is attached to the extremely electronegative elements fluorine, oxygen or nitrogen, there is sufficient separation of charge to give a detectable electrostatic attraction between different parts of adjacent molecules. This attraction is the basis of the *hydrogen bond*.

Electron clouds of adjacent atoms can interact with each other to give electrostatic induction and attraction (rather like the induction and attraction which result from a magnet being brought close to a bar of iron). This is one aspect of the *van der Waals forces* operating between molecules.

There is a wide variation of bond energy associated with these various chemical bonds. Whereas a metallic, ionic or covalent bond can have a bond energy exceeding 400 kJ mol^{-1} a hydrogen bond is usually less than 40 kJ mol^{-1} and a van der Waals bond seldom greater than 8 kJ mol^{-1}.

QUESTIONS

1. Using dots and crosses to indicate valency shell electrons of different atoms, construct structures for the following compounds:

CaO; H_2S; HNO_3; $Cu(NH_3)_4SO_4.H_2O$; $K_3Fe(CN)_6$; CO; CH_3OCH_3

Explain what types of bond are present in each of the compounds.

2. What factors affect the formation of an ionic bond? Why is a C^{4+} ion unknown in any compound?

3. Xenon is known to form a tetrafluoride. Discuss the possible nature of the bonds in this compound.

4. Using the following information, construct a Born–Haber cycle and from it deduce the sublimation energy of iodine:

Heat of formation of sodium iodide $= -$ 291 kJ mol^{-1}
Heat of sublimation of sodium $=$ 109 kJ mol^{-1}
1st ionization energy of sodium $=$ 494 kJ mol^{-1}
Heat of dissociation of iodine $=$ 151 kJ mol^{-1}
Electron affinity of I $= -305$ kJ mol^{-1}
Lattice energy of sodium iodide $=$ 674 kJ mol^{-1}

5. Deduce the shapes of BF_4^-, N_2H_4, H_3O^+, H_2O_2, $BF_3.NH_3^*$.

6. How would you expect the shapes of the molecules to change in the series H_2O, H_2S, H_2Se?

7. What is the oxidation state of the named atom in the following: sulphur in SO_3, nitrogen in NH_2OH, chlorine in $HClO_4$, vanadium in $[V_5O_{14}]^{3-}$, zirconium in $[ZrO(SO_4)_2]^{2-}$?

8. Discuss the various approaches that chemists have made to the concept of electronegativity.

9. Write an account of the hydrogen bond in nature, with particular respect to water and the transmission of information inside the living cell.

10. Write an account of semi-conductors.

11. Comment on *Figure 3.12*.

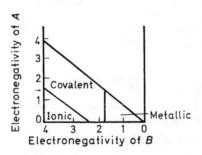

Figure 3.12. Bond type of the compound AB

12. What conclusions can you draw from the following information?

Substance	B.p. K	Molar heat of evaporation (kJ)
Benzene	353	34·8
Caesium chloride	1 573	160
Chlorine	240	20·4
Hydrogen	20	0·92
Potassium	1 033	91·6
Rubidium	952	75·6
Silver chloride	1 837	178
Sodium	1 155	103
Sodium chloride	1 738	170

CHAPTER 4

STATES OF MATTER

INTRODUCTION

Matter exists in three common, easily recognized states—the solid, liquid and gaseous states. (A fourth state, plasma, exists at extremely high temperatures; in this, the energy is such that electrons and nucleons have a largely independent existence.) The most obvious difference between the solid, liquid and gaseous states is in the capacity to maintain a fixed shape and volume. At a given temperature, a solid is rigid and of a definite volume, a liquid occupies a definite volume but it can flow to occupy a vessel of any shape, and a gas has neither shape nor a fixed volume but will expand to fill completely any container into which it is introduced.

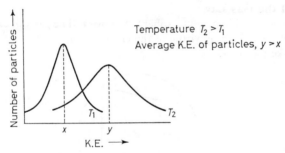

Figure 4.1. Distribution of kinetic energies in a system of molecules

These differences in behaviour can be explained in terms of a kinetic (Greek *kinesis* = motion) theory of matter. This postulates that at all temperatures above absolute zero, 0 K or −273°C, atoms and molecules are in motion. In a solid this motion is very small and probably consists essentially of oscillations of the particles about a mean position. As the temperature is raised (and the kinetic energy increased), this motion becomes more vigorous, until the particles collide with each other and are no longer restricted to particular positions; this represents fusion, and the resultant liquid consists of mobile particles. The interatomic or intermolecular forces, however, are still sufficiently strong to cause the volume at a fixed temperature to remain constant. If still more heat energy is applied, a stage is ultimately reached where the kinetic energy imparted to the molecules is suffi-

59

ciently great to overcome the forces between the molecules, and the result is a gas which can occupy any space available.

For pure substances, melting and boiling occur at definite temperatures; how then can evaporation of liquids be accounted for? At a given temperature it is the *average* kinetic energy of the system which is constant: it does not follow that *all* the individual particles will be endowed with this particular kinetic energy. In fact, the kinetic energies of the molecules will be distributed in accordance with normal probability laws (*Figure 4.1*), and there will be a certain proportion with greater kinetic energy, moving more quickly than the remainder.

It is the faster-moving particles which, when at the surface of the liquid, will be able to overcome the attractive forces exerted by the molecules in the liquid. They will accordingly escape into the atmosphere, and the liquid is said to evaporate. As the molecules which escape are those with greatest energy content, it follows that the average kinetic energy, and hence the temperature, of the liquid will tend to fall. Eventually, if the liquid is in a closed space, the vapour will become saturated and particles which have lost energy by collision will re-enter the liquid at the same rate as fast particles leave; a state of *dynamic* equilibrium will then be obtained.

THE GASEOUS STATE

Deduction of the Gas Laws

Consider a gas enclosed in a hollow sphere of radius r. The gas will exert a pressure on the walls of the sphere by virtue of the change in momentum of the molecules when they strike the sides.

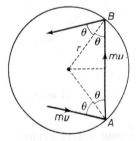

Figure 4.2. Motion of a molecule in a spherical container

Assume that collision between the molecules of both the gas and the wall of the vessel are perfectly elastic, so that there is no loss of momentum. Now consider 1 molecule, of mass m, moving with a speed u; then, if θ is the angle of incidence on collision with the wall, the change in momentum is $2\ mu \cos \theta$ (*Figure 4.2*). The distance, AB, between two impacts with the wall is $2\ r \cos \theta$, and hence the number of impacts per second is $u/2\ r \cos \theta$. Therefore, the rate of change of momentum is

$$2\ mu \cos \theta \times \frac{u}{2\ r \cos \theta} = \frac{mu^2}{r}$$

Now, the rate of change of momentum is equal to the force (Newton's second law of motion), and pressure is equal to force per unit area. Hence the pressure caused by one molecule moving inside the sphere is given by

$$p = \frac{mu^2}{r} \times \frac{1}{4\pi\, r^2} = \frac{mu^2}{4\pi\, r^3}$$

But the volume, v, of a sphere $= \frac{4}{3}\pi r^3$; therefore, $pv = \frac{1}{3}\,mu^2$ and, for n molecules moving with speed u,

$$pv = \tfrac{1}{3}\,nmu^2$$

However, as explained earlier, the molecules in a gas are not all moving with the same speed, and therefore u^2 is the mean of the square velocity of the molecules, i.e. $u^2 = (u_1^2 + u_2^2 + \ldots u_n^2)/n$, and u is the 'root mean square velocity'.

The kinetic theory assumes that the average kinetic energy of the molecules $\frac{1}{2}\,mnu^2$, is proportional to the temperature; thus at a given temperature the kinetic energy is constant. It follows that $pv = \frac{1}{3}\,mnu^2$ is constant or, for a fixed mass of gas at constant temperature, volume is inversely proportional to pressure, which is, of course, a statement of *Boyle's law*.

Rewriting the above equation:

$$pv = \tfrac{2}{3}\left(\tfrac{1}{2}\,mnu^2\right)$$

i.e. $pv = \frac{2}{3}$ (average kinetic energy). Therefore, $pv \propto T$ (where T is the temperature on the absolute scale) so that, for a mass of gas at constant pressure:

$$v \propto T$$

which is the statement of *Charles' law*.

Since $pv \propto T$, pv/T must be a constant. For one mole of gas, the constant is designated R and for n moles of gas, $pv/T = nR$ or

$$pv = nRT$$

which is the *general gas equation*. One mole of gas occupies $22\cdot4$ dm³ (litres) at $1\cdot01 \times 10^5\,\mathrm{N\,m^{-2}}$ (1 atm) and 273K, thus

$$R = \frac{pv}{T} = \frac{1\cdot01 \times 10^5 \times 22\cdot4}{273} = 8\cdot31 \times 10^{-3}\,\mathrm{N\,m^{-2}\,dm^3\,K^{-1}\,mol^{-1}}$$
$$= 8\cdot31\,\mathrm{J\,K^{-1}\,mol^{-1}}$$

If there are equal volumes of two gases at the same temperature and under equal pressure, then $\frac{1}{3}m_1 n_1 u_1^2 = \frac{1}{3}\,m_2 n_2 u_2^2$, and because the average kinetic energies are equal, $\frac{1}{2}\,m_1 u_1^2 = \frac{1}{2}\,m_2 u_2^2$. Thus $n_1 = n_2$, i.e. equal volumes of all gases at the same, temperature and pressure contain equal numbers of molecules (*Avogadro's principle*).

It follows that, at s.t.p., i.e. at 0°C (the standard temperature) and 1.01×10^5 N m^{-2} (1 atm or 760 mmHg) (the standard pressure) the volume occupied by one mole of all perfect gases, the Gramme Molecular Volume, is constant. It has been found experimentally to be 22·4 dm^3 (1).

For example, the density of hydrogen at s.t.p. is 0·09g dm^{-3}. Its molecular weight is 2.016, giving a value for the gramme molecular volume (G.M.V.) at s.t.p. of $2.016/0.09 = 22.4$ dm^3.

Further, $u^2 = 3pv/nm$; but nm/v is the density, d, of the gas and therefore $u^2 = 3p/d$ or, at constant pressure

$$u^2 \propto 1/d$$

$$u \propto \sqrt{(1/d)}$$

This is equivalent to *Graham's law of diffusion* which can thus be derived from a kinetic theory of matter.

The *law of partial pressures* also follows from the kinetic theory; since the molecules are independent of each other, their number alone (and not their nature) is relevant. In a mixture of gases, therefore, the pressure that a particular gas exerts (the partial pressure) is that which it would exert if it alone occupied the volume. The total pressure is then merely the sum of the partial pressures.

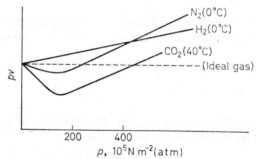

Figure 4.3. Deviations from Boyle's law for some cases

The kinetic theory, and hence the gas laws deduced from it, has been derived on the basis of several assumptions which are known to be oversimplifications. Hence it is not surprising that no real gas perfectly obeys these laws. The variations shown by some common gases are shown in *Figure 4.3*. (Assuming Boyle's law to be correct, a plot of pv against p or v should be a straight line parallel to the axis.)

There are two main reasons for the approximate nature of these laws. The derivation of the kinetic theory assumed that the molecules themselves occupied a negligible volume in comparison with that of the gas as a whole. This becomes increasingly incorrect as the pressure on the gas is increased and the molecules are brought closer together. Thus there will be considerable deviations from the laws at high pressures. Secondly, it has been assumed that there are no forces of attraction between the molecules. But such forces exist and will add to the pressures exerted upon the gas so that the measured, externally applied, pressure will be lower than the true value. This explains the decrease in the product, pv, at low pressures for most gases (*Figure 4.3*). In fact, at sufficiently low temperatures all gases exhibit this phenomenon. Also, because of the molecular attraction, work has to be done by an expanding gas in overcoming this interaction, so that a cooling effect results which is of considerable practical importance in the liquefaction of gases (p. 83).

In an effort to make allowance for these deviations from ideal behaviour, several modifications have been made to the general gas equation, $pv = nRT$. One of the simplest is that due to van der Waals, who suggested the equation

$$\left(p + \frac{a}{v^2} \right)(v - b) = RT$$

where a and b are constants. The former provides for the extra pressure caused by the mutual attraction of the molecules, while the latter makes allowance for the molecular volumes and can be shown to be equal to four times the actual volume of the molecules. As a result of this equation, the intermolecular forces are now referred to as *van der Waals forces*. They are very weak compared with those associated with conventional bonds, seldom exceeding 8 kJ mol^{-1}, whereas the covalent bond can exceed 400 kJ mol^{-1}.

Molecular Weights of Gases

Since the molecular weight of a gas is equal to twice its vapour density (p. 9), measurements of the latter afford a convenient method for the determination of the molecular weight of any gas or of any volatile substance which can easily be converted into the gaseous state, subject to the inevitable slight inaccuracy caused by the non-ideality of the gas.

The most direct method of molecular weight determination will accordingly be by comparison of the weight of a known volume of the gas under consideration with the experimentally observed weight of an equal volume of hydrogen measured under the same conditions. This is the basis of Regnault's method. In theory it is very simple, but in practice, owing to the very small differences in weight involved, accurate results are difficult to obtain. The gas which is being weighed must be perfectly dry, since small amounts of water vapour will cause considerable errors; the globe containing the gas must be repeatedly filled and emptied to ensure that no traces of air remain; precautions must be taken to ensure that the gases are both at atmospheric pressure, and also the decrease in weight caused by the buoyancy of the globe must be balanced by counterpoising a similar globe on the other arm of the balance.

A more convenient method for determining vapour densities is that due to Victor Meyer. A known weight of volatile liquid is made to vaporize and displace its own volume of air. This air is then collected over water and measured at atmospheric pressure and temperature. A conventional apparatus is shown in *Figure 4.4*. The outer tube contains a liquid whose temperature can be maintained at least 20° above the boiling point of the volatile liquid. After this temperature has been maintained for some time, so that equilibrium has been attained, a small sample of the substance whose vapour density is required is quickly introduced into the inner tube and the stopper replaced. The substance immediately volatilizes and expels an amount of air equal to its own volume into the graduated collecting tube.

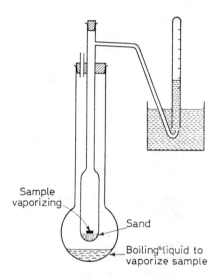

Sample vaporizing

Sand

Boiling liquid to vaporize sample

Figure 4.4. Victor Meyer's apparatus

The volume occupied now will of course be that which the vapour would have occupied if it had been at the temperature and pressure of the receiver. This can be measured, adjusted for the vapour pressure of the water in the graduated tube and converted to s.t.p. Thus the volume occupied at s.t.p. by a known weight of the volatile substance has been determined and hence the vapour density can be deduced from the knowledge that at s.t.p. the density of hydrogen is 0.09 g dm^{-3}. Alternatively, the need to rely upon the vapour density relationship can be avoided by utilizing the fact that the gramme molecular weight of any gas occupies the gramme molecular volume, i.e. 22.4 dm^3 at s.t.p.

Example—0.083 g of bromoethane when volatilized in a Victor Meyer apparatus displaced 18.5 cm^3 of air (adjusted to atmospheric pressure), collected over water at $16°C$. The atmospheric pressure was 755.3 mmHg; the vapour pressure of water at $16°C$ is 13.6 mmHg. Calculate the molecular weight of bromoethane.

The pressure exerted by the bromoethane itself was $(755\cdot3 - 13\cdot6) = 741\cdot7$ mmHg; the volume occupied by 0·083 g of bromoethane at s.t.p. is therefore

$$18\cdot5 \times \frac{741\cdot7}{760} \times \frac{273}{289} \text{ cm}^3$$

and the weight which would occupy 22·4 dm³ (22 400 cm³) is given by

$$\frac{22\ 400 \times 0\cdot083 \times 760 \times 289}{18\cdot5 \times 741\cdot7 \times 273} = 109 \text{ g}$$

Hence, the molecular weight of bromoethane is 109.

The success of this method depends upon strict adherence to certain precautions. The quantity of substance used must be such that none of its vapour escapes from the vessel in which it is volatilized, but only air which has been displaced. If the former occurs, the vapour will condense in the receiver and invalidate the result. It is also important to maintain a constant temperature in the apparatus and to avoid direct contact between the inner tube and the heating liquid; otherwise, superheating might occur and give rise to irregular expansion and contraction of the air in the tube.

Neither of the methods so far mentioned is suitable for compounds—particularly organic ones—which decompose at their boiling point. For such substances, it is necessary to volatilize them at pressures considerably below atmospheric; by this means, their boiling point will be much lower than usual. A suitable apparatus for such determinations is that of Hofmann. The material is introduced into the vacuum above a column of mercury. The tube containing the mercury is surrounded by a jacket containing the vapour of a compound which boils at a temperature higher than that being investigated. The substance vaporizes and the pressure is given by the difference produced in the height of the mercury column.

Example—The mercury in a barometer tube of length 98 cm and cross-sectional area 0·8 cm² fell from a height of 75·9 cm to 30·8 cm after the introduction of 0·077 g of ethoxyethane at 100°C. Calculate the molecular weight of the ether.

The volume of ether at 100°C (373K) and a pressure of (75·9–30·8) cm is $(98\cdot0 - 30\cdot8) \times 0\cdot8$ cm³. Therefore, the volume corrected to s.t.p. is

$$\frac{(98\cdot0 - 30\cdot8) \times 0\cdot8 \times (75\cdot9 - 30\cdot8) \times 273}{373 \times 76} \text{ cm}^3$$

22·4 dm³ at s.t.p. then weighs

$$\frac{0\cdot077 \times 373 \times 76 \times 22\cdot4 \times 1\ 000}{(98\cdot0 - 30\cdot8) \times 0\cdot8 \times (75\cdot9 - 30\cdot8) \times 273} = 74 \text{ g}$$

Assuming the vapour behaves as an ideal gas at 100°C, 74 is the molecular weight of the ether, since it is that which would occupy 22.4 dm^3 (litre) at s.t.p.

A source of error in all molecular weight determinations based on gas densities is the non-ideality of the system. *Figure 4.3* shows that at very low pressures the deviation from ideality is much less; therefore, by referring gases and vapours to conditions of low pressure, more accurate use can be made of the gas laws.

The *limiting density* is defined as

$$w/p_0 v_0$$

where $p_0 v_0$ is the value extrapolated to zero pressure, an imaginary state (see *Figure 4.5*), and w is the weight of the gas.

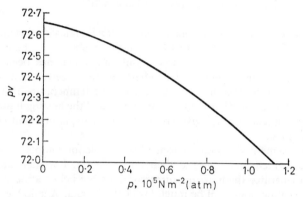

Figure 4.5. Variation of pv with p for hydrogen chloride

The normal density is

$$w/p_1 v_1$$

where $p_2 = 10^5 \text{ N m}^{-2}$ (1 atm) and v_1 is the corresponding volume; therefore

$$\text{limiting density} = \text{normal density} \times (p_1 v_1 / p_0 v_0)$$

The limiting density can then be used to give a more accurate molecular weight for the gas or vapour by using the expression

$$22.4 \times \text{limiting density (g dm}^{-3}) = \text{molecular weight}$$

A further method for the determination of the molecular weight of a gas or vapour is to determine its rate of diffusion relative to that of a substance of known molecular weight and then to make use of Graham's law of diffusion. In practice, it is difficult to determine rates of diffusion and it is more common to study the *effusion* of gases through a small orifice.

The Formulae of Gases

Avogadro's law permits the formulae of many gases to be determined. For the elucidation of formulae it is often vital to know the atomicity of hydrogen and other elements. These have been determined by Cannizzaro's method (see Chapter 1). His reasoning for the diatomicity of hydrogen has since been confirmed by a determination of the specific heat of the element at constant pressure and at constant volume. (The ratio of the specific heat of a gas at constant pressure to that at constant volume, c_p/c_v, is a constant, the value of which depends upon the atomicity of the gas. For monatomic gases its value is 1·67, for diatomic gases, 1·4 and for triatomic gases, 1·33. The determination of this ratio is usually made indirectly by measuring the velocity of sound in the gas, for details of which a physics textbook should be consulted.)

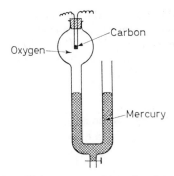

Figure 4.6. Volume composition of carbon dioxide

In general, there are two methods of approach to the problem of determining the formula of a gaseous molecule, namely by synthesis or by decomposition and measurement of the volumes of gases produced.

The first method can be suitably illustrated by the synthesis of hydrogen chloride. It is found that

1 volume of hydrogen + 1 volume of chlorine yields 2 volumes of hydrogen chloride (temperature and pressure being kept constant). Hence, by Avogadro's law

1 molecule of hydrogen + 1 molecule of chlorine yields 2 molecules of hydrogen chloride

Hydrogen and chlorine are both diatomic and therefore

1 atom of hydrogen + 1 atom of chlorine yields 1 molecule of hydrogen chloride

suggesting that the formula of hydrogen chloride is HCl.

Synthesis can also be applied to gases such as carbon dioxide: carbon is heated electrically in a closed tube with oxygen. After adjusting the temperature and

67

pressure to those at the beginning of the experiment, it is found that there is no volume change (*Figure 4.6*):

carbon + 1 volume oxygen yields 1 volume carbon dioxide

∴ xC + 1 molecule oxygen yields 1 molecule carbon dioxide

and so the formula of carbon dioxide is C_xO_2. As is always the case when not all the substances involved are gases, the vapour density is also required. This indicates a molecular weight of 44, hence

$$12x + 32 = 44$$

$$\text{and } x = (44 - 32)/12 = 1$$

i.e. the formula is CO_2.

Synthesis is the method preferred if the gas in question is very stable. Less stable gases are usually decomposed into their elements. For example, if nitrous oxide is heated in a closed tube by an iron wire (*Figure 4.7*), decomposition takes place, with the formation of oxide on the wire; the final volume of gas, nitrogen, is the same as that of the original oxide, i.e.

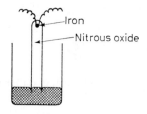

Figure 4.7. Volume composition of nitrous oxide

1 volume nitrous oxide yields 1 volume nitrogen
1 molecule nitrous oxide yields 1 molecule nitrogen ≡ 2 atoms of nitrogen

Hence the formula of nitrous oxide is N_2O_x; vapour density measurements are sufficient to show that $x = 1$. i.e. the formula is N_2O.

THE LIQUID STATE

The relative compactness of the liquid state means that there is considerable resistance to flow; this can be regarded as internal friction or *viscosity*. It is measured in terms of the force required to produce a flow of the liquid. The cause of viscosity is not completely clear, but it appears that liquids consisting of molecules of symmetrical shape are less viscous (or more mobile) than others and also that viscosity depends upon molecular size. But perhaps the major cause of high viscosity is molecular interaction. A liquid can be regarded in one sense as a disordered solid, the disorder being to a large extent a dynamic one. That is, if the system is 'frozen' in terms of time, a liquid, especially if there are large intermolecular attractions, may reveal many regions of an ordered nature. Hydrogen bonding, as for example in ethanediol and propanetriol, is a major cause of high viscosity, and it is doubtless significant that it is also a contributory factor in the ability to attain a high level of order.

Table 4.1. Some typical viscosities

10^{-4} kg m^{-1} s^{-1} (millipoise) at 20°C			
Ethoxyethane	2·3	Ethanol	12·0
Propanone	3·3	Nitrobenzene	20·1
Methanol	5·9	Propanetriol	1×10^4
Water	10·1		

Liquids, unlike gases, possess definite boundary surfaces and the molecules in the surface layer are in a special position in that some of their chemical affinity is unsaturated and the interaction with molecules in the interior is far greater than any attraction towards the few molecules present in the vapour phase (*Figure 4.8*).

Molecule well inside liquid: attractions approximately neutralized

Molecule at surface experiences resultant force inwards

Figure 4.8. The cause of surface tension

Table 4.2. Typical values of surface tension

N m^{-2}			
Water	7·4	Ethoxyethane	1·7
Mercury	54·7	Benzene	2·9

Consequently, the surface of a liquid tends to shrink and is said to be in a state of *surface tension*. If gravitational forces are in any way overcome, it assumes a spherical shape, because this has a minimum surface area for a given mass. The greater the force of molecular or atomic attraction within the liquid, the greater will be the surface tension (*Table 4.2*).

THE SOLID STATE

Broadly speaking, solid substances appear to be either crystalline, i.e. have a definite geometrical form, or amorphous. Substances in the latter category are, however, often microcrystalline and, on the atomic scale, contain regular repeating units.

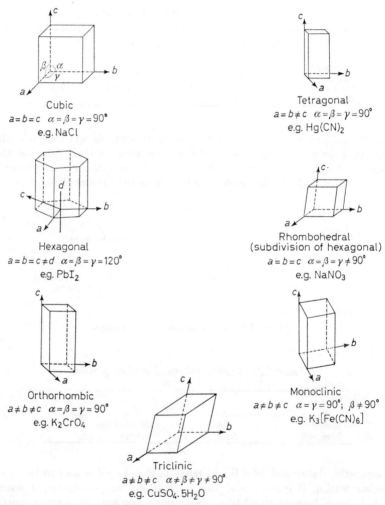

Cubic
$a=b=c \quad \alpha=\beta=\gamma=90°$
e.g. NaCl

Tetragonal
$a=b\neq c \quad \alpha=\beta=\gamma=90°$
e.g. $Hg(CN)_2$

Hexagonal
$a=b=c\neq d \quad \alpha=\beta=\gamma=120°$
e.g. PbI_2

Rhombohedral
(subdivision of hexagonal)
$a=b=c \quad \alpha=\beta=\gamma\neq90°$
e.g. $NaNO_3$

Orthorhombic
$a\neq b\neq c \quad \alpha=\beta=\gamma=90°$
e.g. K_2CrO_4

Monoclinic
$a\neq b\neq c \quad \alpha=\gamma=90°; \quad \beta\neq90°$
e.g. $K_3[Fe(CN)_6]$

Triclinic
$a\neq b\neq c \quad \alpha\neq\beta\neq\gamma\neq90°$
e.g. $CuSO_4.5H_2O$

Figure 4.9. The seven crystal systems

Disorder may still be achieved by having flexible covalent links (as for example in elastomers, which are normally amorphous but on stretching exhibit some crystallinity) or by the distribution of ions, atoms or molecules in a random fashion throughout an irregular crystal structure (e.g. glass).

For crystalline substances, there are seven, and only seven, external shapes which can be fitted together to give a solid structure without voids; accordingly these are the only crystal classes possible (*Figure 4.9*).

Metallic Lattices

The atoms of an element can be superficially regarded as identical spheres, and the structures of metals result from their close packing. *Figure 4.10* shows two layers

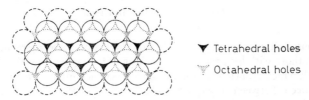

▼ Tetrahedral holes

▽ Octahedral holes

Figure 4.10. Close packing

of such a structure. A third layer of closely-packed atoms may be placed in either of two ways: either directly over the centres of the first layer or over the voids not covered by the previous layers. In the former case, the repeat unit is of the type *XYXYXY*... and is called *hexagonal close packing*, whilst in the latter it is *XYZXYZXYZ*... called *cubic close packing*. In side elevation the latter looks like a face-centred cube (*Figure 4.11*).

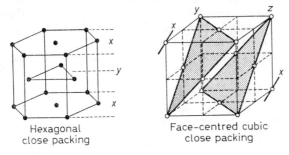

Hexagonal
close packing

Face-centred cubic
close packing

Figure 4.11. Hexagonal and face-centred cubic lattices

In both hexagonal and cubic close packing, each sphere is in contact with twelve others (6 in the same plane and 3 in each plane above and below) and is said to have a *coordination number* of twelve.

71

Some metals form a less well packed structure in which the coordination number of each atom is only eight; this is equivalent to a body-centred cubic structure (*Figure 4.12*).

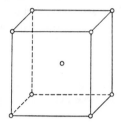

Figure 4.12. Body-centred cube

Metals possessing cubic close packing of their atoms are more malleable and ductile than those with hexagonal close packing. These two properties depend upon the ease with which planes of atoms can glide over each other; such movement can most easily occur between close-packed planes, of which there are more in the cubic than in the hexagonal structure.

A few metals have structures other than those described above; for example, indium has a tetragonal structure.

Some metals exist in more than one crystal form and are then said to exhibit *polymorphism* or *allotropy*. For example, both iron and chromium at normal temperatures exist as body-centred cubic structures but at elevated temperatures are converted into face-centred cubic crystals.

Ionic Lattices

In simple ionic structures, close packing is maintained as far as is possible, consistent with the maintenance of the correct stoichiometry. When equal-sized spheres are closely packed, there will of necessity be some free space (a hole) between them. *Figure 4.10* indicates that in cubic close packing there are two types of such holes. In one type, the hole is equally surrounded by six atoms and is thus

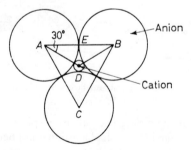

Figure 4.13. The limiting situation for 3-coordination

72

octahedral; in the other it is surrounded by four equidistant atoms and is therefore tetrahedral. In ionic structures, anions are normally arranged in some form of close packing and cations, which are often much smaller than the anions, are placed in the octahedral or in the smaller tetrahedral holes. For this to produce a stable structure, the cations must be sufficiently large to hold the anions in their positions without them coming into too close contact with each other. The structure thus depends upon the ratio cation radius : anion radius. The limiting situation for 3-coordination is shown in *Figure 4.13*.

Let the radius of the cation be r_c and that of the anions, r_a; then $AE = r_a$ and $AD = r_c + r_a$. But $AE/AD = \cos 30°$, therefore

$$\frac{r_a}{r_c + r_a} = \frac{\sqrt{3}}{2}$$

$$\therefore 2r_a = \sqrt{3}r_c + \sqrt{3}r_a$$

hence

$$2 = \sqrt{3}r_c/r_a + \sqrt{3}$$

$$\text{and} \quad \frac{r_c}{r_a} = \frac{0.268}{1.732} = 0.155$$

Similar calculations can be performed for the other possible types of coordination. The results are summarized in *Table 4.3*, whilst the ionic radii of some of the elements are shown in *Table 4.4*.

Table 4.3. Radius ratio limits for different coordination numbers

Co-ordination number	Shape	Radius ratio limits
3	plane triangular	0·155 — 0·225
4	tetrahedral	0·225 — 0·414
6	octahedral	0·414 — 0·732
8	cubic	> 0·732

Table 4.4 (a). Some representative ionic radii (pm)

Group 1		Group 2	The transitional elements					Group 3	Group 4	Group 5	Group 6	Group 7
Li$^+$ 60		Be^{2+} 31						B^{3+} 20		N^{3-} 171	O^{2-} 140	F$^-$ 136
Na$^+$ 95		Mg^{2+} 65						Al^{3+} 50			S^{2-} 184	Cl$^-$ 181
K$^+$ 133		Ca^{2+} 99	Ti^{4+} 68	Cr^{3+} 69	Fe^{3+} 64	Ni^{2+} 72	Zn^{2+} 74	Ga^{3+} 62				Br$^-$ 195
Rb$^+$ 148		Sr^{2+} 113							Sn^{4+} 71			I$^-$ 216
Cs$^+$ 169		Ba^{2+} 135							Pb^{4+} 84			

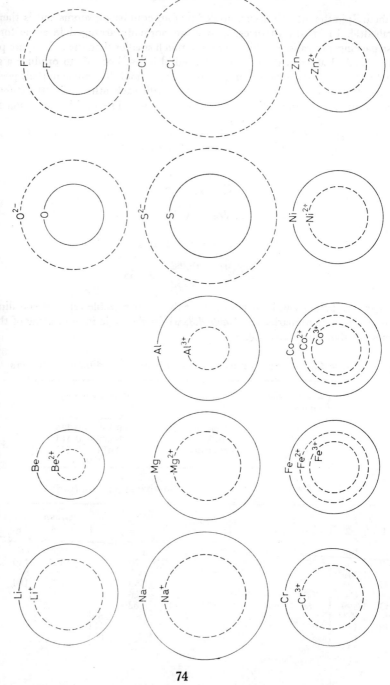

Table 4.4.(b). Comparison of atomic and ionic sizes

A very large number of ionic compounds have crystals based on one of seven different repeating units or *unit cells*. These common structures are described on the following pages.

(i) The cubic sodium chloride and caesium chloride lattices

The radius ratio for sodium chloride is 0·52 and thus octahedral coordination should be exhibited; that this is so is shown in *Figure 4.14*. Each sodium ion occupies an octahedral hole in a face-centred cubic array of chloride ions and vice versa; such coordination is referred to as 6:6.

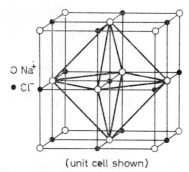

O Na^+

● Cl^-

(unit cell shown)

Figure 4.14. The sodium chloride lattice

In this structure, each corner ion is shared between eight unit cells, each ion on a face of the cell by two cells, each ion on an edge by four cells, while the ion inside the cell belongs only to that particular one. Thus there are

$$\left. \begin{array}{l} \text{8 corner } Na^+ \text{ ions} = \tfrac{8}{8} = 1 \text{ } Na^+ \text{ per cube} \\ \text{6 face } Na^+ \text{ ions} \quad = \tfrac{6}{2} = 3 \text{ } Na^+ \text{ per cube} \end{array} \right\} \text{ 4 } Na^+$$

$$\left. \begin{array}{l} \text{12 edge } Cl^- \text{ ions} = \tfrac{12}{4} = 3 \text{ } Cl^- \text{ per cube} \\ \text{1 centre } Cl^- \text{ ion} = \tfrac{1}{1} = 1 \text{ } Cl^- \text{ per cube} \end{array} \right\} \text{ 4 } Cl^-$$

in agreement with the empirical formula NaCl.

All the alkali halides, except those of caesium, crystallize with the sodium chloride structure. The radius ratios of the caesium halides are too great for octahedral coordination (e.g. $r_{Cs}^+/r_{Cl}^- = 0·93$) and accordingly, the caesium chloride structure (and also that of CsF, CsBr, CsI) has 8:8 coordination (*Figure 4.15*); (in this and subsequent structures, the lines indicated by — show the coordination number of the central atom or ion).

(ii) The fluorite lattice

In order to maintain the correct stoichiometry in compounds of the type MX_2, it is necessary for the coordination number of the cations to be twice that of the

75

anions. An arrangement which satisfies this condition is the fluorite lattice, charac-teristic of calcium fluoride and the fluorides of several other divalent metals. In this, the fluoride ions occupy the tetrahedral holes in a face-centred cubic array of calcium ions, thus giving rise to 8:4 coordination (*Figure 4.16*). When the

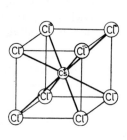

Figure 4.15. The caesium chloride lattice

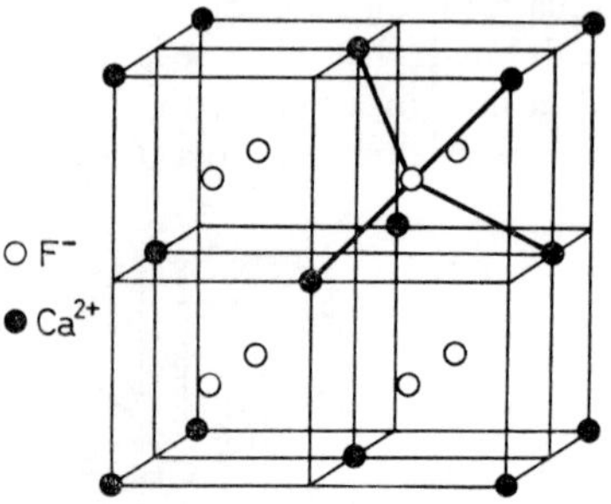

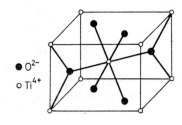

O F⁻

● Ca²⁺

Figure 4.16. The fluorite lattice

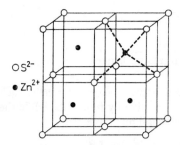

O S²⁻

● Zn²⁺

Figure 4.17. The zinc blende lattice

● O²⁻

o Ti⁴⁺

Figure 4.18. The rutile lattice

positions of the positive and negative ions are reversed, the structure is referred to as an anti-fluorite lattice, as for example in lithium oxide, Li_2O.

(iii) The zinc blende lattice (Figure 4.17)

In the fluorite lattice, the anion occupies *all* the tetrahedral holes of a cubic close-packed array of cations. In the zinc blende lattice, on the other hand, for compounds of the type *MX*, the cation occupies *half* the tetrahedral holes of a cubic arrangement of anions. The sulphides of zinc, cadmium and mercury and the halides of copper, all crystallize in this manner. Several of their compounds do not have the radius ratio 0·225 to 0·414 required theoretically for tetrahedral coordina-tion because there is a considerable degree of covalency, and the simple derivation of limiting radius ratio does not hold.

(iv) The rutile lattice

A less efficient method of packing than those discussed so far is based on a tetragonal prism and 6:3 coordination. Such a structure is that of rutile, TiO_2, in which the titanium ions are surrounded octahedrally by oxide ions, whilst each oxide ion is only surrounded by three titanium ions (*Figure 4.18*). Many compounds in which the radius ratio is about 0·55 crystallize with this structure.

(v) The corundum lattice

An example of a hexagonal lattice is that of corundum (aluminium oxide). Aluminium ions occupy two out of every three octahedral holes in a closely-packed hexagonal lattice of oxide ions, thus giving rise to 6:4 coordination. This type of lattice is adopted by many metal oxides of formula M_2O_3 (*Figure 4.19*).

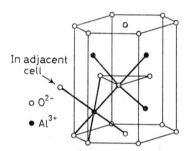

Figure 4.19. The corundum lattice

(vi) The cadmium iodide structure

As a result of the large polarizability of the iodide ions, the cadmium iodide structure is intermediate in character between an ionic and a covalent lattice. The arrangement of ions is hexagonal close packed but the structure forms well-defined layers, each cation being surrounded octahedrally by six iodide ions whilst the three nearest cation neighbours of each anion are all on one side of it (*Figure 4.20*).

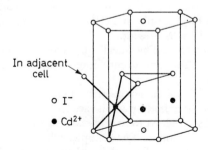

Figure 4.20. Cadmium iodide structcre

77

Covalent Lattices

Covalent structures fall into two types: those in which small individual molecules are held together in a crystal lattice by weak forces such as van der Waals forces or hydrogen bonds, and those in which the covalent bonding continues indefinitely in two or three dimensions. The former are *molecular crystals* whilst the latter are referred to as *macromolecular crystals*. Examples of each type are shown in *Figure 4.21*.

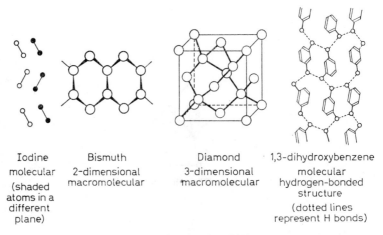

Iodine	Bismuth	Diamond	1,3-dihydroxybenzene
molecular	2-dimensional	3-dimensional	molecular
(shaded	macromolecular	macromolecular	hydrogen-bonded
atoms in a			structure
different			(dotted lines
plane)			represent H bonds)

Figure 4.21. Covalent lattices

Non-stoichiometry

In ionic lattices there are sometimes marked departures from the 'ideal' formulae. Electrical neutrality of such structures is maintained either by the presence of extra electrons held within the lattices or by atoms of the same element exhibiting different valencies, as for example in $Fe^{II}O$ which is deficient in iron and hence has a proportion of Fe^{III} ions present.

Even in ionic lattices possessing 'ideal' formulae, some of the ions may be displaced from their lattice sites and occupy interstitial holes instead, or may indeed be completely absent.

The different types of non-stoichiometric compounds occur when:

(*i*) there is a metal excess caused by anion vacancies. In this case, electrical balance is maintained by electrons (which were originally associated with anions) remaining trapped. This situation is found, for example, when excess sodium is burnt in chlorine. [The sodium chloride formed is yellow because of electrons trapped in what have become known as '*F* centres' (*F* = Farbe (German) = colour).]

78

$$Na^+ \quad Cl^- \quad Na^+ \quad Cl^-$$

$$Cl^- \quad Na^+ \quad e^- \quad Na^+$$

$$Na^+ \quad Cl^- \quad Na^+ \quad Cl^-$$

(*ii*) there is a metal excess caused by extra interstitial cations. In this case, free electrons are trapped in the vicinity of the interstitial cations

$$A^+ \quad B^- \quad A^+$$
$$e^-A^+$$
$$B^- \quad A^+ \quad B^-$$
$$A^+ \quad B^- \quad A^+$$

For example, zinc oxide upon heating loses oxygen (and becomes yellow), producing possibly Zn^+ or Zn at interstitial sites.

(*iii*) there is a metal deficiency caused by cation vacancies. The crystal structure has a complete anion lattice and an incomplete cation lattice, the cation deficiency being made up by oxidation of some of the cations to a higher oxidation state. For example, FeO is metastable but compounds within the range $Fe_{0.91}O$ to $Fe_{0.95}O$ are stable and possess both Fe^{2+} and Fe^{3+} ions:

$$Fe^{2+} \quad O^{2-} \quad Fe^{2+} \quad O^{2-} \quad Fe^{2+} \quad O^{2-}$$

$$O^{2-} \quad Fe^{3+} \quad O^{2-} \quad Fe^{2+} \quad O^{2-} \quad Fe^{2+}$$

$$Fe^{2+} \quad O^{2-} \quad \phantom{Fe^{2+}} \quad O^{2-} \quad Fe^{3+} \quad O^{2-}$$

$$O^{2-} \quad Fe^{2+} \quad O^{2-} \quad Fe^{2+} \quad O^{2-} \quad Fe^{2+}$$

A particularly noteworthy compound is $Ti^{II}O$, which exhibits both anionic and cationic deficiencies (ranging in composition from $Ti_{0.75}O$ through TiO to $TiO_{0.69}$) and also contains a large number of balanced vacant sites, e.g.

Composition	Per cent Ti sites occupied	Per cent O sites occupied
$Ti_{0.75}O$	74	98
TiO	85	85
$TiO_{0.69}$	96	66

In non-stoichiometric compounds there is a small but measurable mobility of trapped electrons and ions; as a result, such compounds are semi-conductors (p. 54). Trapped electrons can sometimes be excited by light energy, the resultant conductivity then being known as *photoconductivity*. The energy acquired from the incident

79

light is sometimes emitted as radiation of somewhat longer wavelength, giving rise to *luminescence*. This property is utilized in zinc sulphide screens for the detection of x-rays and electrons and, when mixed with small amounts of an α-particle emitter as a permanent source of energy, for the luminous hands and numbers of some clocks and watches.

A further consequence of the availability of electrons and of positive holes (i.e. vacant sites where positive ions could be placed) on the surfaces of many non-stoichiometric compounds, particularly oxides, is that they possess valuable catalytic properties.

CHANGES OF STATE

Liquefaction of Gases (*Figure 4.22*)

It was shown in the previous Section that gases obey Boyle's law only to a limited extent; *Figure 4.23* indicates that the deviation from the law is smaller the higher the temperature. This is generally the case; an increase in temperature

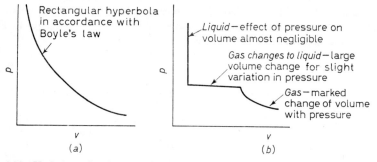

Figure 4.22. Variation of volume with pressure for gas at (*a*) high temperature, (*b*) low temperature

causes molecules to move faster, so that the mutual attraction of one for the other becomes relatively less important. On the other hand, if the temperature is progressively lowered, the attractive forces assume a greater importance, until eventually the gas condenses into a liquid. It was Andrews who, in 1869, carried out a thorough investigation into the conditions under which gases could be converted into liquids. He plotted values of volume against the corresponding pressures at various fixed temperatures. The *isothermals* obtained are shown for carbon dioxide in *Figure 4.23*.

What conclusions can be drawn from these isothermals? At 48·1 °C, carbon dioxide follows Boyle's law closely, but at 35·5 °C pronounced departures from ideal behaviour are apparent. At 21·5 °C, when the pressure on gaseous carbon dioxide is increased to about $60 \times 10^5 \mathrm{N} \ \mathrm{m}^{-2}$, liquid carbon dioxide appears and the pressure remains constant until all the gas has been liquefied. Since liquids are almost incompressible,

80

further increase of pressure has little effect. A similar pattern of behaviour is exhibited at all lower temperatures, the only significant difference being that the pressure needed to liquefy the gas decreases with decreasing temperature. In fact, carbon dioxide can be liquefied by the application of pressure at all temperatures below but not above $31 \cdot 1 °C$. This temperature is therefore known as the 'critical temperature' of carbon dioxide and the corresponding pressure as the 'critical pressure'.

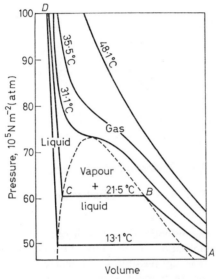

Figure 4.23. Isotherms of carbon dioxide
(From C. W. Wood and A. K. Holliday, *Physical Chemistry*, 2nd ed., London, Butterworth, 1963)

At the critical temperature, the kinetic energy tending to separate molecules equalizes the cohesive forces which maintain the liquid state. If, therefore, a liquid is heated in a closed system (so that the critical pressure can be attained), at the critical temperature the surface tension of the liquid vanishes and with it the meniscus. It is impossible to distinguish liquid from gas and we can say that we have 'continuity of state'.

Carbon dioxide is typical of other gases in its response to pressure at different temperatures, and *Table 4.5* gives the critical temperatures and pressures of other compounds. (Strictly speaking, a substance is only regarded as a gas when above its critical temperature and there is no possibility of liquefying it; below its critical temperature, when it can be liquefied by the application of pressure, it is known as a vapour.)

Table 4.5

Substance	Critical temperature °C	Critical pressure 10^5N m^{-2} (atm)
Helium	−268	2·3
Hydrogen	−234	12·8
Nitrogen	−146	33·5
Oxygen	−118	49·7
Methane	−83	45·6
Carbon dioxide	31	75·0
Ammonia	133	112·0
Water	374	218·5

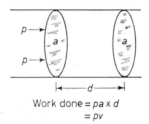

Work done $= pa \times d$
$= pv$

Figure 4.24. Work done by an **expanding** gas at constant pressure

It can be seen that ammonia and carbon dioxide can be liquefied at room temperature merely by the application of pressure. Other gases need to be cooled considerably before there is any possibility of liquefaction. The production of low temperatures is therefore extremely important. There are three fundamental ways of doing this:

(*a*) cooling by bringing into contact with cold substances; the effect can be multiplied by using materials at progressively lower temperatures;

(*b*) by allowing the gas under high pressure to expand into a region of lower pressure. Because intermolecular attraction exists between the molecules of a gas, internal work is done, as the molecules move apart, in overcoming this attraction; the energy required to perform this work can only be obtained from the kinetic energy of the molecules themselves and so the temperature falls, provided the gas is below a certain temperature, called the *inversion temperature.* For most gases, this temperature is so high (for example, for oxygen it is 700°C) that it presents no problem. For gases like hydrogen and helium, however, the inversion temperatures are low, and so initial cooling must be applied before this *Joule–Thomson effect* can be utilized (below −80°C for hydrogen and below −240°C for helium).

(*c*) by causing the gas to perform external work, for example, moving a piston and driving an engine.

Work is done when a force moves through a distance.

Now force = pressure × area, and so work done = pressure × area × distance
= pressure × volume change (*Figure 4·24*)

For an ideal gas $$p_1v_1 = p_2v_2$$

and no work is done when the gas undergoes a change in volume. For a real gas
moving from high to low pressure, work is either done by or on the gas, depending
upon the way in which the product pv varies with p. In *Figure 4.25(a)*, which
applies at lower temperatures, $p_1v_1 < p_2v_2$, and so the gas does external work in
expanding and the cooling due to this reinforces that due to the Joule–Thomson
effect. In *Figure 4.25(b)*, however, which operates at higher temperatures, $p_1v_1 >$

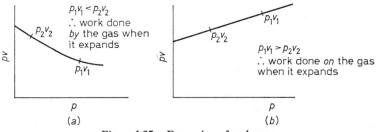

Figure 4.25. Expansion of real gases

p_2v_2, and external work is performed *on* the gas as it expands, increasing the kinetic
energy and thus minimizing or even reversing the Joule–Thomson effect. (It must
be emphasized that all these changes are effected adiabatically, that is, in isolation
from the environment.)

The most important gas liquefaction process in industry today is the production
of liquid air (*Figure 4.26*), followed by fractional distillation into its components.
Air can be liquefied by making use of all the above principles. It is compressed
isothermally to about $150 \times 10^5 \text{N m}^{-2}$ (150 atm) and cooled to $-30°C$ by the first
heat exchanger. About three-quarters of the air is then expanded to $4 \times 10^5 \text{N m}^{-2}$
by doing external work as it drives an engine: the temperature falls to about $-160°C$.
It then goes to the bottom of the fractionator, where it assists vaporization. The
remaining air is liquefied by a second heat exchanger and passed to the fractionator
at X where, as it descends, it 'washes' ascending vapour free from oxygen (b.p.
$-183°C$). The more volatile nitrogen (b.p. $-196°C$) continues to vaporize but is
condensed where it meets the condenser, cooled by cold liquid oxygen. The liquid
nitrogen is then expanded through a valve to 10^5N m^{-2} (1 atm) and enters the top
half of the fractionator Y where, as it 'scrubs' the ascending vapour free from
oxygen, it is vaporized itself and passes out from the plant through heat exchangers
which utilize its low temperature in cooling incoming air. At the same time, liquid
air from the base of the fractionator expands to 10^5N m^{-2} by passage through a
valve and enters the top half of the fractionator at Z; as it falls, nitrogen evaporates,
leaving liquid oxygen which can be removed as required. The intimate contact
between descending liquid and ascending vapour ensures satisfactory fractionation.

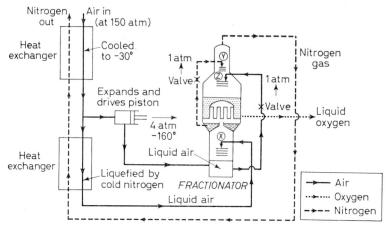

Figure 4.26. Liquefaction of air

The liquid oxygen contains the denser noble gases, notably argon, whilst helium and neon are present in the nitrogen fraction. These gases can be extracted, if required, by further fractionation.

Boiling Point and Melting Point

Although under certain circumstances (at the critical temperature and pressure) the gas and liquid phases merge into one and there is continuity of state, usually the two phases are quite distinct. The more energetic liquid molecules are able to escape from the liquid and become vapour, so giving rise to a vapour pressure. The higher the temperature, the more likely are molecules to possess the necessary kinetic energy to enable them to escape from the liquid phase, so that the vapour pressure is then greater. Eventually, at a sufficiently high temperature, it equals the external pressure and all the liquid becomes vapour, without further increase in temperature. The liquid is said to boil and the heat provided, instead of producing an increase in temperature, produces an increase in disorder (or entropy, p. 120), measured as the latent heat of vaporization. It follows that the boiling point is dependent upon the value of the external pressure, but it is constant at constant pressure. Indeed, a constant boiling point is usually taken as a criterion of purity.

Similarly, when a solid is heated, there is an increase in kinetic energy of the particles, and the temperature rises. As with the liquid, some particles have energy so much in excess of the average that they escape into the vapour phase, so that the solid, too, exerts a vapour pressure, which also increases with temperature. Eventually, vibrations of the ions, atoms or molecules become so extreme that collisions between adjacent particles occur and the crystal lattice breaks down. The melting point at which this occurs is constant at constant pressure for a pure substance, so that this, like the boiling point, is indicative of purity. At the melting point, the heat taken in, instead of producing an increase in temperature, causes an increase in

84

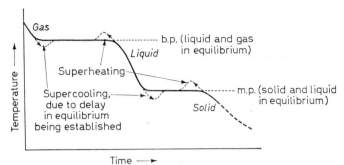

Figure 4.27. Effect of continuous cooling of gas (or heating of solid)

disorder: because the order in the liquid is far higher than in the gas, the latent heat of fusion is less than that of vaporization.

Phase Diagrams

Many of the ideas that have been discussed in this Section can be conveniently summarized in graphical form. In many cases, for pure substances, if vapour presure is plotted against temperature, two distinct curves, *AB* and *AC*, are obtained for solid and liquid, respectively (*Figure 4.28*). The point *B* is known as the triple point and is the temperature where solid, liquid and vapour co-exist. Line *BD* represents the effect of pressure on the melting point and, in the case shown, indicates that an increase in pressure lowers it. Le Chatelier's theorem (p. 133) enables us to predict that in this particular case, when solid melts, there is a diminution in volume.

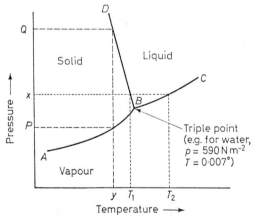

Figure 4.28. Phase diagram for a pure substance with one solid form only

If the pressure is maintained constant at x while the temperature is steadily increased, solid becomes liquid at T_1 and liquid becomes gas at T_2 (at 10^5 N m^{-2} pressure, T_2 represents the normal b.p.) If the temperature is kept constant at y while pressure is steadily increased, vapour is converted into solid at P and the solid melts at Q

85

Polymorphism (Allotropy)

If the solid is capable of existing in different forms (polymorphs or allotropes), *Figure 4.28* requires modification. This depends upon whether the allotropes are each stable under differing conditions or whether one allotrope only is stable. The former, reversible, type of allotropy is called *enantiotropy* and is exemplified by

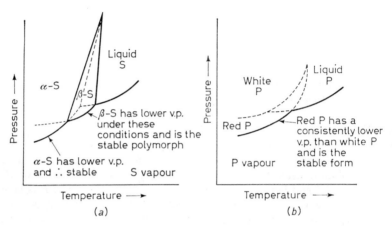

Figure 4.29   Phase diagram for pure substances, with two solid forms
(dotted lines: unstable equilibria)

sulphur (*Figure 4.29(a)*) whilst the other variety is termed *monotropy*. Phosphorus is an example of a monotropic element (*Figure 4.29(b)*).

In principle, the slow cooling of molten sulphur (or its vapour) leads to the formation of β-sulphur (the monoclinic allotrope) and eventually to α-sulphur (the rhombic form). Heating of rhombic sulphur reverses the transitions.

In the case of phosphorus, the vapour pressure of the white allotrope is always greater than that of the red form, and therefore the former is unstable (i.e. has a higher energy) in comparison with the latter. Because of this, the direct transition of the red into the white allotrope is not possible whereas the reverse change is easily produced.

Sublimation

Provided the triple-point pressure is greater than the surrounding pressure, sublimation will occur on heating a solid, i.e. no liquid will form during the transition to the vapour state. This is clearly seen from the phase diagram of carbon dioxide (*Figure 4.30*); at atmospheric pressure, sublimation occurs while liquid carbon dioxide can only be formed above the triple-point pressure.

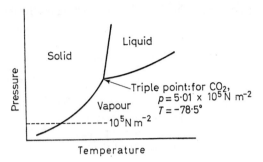

Figure 4.30. Phase diagram of carbon dioxide

SUMMARY

Matter exists between two extremes: that of the perfect gas, where molecules are completely independent of each other and where kinetic energy is high, and the perfect solid at absolute zero, where interaction is at a maximum and kinetic energy zero. The degree of randomness of atoms, molecules and ions consequently decreases from the gaseous state through the liquid to the solid state, where the highest order is attained. Physical constants such as density, melting and boiling points, critical temperature and pressure, surface tension and viscosity are dependent upon the intermolecular and interionic forces as well as the possible geometric arrangement

The structure of solids may be classified in the following terms:

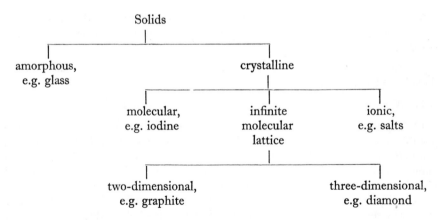

QUESTIONS

1. What properties of the liquid state show that it is intermediate between the solid and gaseous states?

2. What effect has cubic and hexagonal close packing on the density of a metallic element?

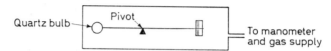

3. The diagram shows a micro-balance, in which the pressures of different gases required to float the quartz bulb are determined. If the beam is balanced in oxygen at a pressure of 5920 N m^{-2} and in argon at a pressure of 4740 N m^{-2}, calculate the atomic weight of argon, given that it is monatomic and the vapour density of oxygen is 16.

4. Explain what precautions are required in using the gramme molecular volume to calculate molecular weights.

If the normal density of hydrogen chloride is 1·64 g dm^{-3}, calculate the molecular weight of the gas, using the values given in *Figure 4.5*.

5. In an experiment, the rate of diffusion of mercury vapour to that of radon was 1·05. Knowing the atomic weight of mercury and the fact that mercury vapour and radon are monatomic, deduce the atomic weight of the latter.

6. Using the Table of ionic radii (*Table 4.4*), suggest what structures you would expect for rubidium bromide and calcium oxide.

7. How would you determine the atomic weights of argon and bromine?

8. Write an essay on the 'defect solid state'.

9. 'Theories are concerned with ideal, non-existent, systems and so are of little use to the practising scientist.' Discuss this statement, with particular reference to gases.

10. Suggest ways of determining the formulae of nitrogen dioxide, sulphur dioxide and ammonia.

11. Write notes on (*a*) thixotropy, (*b*) liquid crystals, and (*c*) glasses.

12. The *Phase Rule* can enable a chemist to make some predictions about the possible outcome of changing conditions. It is expressed as $\mathbf{P + F = C + 2}$ where P is the number of phases, C is the number of components (i.e. the minimum number of substances which are chemically sufficient to define the system), and F is the number of degrees of freedom (i.e. the number of factors which must be fixed to fix the system). It follows that the greater the number of components there are in a system, the greater will be the number of degrees of freedom than a system with one component. Similarly, the greater the number of phases present, the smaller will be the number of degrees of freedom.

Look at the phase diagrams in this book and interpret them in the light of the Phase Rule.

CHAPTER 5

SOLUTIONS

A 'true' solution implies a breakdown at the molecular level; that is to say, the dissolving substance, or *solute*, is separated into individual molecules (or ions) throughout the *solvent*. Such a solution is said to be a *homogeneous* phase, with the composition being uniform and constant throughout. (This description, however, is hardly more than a convenience: a little thought will make it clear that whether such a system appears homogeneous or heterogeneous depends very much upon the scale of dimensions used, i.e. whether atomic or much larger.)

THE COLLOIDAL STATE

If the solute is macromolecular or the size of its aggregates is too great, the average diameter of the solute 'particle' will be much greater than the usual loose solvent aggregates. The mixture then shows deviations from a true solution, and the system breaks down into a solution consisting of a definite disperse phase suspended in a dispersion medium—the continuous phase (*Figure 5.1*). If the particles

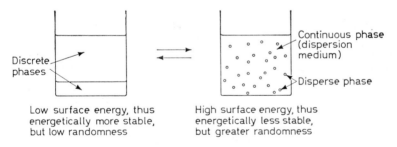

Low surface energy, thus energetically more stable, but low randomness

High surface energy, thus energetically less stable, but greater randomness

Figure 5.1. Colloidal dispersions

of the disperse phase are of the order of about 1–100 nm and remain dispersed, the solution is said to be *colloidal*; if larger and rapidly settling out, the disperse phase is regarded as a coarse suspension. But there are no sharp divisions between the various systems, and it is believed that any solute can be obtained in the form of a colloid (see *Table 5.1*).

Colloidal solutions are either *lyophobic* ('solvent-hating') or *lyophilic* ('solvent-loving'). Broadly speaking, the former owe their stability to the repulsion between similarly-charged aggregates, whilst the latter are prevented from flocculating by having large solvation layers. Many lyophilic colloids, however, are also charged

Table 5.1. Types of Dispersion

	Particle size nm	Optical characteristics
True solution	< 1 (diam.)	clear
Colloids and fine suspensions	1–1 000 (limit of colloidal range)	scatter light, but individual particles invisible
Precipitate (coarse suspension)	>1 000 settle out quickly	opaque: individual particles visible

and some, for example amino acids and proteins, possess both negative and positive charges: the stability of these is then lowest when the algebraic sum of the fixed charges is zero

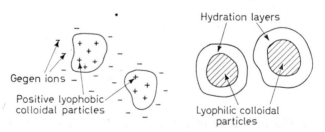

Since the colloidal systems as a whole are electrically neutral, the charged particles are surrounded by mobile counter- (or gegen-)ions (see *Figure 5.2*).

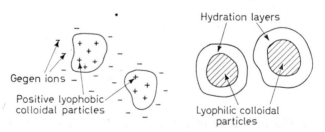

Figure 5.2. Colloidal particles

The presence and type of charge on colloidal particles may be shown by the migration produced under an applied electrical potential, a procedure known as *electrophoresis* (*Figure 5.3*).

90

The stability of a lyophobic colloid can be reduced by adding ions of opposite charge which are more strongly adsorbed than the counter-ions; the net charge may then be so reduced that flocculation occurs. In general, the larger the charge on the ions, the more readily are they adsorbed and the greater their neutralizing power. For negative colloids, the flocculating power of ions is of the order $Th^{4+} >$ $Al^{3+} > Mg^{2+} > Na^+$, and for positive colloids, $Fe(CN)_6^{4-} > Fe(CN)_6^{3-} >$ $SO_4^{2-} > Cl^-$. Flocculation can also be brought about by mixing colloids of opposite charge.

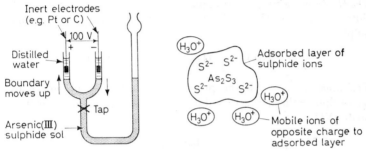

Figure 5.3. Electrophoresis

The stability of lyophilic colloids is not greatly affected by the addition of electrolytes, except in sufficient concentration to compete for the solvent within the solvation layers, when 'salting-out' of the colloid occurs.

The concentration of lyophilic systems can be much higher than that of lyophobic; the taking up of large quantities of solvent often results in the viscosity of lyophilic dispersions differing considerably from that of the dispersion medium (for example, gelatine). Again, because of the solvation layers, lyophilic colloidal particles generally have almost identical refractive indices to the dispersion media, whereas lyophobic particles differ sufficiently to act as scattering surfaces for light waves. This allows a means of observing the latter as a cone of scattered light (the Tyndall cone) when a lyophobic colloid is viewed through the ultramicroscope at right angles to a converging beam of light (*Figure 5.4*).

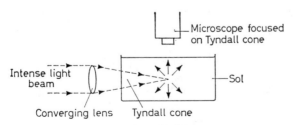

Figure 5.4. The Ultramicroscope

91

Since the colloidal state represents only an intermediate range of particle size, the method of preparation depends either upon the aggregation of ions in solution *up* to, or dispersion of the solid material *down* to, colloidal dimensions.

Examples of methods involving aggregation are:

(*a*) *Hydrolysis*, for example by boiling a very dilute solution of iron(III) chloride in distilled water. A red sol of hydrated iron(III) oxide results:

$$Fe^{3+} + 3 H_2O \rightarrow 3 H^+ + Fe(OH)_3 \rightarrow Fe_2O_3 . xH_2O$$

(*b*) *Double decomposition*, for example by passing hydrogen sulphide through a hot solution of arsenic (III) oxide in distilled water, giving a yellow sol of arsenic (III) sulphide:

$$As_2O_3 + 3 H_2S \rightarrow As_2S_3 + 3 H_2O$$

(*c*) *Oxidation*, for example the production of a white sulphur sol by passing hydrogen sulphide through a very dilute, acidified solution of potassium tetraoxomanganate (VII):

$$2 MnO_4^- + 6 H^+ + 5 H_2S \rightarrow 2 Mn^{2+} + 8 H_2O + 5 S$$

Similarly, a silver sol can be obtained by reducing diammino silver (I) hydroxide with dilute methanal (formaldehyde):

$$2Ag(NH_3)_2^+ + 2OH^- + H.CHO \rightarrow 2Ag + HCOOH + 4NH_3 + H_2O$$

(*d*) *Exchange of solvent:* the addition of water to a solution of phosphorus in ethanol results in the formation of a phosphorus suspension in aqueous ethanol.

Examples of dispersion methods include, for lyophilic colloids, shaking with the dispersion medium ('peptization') or grinding the solid in a ball-mill, perhaps in the presence of dispersing (deflocculating) agents. For lyophobic colloids, electrical methods using spark discharge between electrodes of the metal in the solvent or dispersion with ultrasonic waves can be employed.

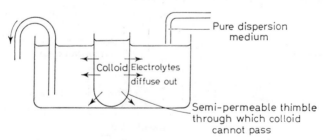

Figure 5.5. Dialysis

92

In order that dispersions, particularly of lyophobic colloids, remain stable, free electrolytes must be removed, usually by *dialysis*. A dialyser, for example parchment paper or cellophane, allows the small particles of impurity to pass through but retains the larger colloidal particles (*Figure 5.5*).

Purified colloidal solutions should remain stable indefinitely, as Brownian motion—the continual agitation of the particles by molecular impact with the dispersion medium—keeps the particles in random motion, preventing their sedimentation under gravity.

Osmotic pressure measurements (p. 109) can be made on the purified colloid and they often reveal, in the case of a lyophilic colloid, that the particle is nothing more than a very large (macro-)molecule.

Factors affecting the Formation of a Solution

A useful guide to the possibility of dissolution is the axiom that 'like dissolves like'. Thus ionic or very polar materials tend to be soluble in polar liquids such as

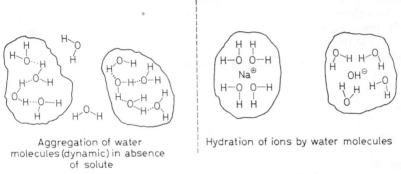

Aggregation of water molecules (dynamic) in absence of solute

Hydration of ions by water molecules

Figure 5.6. Structure of water and aqueous solutions

water, whilst covalent organic molecules are usually soluble in organic solvents of low polarity. What lies behind this generalization?

Two natural tendencies are an increase in randomness and a decrease in energy of a system (see p. 122). The first factor always favours dissolution because there is clearly a greater randomness of structure in a solution than there is in separate, distinct phases. Energy considerations are rather more complex. The lattice energy of ionic solids is considerable, that is to say, ionic lattices are very stable. Increased stability can therefore only come about if there is marked interaction with the solvent. In the case of pure water, the high polarity results in appreciable hydrogen bonding and a loose aggregation of water molecules (*Figure 5.6*). Ions, by virtue of the charge that they carry, can attract the water molecules even more strongly and bring about an arrangement of greater stability. If the free energy (p. 121) decrease accompanying this change exceeds the lattice energy of the solid, then the solid will dissolve. There is no possibility of this sort of interaction between a non-polar covalent molecule and water and so the aggregation of water molecules is unaffected, i.e. the non-polar substance is insoluble.

Electrovalent substances do not generally dissolve in non-polar solvents because there is no prospect of interaction to provide the energy necessary to disrupt the ionic lattice. The fact that there is little lattice energy in a substance of low polarity increases the possibility of this dissolving in a non-polar solvent where there is no molecular interaction and aggregation, especially as dissolution increases the randomness of the system.

1. GAS/LIQUID EQUILIBRIA

Because gases have no boundary surfaces and there is little molecular interaction, energy changes accompanying mixing are usually negligible. The move to greater randomness results in gases being freely miscible and the fact that gaseous molecules are moving rapidly means that solution will be effected rapidly. Furthermore, the rapid movement of the molecules generally offsets any tendency for the denser constituent to settle out. In contrast, equilibrium between a gas and a liquid is reached only slowly and a number of possible equilibria may in fact exist if the gas combines chemically with the solvent. Only in the simplest cases does *Henry's law* apply: the mass, m, of a gas dissolved by a fixed amount of liquid is directly proportional to the gas pressure, p, at constant temperature, i.e.

$$m \propto p \text{ or } m/p = \text{constant} \tag{1}$$

For a perfect gas,

$$pv = \frac{m}{M} \cdot RT$$

i.e.

$$m = \frac{pvM}{RT}$$

where M = molecular weight. Substitution in (1) gives vM/RT = constant, and since the term M/RT is constant, the volume of gas dissolved at a fixed temperature by a fixed amount of liquid is independent of pressure. For this reason, gas solubilities are generally expressed in volume rather than mass units; the 'solubility coefficient' is the volume of gas dissolved in 1 cm^3 of liquid.

For example, the solubility coefficients of two gases, A and B, in water are 0·03 and 0·01, respectively, at 10°. Therefore, if 1 dm^3 of water saturated at 10° by being in contact at 10^5 N m^{-2} with an equimolecular mixture of the two gases is boiled, the volume of each gas (measured at 10°C and $0·5 \times 10^5$ N m^{-2}, the partial pressure) expelled will be $0·03 \times 1000 = 30$ cm^3 of A and $0·01 \times 1000 = 10$ cm^3 of B.

At s.t.p. these volumes become

$$\frac{30 \times 380 \times 273}{760 \times 283} = 14·4 \text{ cm}^3 \text{ of } A$$

$$\frac{10 \times 380 \times 273}{760 \times 283} = 4·8 \text{ cm}^3 \text{ of } B$$

94

Those gases which dissolve exothermically have, in accordance with Le Chatelier's principle (p. 134), negative temperature coefficients; that is, they are less soluble in hot than cold solvent; this is the case for most gases in water.

Departures from Henry's law are caused by:

(a) the non-ideal nature of the gas over the pressure range employed, and
(b) the chemical solution of the gas, e.g.

$$NH_{3(g)} \overset{H_2O}{\rightleftharpoons} NH_{3(soln.)} \rightleftharpoons NH_3 . H_2O \rightleftharpoons NH_4OH \rightleftharpoons NH_4^+ + OH^-$$

The more soluble gases in fact are those which react with the solvent.

In the colloidal range, gases dispersed in liquids constitute *foams*, whilst liquids dispersed in gases are generally termed *aerosols*. Foams are stabilized by lowering the surface tension of the liquid and providing some mechanical strength (for example, by increasing the viscosity) by the addition of foaming agents such as proteins and soaps. Aerosols are inherently unstable and are generally produced by the rapid escape of gas, the aerosol propellant (often an inert fluorohydrocarbon), through a dispersed mixture.

2. GAS/SOLID EQUILIBRIA

The vapour pressure of solvated molecules is determined by the strength of the bonding of the coordinated solvent molecules. For hydrates, the terms 'deliquescent', 'efflorescent' and 'hygroscopic' are only relative and depend on the vapour

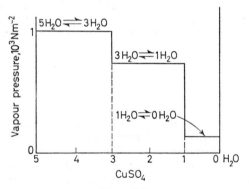

Figure 5.7a

pressure of the water in the air. A substance is hygroscopic and takes in water only if the vapour pressure of the hydrate is less than that of water in the environment. A substance is efflorescent and gives out water vapour only if the vapour pressure of its hydrates exceeds that of the surroundings. The vapour pressure increases, of course, with the extent of hydration. *Figure 5.7a* shows the relationship between the vapour pressure and the degree of hydration of copper(II) sulphate.

The effect of hydration on the behaviour of a substance is well illustrated by sodium carbonate; anhydrous sodium carbonate is hygroscopic, whereas the decahydrate is efflorescent—the stable form is the monohydrate, *under normal atmospheric conditions*, where the vapour pressure is about 2×10^3 N m^{-2} (*Figure 5.7b*)

$$Na_2CO_3 . 10\ H_2O \rightleftharpoons Na_2CO_3 . H_2O \rightleftharpoons Na_2CO_3$$
$$\underset{2\cdot4 \times 10^3\,N\,m^{-2}}{} \qquad \underset{0\cdot66 \times 10^3\,N\,m^{-2}}{}$$

Because of the unsaturated nature of surfaces, due to the presence of residual bonds, adsorption of gases readily takes place, particularly at low temperatures; the adsorbed molecules saturate the residual bonds of the solid and are removed

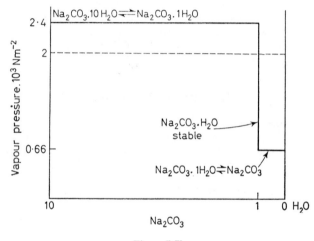

Figure 5.7b

only with some difficulty, for instance in the metallurgical operation of degassing a solid. Solids containing large internal surface areas (that is, having a large porous capillary system) can adsorb very large volumes of gases: a common adsorbent is activated charcoal. In the colloidal range, entrapped gases cause otherwise transparent minerals (by internal reflection) to become opaque, for example milk quartz. Collectively, such gases dispersed in solids are called *solid foams*; they are now commonplace in the form of foam rubbers.

Solids dispersed in gases are referred to as *smokes*. They usually represent a nuisance, both domestically and industrially. It is normally necessary to remove smoke particles from industrial gases such as sulphur dioxide by electrostatic precipitation. The impure gas is passed between insulated metal plates maintained at a high electrical potential: the solid particles become charged under such conditions and are attracted to the plate of opposite charge.

96

3. LIQUID/LIQUID EQUILIBRIA

Liquids of similar polarity are completely miscible, on account of the mutual interaction that is possible between the different molecules (see previous discussion). Thus, ethanoic (acetic) acid is completely miscible with water and methyl benzene with benzene, while water and benzene themselves are immiscible. Between these extremes can be found liquids which are partially miscible, where the two liquid phases consist of solutions of one in the other. The compositions of these liquid phases are often susceptible to temperature change, as shown in *Figure 5.8.*

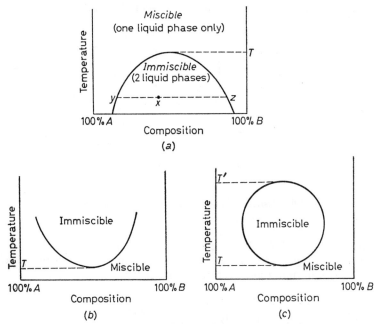

Figure 5.8. Mutual solubilities (*a*) increase with temperature (e.g. phenol and water); (*b*) decrease with temperature (e.g. triethylamine and water); (*c*) can decrease and increase with temperature (e.g. nicotine and water)

(*a*) x = two liquid phases, of composition y and z, present in the proportions

$$\frac{\text{amount of } y}{\text{amount of } z} = \frac{xz}{xy}$$

The *consolute* or *critical solution temperatures*, T, are very sensitive to impurity, and so measurements of miscibilities can often afford information about the condition of binary and more complex mixtures.

The Distribution Law

If a solute is capable of simple solution in two immiscible liquids, then it distributes itself between them in accordance with the respective solubilities, so that the ratio

$$\frac{\text{concentration of solute in liquid } A}{\text{concentration of solute in liquid } B}$$

is constant at constant temperature. This ratio is called the *distribution* or *partition coefficient*.

If a system with a favourable partition coefficient can be found for a solute, then it can be largely transferred from one liquid to the other, possibly leaving impurities behind: this is the basis for *extraction* (*Figure 5.9*).

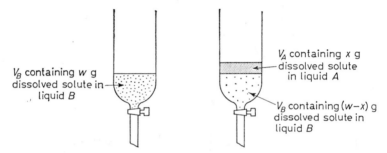

Figure 5.9   Distribution of solute between two immiscible liquids

Suppose V_B to be the volume of liquid B containing W g of solute to which a volume, V_A, of liquid A is added. Then the weight, x g, of solute extracted by liquid A at equilibrium can be calculated as follows:

$$\frac{\text{Concentration in } A}{\text{Concentration in } B} = \frac{x/V_A}{(W-x)/V_B} = K \text{ (the partition coefficient)}$$

$$x = \frac{WK V_A}{(V_B + KV_A)}$$

Example—The solubility of iodine in carbon tetrachloride and water at 20° is 15·8 and 0·29 g dm^{-3}, respectively. Find the volume of carbon tetrachloride required to extract 90 per cent of the iodine from 3 dm^{-3} of water at 20°.

The partition coefficient for $I_{2(CCl_4)}/I_{2(H_2O)} = 15\cdot8/0\cdot29$ and if $V =$ volume of CCl_4 required in dm^3, then

$$\text{Partition coefficient} = \frac{15\cdot8}{0\cdot29} = \frac{90/V}{10/3} \qquad\qquad V = \frac{0\cdot29 \times 90 \times 3}{15\cdot8 \times 10} = 0\cdot50 \text{ dm}^3$$

Three points should, however, be noted:

(*a*) the solute must be in the same physical condition in both solvents; for example, if association or dissociation of solute molecules occurs in one liquid, then the partition coefficient will not remain constant (in fact, departure from constancy can provide significant evidence about such things as ionization);

(*b*) as partition is a manifestation of differential solubility, neither liquid should be saturated with respect to the solute;

(*c*) more efficient extraction is achieved by shaking separate small samples of solvent with the solution and combining the extracts than by using all the solvent in a single extraction.

Applications of Distribution

Ethereal extraction—Diethyl ether (ethoxyethane) is only slightly soluble in water; it dissolves most organic substances, is chemically inert and, being volatile, easily removed from the liquid phase. It is consequently widely used for extracting organic substances from aqueous systems, the more hydrophilic impurities remaining behind.

Desilverization of lead—Zinc, when just above its m.p. (419°), is largely insoluble in molten lead (m.p. 327°). The distribution coefficient for silver between these two liquids at this temperature is

$$\frac{\text{concentration in zinc}}{\text{concentration in lead}} \sim 3\,000$$

Zinc is therefore used to desilverize lead. Relatively small quantities of zinc added to molten argentiferous lead remove most of the silver present; it rises to the surface, where it is skimmed off. The zinc is finally removed from the silver by distillation.

Distillation of Liquid Mixtures

Liquids always have a definite vapour pressure associated with them, the result of certain molecules with excess energy escaping from the liquid phase. At equilibrium, when the 'saturation vapour pressure' prevails, the rate at which liquid molecules leave the liquid equals that at which vapour molecules return (*Figure 5.10*) A liquid of high volatility has a high vapour pressure because its molecules are, on the whole, moving rapidly and are therefore more able to escape into the vapour phase. In the case of an ideal mixture of liquids, where the interactions between molecules are equal and each constituent retains its independent volatility, the total vapour pressure is the sum of the separate vapour pressures (although these are proportional to the relative number of molecules present). It follows that the more volatile component, if of sufficient concentration, contributes the larger proportion to the total. In *Figure 5.11a* the total vapour pressure, *RX*, of the mixture of composition *X* is made up of *XQ*, from the more volatile constituent *A*, and of *XP* from component *B*.

99

As the temperature is increased, so the vapour pressure increases until it equals the external pressure and the liquid boils. If follows from what has been said that the vapour given off is richer in the more volatile constituent than is the liquid left behind. It can be condensed and analysed and a graph such as that in *Figure 5.11b* constructed. A mixture of composition *x* boils at temperature *T'* and its vapour has the composition *y*. If it is now condensed, it gives a liquid that boils at *T'''* giving off vapour of composition *z*. If this process is continued, virtually

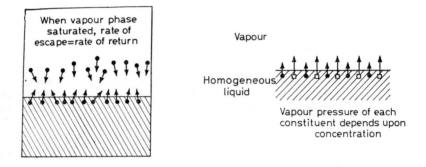

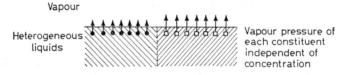

Figure 5.10. Vapour pressure of liquids

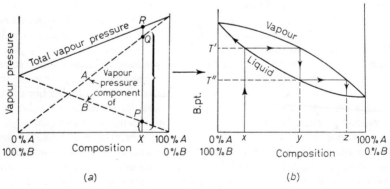

(a)   (b)

Figure 5.11. Vapour pressure and boiling point curves for two ideally miscible liquids

100

pure A can be obtained. This is what happens in *fractional distillation:* a temperature gradient is established in a tall vertical column, sufficient to permit the above
procedure to take place continuously, the vapour that is formed condensing, falling,
warming and reboiling to give off a vapour still richer in the more volatile constituent.

Most liquids are non-ideal in the sense that unequal interaction takes place
between molecules of the different components. If loose compound formation
occurs, for example in hydration, a minimum value for the total vapour pressure
results (*Figure 5.12a*). The mixture with this minimum vapour pressure is called

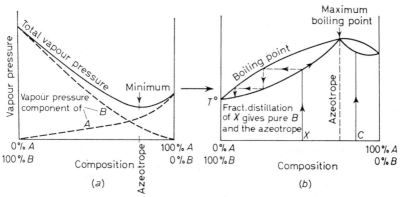

Figure 5.12. Vapour pressure and boiling point curves for two liquids showing negative
deviation

an *azeotrope*. It has the highest boiling point of the system; the vapour and liquid
curves coincide at this point, so the vapour and liquid have the same composition
and in this respect resemble a pure substance (*Figure 5.12b*). Fractional distillation
of any mixture other than the azeotrope gives a liquid of the azeotropic composition
and a vapour of pure A or B (e.g. hydrochloric acid–water).

The existence of repulsive forces between the molecules of different liquids (or,
at least, the non-existence of attractive forces) results in the total vapour pressure
being greater than the sum of the separate ones. This once again leads to the
existence of an azeotrope, this time with maximum vapour pressure and therefore
minimum boiling point. Such a system (e.g. acetone–carbon tetrachloride or
ethanol–water) is shown in *Figure 5.13*. Distillation of the azeotrope gives a vapour
with the same composition as the liquid, so that purification by distillation is
impossible. Fractional distillation of any mixture other than the azeotrope gives a
vapour with the azeotropic composition and residual liquid of A or B.

Repulsive forces between the different molecules result, in the extreme case,
in the liquids being immiscible. Now each liquid phase can exert its own true
vapour pressure (*Figure 5.10*) independent of the overall composition, so that the
boiling point remains constant whilst the two liquids are present, and is not affected

101

by the variation in composition (*Figure 5.14*). Furthermore, because each liquid exerts its normal vapour pressure, the external pressure is reached below the boiling point of either. This fact is utilized in the *steam distillation* of organic substances insoluble in water. Steam is passed in from an external vessel and, by

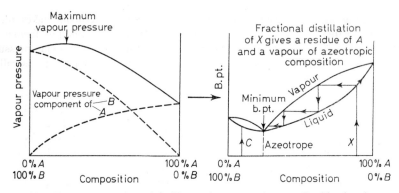

Figure 5.13. Vapour pressure and boiling point curves for two liquids showing **positive deviation**

providing a nucleus for the formation of bubbles of vapour, minimizes 'bumping' and superheating. In the case of nitrobenzene and water, the mixture boils at 99°, where the vapour pressure of water is 0.977×10^5 N m^{-2} (733 mmHg) and of

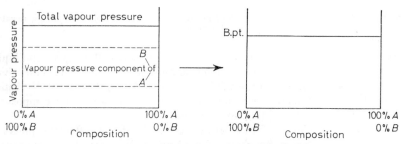

Figure 5.14. Distillation of immiscible liquids

nitrobenzene, 0.036×10^5 N m^{-2} (27 mmHg) Since the vapour pressure is a function of the number of molecules (and, hence, moles) present

$$\frac{\text{molar concn. water}}{\text{molar concn. nitrobenzene}} = \frac{[H_2O]}{[C_6H_5NO_2]} = \frac{0.977 \times 10^5}{0.036 \times 10^5} = \frac{\text{wt. water/18}}{\text{wt. nitrobenzene/123}}$$

(mol. wt. water = 18; nitrobenzene = 123)

Therefore

$$\frac{\text{wt. of nitrobenzene}}{\text{wt. of water}} = \frac{123 \times 0.036}{18 \times 0.977} \sim \frac{1}{4}$$

102

Thus, because of the relatively high molecular weight of the nitrobenzene and despite its low vapour pressure, about one-fifth of the distillate consists of nitrobenzene, and it is distilling at more than 100° below its normal boiling point, so that there is far less risk of decomposition.

4. SOLID/LIQUID EQUILIBRIA

Solutions of solids in liquids provide the most common type of solute/solvent interaction. With water as the solvent, the characteristic solute is an electrovalent compound. Dissolution results in the ions of the crystal lattice being further separated; work is done when ions of opposite charge are moved apart, and unless the energy required is completely forthcoming in the form of hydration energy of the ions, the dissolution will be endothermic. By Le Chatelier's theorem (p. 134), the substance will then be more soluble in hot than cold water and a positive temperature coefficient will result. *Figure 5.15* illustrates the different types of solubility curves possible for saturated aqueous solutions.

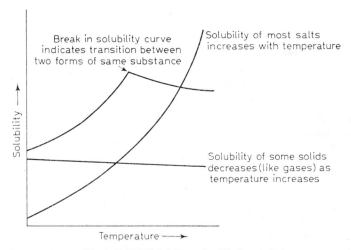

Figure 5.15. Solubility of solids in water

The addition of solid to liquid usually depresses the freezing point of the latter. Two fundamental phase diagrams are possible, depending upon whether the solid deposited on cooling consists of a solution of one component in the other or whether, because of mutual insolubility, two solid phases result. In the case of cobalt and nickel, for example (*Figure 5.16a*), progressive substitution of atoms occurs without significant change in the shape of the crystal lattice, and a solid solution is formed. If a liquid mixture of composition X is cooled, solid first appears at $T°$, being of the composition represented by Y. The remaining liquid consequently becomes richer in component A and the freezing point progressively falls. The solid crystallizing out also becomes richer in A but, of course, it is never as rich in A as the liquid with which it is in contact. (Such a phase diagram is very similar to that for distillation of an ideal liquid mixture.) It should be noted that only liquid exists above the liquidus curve and only solid below it, whilst between the two curves solid and liquid coexist.

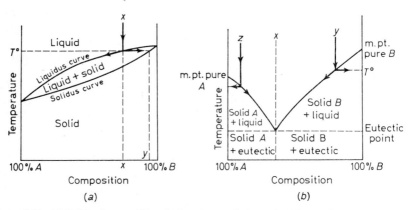

Figure 5.16. (*a*) Solids form solid solution (e.g. cobalt and nickel); (*b*) solids present as separate phases (e.g. ice and salt)

When the atoms of the two components are insufficiently compatible (for example in size) to allow the formation of solid solutions, a phase diagram such as that of *Figure 5.16b* operates. If a liquid of composition Y is cooled, the pure component B begins to settle out when temperature T° is reached. The liquid becomes richer in A as the temperature falls and pure B continually separates out, until the *eutectic* temperature is reached, when the remaining liquid crystallizes out without further change in temperature. The solid so formed, containing fine-grained crystals of both components, is called the eutectic, and embedded in it will be pure B. A similar situation prevails with a mixture of composition Z, but this time it is pure A that settles out until the eutectic is reached. If a liquid mixture of the eutectic composition X is cooled, it freezes at the constant eutectic temperature, giving a solid with the same composition as the liquid (in this respect it resembles a pure substance).

COLLIGATIVE PROPERTIES OF DILUTE SOLUTIONS

For solutions of non-volatile solids in liquids, several relationships can be derived which depend only on the ratio of the number of solute 'particles' to solvent 'particles', provided that the solutions are dilute. Such relationships which depend only on the number ratio and not on the type of 'particles' are termed *colligative*.

(It is most important that the units employed are proportional to the number of particles present; the mole, containing the Avogadro constant of molecules can be used, but dissociation or ionization will lead to anomalous results.)

The number ratio is usually expressed in terms of the molar fraction. If $n =$ number of solute molecules and $n_0 =$ number of solvent molecules, then $n + n_0 =$ total number of molecules, and the mole fractions of solute and solvent are

$$N = \frac{n}{n_0 + n} \quad \text{and} \quad N_0 = \frac{n_0}{n_0 + n} \quad (N + N_0 = 1) \tag{i}$$

Lowering of Vapour Pressure

If a non-volatile solute is added to a liquid solvent, the 'escaping tendency' of the latter is reduced by the presence in the surface of non-volatile molecules: in other words, the vapour pressure of the liquid is reduced. Clearly, the effect is proportional to the relative numbers of molecules present. If

$p\ =$ vapour pressure of the solution

$p_0 =$ vapour pressure of the pure solvent

then

$$p \propto N_0 \text{ or } p = kN_0$$

where k is a constant.

When $N_0 = 1$, i.e. for pure solvent, $p = p_0$, i.e. $p_0 = k$

Therefore

$$p = p_0 N_0 \quad \text{or} \quad \frac{p}{p_0} = N_0$$

and

$$N = 1 - N_0 = \frac{p_0 - p}{p_0} = \frac{n}{n_0 + n} \ (\sim n/n_0 \text{ for a dilute solution}) \text{ [from } (i)\text{]}$$

That is, *the relative lowering of vapour pressure is equal to the molar fraction of the solute*. This statement is known as *Raoult's law*, and it can obviously be used for determining molecular weights.

A dynamic method of measuring the vapour pressure lowering is due to Walker and Ostwald. Air is drawn slowly through tubes containing the solution and then through tubes of pure solvent, all of known weight. The loss in weight caused by solvent being taken up by the air from both sets of tubes is determined. Assuming that the air is saturated as it leaves the last of each set of tubes,

loss in weight from

$$\text{solution} \propto p$$

and from

$$\text{pure solvent} \propto p_0 - p$$

(since the air is already saturated to the value of p).

For example, the loss in weight of a set of bulbs containing 5·04 g of non-volatile solute, S, in 30·0 g of ethanol after a current of air had been drawn through was 0·607 g; the further loss from a second set of bulbs, containing pure ethanol, when the same current of air was passed through, was 0·079 g, i.e.

$$p \propto 0\cdot607 \text{ g}, \quad p_0 - p \propto 0\cdot079 \text{ g}$$

Therefore

$$p_0 \propto 0\cdot079 + 0\cdot607 \propto 0\cdot686 \text{ g}$$

$$\frac{p_0 - p}{p_0} = \frac{0\cdot079}{0\cdot686} = N = \text{mole fraction of solute} = \frac{n}{n_0} \sim \frac{w/m}{W/M}$$

where

$$w = \text{mass of solute} = 5\cdot04 \text{ g}$$

$$m = \text{mol. wt. of solute}$$

$$W = \text{mass of solvent} = 30\cdot0 \text{ g}$$

$$M = \text{mol. wt. of solvent} = 46$$

$$\therefore \frac{5\cdot04/m}{30/46} \sim \frac{0\cdot079}{0\cdot686}$$

i.e.

$$m \sim \frac{5\cdot04 \times 0\cdot686 \times 46}{0\cdot079 \times 30} \sim 68$$

Elevation of Boiling Point and Depression of Freezing Point

Since a solution boils when its vapour pressure is equal to the surrounding atmospheric pressure, a lowering of vapour pressure is equivalent to an elevation of boiling point. This effect is illustrated by *Figure 5.17*.

It can be seen that the relative lowering of vapour pressure is proportional to the relative elevation of boiling point, i.e.

$$\frac{p_0 - p}{p_0} = \text{constant} \cdot \frac{T - T_0}{T_0}$$

$$\therefore \frac{T - T_0}{T_0} = \text{constant} \cdot N$$

$$\sim k \frac{w/m}{W/M} \text{ (for very dilute solutions)}$$

106

Consequently, if the molecular weight of the solvent is known, together with the constant k and the weight of solute and solvent producing the measured elevation, then the molecular weight of the solute can be calculated. It is more usual, however, to introduce the term 'molecular elevation constant' (that is, the elevation produced by one mole of solute in, usually, 1 kg of solvent) and to calculate the molecular weight by proportion.

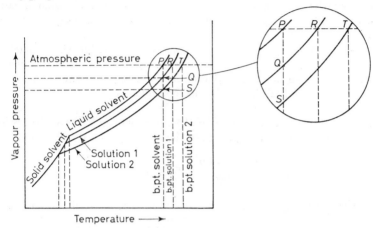

Figure 5.17

For dilute solutions, the lines are close together and virtually parallel:

$$\therefore \frac{PR}{PQ} = \frac{PT}{PS} = \text{constant}$$

i.e. $\dfrac{\text{elevation of boiling point}}{\text{depression of vapour pressure}} = \text{constant}$

For example, the elevation constant for benzene is $2 \cdot 7°$ kg^{-1}. If 1 g of a non-volatile solute in 40 g of benzene produces an elevation of $0 \cdot 20°$, calculate the molecular weight of the solute.

1 g solute in 40 g benzene gives an elevation of $0 \cdot 20°$,

1 g in 1 kg gives an elevation of $0 \cdot 20 \times 40/1000$

$\dfrac{1 \times 1\ 000 \times 2 \cdot 7}{0 \cdot 2 \times 40}$ g solute in 1 kg gives one of $2 \cdot 7°$

$\therefore$ molecular weight of solute $= \dfrac{1\ 000 \times 2 \cdot 7}{0 \cdot 2 \times 40} = 337 \cdot 5$

It is important that the solutions used be always homogeneous and dilute and that the solute be non-volatile and neither associated nor dissociated if the correct molecular weight is to be obtained. It follows that the elevations measured are inevitably small, so that very sensitive thermometers must be used and precautions taken to minimize superheating and fluctuations of temperature.

Reference to *Figure 5.17* will show that similar considerations apply to the effect of non-volatile solute on freezing point. By use of the term 'molecular depression constant', therefore, the molecular weight of the solute can be calculated, provided that it behaves normally and that it is pure solvent which separates on freezing. Various solvents are now known which have extremely large depression constants, so that the use of a very accurate thermometer is obviated; foremost among these is camphor, which is used in the Rast method. As its depression constant is variable, it is usual to calibrate it against a solute of known molecular weight, for example naphthalene.

Example—The melting point of camphor was found to be 178°C. When 0·019 1 g naphthalene was dissolved in 0·401 9 g camphor, the melting point was depressed to 159°C. The melting point of a solution of 0·014 8 g of unknown substance in 0·408 5 g camphor was 162°C. Calculate its molecular weight.

0·019 1 g naphthalene in 0·401 9 g camphor gives 19°C depression. 128 g (1 mole) in 1 kg camphor therefore, gives a depression of

$$\frac{19 \times 128 \times 0.401\,9}{0.019\,1 \times 1000} = 51.2° \text{ kg}^{-1}$$

and the molecular weight of the unknown substance is, by similar reasoning

$$\frac{51.2 \times 0.014\,8 \times 1000}{(178{-}162) \times 0.408\,5} = 116$$

Osmosis

The mixing (diffusion) of a solute and a solvent, until the concentration is uniform throughout, results in an increase in randomness (ΔS is positive, see p. 122) and is therefore often a spontaneous process. If the diffusion is restricted by a semipermeable membrane, so that chiefly solvent molecules pass through, the process is termed *osmosis* (*Figure 5.18*). Osmosis will continue as long as there is a difference in concentration on different sides of the membrane or until a hydrostatic pressure is set up, called *osmotic pressure*, which counteracts the diffusion pressure. Looked at in this way, osmotic pressure can be regarded as negative diffusion pressure. Like the other phenomena discussed in this Section, osmosis is a function of the ratio of solute to solvent molecules and affords another means of calculating molecular weights. In fact, osmotic pressure, π, is found experimentally to obey the laws which hold for gases:

$$\pi \propto 1/v \text{ (or } \pi \propto c) \text{ at constant temperature}$$

$$\propto T°K \text{ at constant concentration}$$

where v = volume containing a fixed weight of solute

c = concentration

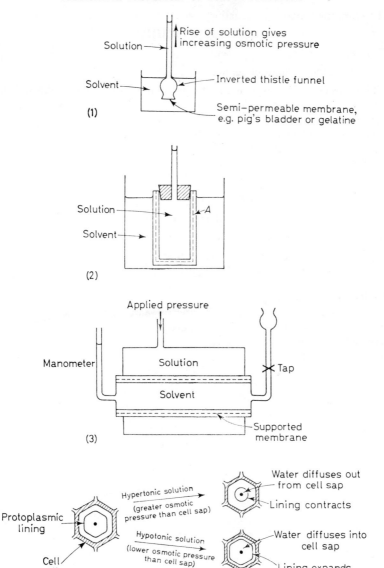

Figure 5.18. Methods for investigating osmosis

(2) A Membrane (A) of $Cu_2[Fe(CN)_6]$ deposited in porous pot by allowing diffusion of $CuSO_4$ and $K_4[Fe(CN)_6]$ in opposite directions (membrane more uniform if diffusion assisted by electrical migration); (3) external osmotic pressure just prevents osmosis (i.e. manometer level remains constant): no unknown dilution of solution occurs; (4) if no diffusion of water either into or out of cell sap, solution has the same o.p. as cell sap, and the two are said to be *isotonic*

Therefore the general gas equation holds

$$\pi v = (m/M)RT$$

where m = mass of solute present

M = mol. weight of solute.

The constant R is even found to be numerically equal to the universal gas constant. (Just as the gas laws are obeyed only by gases at low pressure, so the solutions used must be dilute.)

Example—The osmotic pressure of a solution of 1·82 g of mannitol in $0\cdot1 \times 10^{-3}\,m^3$ (0·1 litres) water is $2\cdot32 \times 10^{-5}$ N m^{-2} (2·32 atm) at 10°C. Calculate the molecular weight of mannitol.

$$\pi v = (m/M)\,RT$$

$$M = \frac{mRT}{\pi v} = \frac{1\cdot82 \times 8\cdot31 \times 283}{2\cdot32 \times 10^5 \times 0\cdot1 \times 10^{-3}} = 180$$

(or alternatively,

$$M = \frac{mRT}{\pi v} = \frac{1\cdot82 \times 0\cdot08 \times 283}{2\cdot32 \times 0\cdot1} = 180)$$

Osmotic Pressure and Vapour Pressure

The relationship between the osmotic pressure of a solution and the relative lowering of vapour pressure can be deduced by considering the equilibrium in the closed system shown in *Figure 5.19*.

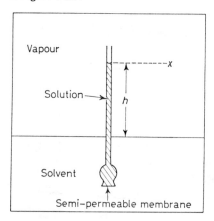

Figure 5.19. Osmotic pressure and vapour pressure

Let p_0 = the vapour pressure of the pure solvent at the surface of the liquid. The vapour pressure, p_0, will decrease with height, h, until at the level x it equals p. Since the system is at equilibrium, this must also be the vapour pressure of the solution at the same height.

110

Now

$$\text{pressure} = \text{height} \times \text{density}$$

$$p_0 - p = \sigma h$$

$$\pi = \rho h$$

where σ = mean density of vapour

ρ = density of solution

Substituting for h

$$p_0 - p = \frac{\sigma}{\rho} \cdot \pi$$

but

$$\sigma = \frac{M}{V} = \frac{M}{RT} \cdot p_0$$

where

$$M = \text{molecular weight of solvent}$$

$$V = G.M.V$$

Therefore

$$\frac{p_0 - p}{p_0} = \frac{M}{\rho RT} \cdot \pi = \text{constant} \cdot \pi$$

That is, the relative lowering of vapour pressure is proportional to the osmotic pressure.

Ionization and Association

It has been pointed out above that the solute must be neither ionized nor associated if the correct molecular weight is to be determined. This is quite obvious when it is realized that the effects described in this Section are functions of only the number but not the nature of the particles. Thus, if a substance ionizes *completely*, each molecule giving rise to two ions, the effect will be doubled; similarly, if a substance associates completely into double molecules, the effect will be halved.

Allowance is made for abnormality by introducing the *van't Hoff factor*, i. If the calculated values for osmotic pressure, vapour-pressure lowering, boiling-point elevation and freezing-point depression (assuming no abnormality) are referred to as c_{calc}, and c_{obs} is the observed colligative value, then

$$i = \frac{c_{obs}}{c_{calc}}$$

A knowledge of i permits the degree of dissociation or association to be evaluated. Thus, for a binary electrolyte, $AB \rightleftharpoons A^+ + B^-$; if α is the degree of ionization, one molecule gives rise, after ionization, to $(1 - \alpha)$ molecules of undissociated AB (in

111

a statistical sense), and a ions each of A^+ and B^-. That is, the number of particles formed from one particle of AB is $1 - a + 2a = 1 + a$, and since c_{calc} and c_{obs} depend only on the number of particles present

$$i = \frac{\text{number of particles due to ionization}}{\text{number of particles assuming no ionization}}$$

$$= \frac{1 + a}{1} \text{ for a binary electrolyte}$$

A similar relationship can be deduced for association. Let substance A associate into double molecules

$$2A \quad \rightleftharpoons (A)_2$$

$$2(1 - a) \quad a$$

If $a =$ degree of association, 2 molecules of A give rise to $(2 - 2a + a)$ particles

$$\therefore i = \frac{2 - a}{2}$$

Example—A solution of a monobasic organic acid contains 0·1 mole in 1 kg of water and depresses the freezing point of the water by 0·223 °C. A solution of the same molarity in benzene depresses the freezing point of benzene by 0·26 °C. What can be deduced? (Freezing-point depression for water, 1·86° kg^{-1}; for benzene, 5·1° kg^{-1}).

Assuming no dissociation 1·0 mole of acid in 1 kg water would produce a depression of 1·86°, and 0·1 mole a depression of 0·186°

hence

$$c_{calc} = 0·186$$

Therefore $\quad i = \dfrac{c_{obs}}{c_{calc}} = \dfrac{0·223}{0·186} = 1·2 = 1 + a_1 \text{ (binary electrolyte)}$

$a_1 = i - 1 = 0·2$. Thus the acid is 20 per cent dissociated in water.

Assuming no association 1·0 mole of acid in 1 kg benzene would produce a depression of 5·1° and so, 0·1 mole, a depression of 0·51°.

Therefore $\quad i = \dfrac{0·26}{0·51} = \dfrac{2 - a_2}{2}$

$a_2 = 0·98$. Thus the acid is 98 per cent associated in benzene.

QUESTIONS

1. Give reasons for the following observations:
(a) iron (III) oxide sol prepared by boiling a few drops of iron (III) chloride in distilled water is positively charged, whilst addition of sodium hydroxide solution under carefully controlled conditions produces a negatively charged colloid;
(b) hydrogen ions are more effective than other univalent ions in causing the flocculation of most negatively-charged colloids;
(c) arsenic (III) sulphide sol is negatively charged;
(d) dilute solutions must be used in preparing lyophobic colloids;
(e) at an oil/water interface, soap molecules orientate in a specific manner.

2. Calculate the weight of naphthalene, $C_{10}H_8$, required to produce a vapour pressure lowering of 264 N m^{-2} (2 mmHg) when dissolved in 50 g methylbenzene at 50° (the vapour pressure of methylbenzene at 50° is 1.22×10^4 N m^{-2} (92.6 mmHg).

3. Derive an expression relating the van't Hoff factor, i, with the degree of dissociation of a molecule producing n ions.

4. A solution of 6.10 g of a monobasic organic acid in 50 g of benzene has a boiling point 1.5° higher than pure benzene. At 17°C, 0.40 g of the acid in 1 dm^3 of water has an osmotic pressure of 7.8×10^3 N m^{-2} (61.4 mmHg). Also 0.50 g of the same acid requires 41.0 cm^3 0.1 M NaOH for neutralization. Given that the molecular elevation constant for benzene is 2.53° kg^{-1}, explain the above data.

5. Describe the precautions necessary in the use of standard types of apparatus for obtaining accurate colligative values and discuss critically the various types of design that you have met.

6. Calculate the proportion by weight of chlorobenzene in the distillate during steam distillation. The mixture boils at 90° at a pressure of 9.78×10^4 N m^{-2} (740 mmHg). The vapour pressure of water at 90° is 7.00×10^4 N m^{-2} (530 mmHg).

7. Find the quantity of solute extracted per dm^3 from an aqueous solution containing 50 g dm^{-3} when shaken with (a) one 200 cm^3 portion of ether, (b) two 100 cm^3 portions, if the partition coefficient (ether)/(water) for the solute is 8.

8. Discuss how Le Chatelier's principle can be applied to the effect of pressure on the solubility of a gas in water and to the effect of temperature on the solubility of a solid in water.
Assuming that air contains 80 per cent of nitrogen and 20 per cent of oxygen by volume and that the solubility coefficients of nitrogen and oxygen in water are 0.02 and 0.04, respectively, at 20°, calculate the volume of each gas expelled by boiling 1 dm^3 of water saturated with these gases at 20°.

113

9. Trace the changes taking place when mixtures of composition C in *Figures 5.12* and *5.13* are heated.

10. Comment on the following flocculation values (arbitrary scale) for silver iodide and aluminium oxide sols:

Silver iodide		Aluminium oxide	
Electrolyte	*Flocc. value*	*Electrolyte*	*Flocc. value*
$NaNO_3$	140	KCl	46·0
$Mg(NO_3)_2$	2·6	K_2SO_4	0·3
$Al(NO_3)_3$	0·07	$K_3[Fe(CN)_6]$	0·08

11. Write an essay dealing with the use of colloids in industry.

12. Interpret the phase diagram, *Figure 5.20*, for steel

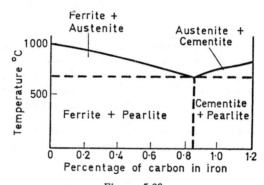

Figure 5.20

13. Comment on the following:
(*a*) although sodium sulphate is soluble in water barium sulphate is insoluble, whilst barium chloride, like sodium chloride, is freely soluble;
(*b*) glucose $C_6H_6(OH)_6$ is soluble in water but insoluble in petrol, whereas hexanol, $C_6H_{13}OH$, is insoluble in water but soluble in petrol.

114

CHAPTER 6

ENERGETICS AND KINETICS

THERMOCHEMISTRY

The law of conservation of energy states that energy can be neither created nor destroyed; that is, when changes take place in which the final energy state is not that of the original, then energy in one form or another must have been given out or taken in. This is the basis of the first law of thermodynamics, which is represented by the equation

$$\Delta U = Q - w$$

where ΔU = the change in the *internal energy* of the system
Q = the heat absorbed by the system
w = the work done by the system.

If, when a chemical reaction takes place, the total energy of the products is less than that of the reactants, then energy must be given out, largely as heat, and the reaction is said to be *exothermic*. If the energy of the reactants is less than that of the products, energy, again usually in the form of heat, is taken in and the reaction is described as *endothermic*. The branch of chemistry dealing with the heat changes accompanying reactions is called thermochemistry. Evidently the amount of heat absorbed or evolved is dependent upon the state of the substances and the quantities reacting. Unless otherwise stated, it is customary in thermochemical equations (i.e. those which take account of the energy changes involved) for the reactants and products to be in their standard (normal) states and for at least one to be of unit molar concentration or, in the case of neutralization reactions, one gramme equivalent. For example, the equation

$$C + O_2 \rightarrow CO_2 \qquad\qquad \Delta H - 394 \text{ kJ}$$

means that when one mole (12 g) of graphite (the standard, most stable, form of carbon) is reacted with one mole (32 g) of gaseous oxygen to produce gaseous carbon dioxide, the heat of the reaction (or heat content change, ΔH) is 394 kilojoules, and since the value is negative, heat is evolved and the reaction is *exothermic*.

The heat of reaction can be redefined in terms of a few specific reactions.

The *heat of formation* of a compound is the heat evolved when one mole of the compound is formed from its elements. In the example given above, the heat of formation of carbon dioxide is 394 kJ i.e. $\Delta H = -394$ kJ.

The *heat of combustion* of an element or compound is the heat evolved when one

115

mole is completely burnt in oxygen (at constant volume). The heat of combustion of carbon is therefore 394 kJ ($\Delta H = -394$ kJ).

The *heat of neutralization* of an acid or base is the heat evolved when a mole of H^+ or OH^- is neutralized. This value is always about 57·5 kJ ($\Delta H = -57·5$ kJ) for strong acids reacting with strong bases (p. 159).

With gaseous reactions in particular the effect of volume change has to be considered. Heats of reaction are generally measured at constant pressure, but where a substantial increase in volume occurs, work has to be performed to cause the gas to expand against the surrounding atmosphere. This means that the heat of reaction at constant pressure*, ΔH_p, will differ from that at constant volume, ΔH_v.

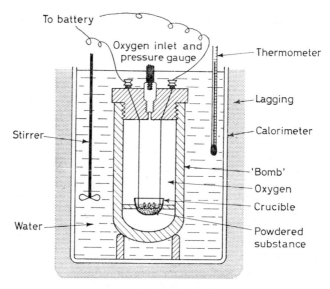

Figure 6.1. Bomb calorimeter

In measuring heats of combustion, the substance is normally ignited by an electric fuse in oxygen under pressure (to ensure complete combustion) in a stout stainless steel vessel called a bomb, so that the reaction takes place at constant volume. The bomb is surrounded by water in an insulated calorimeter. The whole is known as a bomb calorimeter (*Figure 6.1*) and the heat developed is measured by observing the temperature rise and making due allowance for the thermal capacity of the calorimeter and contents. From this, ΔH_v is obtained.

Suppose the heat of combustion of graphite were measured. Then ΔH_v would be

* Where no subscript is employed, it can be taken that the measurement is at constant pressure.

equal to ΔH_p since no change in volume occurs after returning the system to atmospheric temperature

$$C + O_2 \rightarrow CO_2 \qquad \Delta H_v = \Delta H_p = -394 \text{ kJ}$$
$$\text{1 vol.} \quad \text{1 vol.}$$

In a reaction such as

$$C + CO_2 \rightarrow 2CO \qquad \Delta H_p = 173 \text{ kJ}$$

a volume change occurs and for one mole of carbon dioxide used, two moles of carbon monoxide are produced. Assuming ideal behaviour of these gases, $P\Delta V = nRT$. This is the work done in expanding against the pressure and will be equal to $(2 - 1)RT$. If $T = 298K$ (25°C) and R is taken as 8·31 J K^{-1} mol^{-1}, then the work done is $1 \times 8\cdot31 \times 298$ J and, since $\Delta H_p = 173$ kJ

$$\Delta H_v = 173 - \frac{8\cdot31 \times 298}{1\ 000} = 170 \text{ kJ}$$

Hess's law

An extension of the law of conservation of energy is *Hess's law of constant heat summation*. Because the energy associated with a particular substance in a certain condition is constant, there is always the same energy change associated with the conversion of a given quantity of this substance into another, independent of the manner in which the conversion is effected, i.e. whether the change is direct or indirect (*Figure 6.2*).

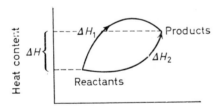

Heat content change, $\Delta H = \Delta H_1 = \Delta H_2$

Figure 6.2. Hess' law

Hess's law is the basis for the calculation of several different energy changes: for example, to calculate the *heat of formation* of ethanol, eqn. (*iv*), from the heats of combustion given in eqn. (*i*)–(*iii*)

		ΔH_v kJ
(*i*)	$C_2H_5OH + 3\ O_2 \rightarrow 3H_2O + 2CO_2 \uparrow$	− 1 430
(*ii*)	$C + O_2 \rightarrow CO_2 \uparrow$	− 394
(*iii*)	$H_2 + \frac{1}{2}O_2 \rightarrow H_2O$	− 285
(*iv*)	$2C + 3H_2 + \frac{1}{2}O_2 \rightarrow C_2H_5OH$	?

This requires that we start with two moles of carbon, three moles of hydrogen and half a mole of oxygen, and that we finish with two moles of carbon dioxide and three moles of water (all in their standard states). From Hess's law the total energy change will be the same whether the change is effected directly or through the intermediacy of ethanol. Since the measurements are made at constant volume, no work is done by or on the system and therefore, from the first law of thermodynamics, $\Delta U = q$. As shown in *Figure 6.3*, $(iv) = 2 \times (ii) + 3 \times (iii) - (i)$ and therefore the heat of formation of ethanol has the value

$$(2 \times - 394) + (3 \times - 285) - (- 1\ 430)\ kJ$$

$$\Delta H_v = - 213\ kJ$$

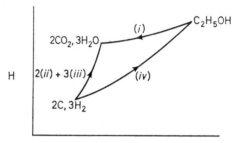

Figure 6.3

Since no absolute heat contents are involved but only changes in their values, the method of *intrinsic energies* can also be used. Elements in their standard states are arbitrarily assigned an intrinsic energy of zero. (Where elements show polymorphism, only one form, the most stable, will have zero intrinsic energy.) The energy associated with a compound relative to its elements is then referred to as its intrinsic energy. For our example

$$C_2H_5OH + 3\ O_2 \longrightarrow 3H_2O \qquad + 2CO_2 \qquad + 1\ 430\ kJ$$
intrinsic energies ? $\qquad$ 0 $\qquad$ $- (3 \times 285) - (2 \times 394)\ kJ$

Therefore the heat of formation of ethanol is
$$+ (1\ 430 - 855 - 788) = - 213\ kJ.$$

Bond Energy

The bond energy is defined as the energy required to break 'one mole's worth' of the bond in question. That is to say, the bond energy of $A—B$ is the heat content of the reaction

$$A—B \rightarrow A + B \qquad\qquad \Delta H = \text{bond energy}$$

It is possible to calculate these values from a knowledge of the relevant heats of

118

dissociation and combustion. For example, in the case of the O—H bond the following figures can be used:

$$2H_2 + O_2 \rightarrow 2H_2O \qquad \Delta H = -486 \text{ kJ}$$
$$2H_2 \qquad \rightarrow 4H \qquad \Delta H = +865 \text{ kJ}$$
$$O_2 \qquad \rightarrow 2O \qquad \Delta H = +495 \text{ kJ}$$
$$4H + 4O \rightarrow 4OH \qquad \Delta H = -4x \text{ kJ}$$

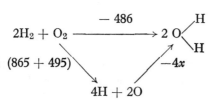

By Hess's law:

$$-486 = 865 + 495 + (-4x)$$
$$x = 1846/4 = 461 \cdot 5 \text{ kJ}$$

Resonance Energy

Evidence for there being non-localization of orbitals (resonance) producing substances which are more stable than the simple structures suggest, is provided from thermochemical data. For example, the 'theoretical' heat of formation of carbon dioxide can be calculated by adding the appropriate heats of dissociation and bond energies, e.g.

heat of atomization of carbon, $C_{crystal} \rightarrow C_{isolated}$ $\Delta H = +713 \text{ kJ}$
heat of dissociation of oxygen, $O_2 \rightarrow 2O$ $\Delta H = +502 \text{ kJ}$
bond energy for $C{=}O$, $C + O \rightarrow C{=}O$ $\Delta H = -741 \text{ kJ}$

Therefore, the energy required to form one mole of carbon dioxide from its elements,

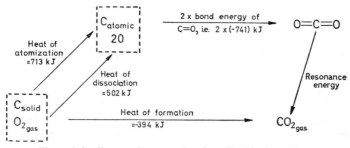

Figure 6.4.   Energy diagram of carbon dioxide formation

119

assuming that the molecule contains two simple $C=O$ bonds, is $713 + 502 + 2(-741) = -267$ kJ but the actual heat of formation, as determined experimentally, is -394 kJ, so that each mole of carbon dioxide is stabilized by $394 - 267)$ kJ, i.e. the resonance energy is 127 kJ and carbon dioxide does not therefore consist of simple multiple bonds. The energy diagram for the formation of carbon dioxide is shown in *Figure 6.4*.

Lattice Energy

The formation of an electrovalent compound from its elements can be regarded as made up of several stages:

(*a*) the dissociation (or sublimation) of the elements into single atoms;

(*b*) the loss or gain of electrons to form ions (ionization energies and electron affinities, respectively);

(*c*) the arrangement of these ions to form one mole of solid. The lattice energy is the energy required to reverse the latter operation, that is, to remove the ions in one mole of solid to an infinite distance where mutual attractions are zero.

Provided that all the other quantities are known, the lattice energy can be calculated by means of Hess's law. *Figure 6.5* shows the sequences for calcium oxide:

$$- 635 + \text{lattice energy} = 199 + 251 + 1\ 730 + 640$$
$$\therefore \text{lattice energy} = 3\ 455 \text{ kJ}$$

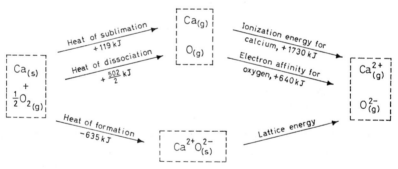

Figure 6.5. Calculation of lattice energy of calcium oxide: the Born–Haber cycle
Positive values indicate that energy is absorbed by, negative values, that energy is lost from, the system

FREE ENERGY, ENTROPY AND THE PROSPECT OF REACTION

It is well known that systems tend to move from high to low energy, that is, from low to high stability. For example, a ball runs of its own volition downhill and not uphill. That is not to say that it could never run uphill, but it would only do so if given enough external energy. We can say that systems tend to move spontaneously downhill, energetically speaking, and in fact most spontaneous reactions are exothermic. But not *all* are and indeed, if they were, reversible reactions would be impossible, for if the change $A \rightarrow B$ were exothermic (i.e. ΔH negative), then the reverse reaction, $B \rightarrow A$, would be endothermic (i.e. ΔH positive).

It is obvious that heat change cannot be the sole criterion of spontaneity in chemical reaction. Spontaneous changes also tend to increase the state of disorder of things. For instance, gases tend to mix completely when brought into contact and solutes tend to dissolve in solvents. The term *entropy*, S, is used to indicate the state of disorder of a system, the entropy being low when the order is high and *vice versa*. The entropy of a perfect crystal at absolute zero, when there is no movement of the atoms, is put at zero, and a glance at *Table 6.1* will reveal that gases, with their more chaotic structure, have higher entropies than liquids, and liquids than solids.

Table 6.1. Standard Entropies (at 298K and 10^5 N m^{-2})

		J k^{-1} mol^{-1}			
C$_{(diamond)}$	2·4	Br$_2$(liq)	152	SO$_2$	249
S$_{(rhombic)}$	32	N$_2$	192	SO$_3$	256
Cu	33	O$_2$	206	NO	211
Zn	42	Cl$_2$	223		

The important equation emerges

$$\Delta G \quad = \quad \Delta H \quad - \quad T\Delta S$$

free energy   heat $- T \times$ (change in
change   change   order)

where $T =$ temperature (in absolute degrees)
$T\Delta S = Q$, the heat absorbed by the system (reversibly)
ΔG, the change in *free energy*, represents the useful work that the system can perform.

The notion that an increase in entropy represents an increase in disorder can be reconciled with the equation $Q = T\Delta S$ by means of a simple calculation. The conversion of liquid water into water vapour is clearly accompanied by an increase in randomness of structure; if the change takes place at the boiling point, the heat required to effect the change reversibly is approximately equal to the latent heat of vaporization ($= 2\,260$ J g^{-1})

$$Q_{\text{revers}} = T\Delta S$$

$$\therefore \Delta S = \frac{Q_{\text{revers}}}{T} \approx \frac{2\,260 \times 18}{373} \approx 109 \text{ J k}^{-1} \text{ mol}^{-1}$$

Consider the reaction between hydrogen and oxygen to give water

$$H_2 + \tfrac{1}{2} O_2 \rightarrow H_2O$$

121

The energy required to electrolyse one mole of water (i.e. to reverse the above reaction) is 237kJ. This is also the electrical work that is obtained when the gases are recombined isothermally and reversibly in a cell, using hydrogen and oxygen electrodes, i.e.

$$\varDelta G = -237 \text{ kJ}$$

(It should be noted that the use of the term 'reversible' in the thermodynamic sense applies to the fact that, depending on the direction and value of the potential difference across the electrolytic cell, the direction of the reaction can be reversed. The word 'isothermal' means that the system is not isolated from its surroundings and that heat is able to flow into or out of the system.)

If, however, the gases are exploded, the amount of heat given out is 286 kJ i.e.

$$\varDelta H = -286 \text{ kJ}$$

The reason for this difference in the values between heat content and free energy is that, in the reversible recombination of the gases, some heat is lost to the surroundings (i.e. Q, which equals $T\varDelta S$, is negative). In general, this occurs whenever a system goes from a *less* to a *more* ordered state: in this case, the change has been from gas to liquid, i.e.

$$T\varDelta S = \varDelta H - \varDelta G = -286 + 237 = -49 \text{ kJ} = Q_{\text{revers}}$$

The conclusion can be drawn that, if a reaction is capable of doing work (i.e. of losing free energy) then it *may* proceed spontaneously; how readily it does, however, depends on other factors (p. 124).

This picture of spontaneous chemical change representing an *overall* increase in stability and disorder can be very useful to the chemist in appraising the prospect of a certain reaction occurring. Often these two factors supplement each other, and there is little difficulty in predicting in which direction the reaction ought to go. In the case of the dissolution of zinc in non-oxidizing acid

$$\text{Zn} + 2\text{H}^+ \rightarrow \text{Zn}^{2+} + \text{H}_2 \uparrow$$

there is a considerable decrease in heat content ($\varDelta H = -152$ kJ), and also an increase in the entropy of the system, since the overall change in state is from a solid to a gas. Therefore, as expected, the reaction is spontaneous, as the entropy increase and heat change both favour a free energy decrease.

In the case of the reaction

$$\text{C} + \text{H}_2\text{O} \rightarrow \text{CO} \uparrow + \text{H}_2 \uparrow \qquad \varDelta H = 132 \text{ kJ}$$

heat is absorbed, but there is a considerable increase in entropy (a solid and a gas producing two gases) which, at least at white heat, more than compensates for the heat absorbed: so the reaction is spontaneous.

Standard molar entropies can be used, in conjunction with the heat content, to calculate the free energy of a reaction, and hence assess the prospect of its occurring. For example, the heat content change for the reaction

$$\tfrac{1}{2}N_2 + \tfrac{1}{2}O_2 \rightarrow NO$$

is 89·0 kJ at 25°C. The standard entropies at this temperature are given in *Table 6.1*, from which it can be seen that the change in entropy accompanying the forward reaction is $211 - \tfrac{1}{2}(206 + 192) = 12$ J K^{-1} mol^{-1}. The change in free energy

$$\Delta G = \Delta H - T\Delta S$$
$$= 89 \cdot 0 - (298 \times 12/1\,000) = 85 \cdot 5 \text{ kJ mol}^{-1}$$

is positive and so the reaction will not take place spontaneously. (This conclusion could also have been arrived at qualitatively by considering the high positive heat of reaction and the fact that there is no change in the volume of gas as the reaction proceeds and thus little change in entropy to counteract the heat of reaction.)

KINETICS

The Effect of Temperature and Catalyst on Reaction Velocity

We have so far considered the probability of spontaneous chemical reaction in terms of a balance between heat of reaction and entropy, represented by the free energy. But not even if the free energy term is favourable need the reaction be readily spontaneous. For example, the free energy change accompanying the reaction

$$H_{2(gas)} + \tfrac{1}{2}O_{2(gas)} \rightarrow H_2O_{(liq)}$$

is -237 kJ mol^{-1}, and yet the reaction, in the absence of a suitable third party, takes place at an incredibly slow rate at ordinary temperature, because of the existence of a high energy barrier which has to be surmounted in the course of the reaction *Figure 6.6* shows that when two reactive species approach one another, an initial decrease in energy occurs as induced dipoles are formed and orientation results to minimize the energy. As the distance of approach becomes smaller the bonds are strained, resulting in an increase of energy for the system. Eventually a transition stage is reached where the combined *activation complex* can break down, either to give back the initial reactants or to form new products, so releasing energy. The energy that has to be acquired in order that the reactants may 'climb' the energy barrier is known as the *activation energy* and is given by the expression

$$k = Ae^{-E/RT}$$

where k = the rate constant

$\quad A$ = a constant, known as the Arrhenius constant

$\quad T$ = the absolute temperature

$\quad E$ = energy of activation.

123

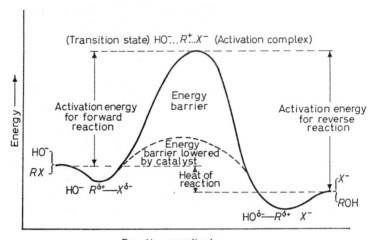

Reaction coordinate

Figure 6.6. A typical reaction path for a reversible reaction

ΔG for conversion of A to B is favourable, but reaction is prevented from being noticeably spontaneous by the energy barrier
For the catalysed reaction, the energy barrier is in some way lowered and there is accordingly a greater probability of some reacting molecules having sufficient activation energy to surmount the barrier

It will be seen from this expression that the rate of reaction increases with an increase in temperature. This is in keeping with the kinetic theory which suggests that the higher the temperature the greater the energy associated with the molecules, and the greater the number of collisions whereby some molecules may acquire the requisite amount of activation energy. In fact, a 10° rise in temperature approximately doubles the rate of reaction, a matter of considerable interest to biologists, affecting as it does the response of cold-blooded animals to a change of temperature, unlike warm-blooded creatures with their constant body temperatures.

In discussing the synthesis of water earlier, reference was made to 'suitable third parties'. These are *catalysts*, which assist reaction by lowering the energy barrier (*Figure 6.6*). Catalysts contribute nothing permanently to the energy of the system and so they cannot affect the position of equilibrium of a reversible reaction (p. 131). Furthermore, they are normally required in small quantities only and can be recovered at the end of the reaction, although they are often changed physically, suggesting that they have become chemically involved at some stage and then reformed.

Catalysts can be classified as *homogeneous*, when both the catalyst and the reactants are in the same phase, or *heterogeneous* when they constitute two phases. There appear to be at least two distinct types of mechanisms in catalysed reactions. The adsorption theory, which is particularly applicable to heterogeneous catalysis, suggests that the reacting species are adsorbed onto the surface of the catalyst,

124

probably in the form of a monomolecular layer, producing an effective increase in the concentration of the reactants and thus enhancing the chances of reaction. Adsorption of this sort would lead to a weakening of bonds inside the molecules of the reactants and also ensure that the molecules were in a particular orientation, both factors affecting the energy of activation. In the case of metallic catalysts, their good thermal conductivity would permit any heat produced to be conducted away before it could bring about the dissociation of the product. Surface catalysts of this type are often highly specific, a fact that is attributed to the difference in the distances between the 'active centres' of different substances. For example, nickel can bring about the dehydrogenation of a primary alcohol to an aldehyde, in accordance with adsorption of the hydroxyl bond on the metal and a consequent weakening of the oxygen–hydrogen linkage; on the other hand, alumina catalyses the dehydration of alcohol to alkene, suggesting the weakening of the carbon–oxygen bond by adsorption of these two atoms because of the more favourable distances between active centres (*Figure 6.7*). There are countless examples of surface catalysis, many of industrial importance and involving transition metals, such as the use of vanadium in the contact process for sulphuric acid and of iron in the Haber process for ammonia. It is believed that those complex proteins called enzymes also function by surface adsorption, a fact that explains their high specificity.

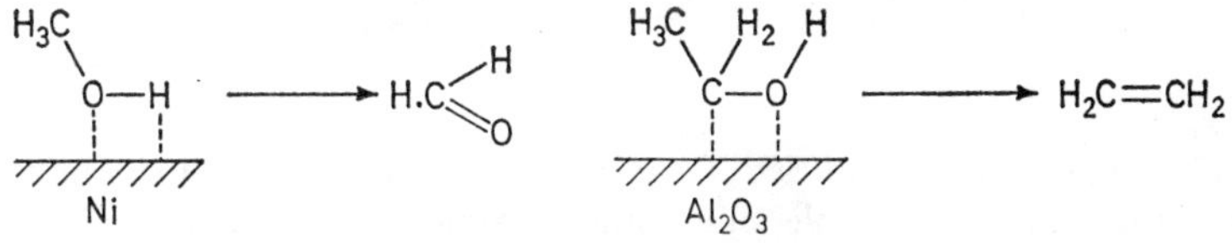

Figure 6.7. Catalytic specificity

An alternative mechanism applicable in some reactions involves the formation and subsequent decomposition of an intermediate compound. If C represents the catalyst and A and B are the reacting species, the reaction can be written as

$$A + C \rightarrow AC \qquad AC + B \rightarrow AB + C$$

(overall reaction, $A + B \rightarrow AB$)

For example, in the Friedel–Crafts reaction between chloroethane and benzene, in the presence of anhydrous aluminium chloride as catalyst, the reaction takes place as follows

$$C_2H_5Cl + AlCl_3 \rightarrow C_2H_5^+(AlCl_4)^-$$
(intermediate compound)
$$C_6H_6 + C_2H_5^+(AlCl_4)^- \rightarrow C_6H_5 . C_2H_5 + HCl + AlCl_3$$

Many other examples of catalysis will be described in later Chapters.

Autocatalysis
Sometimes a catalyst is formed in the course of reaction, so that its rate becomes

125

progressively greater. A common example of this phenomenon is the titration of oxalic acid with potassium permanganate. Manganese(II) ions are formed as a result of the oxidation of the acid, and these catalyse the reaction

$$5 \begin{vmatrix} COOH \\ COOH \end{vmatrix} + 2MnO_4^- + 6H^+ \rightarrow 10\ CO_2 \uparrow + 2Mn^{2+} + 8H_2O$$
$$\text{autocatalyst}$$

Poisoning of Catalysts

Certain substances, often present as impurities, are preferentially adsorbed on to surfaces of catalysts, preventing the latter from acting efficiently. The catalyst is then said to be poisoned, an apt description because in the case of a living organism, such adsorption, for example of arsenic on enzymes, can lead to its death.

Inhibitors

Because of the general lowering of energy barriers, catalysts can allow side reactions to occur; it is thus possible for the presence of certain materials to lead to a reduction in the original reaction rate. Such substances are called *inhibitors* or *negative catalysts*. Thus, whilst benzoyl peroxide provides the free radicles to initiate polymerisation reactions, an inhibitor which destroys free radicals acts to terminate such chain reactions.

Promoters

The addition of small quantities of a second substance may enhance the catalytic activity of the first; for example, aluminium 'promotes' the activity of iron in the Haber process. Most promoters have been discovered empirically—a reminder that our theories of catalysis are far from complete.

Effect of Concentration on Reaction Velocity

Chemical reaction is generally the result of the approach, in a favourably orientated manner, of suitably activated reactants. Clearly, the probability of reaction is raised if the concentration of the molecules is increased. For instance, in a simple reaction between A and B, the rate is increased sixfold if the concentration of A is tripled and that of B doubled:

$$\text{reaction velocity} \propto [A]\,[B]$$

where $[A]$ and $[B]$ represent the 'active masses' of A and B, respectively. Since the number of molecules in one mole of any substance is the same (the Avogadro constant, $6 \cdot 023 \times 10^{23}$), the active mass can be taken as the molar concentration, provided that there is neither association nor dissociation and the substances are not highly concentrated. (Strictly speaking, the term 'activity', taking account of these various factors, should be used, but a discussion of this term is outside the

126

scope of this book.) This is the basis of the *law of mass action* of Guldberg and Waage, which states that the rate of a chemical reaction is directly proportional to the product of the 'active masses' of the reactants.

Order of Reaction

In the example given above, the reaction velocity is dependent on the concentrations of both reactants, i.e. it is a reaction of the second order. Sometimes the rate of a reaction is dependent only on the concentration of one substance, in which case it is a *first-order reaction*

$$A \rightarrow \text{products}$$

The rate will evidently fall off as the reaction proceeds and less of A remains, so that the complete reaction requires an infinitely long interval of time (*Figure 6.8*).

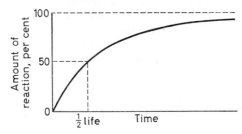

Figure 6.8. Rate of first-order reaction

If, at any one instant, x moles of product have been formed from a moles of reactant, the number of moles of reactant left will be $(a - x)$ and the rate of formation of x will be proportional to this, i.e.

$$\frac{\mathrm{d}x}{\mathrm{d}t} = k(a - x)$$

where k is a constant or

$$\frac{\mathrm{d}x}{a - x} = k \cdot \mathrm{d}t$$

This, on integration, gives

$$- \log_e(a - x) = kt + \text{constant}$$

When $t = 0$, $x = 0$ and therefore the constant $= - \log_e a$. The equation can now be rewritten

$$\log_e \frac{a}{(a - x)} = kt$$

Substituting the *half-life* (i.e. time required to reduce the original concentration

127

to one-half), $t_{\frac{1}{2}}$, for t gives $x = \frac{1}{2}a$ and therefore

$$\log_e \frac{a}{\frac{1}{2}a} = kt_{\frac{1}{2}}$$

$$\log_e 2 = kt_{\frac{1}{2}}$$

$$t_{\frac{1}{2}} = \frac{0 \cdot 693}{k}$$

i.e. the half-life of a first-order reaction is independent of the initial concentration. Consequently, it is a very significant quantity, for example in radioactivity (see p.29)

It is seldom possible to deduce the order of reaction from the overall equation; the rate of a reaction is that of the slowest step, which acts as a 'bottleneck'. For example, kinetic studies have shown that the decomposition of nitrogen(V) oxide by heat is first order. This could be explained if the rate-determining step involved one molecule only (i.e. if it was unimolecular):

$$N_2O_5 \xrightarrow{\text{slow}} NO_2 + NO + O_2$$

The nitrogen(II) oxide formed could then react quickly with another molecule of nitrogen(V) oxide

$$NO + N_2O_5 \rightarrow 3NO_2 \text{ (fast)}$$

giving the overall equation

$$2N_2O_5 \rightarrow 4NO_2 + O_2$$

Sometimes a reaction obeys the first-order equation because one reactant is in such large excess that its diminution in concentration as reaction proceeds is negligible compared to that of the other constituents. An example of this is the hydrolysis ('inversion') of dilute sucrose solution in the presence of acid as catalyst

$$C_{12}H_{22}O_{11} + H_2O \rightarrow 2C_6H_{12}O_6$$

The concentration of water in this dilute solution remains almost constant and its small variation is not sufficient to affect the order of the reaction. Such a reaction is called 'pseudo-unimolecular'.

Second-order Reactions

Consider now a bimolecular reaction of the type

$$2A \rightarrow \text{products}$$

If, after time t, x moles of product have been obtained from an original

128

concentration of a moles of each reactant, the reaction velocity will be

$$\frac{dx}{dt} = k(a-x)^2 \quad \text{or} \quad \frac{dx}{(a-x)^2} = k \cdot dt$$

Integration of this gives

$$kt = \frac{x}{a(a-x)}$$

And this time, for the half-life, $t_{\frac{1}{2}}$,

$$kt_{\frac{1}{2}} = \frac{1}{a}$$

so that, unlike first-order reactions, the half-life is *not* independent of the initial concentration.

An example of such a second-order reaction is the decomposition of hydrogen iodide

$$2HI \rightleftharpoons H_2 + I_2$$

When the prerequisite for most chemical reactions is considered, i.e. collision of the 'right' molecules in the 'right' orientation, it will be realized that reactions higher than second order are very rare.

Determination of Order of Reaction

Before the order can be elucidated, it is imperative that the state of the system be known at various intervals of time. The most obvious, but perhaps least satisfactory, method entails the chemical sampling of the system. This has the disadvantage that the system is disturbed by the sampling and that, even though the sample is 'frozen', i.e. diluted considerably or cooled down as quickly as possible, a certain change after removal of the sample and before analysis is inevitable.

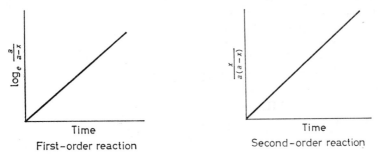

Figure 6.9. Characteristics of reaction kinetics

Far more satisfactory is the measurement of some physical property which is a function of the state of the system. A well known example is the measurement of the optical rotation of sucrose solution in the process of inversion; other physical properties that can be used include the conductivity or volume.

If different concentrations of reactants are employed in different experiments and the time taken for reaction to become half complete is measured in each case, the order can be evaluated by comparison with the various equations for half-life.

Alternatively, the change in concentration can be measured from time to time and the time plotted against $\log_e a/(a - x)$, $x/[a(a - x)]$, etc. A straight line will be given by the appropriate expression (*Figure 6.9*).

Chain Reactions

The order of reaction merely gives the kinetics of the slowest, rate-determining, step; it does not therefore provide an infallible indication of the overall reaction. This is particularly the case with chain reactions, where initiation is the measured stage and where propagation can continue until combination between two free radicals eventually terminates the chain. An example of a chain reaction is that between chlorine and hydrogen in the presence of light, the energy of which dissociates a chlorine molecule into two atoms, one of which can attack a hydrogen molecule to produce a molecule of hydrogen chloride and a reactive hydrogen atom, which can then attack a chlorine molecule and so perpetuate the reaction:

$$Cl_2 + \text{light energy} \rightarrow 2Cl\cdot \qquad \text{Initiation}$$

$$\left.\begin{array}{l} Cl\cdot + H_2 \rightarrow HCl + H\cdot \\[2mm] H\cdot + Cl_2 \rightarrow HCl + Cl\cdot \end{array}\right\} \qquad \text{Propagation}$$

$$2Cl\cdot \rightarrow Cl_2 \qquad \text{Termination}$$

Parallel Reactions (*Figure 6.10*)

A cause of much frustration in practical chemistry is the parallel reaction whereby a reactant can react in different ways, only one of which is perhaps desired. To achieve the maximum yield of the required product, it is important to recognize the situation and control the conditions, e.g. concentration, temperature, catalyst, so that the required reaction is favoured to the partial exclusion of the other. This situation is encountered frequently in organic chemistry, as for example in the treatment of ethanol with concentrated sulphuric acid (see pages 372 and 439)

$$\begin{array}{c} \qquad\qquad \nearrow \quad C_2H_4 + H_2O \\[2mm] C_2H_5OH \\[2mm] \qquad\qquad \searrow \quad C_2H_5OC_2H_5 + H_2O \end{array}$$

OPPOSING REACTIONS: CHEMICAL EQUILIBRIUM

The products formed by a reaction often react to give back the original reactants, i.e. the reaction is *reversible*:

$$A \rightleftharpoons X$$

130

As the rate of formation of X depends on the concentration of A, it will become progressively less as reaction proceeds. On the other hand, as more X is formed, the rate at which this decomposes increases. Inevitably there will be a point at which the concentrations of A and X remain constant: a *dynamic equilibrium* is established. (*Figure 6.10*).

The law of mass action can be applied to both the forward and backward

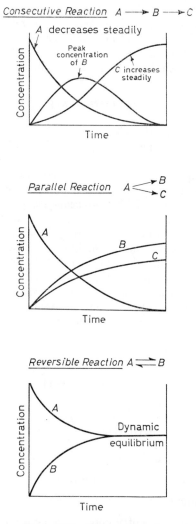

Figure 6.10. Types of reactions

131

reactions for the elementary system

$$A + B \rightleftharpoons C + D$$

The velocity of

$$\text{forward reaction} = k_1 \, [A] \, [B]$$

$$\text{reverse reaction} = k_2 \, [C] \, [D]$$

where [A], [B], [C], [D], represent the active masses (usually molar concentrations) of the constituents, and k_1 and k_2 the velocity constants.

It must be stressed that these equations for the rate reactions apply only to changes involving a direct union of one molecule of A with one of B to produce one molecule each of C and D.

When the rates of forward and backward reactions are equal, i.e. when there is dynamic equilibrium

$$k_1 \, [A] \, [B] = k_2 \, [C] \, [D]$$

$$\frac{[C] \, [D]}{[A] \, [B]} = \frac{k_1}{k_2} = K \text{ (the equilibrium constant)}$$

The value for the equilibrium constant always follows from the overall equation, irrespective of whether the reaction occurs as indicated by this equation.

The above expression, known as the *mass action equation for equilibrium*, is of wide application in chemistry. It should be remembered that it has its roots in the kinetic theory and is the result of considering the effect of differing numbers of molecules of the reactants on the probability of collisions occurring. Therefore, the mass action equation will hold, provided that the concentrations in which the reactants are expressed bear a direct relationship to the actual numbers of molecules. For liquids, molar concentrations are used; for gaseous reactions, partial pressures (p. 62), as these also depend directly on the number of molecules present, and it is more convenient to observe pressure changes than to measure concentrations. The equilibrium constant for molar concentrations is designated K_c, that for partial pressures, K_p.

In the case of the reaction

$$N_2 + 3H_2 \rightleftharpoons 2NH_3$$

$$K_c = \frac{[NH_3]^2}{[N_2] \, [H_2]^3} \qquad K_p = \frac{p^2_{NH_3}}{p_{N_2} \cdot p^3_{H_2}}$$

Let the total volume be v dm³, and

a = initial number of moles of nitrogen

$3a$ = initial number of moles of hydrogen

$2x$ = number of moles of ammonia produced under a pressure P

132

Then, at equilibrium, there are $(a - x)$ moles of nitrogen, $3(a - x)$ moles of hydrogen and $2x$ moles of ammonia, giving a total number of $2(2a - x)$ moles.

$$K_c = \frac{\left(\dfrac{2x}{v}\right)^2}{\left(\dfrac{a - x}{v}\right)\left(\dfrac{3(a - x)}{v}\right)^3}$$

From Dalton's law of partial pressures, the pressure of nitrogen will be

$$p_{N_2} = \frac{(a - x)}{2(2a - x)} \cdot P$$

Similarly,

$$p_{H_2} = \frac{3(a - x)}{2(2a - x)} \cdot P \qquad p_{NH_3} = \frac{2x}{2(2a - x)} \cdot P$$

Therefore,

$$K_p = \frac{p_{NH_3}^2}{p_{N_2} \cdot p_{H_2}^3} = \frac{\left(\dfrac{2x}{2(2a - x)} \cdot P\right)^2}{\left(\dfrac{a - x}{2(2a - x)} \cdot P\right)\left(\dfrac{3(a - x)}{2(2a - x)} \cdot P\right)^3} = \frac{2^4 x^2 (2a - x)^2}{3^3 (a - x)^4 P^2}$$

In practice, x is much smaller than a, and so the equation can be simplified to

$$K_p \sim \frac{64x^2}{27P^2} \quad \text{or} \quad x^2 \sim P^2 \cdot \frac{27}{64} \cdot K_p$$

so that the yield of ammonia is directly proportional to the pressure (approximately).

This conclusion is in accordance with *Le Chatelier's principle* which states that *if a constraint be applied to a system at equilibrium, then the system so shifts as to nullify as far as possible the effect of the constraint*. The formation of ammonia from its elements results in a contraction of volume or pressure. Hence an increase in applied pressure shifts the equilibrium in a forward direction in favour of the formation of ammonia. Also, since the reaction is exothermic, it can be predicted from the same principle that lowering the temperature will favour the forward reaction. Unfortunately, under such conditions, the rate of attainment of equilibrium is very slow; therefore, a catalyst and compromise temperature, giving a reasonable reaction velocity at the expense of the position of equilibrium, are employed in the large-scale manufacture of ammonia.

From the above relationships

$$\frac{K_p}{K_c} = \frac{\dfrac{2^4 x^2 (2a - x)^2}{3^3 (a - x)^4 \, P^2}}{\left(\dfrac{2x}{v}\right)^2 \overline{\left(\dfrac{a - x}{v}\right)\left(\dfrac{3(a - x)}{v}\right)^3}} = \frac{2^2 (2a - x)^2}{(Pv)^2} = \frac{N^2}{(NRT)^2} = (RT)^{-2}$$

where N = number of moles present.

This is an example of the general relationship

$$K_p = K_c (RT)^{\Delta n}$$

where Δn = change in the number of moles of gas.

Calculation of Equilibrium Constants
1. *From molar concentrations*—0·7 mole of acetic acid and 0·8 mole of ethanol were kept in contact until equilibrium was reached, when the amount of acid remaining was found to be 0·2 mole. Calculate the equilibrium constant.
From the equation

$$CH_3COOH + C_2H_5OH \rightleftharpoons CH_3COOC_2H_5 + H_2O$$

it can be seen that for every mole of acetic acid and ethanol that are used up, one mole each of ethyl acetate and water are formed. In this particular case, the amount of acid used, $(0·7–0·2) = 0·5$ mole, must equal the amount of alcohol used and the number of moles of ester and water formed, so that at equilibrium [acetic acid] = $0·2/v$; [ethanol] = $(0·8 - 0·5)/v = 0·3/v$; [ethyl acetate] = [water] = $0·5/v$ where v is the total volume of the system.

Therefore

$$K_c = \frac{[CH_3COOC_2H_5]\,[H_2O]}{[CH_3COOH]\,[C_2H_5OH]} = \frac{0·5 \times 0·5}{0·2 \times 0·3} = 4·2$$

2. *From measurements of density*—The density of iodine vapour at 1473 K and 10^5 N m^{-2} pressure is 1·2 g dm^{-3}. Calculate the equilibrium constant, in terms of partial pressures, for the dissociation

$$I_2 \rightleftharpoons 2I$$

i.e.

$$K_p = \frac{p_I^2}{p_{I_2}}$$

134

Let the degree of dissociation be a. Then the number of moles of undissociated iodine left after dissociation of 1 mole of iodine $= (1 - a)$, and the number of moles of dissociated iodine $= 2a$. Hence the number of 'particles' obtained from one original 'particle' $= (1 - a) + 2a = 1 + a$.

Since at s.t.p. 1 mole (254 g iodine) occupies 22·4 dm³,

at 1 473 K, it will occupy $(22·4 \times 1\ 473)/273$ dm³

and its density will be $(254 \times 273)/(22·4 \times 1\ 473)$

$= 2·1$ g dm⁻³ if there is no dissociation. But

$$\frac{\text{density assuming no dissociation}}{\text{actual density}} = \frac{2·1}{1·2} = \frac{\text{actual number of particles}}{\text{number of particles assuming no dissociation}}$$

$$= \frac{1 + a}{1}$$

$$\therefore \qquad a = \frac{2·1 - 1·2}{1·2} = \frac{0·9}{1·2} = 0·75$$

For a total pressure, P

$$p_{I_2} = \frac{1 - a}{1 + a} . P = \frac{1}{7} \qquad p_I = \frac{2a}{1 + a} . P = \frac{6}{7}$$

therefore

$$K_p = \frac{p_I^2}{p_{I_2}} = \frac{(6/7)^2}{1/7} = 5·14$$

Heterogeneous Equilibria

The law of mass action is essentially concerned with homogeneous systems; this much is clear from its statistical derivation. It can also be applied with some success to a homogeneous phase of a heterogeneous system. For example, in the dissociation of solid calcium carbonate into solid calcium oxide and gaseous carbon dioxide

$$CaCO_3 \rightleftharpoons CaO + CO_2$$

$$K_p = \frac{p_{CO_2} \times p_{CaO}}{p_{CaCO_3}}$$

The values p_{CaCO_3} and p_{CaO} are the vapour pressures of the solids and are constant at constant temperature. The equation can therefore be modified

$$K_p = \frac{k_1 \times p_{CO_2}}{k_2}$$

135

or
$$p_{CO_2} = \text{constant, at constant temperature}$$

This is found to be the case in practice, the constant being known as the *dissociation pressure.*

The same sort of reasoning for the system

$$NH_4HS \rightleftharpoons NH_3 + H_2S$$
$$\text{solid} \qquad \text{gas} \qquad \text{gas}$$

leads to the conclusion that $(p_{NH_3} \times p_{H_2S})$ is constant at constant temperature. How closely theory accords with practice can be seen from *Table 6.2.*

Table 6.2

p_{NH_3} $10^3 Nm^{-2}$	p_{H_2S} $10^3 Nm^{-2}$	$(p_{NH_3} \times p_{H_2S})$ $10^6 N^2 m^{-4}$
33·1	33·1	10 920
27·6	38·8	10 740
59·9	18·9	11 300

Free Energy and Position of Equilibrium

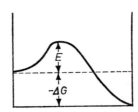

Figure. 6.11

For forward reaction $k_1 = Ae^{-E/RT}$
For backward reaction $k_2 = A_e^{(E-\Delta G)/RT}$

Therefore $K = \dfrac{k_1}{k_2} = \dfrac{Ae^{-E/RT}}{Ae^{+(E-\Delta G)/RT}}$

i.e. $\ln K = \dfrac{-\Delta G}{RT}$

and $\Delta G = -RT\ln K$, i.e. if ΔG is large, the reaction is virtually irreversible.

SUMMARY

In the course of chemical reaction, chemical energy is either lost or gained. There is a tendency for reactions to take place spontaneously if the energy of the products is less than that of the reactants, especially if there is an increase in disorder, or entropy, as the reaction takes place. But even then an energy barrier may limit the progress of the reaction considerably, as only those molecules which have energy in excess of the activation energy will be able to surmount the barrier. An increase in temperature enables more molecules to acquire this activation energy and consequently has a marked effect on the reaction velocity. So also have catalysts, although these function by lowering the energy barrier, either by forming intermediate compounds or adsorption complexes. Other factors which influence the rate of reaction are the physical state of the system (homogeneous systems allowing greater contact between reacting molecules than heterogeneous systems where the only contact is at the junction between the phases) and the concentration of the reactants. In the latter case, the rate of reaction is proportional to the product of the active masses.

At equilibrium, there is a balance of forward and backward reactions and the application of the above idea leads, for the system $aA + bB \rightleftharpoons cC + dD$, to the mass action equation

$$\frac{[C]^c [D]^d}{[A]^a [B]^b} = K \text{ (the equilibrium constant)}$$

QUESTIONS

1. Calculate the heat of formation of carbon disulphide from the following data:

	$\Delta H =$
$CS_2 + 3O_2 \rightarrow CO_2 + 2SO_2$	$- 1\ 110$ kJ
$C + O_2 \rightarrow CO_2$	$- 394$ kJ
$S + O_2 \rightarrow SO_2$	$- 297$ kJ

2. Calculate the bond energy of the C—H link from

	$\Delta H =$
Heat of formation of carbon dioxide	$- 394$ kJ
Heat of formation of water	$- 285$ kJ
Heat of atomization of carbon	$+ 713$ kJ
Heat of dissociation of hydrogen	$+ 435$ kJ
Heat of combustion of methane, CH_4	$- 899$ kJ

3. Suggest the roles of the catalyst in the following:

(a) decomposition of potassium chlorate in the presence of manganese dioxide;

(b) conversion of sulphur dioxide and oxygen to sulphur trioxide in the presence of platinum;

(c) the esterification of ethanol by acetic acid in the presence of mineral acid;
(d) the hydrolysis of ethyl acetate by alkali.
Give any evidence you can to support your mechanisms.

4. Discuss the following statements:

(a) magnesium reacts exothermically with oxygen only when heated;
(b) magnesium oxide is reduced by carbon only at high temperatures;
(c) thionyl chloride reacts instantly and endothermically with acetic acid;
(d) ammonium chloride dissolves in water with the absorption of heat.

5. A mixture of 21 g of acetic acid and 16 g of ethanol were kept at constant temperature until equilibrium was attained; analysis showed that 6 g of acetic acid remained. Calculate the equilibrium constant for the reaction.
What would be the composition of the equilibrium mixture if 5 g of acetic acid, 10 g of ethanol and 8 g each of water and ethyl acetate were kept at the same constant temperature until equilibrium was reached?

6. Nitrogen dioxide contains 20 per cent by volume of NO_2 molecules at a pressure of 10^5 N m^{-2} and a temperature of $30°$, the remainder being dimeric molecules. Calculate the equilibrium constant, K_p, for the reaction

$$N_2O_4 \rightleftharpoons 2NO_2$$

at $30°$ and hence the percentage dissociation at the same temperature when the pressure is increased to 4×10^5 Nm^{-2}

7. The following values represent the change in optical rotation produced in the conversion of a solution of α-glucose containing 10 g dm^{-3}. to β-glucose:

time (min)	5	10	20	30	40	50	60	24h (∞ reading)
rotation($°$)	97·6	96·8	95·2	94·2	93·3	92·5	91·9	87·4

From these values show that the reaction is of first order, and hence evaluate the rate constant.

8. Suggest how you could investigate the effect of temperature and concentration on the reaction between thiosulphate ions and acid:

$$S_2O_3^{2-} + 2H^+ \rightarrow SO_2 \uparrow + S \downarrow + H_2O$$

How could the order of the reaction be confirmed?

9. Derive the relationship $K_p = K_c(RT)^{\Delta n}$.

10. Draw a graph of energy versus internuclear distance, to show the approach of molar quantities of sodium and chloride ions. Indicate the position of internuclear equilibrium.

11. Use the following information (from Moelwyn-Hughes and Johnson, *Trans. Faraday Soc.*, 1940, **36**, 954) to calculate the energy of activation of the first-order decomposition of benzene diazonium chloride:

Temperature °C	k (sec^{-1})
15·1	$9·30 \times 10^{-6}$
19·9	$2·01 \times 10^{-5}$
24·7	$4·35 \times 10^{-5}$
30·0	$9·92 \times 10^{-5}$

(Plot values of log k against $1/T$ to get a straight line of slope $-E/2·30 \, R$: see the Arrhenius equation. Hence calculate E, the energy of activation.)

12. Calculate the equilibrium constant for the reaction

$$C_6H_6 + 3H_2 \rightleftharpoons C_6H_{12}$$

given that the free energy change at 23°C is $-97·4 \, kJ \, mol^{-1}$.

CHAPTER 7

ELECTROCHEMISTRY

A flow of charged particles constitutes an electric current. If the particles are the mobile electrons of the metallic lattice, then the type of conduction is termed *electronic* and is not accompanied by any chemical side effects. If, on the other hand, the charged particles are *ions* that result from the transfer of electrons by electrovalency, then the conduction is *electrolytic*. Provided that the ions are rendered mobile by dissolving in a suitable solvent or by melting, the application of an electric potential causes, by attraction of opposite charges, migration of the negative ions (anions) to the positive electrode (the anode) and of positive ions (cations) to the negative electrode (the cathode). Electrons are transferred at the electrodes. *to* the anode from the anion and *from* the cathode to the cation. That is, chemical changes occur at the electrodes, and this is a characteristic feature of *electrolysis*.

For example, in the electrolysis of sodium chloride, where the crystal lattice is broken down by melting, sodium ions migrate to the cathode and are discharged, whilst chloride ions are discharged at the anode (*Figure 7.1*).

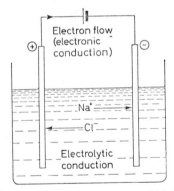

Figure 7.1. Electrolysis of fused NaCl

Cathode	*Anode*
$Na^+ + e \to Na$	$Cl^- - e \to Cl$
	$2\,Cl \to Cl_2 \uparrow$

It is clear that the precursor to electrolysis is the formation of mobile ions which, in the case of strongly electrovalent compounds, occurs either by melting or by solution in a suitable solvent. Evidence that ions are present when the substance is

140

dissolved in a solvent like water is provided by the abnormal colligative properties of solutions of electrolytes (p. 112).

The course of electrolysis is sometimes difficult to predict when there is a mixture of ions present. It is normally true to say that the process which takes place is that which requires least electrical energy. Consequently, the least stable ions are discharged first; the electrochemical series (p. 164) can be used as a rough guide in this respect, but it should be remembered that an excessive potential (the *overpotential*) is required before a substance, especially a gas, is actually discharged: the amount varies from electrode to electrode and sometimes exceeds the electrode potential, preventing discharge. For example, in the electrolysis of sodium chloride solution using a mercury cathode, the following ions are present:

$$Na^+ + Cl^- \text{ (present even in solid lattice)}$$

$$H^+ + OH^- \text{ (from the equilibrium } H_2O \rightleftharpoons H^+ + OH^-)$$

It must be remembered that simple hydrogen ions do not exist in aqueous solution: the polarizing power of a proton is so great that it combines with water to form an oxonium ion, i.e.

$$H^+ + H_2O \rightarrow H_3O^+$$

At the anode, OH^- and Cl^- are attracted. It requires less energy to discharge OH^- than Cl^-, but water is a very weak electrolyte, and so there are far fewer hydroxide ions present than chloride ions. Therefore, unless alkali is added, there are virtually no hydroxide ions available at the electrode and chloride ions are accordingly discharged. Thus, concentration also affects the course of electrolysis; the greater the concentration, the greater the chance of discharge.

At the cathode, there is a surfeit of electrons, and electrolysis proceeds as these are donated to cations. The electrochemical series suggests that the less stable hydrogen ions will be discharged in preference to sodium ions but, if the cathode is mercury, the overpotential is so great for hydrogen that the sodium ions are preferentially discharged. The resulting atoms dissolve in the mercury to form sodium amalgam; the energy released in this latter reaction also favours this particular step.

This is an example of a metal being discharged and reacting with the electrode. More often, reaction takes place between the electrode and a newly-discharged gas; the initial product is a reactive atom, and unless the electrode is particularly inert (e.g. platinum), interaction is a distinct possibility. More often than not, this is a disadvantage, but in the case of the 'anodizing' of aluminium it can be turned to account. Dilute sulphuric acid is electrolysed using an aluminium anode; hydroxide ions are discharged at this electrode to give nascent oxygen, which attacks the aluminium to give a thick protective film of oxide, which can be dyed various colours.

Sometimes, instead of an ion being discharged at the anode to provide the electrons which are essential to a continuous flow of electricity, it is energetically favourable for the anode to dissolve as positive ions, leaving behind electrons. For example, in the electrolysis of copper(II) sulphate solution between copper

141

electrodes, copper(II) ions dissolve from the anode, instead of anions being discharged: as copper is deposited at the cathode, the overall process represents the transfer of copper from anode to cathode and provides a means of purification:

Anode	*Cathode*
$Cu - 2e \to Cu^{2+}$	Cu^{2+} and H^+ attracted
SO_4^{2-} and OH^- attracted but not discharged	Cu^{2+} selectively discharged (less stable and more concentrated than H^+)
	$Cu^{2+} + 2e \to Cu$

Electrolysis can be a very useful industrial process; apart from the above examples of anodizing and purification, it can be the basis for the extraction of elements (e.g. aluminium and chlorine) and for the manufacture of compounds (e.g. sodium hydroxide and potassium chlorate) as well as a means of electroplating.

Faraday's Laws of Electrolysis

The actual quantities of substance liberated during electrolysis are summarized by Faraday's laws:

1. The mass of substance liberated at an electrode is proportional to the quantity of electricity passed (s). i.e. to the number of coulombs (C) = amperes (A) × seconds

2. The masses of different substances liberated by the same quantity of electricity are inversely proportional to the respective ionic charges.

These laws are the consequence of each ion carrying a definite charge and there being a constant number of ions (the Avogadro constant, L, equal to $6 \cdot 023 \times 10^{23}$) in a mole of any ion.

If each ion has a valency of z, then a charge of $\pm z$ units will be associated with each ion. If m is the mass of an ion, then m/z is the mass of the substance which will be liberated for every unit of electricity supplied to or from the cell (First Law). For one ion the actual charge will be $\pm ze$, where e is the electronic charge ($1 \cdot 602 \times 10^{-19}$ C) and for one mole of ion the total charge is $\pm zeL$. Now eL is a constant, known as the Faraday constant, F, and is found experimentally to equal $9 \cdot 649 \times 10^4$ C. Therefore the discharge of one mole of a z-valent ion requires zF coulombs and the number of moles liberated by F coulombs, is $1/z$ (Second law). For example, one mole of copper ions of ionic charge two, requires $2F$ coulombs for discharge and so F coulombs will liberate half a mole of copper. On the other hand, one mole of aluminium ions, of ionic charge three, require $3F$ coulombs and so F coulombs liberate one third of a mole of aluminium.

Conductivity

There is always a resistance to flow of current, which can be expressed in terms of Ohm's law:

$$\text{Resistance} = \frac{\text{voltage}}{\text{current}} \left(\text{ohms} = \frac{\text{volts}}{\text{amps}} \right)$$

142

$$\text{Conductance} = \frac{1}{\text{resistance}} \text{ (in ohm}^{-1} \text{ or } \Omega^{-1})$$

Resistivity is defined as the resistance of a 1 cm cube of material; the total resistance is then given by

$$\text{Resistance} = \frac{\text{length} \times \text{resistivity}}{\text{area}}$$

and hence

$$\text{Resistivity} = \frac{\text{area} \times \text{resistance}}{\text{length}}$$

The electrolytic conductivity (specific conductivity), κ, is the reciprocal of the specific resistance, i.e.

$$\text{Electrolytic conductivity} = \frac{\text{length}}{\text{area} \times \text{resistance}} (\Omega^{-1} \text{ cm}^{-1})$$

In the measurement of conductivities, the solution under test is contained in a conductivity cell (*Figure 7.2a*) the electrodes of which are made of platinum and

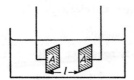

Figure 7.2a. A conductivity cell

rigidly fixed, so that (length)/(area) is constant (the cell constant), and the electrolytic conductivity is calculated simply by dividing the observed resistance into the cell constant.

$$\text{Electrolytic conductivity} = \frac{\text{cell constant}}{\text{resistance}}$$

The resistance itself is determined by using the Wheatstone bridge circuit and an alternating current of such frequency that electrolysis is negligible. The balance point is measured by means of, e.g. a 'magic eye' indicator (*Figure 7.2b*).

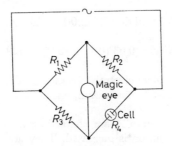

Figure 7.2b. A conductivity bridge

143

When the bridge is balanced

$$\frac{R_1}{R_2} = \frac{R_3}{R_4}$$

$$\therefore \quad R_4 = \frac{R_3 \cdot R_2}{R_1}$$

$$\kappa = \frac{\text{cell constant}}{R_4}$$

As a solution is diluted, the electrolytic conductivity decreases, since the number of conducting particles (ions) between the electrodes diminishes. Therefore, to compare the conducting power of a solution at different concentrations, allowance must be made to keep the same mass of solute under observation. This may be done as follows. If the concentration of the electrolyte is c mol dm^{-3} then the area of each electrode must be $10^3/c$ cm^2 (if the length separating them is 1 cm) in order that one mole is contained between the electrodes, i.e.

$$\kappa = \frac{\text{length}}{\text{area} \times \text{resistance}} = \frac{c}{10^3 \times \text{resistance}}$$

$$\therefore \frac{1}{\text{resistance}} = 10^3 \kappa/c$$

This product is called the molar conductivity, Λ, and is therefore the conductivity, of a solution containing one mole of electrolyte between electrodes 1 cm apart.

Example—The resistance of a 0·1 M solution of an electrolyte in a conductivity cell of cell constant 1·5 cm^{-1} is 50 Ω. Calculate the molar conductivity.

$$\text{Electrolytic conductivity} = \frac{\text{cell constant}}{\text{resistance}} = \frac{1 \cdot 5}{50} = 0 \cdot 03 \ \Omega^{-1} \text{cm}^{-1}$$

A 0·1 M solution contains 0·1 mol dm^{-3}, i.e. 0·1 $\times$ 10^{-3} mol cm^{-3}. Hence

$$\text{molar conductivity} = 0 \cdot 03/(0 \cdot 1 \times 10^{-3}) = 300 \ \Omega^{-1} \text{ cm}^2 \text{ mol}^{-1}$$

If the molar conductivity of an electrolyte is measured at different concentrations and the values plotted against concentration, one of two types of curve is obtained (*Figure 7.3*).

Substances which follow curve A are characterized by a rapid rise in Λ to a limiting value, Λ_0, as the dilution is increased. They are called *strong electrolytes* and are ionic even in the solid state. The increase in conductivity on dilution is

144

merely the effect of rendering the ions more mobile and freer from each other's influence. Substances which give curve *B*, on the other hand, do not give a clear limiting value for Λ; they are *weak electrolytes*, e.g. organic acids, and are only slightly ionized in concentrated aqueous solution. Dilution in this case serves to actually increase the degree of ionization and, indeed, the degree of dissociation

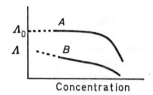

Figure 7.3. Conductivity curves

at concentration c is given by the fraction Λ_c/Λ_0 although Λ_0 can only be obtained indirectly (p. 148). For strong electrolytes this fraction is more correctly called the apparent degree of dissociation, since it really represents a mobility ratio (p. 149).

Law of Independent Migration of Ions (Kohlrausch)

Kohlrausch, in 1876, propounded his law that each ion of an electrolyte contributes a definite, constant amount, the ionic conductance, to the molar conductivity:

$$\Lambda_0 = \Lambda_+ + \Lambda_-$$

where Λ_+ and Λ_- are the ionic conductances at zero concentration

This law was based upon the following sort of evidence:

$$\Lambda_{0(\text{NaCl})} = 126 \cdot 3 \qquad\qquad \Lambda_{0(\text{NaNO}_3)} = 121 \cdot 5$$

$$\dfrac{\Lambda_{0(\text{KCl})} = 149 \cdot 8}{\Lambda_{\text{K}} - \Lambda_{\text{Na}} = 23 \cdot 5} \qquad \dfrac{\Lambda_{0(\text{KNO}_3)} = 145 \cdot 0}{\Lambda_{\text{K}} - \Lambda_{\text{Na}} = 23 \cdot 5}$$

Table 7.1

Ionic conductances at 25°C

Ω^{-1} cm^2 mol^{-1}			
H$^+$	350	K$^+$	73·5
OH$^-$	198·5	NO$_3^-$	71·5
$\frac{1}{2}$SO$_4^{2-}$	80	Na$^+$	50
Cl$^-$	76·3		

Electrolytic conduction can thus be regarded as the sum of the contributions of the separate ions. We now have a picture of ions moving at different speeds

145

and accordingly carrying different amounts of current. In fact, Kohlrausch's law permits Λ_0 for a weak electrolyte to be calculated by adding together the relevant ionic conductances; for example, at infinite dilution that of $H^+ = 315 \cdot 0$, of $CH_3COO^- = 42 \cdot 3$ Ω^{-1} cm^2. mol^{-1}.

Therefore, Λ_0 of acetic acid $= 315 \cdot 0 + 42 \cdot 3 = 357 \cdot 3$ Ω^{-1} cm^2 mol^{-1}.

Solubility of Sparingly Soluble Substances

A further application of Kohlrausch's law is the determination of the solubilities of sparingly soluble salts. A solution of a sparingly soluble salt will be so dilute as to give complete dissociation. Λ_0 may therefore be calculated from Kohlrausch's law. Determination of the electrolytic conductivity permits the dilution to be calculated and hence the solubility.

Example—The ionic conductance at 18°C of $Ag^+ = 54 \cdot 3$ and of $Cl^- = 65 \cdot 5$ Ω^{-1} cm^2 mol^{-1}. Therefore,

$$\Lambda_{0AgCl} = 54 \cdot 3 + 65 \cdot 5 = 119 \cdot 8 \ \Omega^{-1} \ cm^2 \ mol^{-1}$$

But

$$\Lambda_0 = \kappa/c$$

From experiment, the electrolytic conductivity of silver chloride at 18°C $= 1 \cdot 12 \times 10^{-6} \Omega^{-1}$ cm^{-1}.

Therefore $c = \kappa/\Lambda_0 = 1 \cdot 12 \times 10^{-6}/119 \cdot 8 = 9 \cdot 35 \times 10^{-9}$ mol cm^{-3}

But the molecular weight of silver chloride is $143 \cdot 5$. Therefore the solubility of silver chloride is $(9 \cdot 35 \times 10^{-9} \times 143 \cdot 5)$g cm$^{-3} = 1 \cdot 34 \times 10^{-3}$ g dm^{-3} at 18°C.

Conductometric Titration

Reference to *Table 7.1* will reveal that hydrogen has by far the greatest conductance. This fact can be utilized in conductometric titrations, for instance between hydrochloric acid and sodium hydroxide. As hydrochloric acid is added to sodium

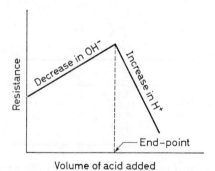

Volume of acid added

Figure 7.4. Conductometric titration

146

hydroxide solution, hydroxide ions are removed in forming water. The resistance will therefore rise until the end point is reached. When excess hydrochloric acid is then added, the resistance will fall again by virtue of the free hydrogen ions; intersection of the two curves gives the end point (*Figure 7.4*).

Transport Numbers

The greater the mobility of a particular ion, the greater will be the fraction of the total current carried by it. If t_+ is the fraction carried by the cation and t_- that carried by the anion, then $t_+ + t_- = 1$. These fractions are called the *transport numbers* of the ions, and they can be determined experimentally by measuring the changes in concentration around the electrodes during electrolysis. Consideration of a hypothetical case will make this clear. Imagine a simple binary electrolyte contained in a voltameter divided into three compartments, A, B, and C, such that there are equal numbers of ions in the anode and cathode compartments (*Figure 7.5a*).

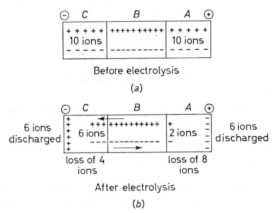

Figure 7.5. Transport numbers

If an electric current is now passed through the apparatus, and if the cations move twice as fast as the anions, the fall in concentration in the anode compartment in a given time will be twice that in the cathode compartment (*Figure 7.5b*). In more general terms

$$\frac{\text{current carried by cation}}{\text{current carried by anion and cation}} = \frac{\text{fall of concentration in anode compartment}}{\text{total fall of concentration}}$$

$$= \text{transport number of the cation, } t_+$$

The concentration of a particular ion, instead of decreasing, may actually increase, due to the possibility of electrodes dissolving. By incorporating a

147

coulometer—a cell used for measuring the quantity of electricity passed—in the circuit, the amount of electrode dissolved can be calculated and allowed for.

Example—There are 0·23 g of silver ions in the anode compartment of a voltameter (with silver anode) before electrolysis and 0·40 g afterwards. In a silver coulometer in series, 0·32 g of silver dissolves from the anode. Calculate the transport number of the silver ions.

If no migration from the anode compartment occurred, the amount of silver ions after electrolysis would be $(0·23 + 0·32) = 0·55$ g

$$\therefore \text{ fall due to migration} = (0·55 - 0·40) = 0·15 \text{ g Ag}$$

$$\text{total current} \equiv 0·32 \text{ g Ag}$$

$$\therefore \text{ transport number of Ag}^+ = 0·15/0·32 = 0·47$$

The total molar conductivity, Λ, is, by Kohlrausch's law, made up of the cationic component, Λ_+, and the anionic component, Λ_-, and it follows from the above that the differences in these values are due to the different mobilities of the ions; for instance, the hydrogen ion has a very high conductance because it is highly mobile and consequently carries a high proportion of the total current. Clearly, then, the transport numbers are related to the conductances in the following manner:

$$t_+ = \frac{\Lambda_+}{\Lambda} \qquad t_- = \frac{\Lambda_-}{\Lambda}$$

Indeed, one purpose of measuring transport numbers is to derive ionic conductances.

APPLICATION OF MASS ACTION: IONIC EQUILIBRIA

Ostwald, in 1888, applied the law of mass action to electrolytic dissociation. For a weak binary electrolyte, AB, in solution of dilution V and with a degree of ionization under these conditions of α, the following equilibrium exists between the undissociated molecule and its ions:

$$AB \rightleftharpoons A^+ + B^-$$

$$\text{Concentrations} \quad \frac{1 - \alpha}{V} \qquad \frac{\alpha}{V} \qquad \frac{\alpha}{V}$$

Applying the law of mass action

$$\frac{[A^+][B^-]}{[AB]} = \frac{(\alpha/V) \times (\alpha/V)}{(1 - \alpha)/V}$$

$$= K, \text{ the dissociation constant, at constant temperature}$$

that is,
$$K = \frac{\alpha^2}{(1 - \alpha)V}$$

For a very weak electrolyte (i.e. when a is small) this can be simplified to

$$K = \frac{a^2}{V}$$

It has been pointed out before that the term 'degree of dissociation' has no real meaning for strong electrolytes, because these are always fully dissociated. In the case of weak electrolytes, however, there is considerable experimental support for the correctness of the above expression, which has come to be known as *Ostwald's dilution law*. For example, in the case of acetic acid:

V $dm^3\,mol^{-1}$	$a = \Lambda_v/\Lambda_0$	$K = a^2/(1-a)V$ $\times 10^{-6}$
5·374	0·009 8	18·1
10·573	0·013 8	18·0
24·875	0·021 6	19·2
63·26	0·033 6	18·5

Example—The molar conductivity of $0·1M$ ammonium hydroxide at $20°C$ is $3·17$. The ionic conductances for the ammonium and hydroxide ions at this temperature are $73·5$ and $198·5$, respectively. Calculate the dissociation constant.

$$\Lambda_0 = \Lambda_+ + \Lambda_- = 73·5 + 198·5 = 272 \ \Omega^{-1} \ cm^2 \ mol^{-1}$$

Therefore

$$\text{degree of dissociation} = \frac{\Lambda}{\Lambda_0} = \frac{3·17}{272} = 0·011\,7$$

$$\text{dissociation constant} = \frac{(0·011\,7)^2}{(1 - 0·011\,7) \times 10} = 1·4 \times 10^{-5} \ M$$

AQUEOUS SYSTEMS: DISSOCIATION OF WATER

As water is progressively purified, its electrolytic conductance approaches a limiting value of $0·54 \times 10^{-7} \ \Omega^{-1} \ cm^{-1}$ at $25°$. Thus water itself is a weak electrolyte by virtue of the tendency towards ionization

$$H_2O \rightleftharpoons H^+ + OH^-$$

By Kohlrausch's law

$$\Lambda_0 = \Lambda_{H^+} + \Lambda_{OH^-}$$

$$= 350 + 198·5 = 548·5 \ \Omega^{-1} \ cm^2 \ mol^{-1}$$

but

$$\Lambda_0 = \kappa/c$$

149

Therefore if V = number of dm^3 containing one mole of ions

$$V = \frac{548 \cdot 5}{1\ 000 \times 0 \cdot 54 \times 10^{-7}} \approx 1 \times 10^7 \, \text{dm}^3$$

$$[\text{H}^+] = [\text{OH}^-] = \frac{1}{1 \times 10^7} = 1 \times 10^{-7} \, \text{M}$$

Applying the mass action equation

$$K_{\text{H}_2\text{O}} = \frac{[\text{H}^+][\text{OH}^-]}{[\text{H}_2\text{O}]}$$

Because the ionization is so small $[\text{H}_2\text{O}]$ is approximately constant; therefore $[\text{H}^+][\text{OH}^-]$ is approximately constant and is called the *ionic product* of water, K_w.

Now $\qquad\qquad [\text{H}^+] = [\text{OH}^-] = 1 \times 10^{-7} \, \text{M}$

thus $\qquad\qquad K_w = (1 \times 10^{-7})^2 = 1 \times 10^{-14} \, \text{M}^2 \text{ at } 25°\text{C}$

One's immediate reaction may be to discount anything as small as this, but to do so would be to ignore the essence of chemical equilibrium: in aqueous systems, the equilibrium will always adjust itself to maintain the ionic product, which can thus exert the most profound effect. For instance, if the hydrogen ion concentration is increased from 1×10^{-7} M (neutrality) to 1×10^{-1} M (decimolar acid), the concentration of hydroxide ions will be decreased from 1×10^{-7} to 1×10^{-13} to maintain the ionic product $(1 \times 10^{-1} \times 1 \times 10^{-13})$ at 1×10^{-14}M^2

Constant use of negative indices like these soon proves tiresome. To avoid this, the term pH has been introduced; it is defined as the negative logarithm of the hydrogen ion concentration. Similarly, pOH is the negative logarithm of the hydroxide ion concentration. In the above case, pH = 1 and pOH = 13. It follows from the definition that pH + pOH = pK_w (= 14 at 25°).

Relationships between $[\text{H}^+]$, $[\text{OH}^-]$, *pH and pOH*

$[\text{H}^+]$ mol dm^{-3}	10^2 10^1 10^0 10^{-1} 10^{-2} 10^{-3} 10^{-4} 10^{-5} 10^{-6} 10^{-7} 10^{-8} 10^{-9} 10^{-10} 10^{-11} 10^{-12} 10^{-13} 10^{-14} 10^{-15}
$[\text{OH}^-]$ mol dm^{-3}	10^{-16} $\qquad\qquad\qquad\qquad 10^{-7} \qquad\qquad\qquad\qquad 10^1$
pH	-2 -1 0 1 2 3 4 5 6 7 8 9 10 11 12 13 14 15
pOH	16 15 14 13 12 11 10 9 8 7 6 5 4 3 2 1 0 -1

Increasing acidity ← 1 mol dm^{-3} of H$^+$ — Neutral — Increasing alkalinity 1 mol dm^{-3} of OH$^-$

Examples

1. Find the pH of $0 \cdot 15$ M hydrochloric acid.

A $0 \cdot 15$ M solution contains $0 \cdot 15$ mol dm^{-3}. Since complete ionization of a strong acid may be assumed, $0 \cdot 15$ mole of hydrogen ions will be present in each dm^3, i.e. $[H^+] = 0 \cdot 15$. Therefore

$$pH = -\log 0 \cdot 15 = -(\bar{1} \cdot 18) = 1 - 0 \cdot 18 = 0 \cdot 82$$

2. Find the pH of a $0 \cdot 1$ M solution of acetic acid, given that the dissociation constant at the temperature in question is $1 \cdot 8 \times 10^{-5}$ M^2.

For this weak electrolyte

$$K \sim a^2/V \text{ and } V = 10$$

Therefore

$$a = \sqrt{(1 \cdot 8 \times 10^{-5} \times 10)} = 1 \cdot 35 \times 10^{-2}$$

If the acetic acid were completely dissociated, the hydrogen ion concentration would be $0 \cdot 1$ mol dm^{-3}; therefore

$$[H^+] = 0 \cdot 1 \times 1 \cdot 35 \times 10^{-2} = 1 \cdot 35 \times 10^{-3}$$

$$\log [H^+] = \bar{3} \cdot 130 \, 3 = -2 \cdot 869 \, 7$$

$$\therefore pH = +2 \cdot 87$$

Changes in pH during Neutralization

Consider the progressive addition of $0 \cdot 1$ M hydrochloric acid to 25 cm^3 of $0 \cdot 1$M sodium hydroxide.

(*a*) after addition of $24 \cdot 0$ cm^3 of acid, $1 \cdot 0$ cm^3 of $0 \cdot 1$ M alkali (i.e. 1×10^{-4} mole) remains in 49 cm^3 of solution.

Therefore $\quad [OH^-] = \dfrac{1\,000 \times 10^{-4}}{49} = 2 \times 10^{-3}$ M

hence $\quad pOH = -(\bar{3} \cdot 301) = +2 \cdot 699$

and $\quad pH = 14 - 2 \cdot 699 = 11 \cdot 3$

(*b*) after addition of $24 \cdot 5$ cm^3 of $0 \cdot 1$ M acid, $0 \cdot 5$ cm^3 of $0 \cdot 1$ M alkali, containing 5×10^{-5} mole, remains in $49 \cdot 5$ cm^3 of solution.

Therefore $\quad [OH^-] = \dfrac{1\,000 \times 5 \times 10^{-5}}{49 \cdot 5} = 1 \times 10^{-3}$ M

and $\quad pOH = 3 \qquad pH = 11$

(*c*) after addition of another $0 \cdot 45$ cm^3 of acid, $0 \cdot 05$ cm^3 of $0 \cdot 1$ M alkali, containing 5×10^{-6} mole, remain in $49 \cdot 95$ cm^3 of solution; hence

$$[OH^-] = \frac{1\,000 \times 5 \times 10^{-6}}{49 \cdot 95} = 1 \times 10^{-4} \text{ M}$$

$$pOH = 4 \qquad pH = 10$$

(*d*) after addition of a further 0·05 cm³ of acid, the alkali is exactly neutralized, i.e.

$$[H^+] = [OH^-] = 1 \times 10^{-7} \text{ M}$$

$$pH = 7$$

A similar situation prevails as excess acid is gradually added. It can be seen, therefore, that, in the titration of strong acid with strong base, there is a considerable change in pH around the end point. In the case of weak electrolytes, the small degree of dissociation results in the initial pH or pOH being less; this, together with the fact that dissociation is encouraged as electrolyte is neutralized, results in the pH change being far less pronounced (*Figure 7.6*). Also, because of hydrolysis of the salt formed (p. 154) when either a weak base or a weak acid is used, the pH at the end point will not be 7.

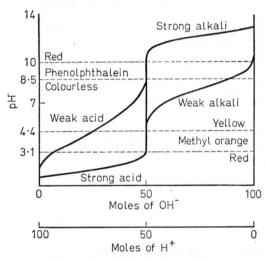

Figure 7.6. Neutralization curves

Indicators

The progress of a neutralization reaction can usually be followed by adding a small quantity of an indicator, often a weak organic acid or base where at least one ion is a different colour to the undissociated molecule. In the case of methyl orange, the undissociated molecule is red and the anion yellow

$$(CH_3)_2\overset{+}{N}=\hspace{-2pt}\langle\hspace{4pt}\rangle\hspace{-2pt}=N-NH-\hspace{-2pt}\langle\hspace{4pt}\rangle\hspace{-2pt}-SO_3^-$$
red

$$\rightleftharpoons (CH_3)_2N-\hspace{-2pt}\langle\hspace{4pt}\rangle\hspace{-2pt}-N=N-\hspace{-2pt}\langle\hspace{4pt}\rangle\hspace{-2pt}-SO_3^- + H^+$$
yellow

152

With phenolphthalein, the undissociated molecule is colourless and the anion red

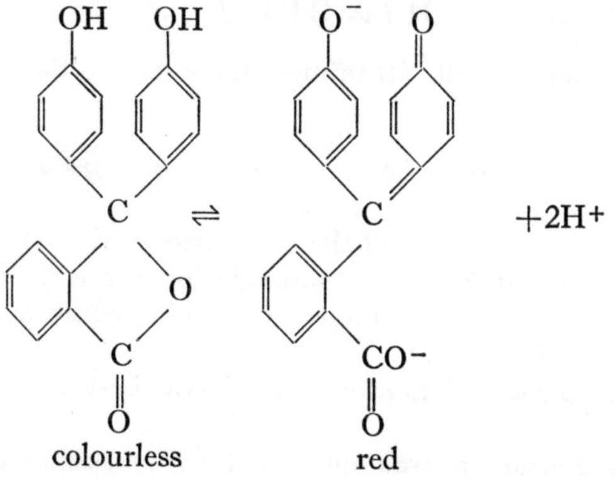

colourless   red

The amount of dissociation of the indicator, and hence the colour, depend upon the pH of the solution; the considerable changes in pH occurring around the end point in all titrations (except those between weak acid and weak base, for which there are no suitable indicators) bring about sharp changes in colour. The equilibrium constant for the reaction is given by

$$HX \underset{acid}{\overset{alkali}{\rightleftharpoons}} H^+ + X^-$$

$$K = \frac{[H^+][X^-]}{[HX]} \quad \text{i.e.} \quad \frac{K}{[H^+]} = \frac{[X^-]}{[HX]}$$

Hence, when $pH = pK^*$ $\dfrac{[X^-]}{[HX]} = 1$

Accordingly, there are different amounts of dissociation at a certain pH, depending on the differing dissociation constants of the indicators. In the case of the titration of strong acid against strong base, the pH changes are such that a wide variety of indicators is available, for strong base against weak acid (when the end point is at pH>7), phenolphthalein is often used whilst in titration of weak base against strong acid, methyl orange can be employed (*Figure 7.6*).

Hydrolysis

Reference has been made in the previous Section to hydrolysis occurring if one of the reactants is a weak electrolyte, and to the subsequent effect on pH. This requires further explanation; if reaction is between strong base and weak acid, there will be a tendency for the anion of the acid to react with water and so disturb the balance:

* By analogy with pH, $pK = - \log_{10} K$

153

Strong base $BOH \rightarrow B^+ + OH^-$ (reaction virtually complete)

Weak acid $HA \rightleftharpoons H^+ + A^-$ (reaction reversible)

Therefore, for the salt AB formed, there will be the following interaction with water

$$B^+ + A^- + H_2O \rightleftharpoons HA + B^+ + OH^-$$

that is, the delicate balance of the water dissociation is disturbed, with the result that the concentration of the hydroxide ion exceeds that of the hydrogen ion, and the aqueous solution, although containing equivalent quantities of acid and base, has an alkaline reaction.

Similarly, aqueous solutions of salts of weak base and strong acid have an acidic reaction.

This phenomenon is well illustrated by potassium cyanide and ammonium chloride solutions. The former, being a salt of the weak acid, hydrogen cyanide, has a pH of 11·1 when in decimolar solution, whilst the latter, the salt of the weak base ammonia, in decimolar solution has a pH of 5·1.

Buffer Solutions

The pH of a solution containing a weak acid or base and one of its salts can remain remarkably constant despite the addition of relatively large quantities of acid or alkali. In the case of ammonium hydroxide containing ammonium chloride, the large ammonium ion concentration resulting from the addition of the fully dissociated salt suppresses the ionization of the ammonium hydroxide (since K must remain constant)

$$NH_4OH \rightleftharpoons NH_4^+ + OH^- \qquad K = \frac{[NH_4^+][OH^-]}{[NH_4OH]}$$

This phenomenon is known as the 'common ion' effect.

If acid is added to this system, the hydrogen ions will be removed by hydroxide ions to form water, and the ammonium hydroxide will dissociate further (and it has considerable potential for further dissociation, so much will its dissociation have been suppressed in the first place) to restore the equilibrium.

Hydroxide ions, on the other hand, are removed by combination with ammonium ions; at the same time, the ionization of ammonium hydroxide is suppressed. It can be seen, therefore, that systems like the above preserve the *status quo*; they are known as *buffer solutions*. Buffer solutions are very important in the metabolic processes of living organisms for the maintenance of a constant pH and for ensuring that the various chemical activities continue unimpaired. The fact that well-chosen buffer solutions provide a constant pH means that they can be used, in conjunction with indicators, for comparison with unknown solutions.

The pH can be calculated as follows. Consider the equilibrium for a weak acid HA

$$HA \rightleftharpoons H^+ + A^-$$

$$K = \frac{[H^+][A^-]}{[HA]}$$

and $\log K = \log[H^+] + \log\frac{[A^-]}{[HA]}$

Therefore, by rearranging and converting to pH

$$pH = pK_a + \log\frac{[A^-]}{[HA]}$$

If now a salt with the common ion A^- is added, it will provide virtually the entire quantity of this anion, so that $[A^-] \sim [salt]$. Furthermore, the ionization of the acid will be so suppressed that $[HA] \sim [acid]$.

The above equation can now be rewritten:

$$pH = pK_a + \log\frac{[salt]}{[acid]}$$

For example, if 50 cm³ of 0·15 M sodium acetate solution is mixed with 20 cm³ of 0·1 M acetic acid, for which $K = 1\cdot7 \times 10^{-5}$ M

$$pH = -\log 1\cdot7 \times 10^{-5} + \log\frac{50 \times 0\cdot15}{20 \times 0\cdot1}$$

$$= 4\cdot76 + \log 3\cdot75 = 5.33$$

Solubility Products

It has been seen earlier that continuing dilution of an aqueous solution of a weak electrolyte encourages ionization. The saturated solution of a sparingly soluble salt (which will inevitably be of extremely high dilution) will consequently contain only ions and no undissociated molecules. In other words, the equilibrium that exists is between the ions in solution and the solid in contact with the solution:

$$AB \rightleftharpoons A^+ + B^-$$
$$\text{solid}$$

The active mass of the solid is assumed to be constant, and use of the mass action equation gives

$$K = \frac{[A^+][B^-]}{[AB]_{solid}} \quad \text{or } K' = S = [A^+][B^-]$$

where S is the *solubility product* (a term reminiscent of the ionic product of water).

155

The solubility of a sparingly soluble salt is therefore seen as an ionic rather than a molecular phenomenon; addition of either ion to a saturated solution of the salt will result in the solubility product being momentarily exceeded, with precipitation of the solid instantly following to restore the constant value for the solubility product.

Example—The solubility of silver chloride is 1.5×10^{-3} g dm^{-3} Calculate the solubility product and hence its solubility in M sodium chloride.

Mol. wt. of silver chloride = 143·5, therefore the solubility, in terms of mol dm^{-3}, will be $1.5 \times 10^{-3}/143.5 \sim 1 \times 10^{-5}$. One molecule of silver chloride gives one ion each of silver and chloride, so

$$[Ag^+] = [Cl^-] = 1 \times 10^{-5} \text{ M}$$

and solubility product $= [Ag^+][Cl^-] = (1 \times 10^{-5})^2 = 1 \times 10^{-10} \text{ M}^2$

A molar solution of sodium chloride, which will be fully dissociated, will contain 1 mol dm^{-3} of both sodium and chloride. Hence the concentration of the common chloride ion will be 1 (neglecting the minute contribution from the silver chloride). As

$$[Ag^+][Cl^-] = 1 \times 10^{-10}$$

$$[Ag^+] = \frac{1 \times 10^{-10}}{1} \text{ M}$$

that is, the amount of silver chloride now dissolved

$$= 1 \times 10^{-10} \text{ mol dm}^{-3}$$

$$= 1 \times 10^{-10} \times 143.5 \text{ g dm}^{-3}$$

$$= 1.435 \times 10^{-8} \text{ g dm}^{-3}$$

Thus, the effect of the molar sodium chloride is to reduce the solubility of silver chloride one hundred thousand-fold.

We have seen how the dissociation of water and the ionic product play a fundamental part in the equilibria of aqueous solutions. Let us now study the role of the solubility product.

Precipitation of Sulphides

Sulphides are on the whole insoluble substances and are therefore readily precipitated by the passage of hydrogen sulphide through a solution of a salt of the metal. We can now say that precipitation will occur only when the solubility product has been exceeded, that is, when the concentrations of metallic ion and sulphide ion are high enough. If the latter can be reduced sufficiently, then there is the possibility that precipitation of the more soluble sulphide will be prevented and some separation of metals thereby effected.

This possibility is realized very simply by passing hydrogen sulphide into an acidified solution of the metal ions. The large concentration of hydrogen ions from the acid will, by the common ion effect, suppress the ionization of the hydrogen sulphide:

$$HCl \rightarrow H^+ + Cl^-$$

$$H_2S \rightleftharpoons H^+ + HS^- \qquad\qquad HS^- \rightleftharpoons H^+ + S^{2-}$$

The presence of molar H^+, in fact, reduces the value of $[S^{2-}]$ from about 10^{-10} to about 10^{-21} M. As the concentration of the metal ion is usually in the range $10^{-2} \sim 10^{-1}$ M, precipitation will only be effected if the solubility product for the sulphide is less than about 10^{-22}. *Table 7.2* indicates which particular metals are most likely to be precipitated in acid solution, i.e. in Group 2 in the conventional scheme of qualitative analysis.

Table 7.2. Some relevant solubility products

Mercury(II) sulphide	(4×10^{-53}) M^2	Nickel(II) sulphide	(1×10^{-19}) M^2
Copper(II) sulphide	$[(8 \times 10^{-45})$ M^2	Cobalt(II) sulphide	(2×10^{-20}) M^2
Cadmium sulphide	(4×10^{-29}) M^2	Iron(III) hydroxide	(1×10^{-36}) M^4
Lead(II) sulphide	(4×10^{-28}) M^2	Chromium(III) hydroxide	(3×10^{-29}) M^4
Zinc sulphide	(1×10^{-20}) M^2	Aluminium hydroxide	(8×10^{-23}) M^4
Manganese(II) sulphide	(1×10^{-15}) M^2	Manganese(II) hydroxide	(4×10^{-14}) M^3

More soluble sulphides will be precipitated more readily using alkaline rather than acid solution. Hydroxide ions from the alkali will remove hydrogen ions (to form largely undissociated water) and thus disturb the equilibrium in favour of the formation of more sulphide ions

$$H_2S \rightleftharpoons H^+ + HS^- \qquad\qquad HS^- \rightleftharpoons H^+ + S^{2-}$$
$$\downarrow OH^- \qquad\qquad\qquad\qquad \downarrow OH^-$$
$$H_2O \qquad\qquad\qquad\qquad\quad H_2O$$

Precipitation of Hydroxides

In Group 3 of conventional qualitative analysis, ammonium chloride and ammonium hydroxide are added. Here, the large concentration of ammonium ions from the ammonium chloride prevents the weak base, ammonium hydroxide, from ionizing sufficiently for the solubility products of the hydroxides, other than those of iron(III), aluminium and chromium(III), to be exceeded (*Table 7.2*)

$$NH_4Cl \rightarrow NH_4^+ + Cl^-$$

$$NH_4OH \rightleftharpoons NH_4^+ + OH^-$$

$$\frac{[NH_4^+][OH^-]}{[NH_4OH]} = K$$

i.e. as $[NH_4^+]$ is increased by the addition of ammonium chloride, $[OH^-]$ is decreased.

157

Solubility of Salts of Weak Acids

Many salts of weak acids which are virtually insoluble in water dissolve readily in acids. This, too, can be explained in terms of the law of mass action: hydrogen ions from the acid combine with the anions of the salt to form undissociated molecules of the weak acid, thus disturbing the equilibrium in favour of further dissolution of the salt; e.g. calcium phosphate

$$Ca_3(PO_4)_2 \rightleftharpoons 3Ca^{2+} + 2PO_4^{3-}$$

$$3H^+ + PO_4^{3-} \rightleftharpoons H_3PO_4$$

NON-AQUEOUS SYSTEMS

By now it should be clear that behind all reactions in aqueous systems lies the equilibrium

$$H_2O \rightleftharpoons H^+ + OH^-$$

The proton so formed has, by virtue of its small size, a considerable polarizing force and attacks a further molecule of water to form the oxonium ion, so that the complete self-ionization of water can be represented as

$$2H_2O \rightleftharpoons H_3O^+ + OH^-$$

Most common chemical reactions revolve around this ionic equilibrium for water, and it is therefore not surprising to find that several concepts have stemmed from this source. Thus, an acid has been defined as a substance that provides hydrogen ions in solution and an alkali as one providing hydroxide ions; neutralization is seen as combination of hydrogen and hydroxide ions to form a molecule of solvent, i.e. the reaction

$$HCl + NaOH \rightarrow NaCl + H_2O$$

is essentially

$$H_3O^+ + OH^- \rightarrow 2H_2O$$

By analogy with this, in liquid ammonia (which ionizes thus: $2NH_3 \rightleftharpoons NH_4^+ + NH_2^-$), ammonium compounds act as acids, and amides (or even imides or nitrides) as bases. They can be titrated together and the result is the formation of solvent, in this case ammonia:

$$NH_4Cl + NaNH_2 \rightarrow NaCl + 2NH_3 \qquad (NH_4^+ + NH_2^- \rightarrow 2NH_3)$$

$$2NH_4NO_3 + PbNH \rightarrow Pb(NO_3)_2 + 3NH_3 \qquad (2NH_4^+ + NH^{2-} \rightarrow 3NH_3)$$

$$3NH_4Br + Li_3N \rightarrow 3LiBr + 4NH_3 \qquad (3NH_4^+ + N^{3-} \rightarrow 4NH_3)$$

158

Nor does the similarity end there: amphoteric substances, e.g. zinc amide, exist

$$Zn(NH_2)_2 + 2KNH_2 \rightarrow K_2Zn(NH_2)_4 \quad (cf.\ Zn(OH)_2 + 2KOH \rightarrow K_2Zn(OH)_4)$$

$$Zn(NH_2)_2 + 2NH_4Cl \rightarrow ZnCl_2 + 4NH_3 \quad (cf.\ Zn(OH)_2 + 2HCl \rightarrow ZnCl_2 + 2H_2O)$$

and liquid ammonia can ammonolyse salts in a manner similar to that in which water can hydrolyse:

$$SbCl_3 + NH_3 \rightarrow SbN + 3HCl \quad (cf.\ SbCl_3 + H_2O \rightarrow SbOCl + 2HCl)$$

Ammoniates also exist and can be compared to hydrates, e.g. $Co(NH_3)_6Cl_3$.

Similar situations are believed to exist in other non-aqueous media. For example, dinitrogen tetroxide self-ionizes

$$N_2O_4 \rightleftharpoons \underset{nitrosyl}{NO^+} + NO_3^-$$

In this particular system, nitrosyl compounds will behave as acids and nitrates as bases, neutralization taking place between them.

As a great number of reactions have been studied in a wider variety of media, so the term 'acid' has been extended. An acid can be defined as a proton-donor or an electron-acceptor (a 'Lewis' acid), whilst a base is a proton-acceptor or an electron-donor (a 'Lewis' base). It follows that every acid has its *conjugate base*, and every base its *conjugate acid*:

$$\underset{\substack{\text{(acid as proton} \\ \text{donor)}}}{HNO_3} \rightleftharpoons H^+ + \underset{\substack{\text{[conjugate base (for it can accept a proton} \\ \text{to yield the original acid)]}}}{NO_3^-}$$

$$\underset{\substack{\text{(acid as} \\ \text{electron acceptor)}}}{BF_3 + F^-} \rightleftharpoons \underset{\substack{\text{(conjugate} \\ \text{base)}}}{BF_4^-}$$

Protophilic Solvents

Certain compounds containing oxygen and nitrogen can, by virtue of the lone pair of electrons on that atom, accept protons. They therefore function as bases and enhance the acidity of weak acids. Examples are ethers and amines:

An ether
$$\begin{array}{c} CH_3 \\ \diagdown \\ O: \\ \diagup \\ CH_3 \end{array} + H-A \rightleftharpoons \left[\begin{array}{c} CH_3 \\ \diagdown \\ O-H \\ \diagup \\ CH_3 \end{array} \right]^+ A^-$$

An amine
$$\begin{array}{c} CH_3 \\ | \\ CH_3-N: \\ | \\ CH_3 \end{array} + H-A \rightleftharpoons \left[\begin{array}{c} CH_3 \\ | \\ CH_3-N-H \\ | \\ CH_3 \end{array} \right]^+ A^-$$

159

These protophilic solvents can actually be used in the titration of a strong base against an acid so weak that it would be difficult to titrate in any other way. (Of course, the basic solvent will reduce the strength of the base, but if a very strong base is used the effect will be of no consequence.) For example, phenols can be titrated with methoxides in the presence of 1,2-diaminoethane as solvent

$$\text{C}_6\text{H}_5\text{—OH} + \text{H}_2\text{NCH}_2\text{CH}_2\text{NH}_2 \rightleftharpoons \text{C}_6\text{H}_5\text{—O}^- + \overset{+}{\text{H}_3}\text{NCH}_2\text{CH}_2\text{NH}_2$$

$$\overset{+}{\text{H}_3}\text{NCH}_2\text{CH}_2\text{NH}_2 + \text{CH}_3\text{O}^- \rightleftharpoons \text{CH}_3\text{OH} + \text{H}_2\text{NCH}_2\text{CH}_2\text{NH}_2$$

i.e. the overall reaction is

$$\underset{\text{acid}}{\text{C}_6\text{H}_5\text{—OH}} + \underset{\text{base}}{\text{CH}_3\text{O}^-} \rightleftharpoons \text{C}_6\text{H}_5\text{—O}^- + \text{CH}_3\text{OH}$$

Protogenic Solvents

These are solvents with the opposite quality to the protophilic; that is, they are proton donors; e.g. acetic acid, CH_3COOH

$$CH_3COOH \rightleftharpoons CH_3COO^- + H^+$$

The strength of an acid will decrease if it is transferred from a weaker to a stronger protogenic solvent. For example, nitric acid is still acidic, albeit more weakly, in acetic acid as well as in water because the solvent is still able to accept protons from it

$$HNO_3 + CH_3COOH \rightleftharpoons CH_3COOH_2^+ + NO_3^-$$

but it is basic in sulphuric acid, because the latter is the stronger proton donor

$$HNO_3 + H_2SO_4 \rightleftharpoons H_2NO_3^+ + HSO_4^-$$

The strengths of strong acids in a solvent such as water cannot be compared because dissociation is virtually complete in all cases, but by using a protogenic solvent, the strengths will be reduced (by different amounts), and a comparison becomes possible.

It follows that protogenic solvents, as well as suppressing the dissociation of acids, amplify the strengths of bases; weak bases can therefore be titrated against strong acids by using such a solvent; e.g. amines against perchloric acid, with acetic acid as solvent

$$\text{CH}_3\text{—}\underset{\underset{\text{H}}{|}}{\overset{\overset{\text{H}}{|}}{\text{N}}}\text{:} + \text{H}\text{—OOC}\cdot\text{CH}_3 \rightleftharpoons \left[\text{CH}_3\text{—}\underset{\underset{\text{H}}{|}}{\overset{\overset{\text{H}}{|}}{\text{N}}}\text{—H}\right]^+ + \text{CH}_3\text{COO}^-$$

$$CH_3COO^- + HClO_4 \rightleftharpoons CH_3COOH + ClO_4^-$$

The overall reaction is

$$\underset{\text{base}}{\overset{\overset{\displaystyle H}{|}}{\underset{\underset{\displaystyle H}{|}}{CH_3\!-\!N\!:}}} + \underset{\text{acid}}{HClO_4} \rightleftharpoons \left[\underset{\underset{\displaystyle H}{|}}{\overset{\overset{\displaystyle H}{|}}{CH_3\!-\!N\!-\!H}}\right]^{+} + ClO_4^{-}$$

ELECTRODE POTENTIALS

When most metals are placed in water, the tendency which they have to form positive ions results in the metal becoming very slightly negatively charged with respect to the water. The negative charge residing on the metal attracts the positive metal ions towards it, giving rise to a 'double layer' (*Figure 7.7a*).

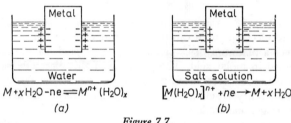

$$M + xH_2O - ne \rightleftharpoons M^{n+}(H_2O)_x \qquad \left[M(H_2O)_x\right]^{n+} + ne \rightarrow M + xH_2O$$

(a) (b)

Figure 7.7

If the metal is placed in a concentrated solution of its ions, there is an opposing tendency for positive ions to leave the solution and to be deposited on the metal; if this 'condensing pressure' of the solution exceeds the 'solution pressure' of the metal, the result will be that the metal acquires a positive charge. This will attract negative ions from solution, and again a double layer will be formed (*Figure 7.7b*).

The actual potential between the metal and the solution will thus depend on two factors: the tendency of the metal to form ions (the less the electronegativity the greater the tendency to ionize) and the concentration of the metal ions in solution. The *standard electrode potential* is the potential developed between the metal and a molar solution of its ions.

Measurement of Electrode Potentials
The e.m.f. of a cell can be regarded as the potential difference between the two electrodes comprising the cell. Therefore, if the e.m.f. of the cell is known, together with one electrode potential, then the other electrode potential can be calculated.

The e.m.f. of the cell is determined by means of a potentiometer which has been calibrated by finding the balance point for a Weston standard cell, whose e.m.f. at

161

$25°$ is $1\cdot018\ 1$ V. If this balance length is L_1 cm, then e, the drop in potential per cm of the potentiometer wire, is given by the expression

$$e = \frac{1\cdot018\ 3}{L_1}\ \mathrm{V\ cm^{-1}}$$

The e.m.f. of the cell will then be eL_2, where L_2 is the new balance point, using the cell under test instead of the standard cell (*Figure 7.8*)

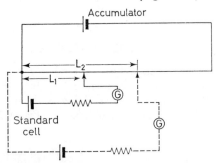

Figure 7.8. Circuit for measurement of electrode potentials

The standard electrode used can be the hydrogen electrode, in which hydrogen is bubbled over a platinum black electrode (which catalyses the dissociation into atoms) immersed in molar acid. The potential of this electrode is, by definition, zero. A more common electrode is the calomel electrode in which mercury is in contact with dimercury(II) chloride; the potential of this electrode can be calibrated against the hydrogen electrode and is found to be $0\cdot242$ V at $25°$.

In the measurement of electrode potentials, two electrode half-cells are connected by a salt bridge which permits the flow of ions to complete the circuit (*Figure 7.9*).

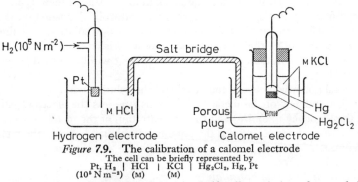

Figure 7.9. The calibration of a calomel electrode

The cell can be briefly represented by

Pt, H_2 | HCl | KCl | Hg_2Cl_2, Hg, Pt
($10^5\,\mathrm{N\,m^{-2}}$) (M) (M)

If the calomel electrode is combined with a half-cell consisting of a metal electrode and an electrolyte containing the metal ions, then the potential difference will be given by

$$\mathrm{e.m.f.} = E_M - E_{cal}$$

where E_M is the potential of the electrode in question.

162

Electrochemical Series

The chemistry of a metal is to a large extent determined by its electronegativity (the readiness with which it attracts electrons). Thus, the arrangement of the metals in order of the standard electrode potentials is a matter of great importance. It is this arrangement which comprises the *electrochemical series* (*Table 7.3*). (Strictly speaking, the electrode potentials are the products of 'wet' systems wherein the ions of the metal are probably hydrated; for 'dry' systems, the ionization energy may be a more relevant quantity.)

Table 7.3. The electrochemical series
Electrode potentials (at 25°, referred to a standard hydrogen electrode)

Metals ($M^{n+}_{(aq)} + ne \rightarrow M$) V		Non-metals ($X + ne \rightarrow X^{n-}_{(aq)}$) V	
Li^+/Li	− 3.02	O_2/OH^-	+ 0.40
K^+/K	− 2.92	$I_2/2I^-$	+ 0.54
Ba^{2+}/Ba	− 2.90	$Br_2/2Br^-$	+ 1.07
Ca^{2+}/Ca	− 2.87	$Cl_2/2Cl^-$	+ 1.36
Na^+/Na	− 2.71	$F_2/2F^-$	+ 2.87
Mg^{2+}/Mg	− 2.38		
Al^{3+}/Al	− 1.67		
Zn^{2+}/Zn	− 0.76		
Fe^{2+}/Fe	− 0.44		
Sn^{2+}/Sn	− 0.14		
Pb^{2+}/Pb	− 0.13		
Fe^{3+}/Fe	− 0.04		
$(2H^+/H_2)$	0.00		
Cu^{2+}/Cu	+ 0.34		
Cu^+/Cu	+ 0.52		
Ag^+/Ag	+ 0.80		

The high potential of metals at the top of the series is a measure of the readiness with which they lose electrons to form positive ions in aqueous solutions. It follows that these metals are very reactive and that their compounds are correspondingly stable, so much so that reduction to the metal can often prove a difficult operation. Consequently, many of these metals have been known in the elemental state only since the last century and the work of Davy on electrolysis.

From *Table 7.3*, reactions can be constructed by algebraic addition of the separate 'half-reactions', e.g.

$$Zn \rightarrow Zn^{2+} + 2e \qquad E = + 0.76 \text{ V}$$

$$Pb^{2+} + 2e \rightarrow Pb \qquad E = - 0.13 \text{ V}$$

$$\overline{Zn + Pb^{2+} \rightarrow Pb + Zn^{2+} \qquad E = + 0.63 \text{ V}}$$

The fact that the resultant potential is positive is indicative of the practicality of the reaction. Indeed, it is well known that if a metal higher in the Series is placed in contact with the ions of a metal lower in it, then the reaction is spontaneous, i.e.

163

ΔG is negative. Now, since one volt is the potential difference when one joule of work is done in transferring one coulomb of electricity across it, E joules of work are performed per coulomb for a potential difference of E volts. For a valency change of z, the number of coulombs per mole $= zF$, where F is the faraday, and so the work done $= zFE$ joules per mole. That is,

$$\text{free energy change} = \Delta G = -zFE$$

It follows that, if the algebraic addition of electrode potentials gives a positive result, the free energy change for the reaction will be negative and the reaction will tend to be spontaneous (p. 122).

Metals high in the Series react vigorously with cold water to give the hydroxide, metals lower down tend to react when heated in steam, whilst metals at the bottom are virtually unaffected

cold water $\quad 2Na + 2H_2O \rightarrow 2NaOH + H_2 \uparrow$

steam $\quad\quad 3Fe + 4H_2O \rightleftharpoons Fe_3O_4 + 4H_2 \uparrow$

Metals above hydrogen react in non-oxidizing acid (unless the overpotential exceeds the electrode potential, as is the case with lead) whilst those below are usually not reactive, e.g.

$$
\begin{array}{ll}
& \text{V} \\
Mg \rightarrow Mg^{2+} + 2e & E = +2.38 \\
2H^+ + 2e \rightarrow H_2 & E = 0.00 \\
\hline
Mg + 2H^+ \rightarrow Mg^{2+} + H_2 & E = +2.38 \\
\text{i.e. } \textit{Reaction spontaneous}
\end{array}
\qquad
\begin{array}{ll}
& \text{V} \\
Cu \rightarrow Cu^{2+} + 2e & E = -0.34 \\
2H^+ + 2e \rightarrow H_2 & E = 0.00 \\
\hline
Cu + 2H^+ \rightarrow Cu^{2+} + H_2 & E = -0.34 \\
\textit{Reaction not spontaneous}
\end{array}
$$

Corrosion and Sacrificial Protection

An electrical cell can be constructed by using two metals of different nobility placed in a salt solution. The potential of the cell will increase as the gap between the metals in the Series widens. Always the less noble metal dissolves, and such dissolution can be regarded as a form of corrosion, although perhaps it would not always be recognized as such. But if a metal is contaminated with another, more noble, metal, the phenomenon takes on the more familiar aspect of corrosion. For example, zinc can be contaminated by momentary immersion in copper(II) sulphate

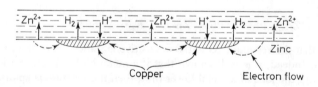

164

solution—it then consists of countless tiny cells and when placed in dilute sulphuric acid corrodes with great rapidity with evolution of hydrogen at the surface of the copper.

The fact that it is the less noble metal that dissolves can be utilized in the sacrificial protection of metals. In the case of galvanized iron, iron is covered by the less noble zinc: if the coating is not complete, then the formation of electrical cells is possible, and if the surface is covered with electrolyte such as rain water (containing carbonic acid), zinc passes into solution. This is important, because it means that the metal dissolves from the bulk of the surface and not from the spot of exposed iron. On the other hand, in the case of tinned iron, it is the iron that is the less noble metal so that, if a flaw develops and a small area of iron is exposed, contact with an electrolyte this time results in the iron, and not the tin, dissolving. As the dissolution takes place from a small area of metal, rapid 'pitting' occurs, with the eventual appearance of a hole.

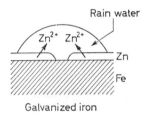
Galvanized iron

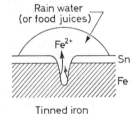
Tinned iron

Concentration Cells

The electrode potential is a measure of the 'solution pressure' of the substance of which the electrode is made and the 'condensing pressure' of the ions of the substance present in the solution. The potential will therefore depend on the concentration of these ions, and so a potential difference will exist between two electrodes of the same material in solutions of different concentrations. This is the principle of the concentration cell. It can be shown that the e.m.f. of such a cell is given by

$$E = \frac{RT}{zF} \ln \frac{c_1}{c_2}$$

where z = charge on the ions, c_1, c_2 = concentrations ($c_1 > c_2$).

Measurements of pH can be carried out by using a concentration cell, for example with the standard hydrogen electrode combined with a hydrogen electrode dipping into a solution whose pH is being determined. As c_2, the concentration of the standard electrode solution, is 1, the above expression simplifies to

$$E = (RT/F) \ln c_{H^+}$$

$$pH = - \ln c_{H^+} = - FE/RT$$

so that, in the pH meter, pH can be calibrated in terms of the potential developed.

165

Reduction–Oxidation Potentials

An oxidizing agent removes electrons from a system whilst a reducing agent adds electrons. An inert electrode will therefore tend to acquire a positive charge when in contact with an oxidizing agent and a negative charge when in contact with a reducing agent. The resultant potential is known as the *redox potential*. *Table 7.4* shows the potential acquired by a platinum electrode when placed in an equimolecular solution of the two ions.

Table 7.4. Redox potentials

System $M^{m+}_{(aq)} + ne \rightarrow M^{(m-n)+}_{(aq)}$	Potential V
Co^{3+}/Co^{2+}	$+ 1 \cdot 82$
Ce^{4+}/Ce^{3+}	$+ 1 \cdot 57$
$MnO_4^-, H^+/Mn^{2+}$	$+ 1 \cdot 52$
$Cr_2O_7^{2-}, H^+/2Cr^{3+}$	$+ 1 \cdot 36$
Fe^{3+}/Fe^{2+}	$+ 0 \cdot 76$
$Fe(CN)_6^{3-}/Fe(CN)_6^{4-}$	$+ 0 \cdot 36$
Cu^{2+}/Cu^+	$+ 0 \cdot 17$
H^+/H	$0 \cdot 00$
Ti^{4+}/Ti^{3+}	$- 0 \cdot 06$
V^{3+}/V^{2+}	$- 0 \cdot 2$
Cr^{3+}/Cr^{2+}	$- 0 \cdot 4$

Systems high in the series oxidize the electrode more strongly than those below; it therefore follows that if two systems are brought into contact, the one higher will oxidize the one below (i.e. the one below will reduce the one above). For example, nascent hydrogen will reduce iron(III) to iron(II)

$$Fe^{3+} + e \rightarrow Fe^{2+} \qquad E = 0.76 \text{ V}$$
$$H - e \rightarrow H^+ \qquad E = 0 \cdot 00 \text{ V}$$

$$\overline{Fe^{3+} + H \rightarrow Fe^{2+} + H^+} \qquad E = 0 \cdot 76 \text{ V}$$

Acidified dichromate oxidizes iron(II) to iron(III)

$$Cr_2O_7^{2-} + 14H^+ + 6e \rightarrow 2Cr^{3+} + 7H_2O \qquad E = 1 \cdot 36 \text{ V}$$
$$6\,(Fe^{2+} \rightarrow Fe^{3+} + e) \qquad E = -0 \cdot 76$$

$$\overline{Cr_2O_7^{2-} + 14H^+ + 6Fe^{2+} \rightarrow 2Cr^{3+} + 6Fe^{3+} + 7H_2O} \qquad E = 0 \cdot 60$$

SUMMARY

The conductance of a solution of an electrolyte is the sum of two parts: the conductance of the anion plus that of the cation. These ion conductances are, in turn, dependent on the speed or mobility of the ions: the faster the ion moves, the greater its transport number or the fraction of the total current carried by it. During electrolysis, the operation requiring least energy is normally carried out. Factors favouring discharge of an ion include high concentration, instability and exothermic reaction with the electrode after discharge. On the other hand, the existence of a high overpotential, high stability and low concentration reduce the possibility of discharge of that ion. One further possibility to consider is the dissolving of the anode, the electrons being released by this process replacing those that would have otherwise been lost by anions.

Electrolysis is dealt with quantitatively by Faraday's laws. Electrolytes can be divided into two classes: strong electrolytes, completely ionized even in the solid state, and weak electrolytes, the ionization of which is increased by dilution.

Application of the law of mass action to weak electrolytes gives rise to the notion of a dissociation constant, which is related to the degree of dissociation in the case of a binary electrolyte by the expression $K = \alpha^2/[(1 - \alpha)V]$.

The law of mass action provides an explanation for phenomena such as the common ion effect, solubility product and hydrolysis. Although these terms are normally used in aqueous systems (where the ionic product is 10^{-14} M^2 and, therefore, pH + pOH = 14), there is no reason why they should not also be applied to non-aqueous systems; fruitful comparisons can be drawn between reactions in water and in solvents such as liquid ammonia.

The potential acquired by a metal when placed in a molar solution of its ions is a measure of its readiness to form hydrated ions. The electrochemical series is an arrangement of elements in order of their electrode potentials and provides a means of comparing the elements with regard to their reactivities, including their tendency to displace each other from solution; the energy liberated in such reactions can be converted directly to electricity by making two metals the electrodes of a cell. A potential is also developed when an inert electrode is placed in a solution of an ion with reducing or oxidizing properties.

QUESTIONS

1. Cite the evidence for the existence of ions.

2. Using specific examples describe electrolytic methods for:
   (a) the isolation of a metal,
   (b) the isolation of a non-metal,
   (c) the formation of an oxidizing compound,
   (d) the formation of an organic compound.

3. Discuss the various possibilities during the electrolysis of sodium chloride solution: (a) with carbon electrodes far apart; (b) with a mercury cathode; (c) hot and concentrated and with carbon electrodes placed close together.

4. The following values for the resistance of potassium oxalate solution of differing dilutions have been obtained, using a conductivity cell of cell constant $1 \cdot 4$ cm^{-1}:

	M/8	M/16	M/32	M/64	M/128	M/256	M/512	
Resistance	215	375	710	1 390	2 700	5 220	10 430	ohms

The transport number of the potassium ion in potassium oxalate is 0·55. If the resistance of a M/32 solution of oxalic acid in the same conductivity cell is 595 Ω, calculate the degree of dissociation and hence the dissociation constant of the acid, given that the molar conductivity of the hydrogen ion is 350 Ω^{-1} cm^2 mol^{-1}.

5. A saturated solution of silver chloride solution has an electrolytic conductivity of 1·5 $\times$ 10^{-6} Ω^{-1} cm^{-1}. If the molar conductivity of the silver ions is 54 and of chloride 65 Ω^{-1} cm^2 mol^{-1} at this temperature, calculate the solubility product for silver chloride and hence the solubility in 3 M sodium chloride solution.

6. Explain the following:

(a) aluminium hydroxide is precipitated if ammonium sulphide is added to aluminium chloride solution;

(b) magnesium hydroxide is not precipitated if ammonium hydroxide is added to a solution of magnesium sulphate and ammonium chloride;

(c) sodium chloride is less soluble in concentrated hydrochloric acid than in water but lead chloride is more soluble;

(d) zinc sulphide is precipitated from an alkaline but not from an acidic solution of hydrogen sulphide whilst tin(II) sulphide is precipitated from an acidic solution;

(e) aluminium chloride solution is acidic and sodium carbonate solution is alkaline.

7. Write an account of the development of the terms 'acid' and 'oxidation'.

8. (a) If the e.m.f. of a cell with zinc and lead as the two electrodes is 0·63 V, calculate the free energy change associated with the replacement of one mole of lead by zinc.

(b) Show from the values in *Table 7.3* which of the following reactions is feasible:

$$2Br^- + I_2 \to Br_2 + 2I^-$$

$$2Br^- + Cl_2 \to Br_2 + 2Cl^-$$

9. (a) Calculate the pH of a 0·01 M solution of an acid whose dissociation constant is 4 $\times$ 10^{-6} M.

(b) If the osmotic pressure of a solution of a compound AB_2 is 2·2 times that expected in the absence of ionization, calculate the degree of ionization.

10. If a solution containing c mol cm^{-3} of a uni-univalent electrolyte is confined between electrodes of area 1 cm^2 and separated by a distance of 1 cm, show that for the application of a potential difference of 1 V, the current is given by cLe $(u_+ + u_-)$, where u_+ and u_- are the velocities (mobilities) of the cations and anions respectively under a potential gradient of 1 V cm^{-1}, L is the Avogadro constant and e the elementary charge. Show also that under these conditions the

current can be equated to the electrolytic conductivity. Hence deduce a relation for the molar conductivity and show that $\Lambda_+ = Fu_-$. What further assumptions have to be made to bring these equations into line with the experimental results shown in *Figure 7.3*?

11. 300 V d.c. was applied across a U-tube containing an agar gel of sodium sulphate and phenolphthalein coloured with a few drops of alkali. Describe what would be observed and state what quantitative values could be obtained from such an experiment.

12. Explain the following results:
(a) Zinc blocks attached to the side of a ship's steel hull prevent corrosion.
(b) The smaller the area at which corrosion occurs, the more intense the effect.
(c) Corrosion occurs where there is a lower concentration of oxygen.
Suggest experiments to investigate these effects quantitatively.

13. (a) Show that for transfer between two half-cells of a mole of a z-valent ion under a potential difference of E volts, the free energy change is given by: $\Delta G = -zFE$.
(b) The osmotic work done in diluting a solution is given by $\int \pi dv$ (cf. *Figure 4.25*). Assuming that the solutions are dilute, substitute for v, using the general gas equation and integrate between pressures P_1 and P_2, corresponding to concentrations c_1 and c_2. Hence derive the e.m.f. for a concentration cell (p 165).
14. Write an account of some of the research being carried out into the development of new cells for powering an electric car.

15. Calculate the transport number of the copper ion in copper(II) sulphate solution from the following data:
Current = 0·015 A; Time for which current flows = 17 580 s
Concentration of copper(II) sulphate solution before electrolysis = 0·1 M
Volume of copper(II) sulphate solution in anode compartment = 40 cm³
After electrolysis, titration of this solution against 0·1 M sodium thiosulphate solution in the presence of excess iodide ion produced the following result:
25 cm³ of anode solution = 29·0 cm³ of the thiosulphate solution.

16. Write an essay on corrosion.

CHAPTER 8

GENESIS, DISTRIBUTION AND EXTRACTION
OF THE ELEMENTS

GENESIS

Any theory about the evolution of the chemical elements must take account of the following facts:

1. As far as is known, hydrogen and helium comprise about 76 and 23 per cent, respectively, of the total weight of the universe.

2. Thermonuclear reactions only take place at very high temperatures.

3. The 'iron group' of elements are about 10 000 times more plentiful than their neighbours and possess very stable nuclei.

Nonetheless there is no shortage of theories of the universe, and the one to be described is not without its competitors.

Gravitational attraction between hydrogen atoms produces a volume contraction and the conversion of potential energy into kinetic energy. When a temperature of about 5×10^6 K is reached, hydrogen atoms fuse together to form helium by some such sequence as

$$2{}^1_1\text{H} \rightarrow {}^2_1\text{H} + {}^{\ 0}_{+1}e$$

$$ature{}^2_1\text{H} + 2{}^1_1\text{H} \rightarrow {}^4_2\text{He} + {}^{\ 0}_{+1}e$$

Cooling occurs as the hydrogen is used up, but this is followed by a further contraction and a resultant increase in the central temperature to about 100×10^6 K, accompanied by the formation of a much expanded envelope of gas. This situation is characteristic of the stars known as 'Red Giants', in which the following reactions probably occur

$$2{}^4_2\text{He} \rightarrow {}^8_4\text{Be} \xrightarrow{\ {}^4_2\text{He}\ } {}^{12}_6\text{C} \xrightarrow{\ {}^4_2\text{He}\ } {}^{16}_8\text{O} \xrightarrow{\ {}^4_2\text{He}\ } {}^{20}_{10}\text{Ne}$$

When most of the helium is used up, contraction occurs again and sufficient heat is generated to permit the products of the above reaction to interact, with the formation of elements up to and including iron and its neighbours, by which time a temperature of about 5×10^9 K has been reached (a 'White Dwarf'). It is the stability of the iron group of elements which is responsible for terminating this sequence of events, and with it the life of a 'first-generation' star.

Disintegration of such a star results in the mixing of the various elements with the debris of other, possibly younger, stars and with interstellar hydrogen ('Supernovae'). The operations described above can now be repeated with the 'second-generation' star that is formed from coalescence of this mixed material into a dense hot mass. This time, however, helium can be formed by the carbon–nitrogen cycle

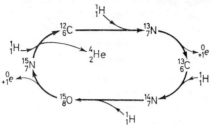

The ^{13}C isotope can also capture a proton and lose a neutron. Neutrons so released can then be captured by other elements, and it is believed that by neutron capture the stable iron group can be converted into the heavier elements. Such is one possible theory for the genesis of the elements in our own Sun.

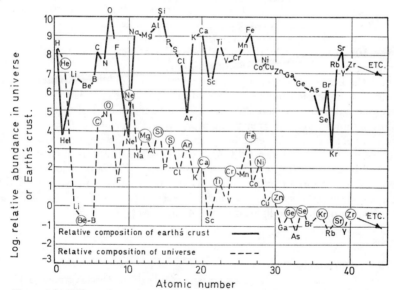

Figure 8.1. Comparison of composition of the universe and of the Earth's crust

Note (*i*)   The relative abundance in Universe of elements of even atomic number (ringed)

(*ii*)   Relative rarity in Earth's crust of noble gases

DISTRIBUTION

For some reason—it may have been the result of intense gravitational attraction as another star passed close by—matter was removed from the Sun to form the planets and their satellites. The relative increase in surface area soon produced sufficient cooling to bring thermonuclear reactions to an end. As cooling of the

Earth continued, less volatile elements tended to liquefy and chemical bonds assumed some measure of permanence. At the same time, the small gravitational attraction due to the small size of the Earth allowed the lighter materials to escape into space. Although some hydrogen may have been absorbed into the molten metallic core, much must have been lost in this manner. Oxygen, being a far denser gas, would be retained to a much larger extent, a very important factor in subsequent development. The escape velocity from the earth is 11 km sec^{-1}, and hydrogen molecules have an average velocity 1/4 of this value at 250°C. According to Jeans, a gas having 1/4 escape velocity would take 5×10^4 years to escape,

Table 8.1. Free Energies of Formation of Oxides and Sulphides

Oxides kJ mol^{-1}		Sulphides kJ mol^{-1}	
CaO	−603	MnS	−189
MgO	−569	ZnS	−167
Al$_2$O$_3$	−527	MoS$_2$	−113
SiO$_2$	−398	PbS	−96
FeO	−251	FeS	−92
PbO	−189	As$_2$S$_3$	−38

whilst if it has 1/5 escape velocity the time required for complete escape increases to $2 \cdot 5 \times 10^9$ years. Clearly the mean molecular velocity is a crucial factor in determining the course of events.

In fact, the Earth can be regarded as an incompletely oxidized mixture of metals and silicon. Competition must have existed between the elements for the available oxygen (and sulphur, chlorine, etc.), and the greater the heat of formation of the oxide (or sulphide, etc.) the greater the prospect of its formation (*Table 8.1*).

There is believed to be an excess of iron in the core of the Earth, and therefore only those oxides or sulphides with heats of formation in excess of the values for iron would be expected to be formed in any quantity. The large amount of silicon present had an important influence on events: there tended to be formed three liquid phases, *molten iron*, on which floated the second phase of *iron*(II) *sulphide*, and finally, floating on this, the lightest phase of all, *silica* (*Figure 8.2*).

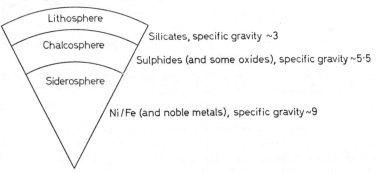

Figure 8.2. The structure of the Earth

172

Table 8.2. Geochemical distribution of the elements

Iron core (siderophil)	In sulphide layer (chalcophil)	In silicate layer (lithophil)	As gases (atmophil)	In organisms (biophil)
Transition metals, Ge, Sn	Ga, In, Tl, Pb, As, Sb, Bi, S, Se, Te, Transition metals	Group I metals Group II metals B, Al, Lanthanides Si, Ti, Zr, Hf, Th, Group VII elements	H C N O Noble gases	H C N, P O, S K, I

Distribution of elements and compounds (*Table 8.2*) would then take place between these three phases. Noble metals (siderophiles), unsuccessful in their quest for oxygen and sulphur, would dissolve in the iron core. Metallic sulphides (chalcophiles) would dissolve in the iron sulphide phase and oxides (lithophiles) in the silica phase. Those oxides which were basic enough would do more than dissolve: they would react with the molten acidic silica to form silicates, which could subsequently solidify under vastly differing conditions. Rapid cooling at the Earth's surface produces fine-grained crystals, such as those present in basalt,

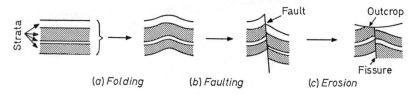

(a) Folding   (b) Faulting   (c) Erosion

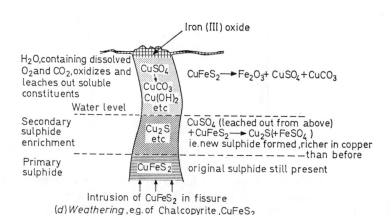

(d) Weathering, e.g. of Chalcopyrite, $CuFeS_2$

Figure 8.3. Changes in the Earth's crust

whilst slow cooling in some intrusion beneath the surface gives the large crystals familiar to us in granite.

Eventually, when the surface of the Earth and its atmosphere had cooled sufficiently, water vapour condensed and fell as rain. Rivers appeared, to grow into oceans. Evaporation of water from the ocean surface maintained a humidity in the atmosphere and perpetuated the 'water cycle', with more water being precipitated as rain to replenish the rivers and maintain the oceans. But the picture is not as

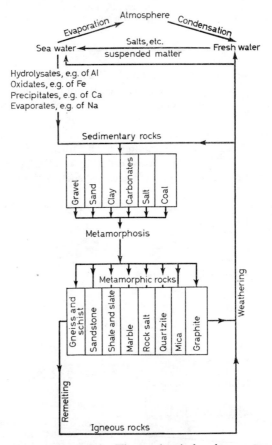

Figure 8.4. The geochemical cycle

symmetrical as it may at first appear. Soluble constituents in the Earth's crust were steadily dissolved, and sea water now contains 1·94 per cent chloride, 1·08 per cent sodium, 0·27 per cent sulphate, 0·13 per cent magnesium, 0·04 per cent each of potassium and calcium and smaller quantities of most other elements.

It was in this 'primeval soup', as it has been called, that life first appeared. There is some evidence for supposing that amino acids were formed by reaction between substances such as methane, water and ammonia, activated by electrical discharge. Amino acids were presumably then condensed into proteins, but it is difficult to see how the various substances could be integrated and coordinated into the activity of the living cell without the control and intervention of something akin to DNA (p. 427). The origin of life may always remain a mystery, but there is no doubting the marks that living things have made on the Earth's crust. Plants, by their photosynthetic processes, provide the atmosphere with oxygen; shells of aquatic animals are deposited on the ocean floor as carbonates which, under pressure and heat, can undergo metamorphosis into substances such as marble. Plant and animal remains can also be converted into coal and petroleum; coal itself can undergo metamorphosis, if exposed to sufficient heat, into coke and graphite. Humic acids, arising from plant decay, are even able to effect the dissolution of several metals whilst bacterial activity in tropical regions may result in the conversion of clays into bauxite.

Even now the story is far from complete. As the Earth diminished in size, the solid crust prevented the gradual and continuous readjustment necessary for the alleviation of stress. Growing strain was relieved dramatically by volcanic eruptions, earthquakes, folding and faulting; fresh material was thus brought to the surface. Seas were isolated and sedimentary rocks formed by evaporation of the water. The contours of the hydrosphere are dependent upon another variable factor: the amount of water fixed as ice at the polar caps. The increased melting of ice during warmer periods causes the oceans to rise, and the varying level of the sea may have played a part in the periodic flooding of tropical forests in the Carboniferous Period and so in the formation of coal seams.

As well as dissolving soluble minerals, water can effect physical 'weathering' of rock when it freezes; the force of expansion accompanying the solidification can be sufficient to break up the largest boulders. Oxygen dissolved in water oxidizes some sulphides as it passes through the surface layers of the Earth, whilst dissolved carbon dioxide converts insoluble carbonates into soluble bicarbonates. Wind, as well as frost, is another important weathering agent and plays its part in reducing massive rock to the fine particles capable of supporting simple plants. Decay of these plants has provided the organic matter necessary for the sustenance of a wider range of organisms, from microbes on the one hand to sophisticated plants on the other. The proliferation of both terrestrial plants and animals has also had its effect on the distribution of certain elements, notably carbon, nitrogen and phosphorus. So it can be seen that the picture constantly changes (*Figure 8.4*).

EXTRACTION

It has been estimated that the major constituents of the Earth as a whole are

	per cent		per cent
Iron	39·76	Aluminium	1·79
Oxygen	27·71	Sulphur	0·64
Silicon	14·53	Sodium	0·39
Magnesium	8·69	Potassium	0·14
Calcium	2·52	Phosphorus	0·11

Fourteen elements make up 99·62 per cent of the Earth's crust (which is taken to comprise the atmosphere, hydrosphere and ten miles of lithosphere. The latter is believed to be made up of 95 per cent igneous rock, 4 per cent shale, 0·75 per cent sandstone and 0·25 per cent limestone):

	per cent		per cent
Oxygen	48·60	Magnesium	2·00
Silicon	26·30	Hydrogen	0·76
Aluminium	7·73	Titanium	0·42
Iron	4·75	Chlorine	0·14
Calcium	3·45	Phosphorus	0·11
Sodium	2·74	Carbon	0·09
Potassium	2·47	Sulphur	0·06

It follows that many of the well known metals are in reality very scarce. They owe their familiarity to the conspicuous nature of their ores, which are concentrated in only a few places, and also to the fact that these ores are often readily reduced to the element. On the other hand, many elements, although not found naturally concentrated, are present as minor constituents of common rocks and so are really quite abundant but too expensive to isolate because of their low concentration. (It should be noted that elements of even atomic number are more abundant than the adjacent elements of odd atomic number.)

Many of the ores used nowadays are of low grade, and the processes involved in extraction are sufficiently costly to warrant the concentration of the mineral prior to further treatment. This process is referred to as *mineral dressing* and consists of an initial crushing and grading of the ore, followed, if necessary, by making use of either density or surface differences to achieve greater concentration. In the former case, the crushed ore is washed in a suspension of suitable density to allow the separation of the mineral from the worthless rock (gangue) accompanying it. Surface differences are made use of in *froth flotation*, which depends on the difference in hydrophobic nature of the surfaces of the mineral and gangue particles. Even if only a small difference in 'wettability' exists, suitable surface-active substances (such as eucalyptus oil) can be added to enhance this effect; if air is then blown through the suspension of the finely crushed ore, the air bubbles become attached to the hydrophobic surfaces and carry the mineral particles to the top of the medium, where they pass into a stable foam which is continually removed (*Figure 8.5*).

More specialized physical methods of concentrating minerals depend on (*a*)

176

Air bubbles adhere to hydrophobic surfaces and make them more buoyant. Surface-active agents are widely used to render a hydrophilic surface hydrophobic and so attractive to air bubbles; selectivity can often be achieved, e.g. by using sodium xanthate with galena

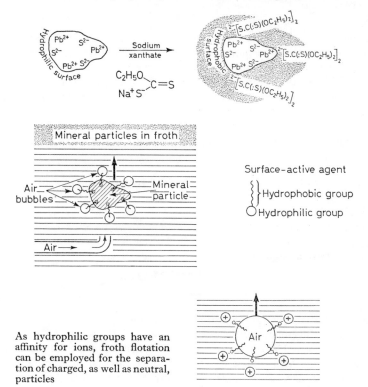

Figure 8.5. Froth flotation

magnetic separation, e.g. for iron(II) (III) oxide, and (*b*) the readiness with which some substances acquire and retain electrostatic charges. This method is used in separating cassiterite, tin(IV) oxide, from gangue. Chemical methods involve separation by (*i*) dissolution, in which, for example, bauxite can be purified to aluminium oxide by dissolving in caustic alkali

$$\text{(impure) } Al_2O_3 \xrightarrow{OH^-} Al(OH)_4^- \xrightarrow{CO_2} Al_2O_3 \text{ (pure)}$$

and (ii) precipitation, whereby, for example, magnesium can be removed from sea water by the addition of hydroxide ions

$$Mg^{2+} \xrightarrow{OH^-} Mg(OH)_2 \downarrow$$

177

Principles Governing the Extraction of the Elements

In the extraction of an element from one of its compounds, the change nearly always involves reduction

$$M^{n+} + ne \to M^0$$

This reduction can be effected by chemical reagents, by electrolysis or, in a few cases, by thermal decomposition. The method employed depends on a number of factors, the two most important of which are the purity of the product required and the stability of the compound containing the element. Metals at the top of the electrochemical series are generally isolated electrolytically, whilst those lower down are prepared by chemical reduction.

A few elements, notably the halogens, are normally present in an ore in a reduced state and must be extracted by oxidative processes

$$X^{n-} \to X^0 + ne$$

although even of these, iodine is often obtained by reduction of sodium iodate.

Chemical reduction methods

The heat of formation of a chemical bond depends on the electronegativity difference between the elements involved in the linkage, and therefore the majority of metal oxides have a large negative heat of formation, e.g.

$$Mg + \tfrac{1}{2}O_2 \to MgO \qquad \Delta H = -622 \text{ kJ}$$

$$2Al + \tfrac{3}{2}O_2 \to Al_2O_3 \qquad \Delta H = -1\,677 \text{ kJ}$$

This implies that the energy required to reduce most of the metal oxides is large. However, from the equation $\Delta G = \Delta H - T\Delta S$ (p. 121) it can be seen that the free energy change, ΔG, associated with the formation of a solid oxide becomes *less* negative as the temperature is increased, since the system is going to a more ordered state, i.e. ΔS, the change in entropy or degree of disorder is negative, and therefore $- T\Delta S$ is positive.

On the other hand, for the reaction

$$2C + O_2 \to 2CO$$

an increase of one mole of gas is experienced and, accordingly, it is accompanied by a large increase in entropy (ΔS positive); thus, as the temperature is increased, ΔG will become *more* negative. At a certain temperature, therefore, the changes in free energy of the two systems

metallic oxide → metal + oxygen	$\Delta G = + x$ kJ
carbon + oxygen → carbon monoxide	$\Delta G = - x$ kJ
metallic oxide + carbon → metal + carbon monoxide	$\Delta G = \quad 0$ kJ

will balance. At, and above, this temperature carbon will be capable of reducing the oxide under consideration. This is one reason for working at elevated temperatures in most metallurgical operations. These results are conveniently

178

shown graphically, as for example in *Figure 8.6*, and the temperatures above which it is possible to reduce different oxides with carbon are then easily determined.

Besides carbon, other reducing agents are used, but to a much smaller extent, e.g. aluminium in the Thermit process (see p. 327) and sodium, calcium or magnesium for reducing certain transition metal halides, e.g.

$$TiCl_4 + 2Mg \rightarrow Ti + 2MgCl_2$$

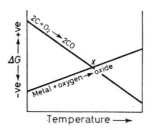

Figure 8.6a. Effect of entropy on temperature coefficient of free energy

$$\Delta G = \Delta H - T\Delta S$$

If ΔS is positive, ΔG becomes more negative at higher temperature, e.g.

$$2C + O_2 \rightarrow 2CO$$
(solid) (1 vol. gas) (2 vol. gas)

If ΔS is negative, i.e. the system is going to a more ordered state, ΔG becomes less negative at high temperature,
e.g.
metal + oxygen → oxide
(solid) (gas) (solid)

To the left of **X**, the system is more stable (minimum ΔG), with metal in form of oxide and carbon in elemental condition. To the right of **X**, the system is more stable, with metal in elemental condition and carbon in form of carbon monoxide, i.e. metal oxide can be reduced.

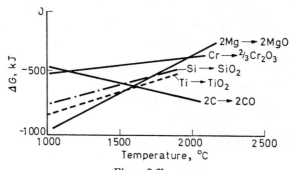

Figure 8.6b

Some sulphides are reduced *in situ* by reaction with part of the ore which has previously been oxidized (see later).

The chief difficulty which makes chemical reduction either impossible or uneconomic is the reaction of the metal produced with the reducing agent; e.g. if potassium is extracted with carbon, the metal reacts with the carbon monoxide formed to give the explosive compound

$$
\begin{array}{c}
\overset{\displaystyle OK \qquad\qquad OK}{\underset{\displaystyle \quad}{\diagdown \diagup}} \\
C\!\!=\!\!C \\
KO\!-\!C \qquad\qquad C\!-\!OK \\
C\!-\!C \\
\overset{}{\underset{\displaystyle OK \qquad\qquad OK}{\diagup \diagdown}}
\end{array}
$$

Similarly, if alumina is reduced at high temperature by carbon, some reaction occurs between the aluminium and the carbon, forming aluminium carbide, Al_4C_3. As a result, this method has not been successful in replacing the electrolytic process.

(i) Elements Obtained by Reduction with Carbon (Coke)

Foremost among these is iron (*Figure 8.7*). The chemistry of the blast furnace is very complex but the overall reactions can be simply represented by the equations

$$Fe_2O_3 + 3C \rightarrow 2Fe + 3CO \uparrow$$

$$Fe_2O_3 + 3CO \rightarrow 2Fe + 3CO_2 \uparrow \qquad Fe_2O_3 + CO \rightarrow 2FeO + CO_2 \uparrow$$

$$CO_2 + C \rightarrow 2CO \uparrow \qquad\qquad FeO + C \rightarrow Fe + CO \uparrow$$

A slag former, limestone, is also intimately mixed with the charge and the slag is removed from above the molten pig iron, which still, however, contains carbon, phosphorus and silicon. Purification by oxidation of iron oxides, for example, is used in producing wrought iron and steel (which contains other added elements such as manganese and chromium).

The oxides of manganese, chromium and zinc are all conveniently reduced by carbon. (When chromite, $FeCr_2O_4$, is so reduced, a series of chromium–iron alloys is obtained, which are collectively referred to as 'ferro-chromium' and which are used for making stainless steel).

Reference to *Figure 8.6b.* shows that magnesium oxide can be reduced at temperatures of about 2 000°C:

$$MgO + C \rightarrow Mg + CO \uparrow$$

The products must be rapidly quenched to prevent the reverse reaction occurring, which at low temperatures is the thermodynamically favoured one.

Tin(IV) oxide is readily reduced by carbon, as are also the oxides of other elements at the lower end of the electrochemical series, such as those of lead, copper, mercury and the 'noble' elements. In these latter cases, the oxides first have to be obtained by roasting sulphide ores.

180

Two important non-metals extracted from their sources by carbon reduction are phosphorus and sulphur. The former is obtained by interaction between calcium phosphate, sand and coke in an electric furnace:

$$2Ca_3(PO_4)_2 + 2SiO_2 \rightarrow 2CaSiO_3 + P_4O_{10}$$

$$P_4O_{10} + 10C \rightarrow P_4 + 10CO \uparrow$$

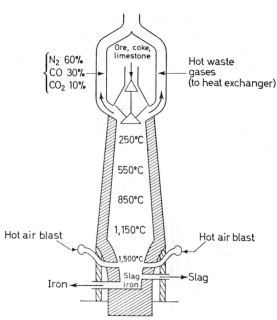

Excess carbon reacts with hot oxygen

$$C + O_2 \rightarrow CO_2 \qquad CO_2 + C \rightarrow 2\,CO$$

Carbon monoxide reduces oxide to metal

$$Fe_2O_3 + 3CO \rightarrow 2Fe + 3CO_2 \uparrow \qquad \Delta H = -23{\cdot}9\,kJ$$

Because of its exothermicity, this reaction takes place from left to right only in the cooler parts of the furnace. Also in this region, partial reduction occurs

$$Fe_2O_3 + CO \rightarrow 2FeO + CO_2 \uparrow$$

whilst in the hottest part of the furnace

$$FeO + C \rightarrow Fe + CO \uparrow$$

Figure 8.7. The blast furnace

181

whilst sulphur can be obtained from calcium sulphate via the intermediate calcium sulphide:

$$CaSO_4 + 4C \rightarrow CaS + 4CO \uparrow$$

$$CaS + 2H_2O \rightarrow Ca(OH)_2 + H_2S \uparrow$$

$$2H_2S + O_2 \rightarrow 2S \downarrow + 2H_2O$$

(ii) Elements Obtained by Reduction with Metals

Manganese and chromium are conveniently prepared by reduction of manganese(II, III oxide and chromium(III) oxide, respectively, with aluminium. The former is obtained from manganese(IV) oxide by the action of heat:

$$3MnO_2 \rightarrow Mn_3O_4 + O_2 \uparrow$$

$$3Mn_3O_4 + 8Al \rightarrow 9Mn + 4Al_2O_3 \qquad \Delta H = -2510 \text{ kJ}$$

Titanium is obtained by reduction of its tetrachloride using magnesium or sodium (*Figure 8.8*) whilst silicon can be obtained from its oxide by reaction with magnesium:

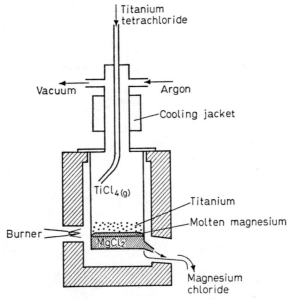

Figure 8.8. Extraction of titanium

$$TiCl_{4(liq)} + 2Mg \rightarrow Ti_{(s)} + 2MgCl_{2(s)} \qquad \Delta G = -452 \text{ kJ}$$

i.e. the free energy change is very favourable. The reaction velocity is increased by bringing $TiCl_4$ vapour into contact with molten magnesium. The lower layer of $MgCl_2$ is run off as necessary, the reaction being continued until there is 85 per cent conversion. Magnesium is then removed from the product by vacuum distillation.

182

$$SiO_2 + 2Mg \rightarrow Si + 2MgO$$

Wet methods of reduction, making use of the order of the elements in the electrochemical series, are employed for the recovery of some metals from low-grade ores. For example, after leaching copper residues with sulphuric acid in the presence of air, the copper is precipitated by metallic iron

$$Fe + Cu^{2+} \rightarrow Cu \downarrow + Fe^{2+}$$

(iii) Elements Obtained by Other Reductive processes

Lead and copper are extracted from their sulphide ores by first partially oxidizing the ore and then heating strongly in the absence of air to effect a mutual reduction. The processes can be summarized by the following sequences of reactions:

$$PbS + 2O_2 \rightarrow PbSO_4$$

$$PbS + PbSO_4 \rightarrow 2Pb + 2SO_2 \uparrow$$

and

$$CuS \xrightarrow{air} CuO \rightarrow Cu_2O$$
$$\xrightarrow{air} Cu_2S + SO_2 \uparrow$$
$$2Cu_2O + Cu_2S \rightarrow 6Cu + SO_2 \uparrow$$

Mercury(II) oxide decomposes so readily on heating (ΔG is negative at temperatures above 500°) that no mutual reduction between sulphide ore and oxidized product is necessary. Instead, the sulphide can be completely converted to the oxide and hence directly to the metal

$$HgS + O_2 \rightarrow Hg + SO_2 \uparrow$$

An interesting reductive extraction of a non-metal is that of iodine from the sodium iodate of Chile. There is partial reduction of iodate to iodide by sodium hydrogen sulphite; the iodide formed then reacts with the remaining iodate to liberate iodine

$$3OH^- + IO_3^- + 3HSO_3^- \rightarrow I^- + 3H_2O + 3SO_4^{2-}$$
$$5I^- + IO_3^- + 6H^+ \rightarrow 3I_2 + 3H_2O$$

Thermal Decomposition

This is a method of preparing some metals in a high state of purity; for example, the decomposition of nickel tetracarbonyl as a means of purifying nickel, and the deposition of a metal from a halide—usually an iodide—by the method of van Arkel, in which the vapour of the halide is decomposed by a red-hot wire. This method is of special use for the purification of titanium and zirconium and is superior to the reduction with sodium or magnesium, as the metals are formed in a pure state and, although in the form of a powder, they can still be moulded and sintered by the application of pressure (the powder-metallurgical process).

Electrolytic Reduction and Oxidation

By using electrolytic methods, an unlimited amount of energy is available for even those reactions which are not thermodynamically favoured. Thus the reductive process

$$2Al_2O_3 + 3C \rightarrow 4Al + 3CO_2 \uparrow \qquad \Delta G = + 1\ 465\ kJ$$

is not a feasible reaction, unless a considerable amount of energy is supplied to overcome the high positive value of the free energy change. The reaction can be accomplished electrolytically using carbon anodes and a current of several thousand amperes at about 1 volt (*Figure 8.9*). Electrolytic methods can be used, provided that

(*a*) the ions in the crystal lattice of the mineral can be rendered mobile or, in the case of covalent molecules, the atoms can be made susceptible to electrolysis by dissolution in a suitable electrolyte;

(*b*) the required element can be discharged in a pure form. This means that either the mineral must be pure or that the discharge potential of the element must be lower than those of the impurities. Since the hydroxides of many metals are precipitated even in acid solutions, a low pH would often have to be maintained in order to prevent the formation of such precipitates; unfortunately, such a condition favours the discharge of hydrogen rather than other cations. Nevertheless, copper and zinc can be discharged in preference to hydrogen, particularly from strong solutions, because of the large overpotentials for hydrogen on these elements.

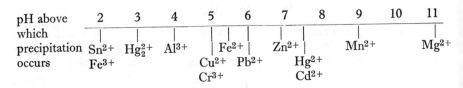

pH above which precipitation occurs	2	3	4	5	6	7	8	9	10	11
	Sn^{2+} Fe^{3+}	Hg_2^{2+}	Al^{3+}	Fe^{2+} Cu^{2+} Cr^{3+}	Pb^{2+}	Zn^{2+} Hg^{2+} Cd^{2+}		Mn^{2+}		Mg^{2+}

Elements obtained by electrolysis are generally those with large positive or negative electrode potentials (p. 164). The electrolyte most commonly used is the molten halide, containing a small amount of impurity to lower the melting point. By this means both a metal and a halogen are obtained at the electrodes.

The cathode is generally iron and the anode graphite, but where possible the electrolytic vessel is made to act as one of the electrodes. The currents used are sufficiently great to keep the electrolyte molten once the external heating has lowered the resistance to such a value that a current can start to flow.

Elements extracted by this method are those of Groups I and II, aluminium, copper, zinc, the halogens and hydrogen (*Figures 8.9* and *8.10*).

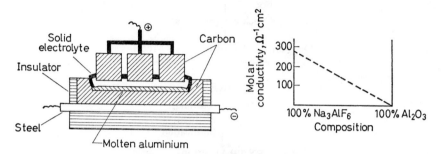

Figure 8.9. Extraction of aluminium

Although the overall reaction can be written
$$Al_2O_3 + 3C \rightarrow 2Al + 3CO$$
the free energy change is highly unfavourable and electrical energy is therefore necessary. This is supplied at about 20 kW (4 000 A at 5 V).

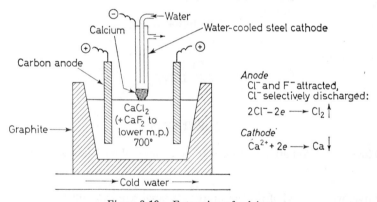

Anode
Cl⁻ and F⁻ attracted, Cl⁻ selectively discharged:
$$2Cl^- - 2e \longrightarrow Cl_2 \uparrow$$

Cathode
$$Ca^{2+} + 2e \longrightarrow Ca \downarrow$$

Figure 8.10. Extraction of calcium

SUMMARY

The elements present on earth were made by nuclear fusion from hydrogen in a star at extremely high temperatures. Much of the hydrogen and helium originally present escaped into space as the earth cooled down. Reaction took place between metals and non-metals to give chiefly oxides and sulphides, with excess metal forming the core of the earth.

185

Extraction of the elements from the compounds formed by these and subsequent weathering processes involves the reversal of the original reactions:

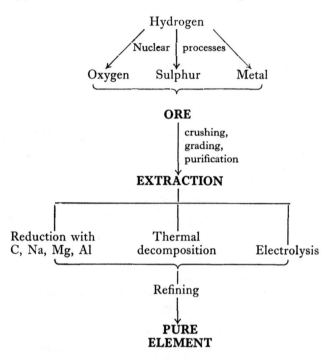

QUESTIONS

1 How would you attempt to isolate potassium, aluminium, zinc and platinum from their chlorides?

2. Give, with the aid of a diagram, the essential requirements for the isolation of sodium from sodium chloride electrolytically.

3 (a) With reference to *Figure 8.6b*, discuss the equilibrium between chromium (III) oxide, chromium, carbon and carbon monoxide.

(b) Describe the electrode reactions involved in the electrolysis of aluminium oxide dissolved in molten sodium hexafluoroaluminate between carbon electrodes.

4. 'Most of the heat lost by radiation from the Earth is accounted for by the disintegration of natural radioactive sources'. Discuss this statement in relation to cosmological theories.

5. What further evidence would you seek to test the theories of evolution and distribution described in this Chapter?

6. What can you deduce from the following information?

Mass of the Earth $\sim 1 \times 10^{25}$ kg
Radius of the Earth $\sim 6 \times 10^6$ m
Density of the Earth's crust ~ 3 gcm^{-3}

7. It has been assumed in this chapter that the Earth reached its present condition by cooling down. Describe a 'cold' theory for the evolution of the Earth.

8. Comment on the following figures, in mol cm^{-2} of total Earth surface, for the amounts of different ions present in the oceans and for the amount added by rivers per 10^8 years:

	Na$^+$	Mg^{2+}	Ca^{2+}	K$^+$	Cl$^-$	SO$_4{}^{2-}$	CO$_3{}^{2-}$	NO$_3{}^-$
Present in oceans	129	15	2·8	2·7	150	8	0·3	0·01
Added in 10^8 years	196	122	268	42	157	84	342	11

9. It is believed that, in the earlier days of the earth, there was insufficient oxygen to react with all the oxidizable material present. On this view, the primeval atmosphere would have been devoid of oxygen. If it has since appeared through photosynthesis, calculate the possible amount of carbonaceous fuel (expressed as carbon) present in the Earth's crust, from the quantity of oxygen now present in the atmosphere.

CHAPTER 9

HYDROGEN

	Atomic weight	Atomic number	Ionization energy kJ mol⁻¹	Electron affinity	Atomic radius, pm	m.p.	b.p.	Latent heat of fusion, J mol⁻¹
						\multicolumn{2}{c}{K}	
Hydrogen, H	1·008		1315	75	30	14	20	118
Deuterium, D	2·015	} 1				18	23	197
Tritium, T	3·006							

The hydrogen atom is the simplest of all atoms. The most abundant isotope (protium) consists simply of a proton and an extranuclear electron. Two further isotopes are known: *deuterium*, occurring to the extent of about one part in 5 000, has a neutron as well as a proton in the nucleus, and *tritium*, present only in minute quantities, contains two neutrons. As is always the case with isotopes, these three forms of hydrogen are virtually identical in chemical properties, but there are significant differences in physical properties owing to the large percentage changes in mass.

The element occurs widely combined with other elements in the earth's crust, but not to any significant extent in elemental form because the earth's gravitational field is too small to retain such a light molecule for any length of time. It accounts for about 1 per cent of the lithosphere by weight (e.g. as petroleum and other organic remains), but on the basis of the number of atoms it is one of the most abundant of elements. As a constituent atom of water it forms about 10 per cent by weight of the hydrosphere.

Preparation

(*i*) *From acids*—Hydrogen can be very conveniently prepared, albeit in a somewhat impure state, by the action of a non-oxidizing acid on any metal above hydrogen in the electrochemical series, provided that the electrode potential is greater than the overpotential. In the case of zinc, for example, the overpotential is almost as great as the electrode potential, and accordingly pure zinc is attacked only very slowly. However, if the metal surface is contaminated by a substance of low overpotential (for example, with copper by treatment with copper(II) sulphate solution), reaction is brisk, the hydrogen being evolved at the impurity

$$Cu^{2+} + Zn \rightarrow Cu\downarrow + Zn^{2+}$$
$$Zn + 2H^+ \rightarrow Zn^{2+} + H_2 \uparrow$$

(*ii*) *From water*—Very pure hydrogen is obtained when barium hydroxide is electrolysed, using nickel electrodes. Hydrogen ions from the water present are selectively discharged at the cathode

$$H_2O \rightleftharpoons H^+ + OH^-$$

$$H^+ + e \rightarrow H\cdot$$

$$2H\cdot \rightarrow H_2$$

Any carbon dioxide that may be present is absorbed by the barium hydroxide, whilst any traces of oxygen are removed by passing the gas over heated platinum gauze which catalyses the reduction of the oxygen by some of the hydrogen to water. The remaining gas can be subsequently dried over phosphorus(V) oxide.

The more electropositive metals will react with water or steam with the evolution of hydrogen, for example, calcium with cold water and iron with steam

$$Ca + 2H_2O_{(liq)} \rightarrow Ca(OH)_2 + H_2 \uparrow$$

$$3Fe + 4H_2O_{(gas)} \rightleftharpoons Fe_3O_4 + 4H_2 \uparrow$$

(*iii*) *From alkalis*—Amphoteric metals such as zinc and aluminium react with strong alkali to produce hydrogen:

$$Zn + 2OH^- + 2H_2O \rightarrow Zn(OH)_4^{2-} + H_2 \uparrow$$
$$\text{zincate}$$

$$2Al + 2OH^- + 6H_2O \rightarrow 2Al(OH)_4^- + 3H_2 \uparrow$$
$$\text{aluminate}$$

Industrial Methods

In industry, hydrogen is obtained by the electrolysis of sodium hydroxide or brine (p. 202), or from water gas by the Bosch process. The water gas is produced by passing steam over white-hot carbon; the reaction is endothermic and the temperature rapidly falls below the optimum

$$C + H_2O \rightarrow CO \uparrow + H_2 \uparrow \qquad \Delta H = + 131 \text{ kJ}$$

If, however, the hydrogen is required with nitrogen for the manufacture of ammonia, the current of steam is replaced by air at this juncture and an exothermic reaction takes place, with the formation of producer gas

$$C + \tfrac{1}{2}O_2 + 2N_2 \rightarrow CO \uparrow + 2N_2 \uparrow \qquad \Delta H = - 110 \text{ kJ}$$

When the temperature is once again high enough, steam replaces air, and the procedure is repeated. If now the mixture of gases, containing hydrogen, nitrogen and carbon monoxide, is passed, together with more steam, over iron(III) oxide at about 500°, the carbon monoxide reduces the steam to hydrogen and is itself oxidized to carbon dioxide

$$CO + H_2O + N_2 + H_2 \rightarrow CO_2 \uparrow + N_2 \uparrow + 2H_2 \uparrow$$

The carbon dioxide produced can be removed by passing through water under pressure and any remaining carbon monoxide absorbed in ammoniacal copper(I) formate.

Hydrogen is also formed in large quantities in the petroleum industry as a by-product in the catalytic dehydrogenation of alicyclic compounds to aromatic structures of higher octane number (p. 380), e.g.

$$C_6H_{12} \rightarrow C_6H_6 + 3H_2$$
Cyclohexane   Benzene

A further by-product of the petroleum industry is methane, and in recent years methods have been developed for the use of this as a feedstock for water gas processes, e.g.

$$CH_4 + H_2O \xrightarrow[900°]{Ni} CO + 3H_2$$

Uses

The hydrogen produced commercially is used chiefly in the Haber process for the synthesis of ammonia (p. 255), for the catalytic reduction of carbon monoxide to methanol (p. 442), for the manufacture of hydrochloric acid and for the hydrogenation of fats and coal. Long-term interest centres on the possibility of the controlled fusion of hydrogen atoms at very high temperatures, with the consequent production of much energy.

Properties

Hydrogen is colourless, odourless and tasteless and is the lightest gas known; its low mass is responsible for its rapid diffusion.

It is possible for the two protons in the diatomic molecule to spin either in the same direction (*ortho*-hydrogen) or in opposite directions (*para*-hydrogen). At room temperature, the ratio of ortho-(*o*-) to *para*-(*p*-)hydrogen is about three to one but, because the internal energy of the latter is the lower, as the temperature falls, the amount of *p*-hydrogen increases until, at around absolute zero, the amount of *o*-hydrogen is negligible.

The chemical versatility of hydrogen is indicated by its forming more compounds than any other element.

If the hydrogen atom loses an electron (i.e. is oxidized), a proton is left. The ionization energy is fairly high, so that the change is not brought about very easily; however, the large polarizing power of the proton, consequent upon its small size, results in its reacting with easily polarizable substances like water, and this subsequent reaction assists the ionization of hydrogen:

$$H_{(gas)} \rightarrow H^+_{(gas)} + e \qquad \Delta H = + 1\,315\,kJ$$

$$H^+ + H_2O \rightarrow H^+_{(aq)} \qquad \Delta H = - 1\,070\,kJ$$

This means that hydrogen is a reducing agent, particularly when in the atomic form. (Several reactions can be brought about by reagents which produce hydrogen *in situ*, but not by hydrogen which has been prepared separately. It is possible that this reactive 'nascent' hydrogen consists of atomic hydrogen of momentary existence.)

The hydrogen atom can also achieve stability by gaining an electron to give the helium configuration, either by electrovalency or by covalency:

(*i*) *Electrovalency*—It is clear that hydrogen could only gain an electron by chemical means from a metal more electropositive than itself. Thus, if it is passed over heated calcium, an electrovalent hydride is formed

$$Ca + H_2 \rightarrow Ca^{2+}H_2^-$$

(*ii*) *Covalency*—This is undoubtedly the chief mode of reaction of the hydrogen atom, and the covalent link with the carbon atom forms the basis of organic compounds. It is also the type of link present in the hydrogen molecule, the bond being formed by the interaction of the *s* orbitals:

or

$$\text{(H)} + \text{(H)} \rightarrow \text{(H H)}$$

$$H\cdot + \cdot H \rightarrow H:H$$

An indication of the strength of this bond is provided by the heat of dissociation

$$H_{2(gas)} \rightarrow 2H\cdot \qquad\qquad \Delta H = +\ 435 \text{ kJ}$$

An electric arc between tungsten electrodes produces sufficient energy to dissociate hydrogen molecules into individual atoms. In the presence of a third body (to absorb the energy given out) these atoms recombine, with the evolution of much heat. This reaction has been utilized in welding, where the material to be welded constitutes the third body; there is the further advantage that the area to be welded is surrounded by an inert atmosphere of hydrogen which prevents oxidation.

Hydrogen reacts directly with many non-metals on the right of the Periodic Table. With fluorine, reaction takes place spontaneously and explosively in the cold, but the vigour declines with the other Group 7 elements. Thus, light is

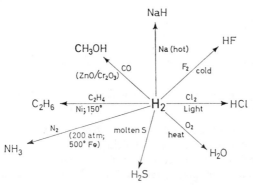

Figure 9.1. Reactions of hydrogen

191

required to activate reaction between hydrogen and chlorine and a catalyst is needed for iodine to bring about even partial reaction.

In the case of Group 6 elements, reaction occurs with oxygen on heating, provided moisture is present, and hydrogen sulphide is formed if hydrogen is bubbled through molten sulphur, although the latter is a reversible reaction. The strong affinity between hydrogen and oxygen is the basis for many reducing actions of hydrogen; for example, the reduction of hot copper(II) oxide to copper

$$CuO + H_2 \rightarrow Cu + H_2O$$

Direct reaction with other non-metals tends to be slight. For instance, nitrogen reacts with hydrogen only under carefully controlled, specialized conditions (p. 255), cf. *Figure 9.1.*

HYDRIDES

Hydrides are binary compounds containing hydrogen and one other element; if the latter is less electronegative than hydrogen, then the name of the compound ends in 'hydride'; for example, NaH is called sodium hydride. On the other hand, if the second element is more electronegative than hydrogen, 'hydrogen' usually forms the first part of the name; thus HCl is hydrogen chloride rather than chlorine hydride.

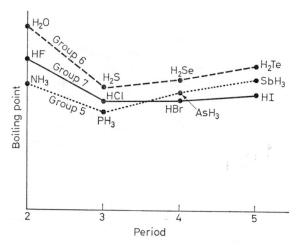

Figure 9.2. Boiling points of covalent hydrides

If the second element is a metal of Group 1 or 2, its valency electrons are transferred to the hydrogen, and salt-like hydrides containing the negative hydride ion result. As the second element becomes more non-metallic (i.e. increases in electronegativity), the hydrogen atom carries a relatively greater positive charge until, in the case of the very electronegative fluorine, oxygen and nitrogen, there is sufficient

192

separation of charge to give rise to hydrogen bonding (e.g. . . . . H—F . . . H—F . . .)
and an approach towards a macromolecular state, so that the hydrides NH_3, H_2O
and HF are *less* volatile than those of their group members of higher atomic weight
(*Figure 9.2*).

It is convenient to classify hydrides as ionic (with elements of Groups 1 and 2)
and covalent (with elements of Groups 3 to 7), although there are also interstitial
hydrides, of vague composition, formed by many of the transitional metals, in which
the small hydrogen atoms occupy interstices in the crystal lattice of the metal.
Hydrogen is therefore occluded, often in very large amounts, on coming into
contact with the metal and is expelled on heating.

Table 9.1 illustrates the typical hydrides formed by members of the various
groups of the Periodic Table.

Table 9.1

1	2		3	4	5	6	7	
			$\delta-$	$\delta+$	$\delta+$	$\delta+$		
Li^+H^-	$Be^{2+}H^-_2(?)$		B_2H_6	CH_4	NH_3	H_2O	$\delta+HF$	↑
			$\delta-$	$\delta-$	$\delta-$	$\delta+$		*Positive*
Na^+H^-	$Mg^{2+}H^-_2(?)$		$(AlH_3)_x$	SiH_4	PH_3	H_2S	$\delta+HCl$	*charge on*
			$\delta-$	$\delta-$	$\delta-$	$\delta+$		*hydrogen*
K^+H^-	$Ca^{2+}H^-_2$		Ga_2H_6	GeH_4	AsH_3	H_2Se	$\delta+HBr$	*increasing*
ionic; solid		Transitional metals; non-stoichiometric, interstitial hydrides		$\delta-$ SnH₄	$\delta-$ SbH₃	$\delta+$ H₂Te	$\delta+HI$	
				$\delta-$ PbH₄	$\delta-$ BiH₃			

covalent; gases (except $(AlH_3)_x$)

→

Positive charge on hydrogen increasing

Ionic Hydrides

These are normally prepared by passing a stream of dry hydrogen over the
heated metal. That the saline product contains the hydride ion, H^-, is shown by
the fact that hydrogen is evolved at the *anode* during the electrolysis of the fused
compound

$$Ca + H_2 \rightarrow CaH_2$$

At anode $\qquad\qquad\qquad H^- - e \rightarrow H; \quad 2H \rightarrow H_2$

The hydride ion instantly reacts with water to liberate **hydrogen**

$$H^- + H_2O \rightarrow OH^- + H_2 \uparrow$$

The readiness with which the hydride ion relinquishes its negative charge is
related to its reducing power; it will, for instance, reduce carbon dioxide to
methanoates (formates):

$$H^- + O{=}C{=}O \rightarrow {}^-O{-}C{=}O$$
$$|$$
$$H$$

193

The complex hydrides, sodium tetrahydridoborate, $NaBH_4$, and lithium tetra-hydridoaluminate, $LiAlH_4$, are particularly important reducing agents in organic chemistry on this account.

Covalent Hydrides

A general method of preparation of these compounds is the action of water or acid on binary compounds of the elements with metals:

$$Mg_2Si + 2H_2O \rightarrow 2MgO + SiH_4 \uparrow \text{ (and other silicon hydrides)}$$
$$AlN + 3H_2O \rightarrow Al(OH)_3 + NH_3 \uparrow$$
$$FeS + 2HCl \rightarrow FeCl_2 + H_2S \uparrow$$
$$PI_3 + 3H_2O \rightarrow H_3PO_3 + 3HI \uparrow$$

Group 3—Boron forms a very interesting series of hydrides when magnesium boride is treated with dilute hydrochloric acid. All of the products are electron-deficient, i.e. it is impossible to formulate conventional structures in which the atoms of boron have an octet of electrons in their outer quantum shells. The parent member of the series is diborane, B_2H_6, which is a gas rapidly hydrolysed to boric acid and hydrogen by the action of water

$$B_2H_6 + 6H_2O \rightarrow 2H_3BO_3 + 6H_2 \uparrow$$

The shape of the molecule is known to be as shown, but the method of bonding in the bridge between the boron atoms is uncertain.

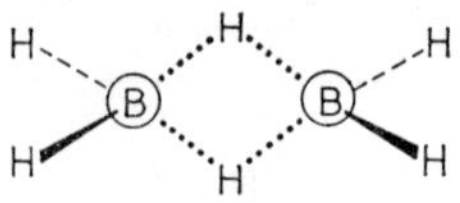

Group 4—Carbon exhibits to a unique degree the ability to form chains of atoms, which can be 'straight', branched or cyclic:

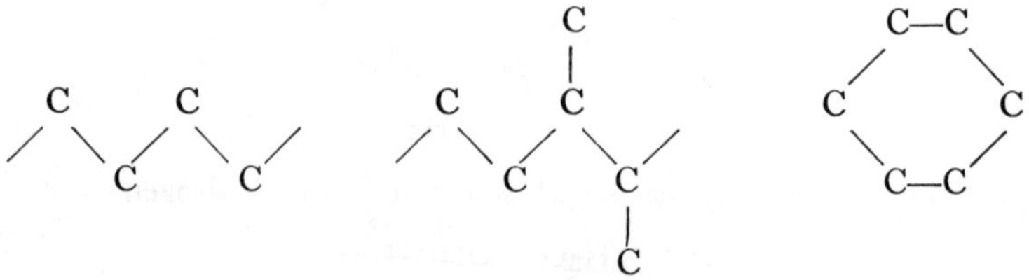

straight   branched   cyclic

Furthermore, carbon can form multiple bonds with ease. These factors give rise to a very large number of hydrides of carbon (discussed in Chapter 19).

Silicon chains are also possible but are not so stable; as a result, the number of hydrides of silicon is very limited, the more so as multiple bonds between silicon atoms are not stable.

194

Most of the hydrides of silicon—the silanes—are more readily inflammable than hydrocarbons in air; they are also rapidly hydrolysed by water in the presence of alkali, whilst the hydrocarbons are inert to water. This can be attributed to the presence of unoccupied $3d$ orbitals in the silicon atom, with which the water molecules can coordinate to initiate reaction:

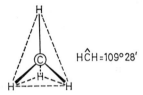

$$SiO_2 + 4H_2\uparrow$$

The structures of all the hydrides of the Group 4 elements are based on a tetrahedron, providing that no multiple bonds are present, e.g. methane, CH_4

$$H\hat{C}H = 109°28'$$

Group 5—The representative hydrides of nitrogen and phosphorus are ammonia, NH_3, and phosphine, PH_3, respectively. Both can be prepared by the decomposition of appropriate salts

$$NH_4^+ + OH^- \rightarrow NH_4OH \rightarrow NH_3\uparrow + H_2O$$

$$PH_4^+ + H_2O \rightarrow H_3O^+ + PH_3\uparrow$$

The structures of ammonia and phosphine are both pyramidal, with the lone pair directed to the corner of a tetrahedron.

$$H\hat{N}H = 106°45'$$

The nitrogen atom of the ammonia molecule donates its lone pair far more readily than the phosphorus atom of phosphine, so that ammonia is far more basic—the ammonium ion formed by coordination of ammonia with a proton closely resembles the ions of the alkali metals:

$$H^+ + :NH_3 \rightarrow [NH_4]^+$$

Hydrogen bonding in ammonia results in it having a lower volatility than phosphine. Ammonia is also far more soluble than phosphine in water because of the ability of the nitrogen atoms to form hydrogen bonds with water molecules, giving rise to molecules of NH_4OH. Ammonia does not burn in air, whereas phosphine readily burns to phosphorus(V) oxide

$$4PH_3 + 8O_2 \rightarrow P_4O_{10} + 6H_2O$$

Group 6—Both water, H_2O, and hydrogen sulphide, H_2S, can be synthesized by direct combination of the elements. Their structures are angular, with bond angles rather less than tetrahedral because of the repulsive force exerted on the bond pairs by the lone pairs

$$104°31'\qquad\qquad 92°20'$$

As a result of association caused by hydrogen bonding, the melting and boiling points of water are much larger than would normally be expected from a molecule of such a size. That oxygen can form these bonds with hydrogen more readily than nitrogen can be seen from the differences betweeen the physical constants of water and of ammonia. Comparable information for phosphine and for hydrogen sulphide shows the considerable part played by the hydrogen bonding in the former two compounds (*Table 9.2*).

Table 9.2

	Molecular weight	Melting point, °C	Boiling point, °C
Ammonia	17	−78	−33
Phosphine	34	−132	−87
Water	18	0	100
Hydrogen sulphide	34	−85	−60

The appreciable separation of charge in the water molecule, together with the asymmetric, angular structure, results in considerable polar character and a very high dielectric constant. Water is also amphoteric, able both to accept and to donate protons:

$$H_2O + H^+ \rightleftharpoons H_3O^+$$

$$H_2O \rightleftharpoons H^+ + OH^-$$

Consequently, water has remarkable solvent powers. In contrast, hydrogen sulphide is able only to donate protons and is only mildly polar

$$H_2S + H_2O \rightleftharpoons H_3O^+ + HS^- \qquad pK_a = 7$$

Group 7—All the halogen elements form hydrides of reasonable stability. Hydrogen fluoride and chloride are prepared by the action of hot, concentrated sulphuric acid on fluorides and chlorides of metals, e.g.

$$F^- + H_2SO_4 \rightarrow HSO_4^- + HF$$

196

Hydrogen bromide and iodide are less stable and are oxidized by concentrated sulphuric acid; they are therefore seldom made by methods analogous to the above but by the action of water on the bromide or iodide of phosphorus, prepared *in situ* from the elements:

$$2P + 3Br_2 \rightarrow 2PBr_3$$

$$PBr_3 + 3H_2O \rightarrow H_3PO_3 + 3HBr$$

All four hydrides ionize exothermically in the presence of water to form acidic solutions

The strength of the acid solutions is in the order HF ≪ HCl < HBr < HI, in accordance with the decreasing bond strength on going from HF to HI.

(Hydrofluoric acid is a particularly weak acid, because of the occurrence of hydrogen bonding which tends to prevent hydrogen escaping as a solvated proton.)

The hydrides of Group 7 elements are all, to a greater or lesser extent, reducing agents. It also follows from the respective bond energies that hydrogen iodide is the strongest reducing agent (as least energy is required to break the bond and provide hydrogen) and hydrogen fluoride the weakest. Hydrogen iodide is also the least stable thermally; it is about 20 per cent dissociated at 350° and becomes progressively more so as the temperature is increased

$$2HI \rightleftharpoons H_2 + I_{2(gas)} \qquad \Delta H = 2 \times 5 \cdot 9 \text{ kJ}$$

QUESTIONS

1. The bond angles between the constituent atoms in water, hydrogen sulphide, ammonia and phosphine are 104°31′, 92°20′, 106°45′, 93°50′, respectively. Comment on the differences in these angles.

2. What conclusions can be drawn from the fact that hydrogen chloride is evolved when sodium chloride is heated with concentrated sulphuric acid, but sulphuric acid is not evolved when sodium sulphate is heated with concentrated hydrochloric acid?

3. Write down reactions which illustrate the reducing power of hydrogen, hydrogen iodide and lithium tetrahydridoaluminate.

4. In what reactions does hydrogen function as an oxidizing agent?

5. Write an essay on the isotopy of hydrogen.

6. Write an account of the role of water as a solvent.

7. Where do you think hydrogen should be placed in the Periodic Table; in Group 1, Group 7, or neither?

197

CHAPTER 10

GROUP 7: THE HALOGENS

Group 6 ns np Group 0

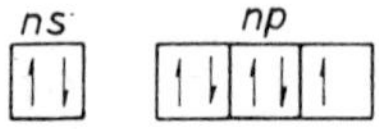

O	Fluorine, F	2.7	Ne
S	Chlorine, Cl	2.8.7	Ar
Se	Bromine, Br	2.8.18.7	Kr
Te	Iodine, I	2.8.18.18.7	Xe
Po	Astatine, At	2.8.18.32.18.7	Rn

Electronic Configuration and Oxidation States
The feature common to all elements of this group is an outer electronic structure of s^2p^5, i.e. one electron is needed to completely fill the p orbitals and thus to attain the noble-gas type configuration either through electrovalent or covalent bond formation. Thus the elements all exhibit an oxidation state of -1. Positive oxidation states are also known for all the halogens except fluorine, and in accordance with the decrease in ionization energies this feature is most pronounced with iodine, which can even completely lose an electron in certain chemical combinations.

Table 10.1

	Ionization energy kJ mol^{-1}	Oxidation states
F	1 685	−1, 0
Cl	1 255	−1, 0, 1, 3, 4, 5, 6, 7
Br	1 142	−1, 0, 1, 5, 6
I	1 010	−1, 0, 1, 3, 5, 7

Atomic and Ionic Radii
The atomic and ionic radii increase in the manner expected as the Atomic Number increases:

	Fluorine	Chlorine	Bromine	Iodine
Atomic radii (pm)	72	99	114	133
Ionic radii (pm), (X^-)	136	181	195	216

198

The ionic radii are much greater than the corresponding atomic radii because of the extra electron accommodated and are larger than the alkali metal (Group 1) ions from the corresponding Periods, since the latter have lost the single electron from their outer shells (p. 74).

Electron Affinities and Electronegativities

As expected from the electronic configuration, the halogens have high electron affinities, the maximum being at chlorine

	F	Cl	Br	I
Electron affinity ($kJ\ mol^{-1}$)	−335	−381	−331	−305

Similarly, the electronegativities are large, fluorine, with its small atomic volume, being the most electronegative of all the elements

	Fluorine	Chlorine	Bromine	Iodine
Electronegativities	3·9	3·0	2·8	2·5

As the sizes of the atoms increase, so the influence of the positively charged nucleus on the periphery of the atoms decreases, and electrons are then less readily attracted. That this is so is indicated by the decrease in electronegativity as one passes down the Group, and by the accompanying appearance of some metallic properties with iodine.

It is clear that the halide ions, X^-, having the configuration s^2p^6, are readily formed and that the elements normally exist as diatomic molecules in which the s^2p^6 configuration is attained by electron sharing

$$\overset{\Large\cdot\ \cdot}{\underset{\Large\cdot\ \cdot}{\cdot\ \underset{}{\times}\ \cdot\ \underset{}{\times}\ \cdot}}$$

or in terms of orbitals

The stability of this configuration is indicated by the large heats of dissociation (bond energies) of the molecules, although that of fluorine is relatively low because its small size cannot prevent some repulsion between pairs of non-bonding electrons:

	F_2	Cl_2	Br_2	I_2
Heat of dissociation ($kJ\ mol^{-1}$)	159	239	188	147

The melting and boiling points of the halogens are low, as expected of structures consisting of covalent diatomic molecules held together by weak van der Waals forces:

	Fluorine	Chlorine	Bromine	Iodine
M.p. (°C)	-223	-102	-7	114
B.p. (°C)	-187	-35	59	183
Colour in gaseous state	pale yellow	green	red	violet

The rise in the melting points is brought about mainly by the increase in the volume of the molecules (and, hence, in van der Waals bonding) and partly through the greater availability of orbitals as the atomic number increases, tending to give stability to the diatomic state as a result of possible electron delocalization. This latter phenomenon is also indicated by the increase in intensity of the colours of the halogen vapours on passing from fluorine to iodine, showing the increasing ease with which electrons are excited.

Electrode Potentials

The reactivities of the elements are related to their redox potentials,

$$X_2 + 2e \rightleftharpoons 2X^- \text{ aq.}$$

Fluorine	Chlorine	Bromine	Iodine
2·87	1·36	1·06	0·54 V

Fluorine is readily reduced to the fluoride ion and is consequently a very powerful oxidizing agent. The redox potentials of the remaining elements decrease rapidly; thus the oxidizing power of chlorine, bromine and iodine gets progressively less.

OCCURRENCE AND EXTRACTION OF THE HALOGENS

Fluorine

This element occurs principally as the insoluble calcium salt, $Ca^{2+}F_2^-$, *fluorspar*, from which hydrogen fluoride is obtained by treatment with concentrated sulphuric acid

$$2F^- + H_2SO_4 \rightarrow SO_4^{2-} + 2HF \uparrow$$

As fluorine is the most electronegative of all elements known, it is impossible to oxidize hydrogen fluoride to the free element by any other element or compound. Sufficient energy can be supplied electrically, however, and the element may therefore be obtained by an electrolytic process in which the non-conducting hydrogen fluoride, after reaction with potassium fluoride to produce potassium hydrogen difluoride, $K^+HF_2^-$, acts as an electrolyte in the molten state (*Figure 10.1*).

The reaction at the carbon anode is

$$F^- \rightarrow \tfrac{1}{2}F_2 \uparrow + e$$

and that at the steel cathode

$$H^+ + e \rightarrow \tfrac{1}{2}H_2 \uparrow$$

(Periodical additions of anhydrous hydrogen fluoride must be made to the electrolyte to replace that decomposed.)

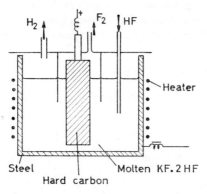

Figure 10.1. Electrolytic preparation of fluorine

Fluorine has not found many uses until recent years, but it is rapidly assuming a considerable importance in the preparation of *fluorocarbons*, in which the fluorine atoms, which are much larger than those of hydrogen, protect the carbon chains from attack, making such compounds particularly valuable as lubricants and refrigerants (p. 408). Fluorine also finds application in uranium technology, partly because uranium(VI) fluoride is a convenient compound for the separation of the isotopes of the metal by diffusion processes (p. 62), and partly because the metal itself is obtained by the reduction of the tetrafluoride.

Chlorine

Chlorine, bromine and iodine exist as salts of the Group 1 metals and magnesium, the extreme solubility of which results in their main source being either

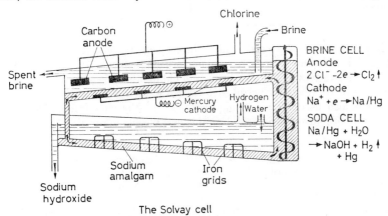

Figure 10.2. Manufacture of chlorine

the sea or salt deposits on the sites of dried-up seas. The main sources of chlorine are *rock salt*, Na^+Cl^-, and *carnallite*, $K^+Mg^{2+}Cl_3^-$. $6H_2O$, from which it is obtained by electrolysis. A strong solution of brine can be electrolysed in a Solvay cell (*Figure 10.2*), the position of the anodes being adjustable so that the anode–cathode distance can be maintained at about 2 mm despite the removal of the anode surface by oxidation.

The reaction at the mercury cathode is

$$Na^+ + e \rightarrow Na \text{ (which dissolves to form an amalgam)}$$

and at the carbon anode

$$2Cl^- \rightarrow Cl_2 \uparrow + 2e$$

In the U.K., 75 per cent of the annual chlorine output is at present obtained by this method. On the other hand, in the U.S.A. about 80 per cent is produced in diaphragm or Downs cells (p. 310), and only 20 per cent by the Solvay process.

Production of chlorine (1965)	10^6 tonnes	per cent	Extraction of salt (1961)	10^3 tonnes
World	15·0		U.S.A.	23 325
U.S.A.	6·6	44	China	14 000
West Germany	1·1	7	U.S.S.R.	7 500
Japan	1·0	7	U.K.	5 760
U.K.	0·7	5	West Germany	4 680

Uses of Chlorine (U.K.)		Manufacture of hydrochloric acid (1962)	
Metallurgy, etc.	16	U.S.A.	960
Car industry (fuel additives, etc.)	15	West Germany	283
Plastics	12	Japan	185
Chemicals (incl. HCl)	12	France	111
Domestic	12	Italy	79

Chlorine is extensively used in metallurgical operations, e.g. for the recovery of tin and aluminium from scrap. Many organic solvents and plastics (p. 406) are chloro-derivatives of hydrocarbons, and chloroethane is an intermediate in the preparation of additives for petrol such as lead tetraethyl for 'anti-knock' (p. 380). Chlorine is also used for sterilizing water and in the chemical industry for the manufacture of hydrochloric acid, bromine and bleaching powder:

$$Ca(OH)_2 \xrightarrow[\substack{\text{cold, solid} \\ \text{current} \\ \text{of } Cl_2}]{\text{counter-}} Ca(OCl)_2 + \text{basic chloride} \\ \text{bleaching powder}$$

Bromine

Bromine is chiefly obtained from sea water, in which it occurs as bromides to the extent of about 0·015 per cent. It is liberated by passing chlorine into the water, which is maintained at a pH of 3·5 by the addition of a suitable quantity of sulphuric acid

$$2Br^- + Cl_2 \rightarrow 2Cl^- + Br_2 \uparrow$$

(acid is necessary to minimize the loss of bromine by reaction with the water, as indicated by the equation $Br_2 + H_2O \rightarrow HBr + HOBr$). The bromine is absorbed in a suitable solvent such as benzene, from which it is subsequently extracted by fractional distillation.

Bromine is used for making 1,2-dibromoethane, an additive for petrol containing lead tetraethyl; it ensures that after combustion the lead is removed in the exhaust as volatile lead tetrabromide instead of forming a deposit on the cylinder walls of the engine. The element is also extensively used in photography, in the form of its silver salt, photochemical reduction of silver bromide to metallic silver being particularly easily accomplished. The manufacture of potassium bromide for medicinal purposes is another outlet for the element, and it is also used in the dyestuffs industry.

Iodine

The main sources of iodine are certain seaweeds in which it is present to the extent of about 3 per cent by weight, and *caliche*, which is impure sodium nitrate containing about 0·2 per cent of sodium trioxoiodate(V) ('iodate').

Extraction from seaweed is effected by burning and leaching the ash, which contains alkali metal iodides, with water. The solution so obtained is evaporated to remove the chlorides and sulphates of sodium and potassium by crystallization, after which sulphuric acid is added and any sulphur precipitated (from sulphides) removed. Finally, iodine is liberated by distillation of the liquor in the presence of manganese(IV) oxide

$$2I^- + Mn^{4+}O^{2-}{}_2 + 4H^+ \rightarrow Mn^{2+} + 2H_2O + I_2 \uparrow$$

Most iodine is obtained from caliche. The ore is leached with water and pure sodium nitrate is crystallized out. The mother liquors are then concentrated and treated with the amount of sodium hydrogen sulphite necessary to reduce five-sixths of the iodate to hydriodic acid. This then reacts with unchanged iodate to liberate iodine. The process can be summarized by the equations

$$IO_3^- + 3HSO_3^- \rightarrow HI + 3SO_4^{2-} + 2H^+$$

$$5HI + IO_3^- + H^+ \rightarrow 3I_2 \uparrow + 3H_2O$$

The iodine so obtained is purified by sublimation over potassium iodide, which reacts with any iodine monochloride present

$$ICl + I^- \rightarrow Cl^- + I_2 \uparrow$$

Iodine and many of its compounds find use in medicine, both as antiseptics and for the prevention of goitre, an abnormal growth of the thyroid gland caused

203

by iodine deficiency. Iodides and iodates are also used in photography and in analytical chemistry.

Astatine

This element does not occur in nature to any appreciable extent but has been synthesized by bombardment of bismuth with α-particles:

$$^{209}_{83}Bi + ^{4}_{2}He \rightarrow ^{211}_{85}At + 2^{1}_{0}n \qquad (t_{\frac{1}{2}} = 2 \cdot 7 \times 10^4 \, s)$$

Laboratory Preparation of the Halogens

Chlorine, bromine and iodine can all be prepared by oxidation of the corresponding anion with a vigorous oxidizing agent, such as manganese(IV) oxide, in the presence of concentrated sulphuric acid:

$$2X^- + Mn^{4+}O^{2-}_2 + 4H^+ \rightarrow Mn^{2+} + 2H_2O + X_2 \uparrow$$

In the case of chlorine, it is often more convenient to drop cold, concentrated hydrochloric acid on to crystals of potassium tetraoxomanganate(VII) ('permanganate'):

$$2MnO_4^- + 16HCl \rightarrow 6Cl^- + 2Mn^{2+} + 8H_2O + 5Cl_2 \uparrow$$

REACTIONS OF THE HALOGENS

As the atomic number increases, so the chemical reactivity decreases; this is particularly so with regard to the -1 oxidation state and is in agreement with the calculated decrease in electronegativity.

Fluorine will react directly with all elements except nitrogen, oxygen and the noble gases, most non-metals igniting spontaneously in the gas. Chlorine also reacts with most elements except carbon, nitrogen and oxygen, although not as readily as fluorine; with both fluorine and chlorine, the highest valency state of the combining metal is usually involved. Bromine often appears to be more reactive than chlorine because of its concentration in the liquid state, although it is really less reactive and combines spontaneously only with a few elements. Iodine is the least reactive of the halogens but it will combine directly with many metals on warming.

Reaction with Hydrogen

The reactions represented by the equation

$$H_2 + X_2 \rightarrow 2HX$$

can be summarized as

Fluorine	Chlorine	Bromine	Iodine
explosive even at 20K	explosive in direct sunlight	requires a catalyst (Pt) at 500K	reversible requires a catalyst and temperatures $> 500K$

The heat of formation of hydrogen iodide is positive for solid iodine but slightly negative for gaseous iodine; in the latter case, therefore, the reaction is exothermic. By Le Chatelier's principle the degree of thermal dissociation should increase with rise in temperature (*Figure 10.3*). The ready reversibility of this reaction makes hydrogen iodide, alone of the hydrogen halides, a useful reducing agent.

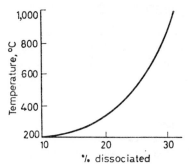

Figure 10.3. Dissociation of hydrogen iodide

Reaction with Water

The reactions of the halogens with water can be considered in terms of the redox potentials involved:

(*i*) $2H_2O \rightarrow 4H^+ + O_2 \uparrow + 4e$ $\qquad\qquad\qquad E = -0.81$ V

	Fluorine	Chlorine	Bromine	Iodine
		V		
(*ii*) $2X_2 + 4e \rightarrow 4X^-$ $\quad E =$	2·87	1·36	1·06	0·54
and, by addition of (*i*) and (*ii*) $2H_2O + 2X_2 \rightarrow 4H^+ + O_2 \uparrow + 4X^-$	2·06	0·55	0·25	−0·27

Therefore, the first three halogens can liberate oxygen from water, and in fact fluorine cannot be kept in aqueous solution, as immediate reaction occurs. When chlorine and bromine liberate oxygen—with much less vigour than fluorine—a different equilibrium is normally established, particularly in diffuse daylight

$X_2 + 2H_2O \rightarrow H_3O^+ + X^- + HOX$ (hydrogen oxohalate(I) or hypohalous acid)

Both reactions will be encouraged by the presence of hydroxide ions, which can then remove the oxonium ions (H_3O^+) as neutral water molecules. Thus, fluorine reacts with dilute alkali to give oxygen difluoride

$$2F_2 + 2OH^- \rightarrow F_2O + H_2O + 2F^-$$

which decomposes in strong alkali to give oxygen

$$F_2O + 2OH^- \rightarrow 2F^- + H_2O + O_2 \uparrow$$

205

Chlorine with cold alkali gives the chloride and oxochlorate(I) ('hypochlorite') ions, which latter disproportionate in hot alkali to form chloride and trioxochlorate(V) (chlorate) ions

$$Cl_2 + 2OH^- \rightarrow Cl^- + OCl^- + H_2O$$

$$3OCl^- \rightarrow 2Cl^- + ClO_3^-$$

Bromine likewise gives hypobromites and bromates.

Because of the unfavourable electrode potentials, iodine will produce any significant amount of hypoiodite ions only in the presence of dilute alkali, and even then they are so unstable that disproportionation readily occurs to give iodide and iodate ions.

Halogens as Oxidizing Agents

Besides the above oxidizing reactions, the halogens find several applications as oxidizing agents, as the following examples indicate:

$$2Fe^{2+} + Cl_2 \rightarrow 2Fe^{3+} + 2Cl^-$$

$$Sn^{2+} + I_2 \rightarrow Sn^{4+} + 2I^-$$

$$4Br_2 + S_2O_3^{2-} + 5H_2O \rightarrow 2SO_4^{2-} + 8Br^- + 10H^+$$

$$I_2 \quad + \quad 2S_2O_3^{2-} \quad\quad \rightarrow 2I^- + \quad\quad S_4O_6^{2-}$$

Trioxothiosulphate μ-Dithio-bistrioxosulphate(VI)
('thiosulphate')   ('tetrathionate')

$$AsO_3^{3-} \quad\quad + I_2 + H_2O \rightleftharpoons \quad AsO_4^{3-} \quad + 2HI$$

Trioxoarsenate(III)   Tetraoxoarsenate(V)
('arsenite')   ('arsenate')

The reactions involving iodine are particularly useful for quantitative estimations because of the colour changes involved and of the ready detection of iodine by a blue-black complex it forms with starch.

The Hydrogen Halides

The more important reactions of these compounds have been considered already in the comparative study of hydrides (p. 193 *et seq.*); it remains to describe their preparation.

(*a*) *Hydrogen fluoride* is prepared by distilling finely ground fluorspar, calcium fluoride, with concentrated sulphuric acid in a lead retort

$$2F^- + H_2SO_4 \rightarrow 2HF \uparrow + SO_4^{2-}$$

Hydrogen fluoride is evolved and absorbed in water until about an 80 per cent solution of hydrofluoric acid is obtained. Alternatively, anhydrous hydrogen fluoride may be obtained by condensation of the vapour below 19°C to a volatile liquid. The aqueous solution attacks glass and is therefore usually kept in gutta-percha or polythene bottles:

$$SiO_2 + 4HF \rightarrow SiF_4 \uparrow + 2H_2O$$
(from glass)

$$SiF_4 + 2HF \rightarrow H_2SiF_6$$
Hexafluorosilicic acid

(Hydrogen fluoride should be handled with the utmost care as it causes unpleasant wounds on the skin.)

(b) *Hydrogen chloride* is prepared by the action of hot, concentrated sulphuric acid on sodium chloride, dried by passing through concentrated sulphuric acid and then collected by the upward displacement of air. In the laboratory it is usual to utilize only half of the available acid:

$$Cl^- + H_2SO_4 \rightarrow HSO_4^- + HCl \uparrow$$

but on the industrial scale this is not sufficiently economical and, by heating the mixture further, the whole of the available acid is utilized:

$$Cl^- + HSO_4^- \rightarrow SO_4^{2-} + HCl \uparrow$$

Because of the very great solubility of hydrogen chloride in water (500 volumes of gas in one volume of water at s.t.p.), care must be taken when the gas is dissolved to prevent water from sucking back on to the concentrated sulphuric acid as the pressure is reduced.

Hydrogen chloride is a by-product of the chlorination of organic compounds and is synthesized industrially by direct combination of the two elements in the presence of activated charcoal:

$$H_2 + Cl_2 \rightarrow 2HCl$$

(c) *Hydrogen bromide* is conveniently prepared by dropping bromine on to red phosphorus in a little water. Bromides of phosphorus are immediately formed and hydrolysed, e.g.

$$2P + 3Br_2 \rightarrow 2PBr_3$$

$$PBr_3 + 3H_2O \rightarrow H_3PO_3 + 3HBr$$

Traces of bromine carried over with the hydrogen bromide are removed by passing the mixture over glass beads coated with a paste of red phosphorus and water.

(d) *Hydrogen iodide* can be obtained in a similar manner, except that in this case water is dropped on to a mixture of red phosphorus and iodine. The most convenient way to prepare the aqueous solution, hydriodic acid, is to pass hydrogen sulphide into a suspension of iodine in water

$$I_2 + H_2S + aq \rightarrow S \downarrow + 2HI_{(aq)}$$
Hydriodic acid

207

All of the hydrogen halides dissolve readily in water and produce azeotropes of maximum boiling point (p. 102); with the exception of hydrofluoric acid, the solutions are strongly acidic:

Halides

The metals of Groups 1 and 2 (excluding beryllium) form ionic halides. Most other elements, particularly in their highest valency states, give halides which are predominantly covalent. As expected, the degree of covalence increases with the size of the halogen atom, fluorides having considerably more ionic character than the other halides.

Preparation—The most versatile method of preparation is by direct halogenation, when, because of the high electronegativity of fluorine and chlorine and, to a lesser extent, bromine, the maximum valency state of a metal of variable valency is exerted, e.g.

$$2Fe + 3Cl_2 \rightarrow Fe_2Cl_6$$

$$2Al + 3Cl_2 \rightarrow Al_2Cl_6$$

$$Sn + 2Br_2 \rightarrow SnBr_4$$

$$C + 2F_2 \rightarrow CF_4$$

Halides which are mainly covalent can also be prepared from the oxide by mixing with carbon as reducing agent and heating in the presence of the halogen, e.g.

$$Al_2O_3 + 3C + 3Cl_2 \rightarrow Al_2Cl_6 + 3CO \uparrow$$

Dry methods such as these are essential for the preparation of easily hydrolysed halides, particularly when they are required in the anhydrous form.

Wet methods can be used when the product is not susceptible to hydrolysis, (either because it is highly ionic or insoluble). The metal, its oxide, hydroxide or carbonate can be dissolved in aqueous halogen acid and the resultant solution evaporated, e.g.

$$Fe + 2HCl \rightarrow FeCl_2 + H_2 \uparrow$$

$$MgO + 2HBr \rightarrow MgBr_2 + H_2O$$

$$Na_2CO_3 + 2HI \rightarrow 2NaI + H_2O + CO_2 \uparrow$$

Insoluble halides, which include the fluorides of calcium, strontium and barium together with the chlorides, bromides and iodides of silver, copper(I) and lead,

208

the chloride of dimercury(II) and the iodides of mercury, can all be made by precipitation reactions of the type

$$M^{n+} + nX^- \to MX_n \downarrow$$

Hydrolysis—Most of the covalent halides are hydrolysed to a greater or lesser extent by water or alkali. There is no single explanation for this, but the likely reaction in individual cases can often be deduced. Some different possibilities are discussed below.

(*i*) Beryllium chloride, $BeCl_2$, has considerable covalent character, since the small size and double charge of the beryllium ion confer a large polarizing power, which gives rise to the reaction between the ion and water molecules, producing an equilibrium of the type

$$Be^{2+} + 4H_2O \to [Be(H_2O)_4]^{2+} \rightleftharpoons [Be(OH)(H_2O)_3]^+ + H^+$$

Similar reasoning applies to the hydrolysis of silicon tetrachloride, except that here the oxygen atoms of the water molecules can donate electrons to the favourably situated $3d$ orbitals of the silicon atom, as is the case with silane (p. 195). In contrast to this, carbon tetrachloride is not susceptible to hydrolysis under normal conditions because the outer (second) quantum level of the carbon atom is already full, and accordingly there are no empty orbitals of suitably low energy available for coordination of water molecules.

The more the covalent character exhibited by the original compound, the greater is the extent of the hydrolysis, provided that suitable empty orbitals are available to accept the electrons of the oxygen in potentially coordinated water molecules. An excellent example of a readily established equilibrium is that afforded by bismuth trichloride

$$BiCl_3 + H_2O \rightleftharpoons BiOCl \downarrow + 2HCl$$

(*ii*) The halides (excluding fluorides) of nitrogen and oxygen are readily hydrolysed by a different mechanism. The lone pair(s) of electrons associated with these atoms can be donated to one of the hydrogen atoms of the water molecule. This results in an electronic rearrangement to yield the products of hydrolysis, e.g.

Oxygen Compounds of the Group 7 Elements

There are several binary compounds of oxygen and the halogens, although very few of them are of any great importance. Compounds of oxygen with fluorine are correctly known as fluorides, since fluorine is more electronegative than oxygen, whilst the remainder (other than I_2O_4 and I_4O_9, see p. 213) are classed as oxides. *Table 10.2* lists all the known oxygen compounds of the halogens (important ones: bold)

Table 10.2. Binary compounds of oxygen and the halogens

OF_2			
O_2F_2			
O_3F_2	**Cl_2O**	Br_2O	
	ClO_2	BrO_2	
		BrO_3	
O_4F_2		(or Br_3O_8)	I_2O_4
	Cl_2O_6		I_2O_5
	Cl_2O_7		I_4O_9

Oxygen difluoride is prepared by the action of fluorine on dilute sodium hydroxide solution (p. 205). There is no oxyacid corresponding to this compound nor, indeed, to any of the remaining oxygen fluorides, as such a compound would require some donation of electrons from the halogen to the oxygen and, owing to the great electronegativity of the fluorine atom, this is unlikely.

The oxides of chlorine are all endothermic, unstable and explosive. Chlorine(I) oxide, Cl_2O, chlorine(IV) oxide, ClO_2, and chlorine(VII) oxide, Cl_2O_7, are all gases at room temperature whilst chlorine(VI) oxide, Cl_2O_6, is a liquid.

Chlorine dioxide is of some importance as an industrial oxidizing agent and is also of theoretical interest, as the molecule contains an odd electron and is accordingly paramagnetic

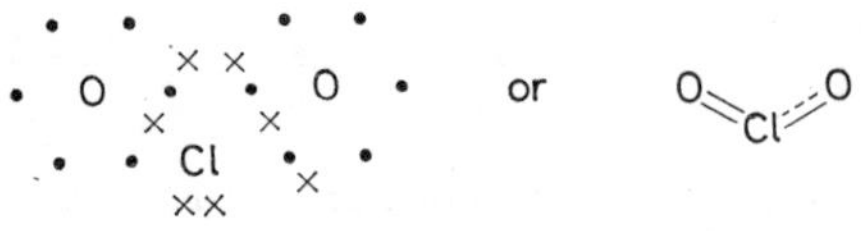

All the oxides of chlorine are acid anhydrides and react in the following way with water:

$$Cl_2O + H_2O \rightarrow 2HOCl$$

$$2Cl^{IV}O_2 + H_2O \rightarrow HCl^{III}O_2 + HCl^{V}O_3$$

$$Cl_2^{VI}O_6 + H_2O \rightarrow HCl^{V}O_3 + HCl^{VII}O_4$$

$$Cl_2O_7 + H_2O \rightarrow 2HClO_4$$

The structures of the acids and of the corresponding anions are shown in *Table 10.3*. Like the oxides, the acids and their salts have powerful oxidizing properties,

210

the sodium and calcium salts of hypochlorous acid being particularly well known as bleaching agents. A notable exception to this oxidizing character is afforded by aqueous solutions of perchloric acid: while the pure covalent acid is an exceptionally powerful oxidizing agent, solutions of the acid are quite stable, possibly because of the perfect symmetry of the ion and the completed orbitals of its constituent atoms.

Table 10.3. The oxyacids of chlorine

Acid	Structure	Stability	Structure of anion
Hydrogen oxochlorate(I) (hypochlorous acid)	H—O—Cl	unstable	$(O—\ddot{C}l\!:)^-$
Hydrogen dioxochlorate(III) (chlorous acid)	H—O—Cl→O		(see structure)
Hydrogen trioxochlorate(V) (chloric acid)	(see structure)		(see structure)
Hydrogen tetraoxochlorate(VII) (perchloric acid)	(see structure)	isolable but explosive on heating	(see structure)

All the structures are based on a tetrahedral arrangement of electron pairs around the central chlorine atom.

Only two oxyacids of bromine are definitely known: hypobromous acid, HOBr, and bromic acid, $HBrO_3$, both similar to their chlorine analogues.

The only simple oxide of iodine is the one of empirical formula I_2O_5. This is a stable white solid and apparently has a polymeric structure. It is easily obtained by direct oxidation of iodine with concentrated nitric acid, the initial product, iodic acid, being easily dehydrated by the application of heat:

$$I_2 \xrightarrow{HNO_3} HIO_3 \xrightarrow{-H_2O} I_2O_5$$

It is a fairly strong oxidizing agent and is used as such for the quantitative estimation of even small amounts of carbon monoxide (and, hence, of oxygen in organic compounds)

$$I_2O_5 + 5CO \rightarrow I_2 + 5CO_2$$

The iodine liberated is estimated in the usual way with sodium thiosulphate.

Characteristically, because of its larger size, iodine also forms 'periodates', the parent acid of which may be regarded as a derivative of $I(OH)_7$. Strong oxidation

211

of iodate(V) by oxochlorate(I) yields, on acidification, an acid of formula H_5IO_6 ($I(OH)_7$–H_2O), commonly known as *ortho*periodic acid and having the structure

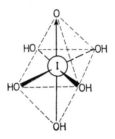

Dehydration of this compound can be carried out in stages to give, first, $H_4I_2O_9$ (i.e. $2H_5IO_6$–$3H_2O$), and then HIO_4 ($H_4I_2O_9$–H_2O), periodic acid. It should be noted that in all these compounds the iodine is present in an oxidation state of + 7.

Cationic Halogen Compounds

Several compounds of iodine (and a few of bromine and chlorine) are known in which the halogen is in the form of a positively charged ion, typifying the increase in metallic character on descending a group. The two 'oxides', I_2O_4 and I_4O_9, are thought to contain such ions, the former being formulated as $IO^+IO_3^-$, iodyl(III) iodate, and the latter as $I^{3+}(IO_3^-)_3$, iodine(III) iodate. Iodine(III) phosphate, ethanoate(acetate) and a few other salts are also known.

Iodine also occurs as a unipositive ion. Thus the unstable hypoiodous acid may better be regarded as iodine hydroxide, since the following equilibria have been determined

$$HOI \rightleftharpoons H^+ + OI^- \quad K \sim 10^{-13} \text{ M}$$

$$HOI \rightleftharpoons HO^- + I^+ \quad K \sim 10^{-10} \text{ M}$$

However, it appears that the ion I^+ is stable only when complexed, e.g. with pyridine (py), (p. 423). In this state several salts are known, such as pyridyl iodine(I) nitrate and perchlorate

$$[Ipy]^+NO_3^- \qquad\qquad [Ipy]^+ClO_4^-$$

(The coordination of a pyridine molecule to the iodine cation stabilizes it by completing an octet of electrons in the outer electronic shell.)

The Interhalogens

Compound formation between halogen atoms themselves results in formation of the interhalogens (*Table 10.4* and *Figure 10.4*).

Table 10.4. The interhalogens

ClF	BrF		BrCl	ICl	IBr	AtI
ClF$_3$	BrF$_3$	IF$_3$		ICl$_3$		
ClF$_5$	BrF$_5$	IF$_5$				
		IF$_7$				

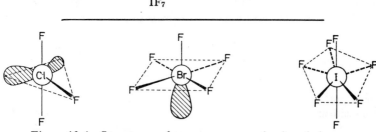

Figure 10.4. Structures of some representative interhalogens

All these compounds can be prepared by direct combination of the elements in appropriate proportions under suitable conditions. Perhaps the most important at present is bromine trifluoride. This exists to a large extent in the form of associated molecules linked by fluorine bridges

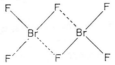

but it also ionizes to some extent in accordance with the equation

$$2BrF_3 \rightleftharpoons BrF_2^+ + BrF_4^-$$

Bromine trifluoride provides a very useful medium for fluorinating reactions, and many metals, oxides and even other halides are readily converted to fluorides by this reagent.

ANALYSIS OF HALOGENS AND THEIR COMPOUNDS

Halogens are detected in their elementary state by their characteristic colours and odours. With the exception of iodine they will also bleach damp litmus paper.

The halide ions are recognized by the formation of the fuming, pungent hydrogen halides on treatment with concentrated sulphuric acid. Because of the increasing reducing power of these latter compounds as the atomic number of the halogen increases, those of bromine and iodine produce other products also in the presence of sulphuric acid:

halide	F$^-$	Cl$^-$	Br$^-$	I$^-$
products with	HF	HCl	HBr,	HI
conc. H$_2$SO$_4$	(attacks		Br$_2$,	I$_2$
	damp glass)		SO$_2$	SO$_2$ + H$_2$S
				(producing S)

Alternatively, the halide ions (except fluoride and chloride) can be directly oxidized using, for example, sodium hypochlorite. The free halogen produced can be extracted by carbon tetrachloride and identified by the colour imparted—yellow indicating bromine and violet indicating iodine.

Halide ions other than fluoride can also be confirmed by treatment with nitric acid, followed by addition of silver nitrate solution

$$Ag^+ + X^- \rightarrow AgX \downarrow$$

The silver halides have the properties:

	AgCl	AgBr	AgI
colour of ppt.	white	cream	yellow
reaction with ammonia	soluble	soluble with difficulty	insoluble

(The difference in the ability of the precipitates to dissolve in ammonia in accordance with the equation $Ag^+ + 2NH_3 \rightleftharpoons Ag(NH_3)_2^+$ is caused by the decreasing solubility products of the halides, in the order chloride > bromide > iodide.)

Halogens in higher oxidation states may be identified by reducing to the halide, for example with sulphur dioxide, followed by application of one of the above techniques.

Quantitatively, fluorides may be estimated gravimetrically as the insoluble calcium fluoride, whilst the other halides are determined by titration with silver nitrate solution, using sodium chromate (or an adsorption indicator) in neutral conditions

$$Ag^+ + X^- \rightarrow AgX \downarrow$$

Then, at the end point,

$$2Ag^+ + CrO_4^{2-} \rightarrow Ag_2CrO_4 \downarrow$$
$$\text{yellow} \qquad \text{red}$$

QUESTIONS

1. In which respects is fluorine not a typical member of Group 7?

2. Boron trifluoride is acidic in anhydrous hydrofluoric acid, whilst ethanol is basic. Suggest reasons for these observations.

3. What sort of compound would you expect to contain halogen in a cationic form? How would you show the existence of halogen in this condition?

4. In preparing chlorine(I) oxide, chlorine is passed at $0°C$ over dry mercury(II) oxide, and for preparing aqueous hypochlorous acid, chlorine is shaken with a suspension of mercury(II) oxide. Discuss the use of the metal oxide in these reactions.

214

5. Comment on the structure and reactions of bleaching powder.

6. Suggest reasons for the existence of SF_6, whereas the highest chloride of sulphur is SCl_4.

7. What would you expect to be the chemical properties of astatine?

8. Under what conditions could an arsenite be estimated volumetrically?

9. Discuss the merits and limitations of various methods of classification of the halides.

10. Write an account of the functions of halogen compounds in the human body.

11. The average concentration of bromide in sea-water is $0\cdot000\ 8$ mol kg^{-1}. Calculate the volume of sea-water of relative density $1\cdot03$ required to produce $1\ 000$ kg of bromine, if the bromine is completely extracted. How much chlorine would be needed to displace this amount of bromine?

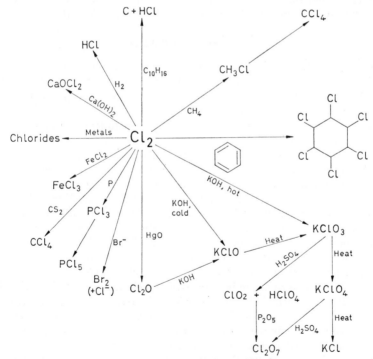

Figure 10.5. The reactions of chlorine

215

CHAPTER 11

GROUP 6 ELEMENTS

Group 5 $\overset{ns}{\boxed{\uparrow\downarrow}}$ $\overset{np}{\boxed{\uparrow\downarrow|\uparrow|\uparrow}}$ Group 7

Group 5			Group 7
N	**Oxygen, O**	2.6	F
P	**Sulphur, S**	2.8.6	Cl
As	**Selenium, Se**	2.8.18.6	Br
Sb	**Tellurium, Te**	2.8.18.18.6	I
Bi	**Polonium, Po**	2.8.18.32.18.6	At

The first member of this group is *oxygen*, by far the most abundant element in the earth's crust, and the last *polonium*, occurring only to the extent of one part in $2 \cdot 5 \times 10^{10}$ parts of pitchblende and with a half-life of $1 \cdot 19 \times 10^7 \, s$

$$^{210}_{84}\text{Po} \rightarrow \,^{206}_{82}\text{Pb} + \,^{4}_{2}\text{He} \; (\alpha\text{-particles})$$

Not surprisingly, then, polonium is the most recent member of this group to have been discovered. Oxygen has been known since the eighteenth century, although it was first regarded as dephlogisticated air. In view of its abundance, it might be argued that the discovery of oxygen was very late; doubtless, the fact that it is an invisible gas and is not the major constituent of the atmosphere is largely responsible. In contrast, sulphur, existing as a solid in a fairly pure state, has been known since well before the time of Christ.

As the elements of this group are just two electrons short of the nearest noble gas and the stable electronic configuration which that implies, much of their chemistry is associated with the acquisition of two electrons, either by transfer or sharing. In addition, compounds are formed in which the coordination is in excess of two: oxygen can have a maximum coordination number of four only, whereas sulphur can have one of six and tellurium even of eight. This difference between the elements is a function of the outer shell; in the case of oxygen, this is complete with eight electrons, i.e. there are no d orbitals of suitable energy available. But the outer shell of tellurium can be expanded to include five d orbitals. The absence of d orbitals in the second shell is responsible for oxygen standing apart from the other elements of the Group, and sulphur is usually preferred as the typical element.

The increase in the number of electron shells on passing from oxygen to polonium results in a progressive increase in atomic volume. There is consequently less attraction between the nucleus and the outer electrons with the latter elements, so that there is a fall in both the ionization energy and electronegativity on descending the series (*Table 11.1*). Tellurium and polonium even exhibit some metallic behaviour, both possessing high densities and forming, for example, basic nitrates and sulphates.

Table 11.1

	Atomic radius, pm	Ionic radius, X^{2-}(pm)	First ionization energy (kJ mol^{-1})
O	74	140	1 318
S	104	180	1 001
Se	117	200	940
Te	137	220	871

REACTIONS

The elements of this group react with most metals. Only the most electropositive metals give ionic compounds and these are hydrolysed in aqueous solution

$$O^{2-} + H_2O \rightarrow 2OH^- \qquad S^{2-} + 2H_2O \rightarrow H_2S + 2OH^-$$

Transitional metals give compounds of appreciable covalent character. Frequently non-stoichiometric (Berthollide) compounds exist: iron(II) oxide can vary in composition between $Fe_{0.91}O$ and $Fe_{0.95}O$, with iron positions in the lattice being occasionally empty and with some iron atoms being in oxidation state $+3$ to maintain electrical neutrality. Perhaps the most striking example of non-stoichiometry is provided by cobalt telluride which can exist between the composition CoTe and CoTe$_2$.

In reactions with halogens, oxygen is clearly atypical. Although hemioxides exist (e.g. Cl$_2$O), the more stable compounds contain more oxygen than halogen (e.g. I$_2$O$_5$). On the other hand, tellurium and polonium form tetrahalides, even with iodine. Sulphur, selenium and tellurium even give hexafluorides, and it is significant that, whilst SF$_6$ and SeF$_6$ (maximum coordination number 6) are very stable, TeF$_6$ (maximum coordination number 8) is rapidly hydrolysed by water. In several of these and other compounds, the electronegativity of the later elements is less than that of the second constituent, so that a positive oxidation number results. For example, in the hexafluorides, sulphur, selenium and tellurium all exhibit an oxidation state of $+$ 6.

The elements of this group react together to some extent. They all react with oxygen to form di- or trioxides. Furthermore, sulphur trioxide can dissolve selenium to give SeSO$_3$ and tellurium to give TeSO$_3$.

STRUCTURES

Polymorphism (or allotropy) is very conspicuous in this Group.

217

Oxygen exists both as the dimer, O_2, and trimer, ozone, O_3 (and, at low temperatures, as O_4). The small mass of the molecules results in these polymorphs being volatile. In the case of the remaining elements, however, the tendency towards chain formation or metallic lattices means that they are all solid at room temperature (*Table 11.2*).

Table 11.2

Element	Structure	Volatility; m.p., °C	Density (g cm^{-3})
Oxygen	diatomic	gas; −220	1·27 (of solid)
Sulphur	puckered 8-ring	solid; 114 (α-)	2·06 (α-)
Selenium	zig-zag chains	solid; 217 (grey)	4·80 (grey)
Tellurium	zig-zag chains	solid; 450 (metal)	6·24 (metal)
Polonium	cubic lattice	solid; 254 (β-)	9·51 (β-)

Oxygen

The diatomic molecule, O_2, is paramagnetic; the orbitals fill with pairs of electrons until the last two are reached when, because of the lower energy content, they remain single and unpaired.

The trimer, *ozone*, on the other hand, has no unpaired electrons and is therefore diamagnetic. Its structure can be regarded as a hybrid of

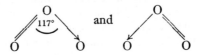

It is prepared endothermically from oxygen by photochemical or electrical stimulus

$$3O_2 \rightarrow 2O_3 \qquad \Delta H = 2 \times 142\,kJ$$

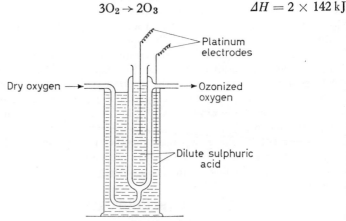

Figure 11.1. Preparation of ozone: Brodie's apparatus

218

Ozone is present in the upper atmosphere, where ultra-violet light is probably the activating agent. Commercially, it is produced by the action of a silent electrical discharge on air or oxygen (*Figure 11.1*). The yield is never higher than 80 per cent and often much lower. Pure ozone is obtained as a dark-blue liquid when ozonized oxygen is liquefied and fractionally distilled.

Ozone is a powerful oxidizing agent, especially in acid solution:

neutral $O_3 + H_2O + 2e \rightarrow O_2 \uparrow + 2OH^-$ $E^\circ = + 1\cdot2$ V

acid $O_3 + 2H^+ + 2e \rightarrow O_2 \uparrow + H_2O$ $E^\circ = + 2\cdot01$ V

As the redox potential for iodine/iodide is only 0·54 V

$$2I^- \rightarrow I_2 + 2e \qquad\qquad E^\circ = - 0\cdot54 \text{ V}$$

it follows that potassium iodide is oxidized to iodine by ozone, even in neutral solution:

$$2I^- + O_3 + H_2O \rightarrow I_2 + O_2 \uparrow + 2OH^- \qquad E^\circ = 1\cdot2 - 0\cdot54 = 0\cdot66 \text{ V}$$

Oxygen is usually evolved when ozone reacts, for example

$$H_2O_2 + O_3 \rightarrow H_2O + 2O_2 \uparrow \qquad PbS + 4O_3 \rightarrow PbSO_4 + 4O_2 \uparrow$$

The oxidizing power of ozone is utilized in bleaching and sterilization. On a more academic level, it is important because of its ability to attack unsaturated carbon–carbon bonds to form ozonides (p. 386). Turpentine contains such unsaturated links and, because it absorbs ozone, is used in the determination of its molecular formula (*Figure 11.2*).

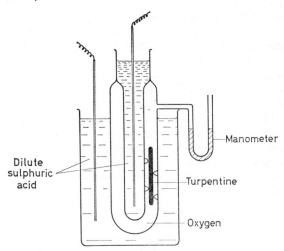

Figure 11.2. Determination of ozone formula

A fixed volume of air is ozonized and the diminution in volume on cooling to the original temperature is observed. The ozone present is then removed by breaking

219

the phial of turpentine. The subsequent contraction as the ozone is absorbed by the turpentine is found to be twice the original contraction accompanying ozonization.

Let the formula of ozone be O_n. Then, on ozonization

$$n\,O_2 \rightarrow 2O_n$$

or n vols. of oxygen give 2 vols. of ozone (by Avogadro's hypothesis). But there is a contraction of 1 volume when ozone is formed

i.e. $n - 2$ vols. $= 1$ vol. or $n = 1 + 2 = 3$

Hence the formula for ozone is O_3.

Sulphur

Sulphur illustrates its power of catenation by forming the S_8 molecule:

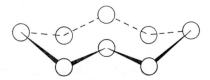

Below 96°, the most stable arrangement of these molecules is in the form of a rhombic crystal (α-S); above 96°, in the form of a monoclinic crystal (β-S). Consequently, the rhombic variety is usually obtained when sulphur crystallizes from a cold solution (for example, in carbon disulphide) and monoclinic crystals result from the solidification of molten sulphur. This type of allotropy, where a definite transition temperature exists between the two allotropes, is called *enantiotropy*. (The equilibrium between the two allotropes and liquid sulphur is represented diagrammatically on page 85.)

Rhombic sulphur melts at 114°, monoclinic sulphur at 120°, to give a honey-coloured liquid consisting of mobile S_8 molecules (S_λ). At about 190°, however, the liquid becomes dark and viscous as the molecules open and form long, tangled chains (S_μ). Sudden cooling of this variety gives plastic sulphur (γ-S). Eventually, at 444°, the liquid boils, giving a vapour consisting initially of S_8 molecules; these dissociate progressively at higher temperatures until, at 2 000°, they are monatomic. (Diatomic sulphur, formed in the course of this heating, resembles oxygen in being paramagnetic.) These changes can be summarized

$$\alpha\text{-S (rhombic)} \underset{< 114°}{\overset{> 114°}{\rightleftharpoons}}$$

> 96° | < 96°

$$\beta\text{-S (monoclinic)} \underset{< 120°}{\overset{> 120°}{\rightleftharpoons}} S_\lambda \text{ (liquid)} \underset{< 190°}{\overset{> 190°}{\rightleftharpoons}} S_\mu \text{ (chains)} \underset{< 444°}{\overset{> 444°}{\rightleftharpoons}} S_8 \text{ (vapour)} \underset{< 2,000°}{\overset{> 2,000°}{\rightleftharpoons}} S$$

quench suddenly

γ-S (plastic)

Selenium, Tellurium, Polonium

There are at least five polymorphs of selenium, two of which, like sulphur, are composed of octatomic molecules and exist as rhombic and monoclinic crystals. The most stable form, however, is the grey 'metallic', which appears to be made up of infinite chains arranged spirally.

At least one crystalline form of *tellurium* is known, a 'metallic' type corresponding to grey selenium.

Two forms of *polonium* have so far been discovered, cubic (α-) and rhombohedral (β-).

OCCURRENCE AND EXTRACTION

Oxygen

There is not sufficient oxygen available to give complete oxidation of all the elements present on earth. It is reasonable to suppose that most of it was initially 'fixed' by reaction with more electropositive elements. Subsequent photochemical decomposition of water vapour and photosynthetic processes of plants have released large quantities of oxygen into the atmosphere, from which it is removed by respiration and other combustion processes, as well as by the weathering of rocks (particularly when the oxygen is dissolved in water). There are, then, two opposing influences tending to maintain a rough balance of atmospheric oxygen. At present, elemental oxygen comprises about 20 per cent of the atmosphere; in combined form, it accounts for about 50 per cent of the lithosphere, chiefly as silicates, whilst, in the form of water, it makes up about 90 per cent of the hydrosphere.

Oxygen is obtained commercially by the fractional distillation of liquid air or by the electrolysis of sodium hydroxide solution, using iron electrodes and a high current density. In the case of the latter method, hydroxide ions are attracted to the anode where they relinquish their valency electrons and decompose into water and oxygen

$$OH^- - e \rightarrow OH \qquad\qquad 4OH \rightarrow 2H_2O + O_2 \uparrow$$

It can be prepared in small quantities in the laboratory either by the electrolysis of alkali solutions or aqueous solutions of electrolytes containing anions more resistant to discharge than the hydroxide ion, for example dilute sulphuric acid

$$H_2SO_4 \rightleftharpoons 2H^+ + SO_4^{2-}$$
$$H_2O \rightleftharpoons H^+ + OH^-$$

Sulphate and hydroxide ions are attracted to the anode and the hydroxide ions are selectively discharged. An inert anode must be used, for otherwise it may react with the nascent oxygen or even dissolve itself to provide the circuit with the electrons that would otherwise be provided by the hydroxide ions.

Oxygen is also liberated by the action of heat on various oxycompounds. *Chlorates*: e.g. potassium chlorate (preferably with manganese dioxide as catalyst)

$$2KClO_3 \rightarrow 2KCl + 3O_2 \uparrow$$

(chlorine and chlorine dioxide are also formed in small quantities)

Nitrates: e.g.

potassium nitrate, $2KNO_3 \rightarrow 2KNO_2 + O_2 \uparrow$

copper nitrate, $2Cu(NO_3)_2 \rightarrow 2CuO + 4NO_2 \uparrow + O_2 \uparrow$

Oxides of noble metals, e.g. mercury (II) oxide, $2HgO \rightarrow 2Hg + O_2 \uparrow$

Peroxides, e.g. barium peroxide, $2BaO_2 \rightarrow 2BaO + O_2 \uparrow$

Peroxides also readily release oxygen on treatment with acidified potassium permanganate

$$2MnO_4^- + 5O_2^{2-} + 16H^+ \rightarrow 2Mn^{2+} + 8H_2O + 5O_2 \uparrow$$

Oxygen is used in large quantities for oxyhydrogen and oxyacetylene welding, and to assist respiration in illness, mountaineering and space and underwater exploration.

Sulphur

Sulphur is found abundantly as sulphides, the high density of which has caused them to sink to low levels in the silicate phase. Where they have come into contact

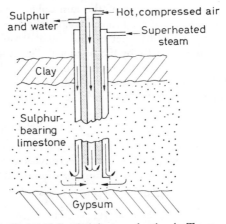

Figure 11.3. Sulphur production in Texas

with water containing dissolved oxygen, oxidation to sulphates has taken place and this has often been followed by leaching out into the oceans of the world. Subsequent disturbance of the earth's crust has in some cases resulted in seas being isolated and sulphates deposited. There is the further possibility of sulphates being reduced by bacterial activity or by carbonaceous matter to elementary sulphur, which is also deposited by volcanic activity. Being a constituent of some proteins, sulphur is found in organic residues such as petroleum and coal.

Large deposits of sulphur occur in Texas and Sicily. In Texas, the sulphur is melted by superheated steam and forced to the surface by compressed air (*Figure 11.3*).

In Sicily, the sulphur-bearing rock is ignited. The heat generated causes the remaining sulphur to melt and run into moulds; it is then purified by distillation. Sulphur can also be extracted from petroleum and from the 'spent oxide' of the gas works (p. 279).

Sulphur finds many uses, for a variety of reasons. Its toxic nature, particularly towards lower organisms, renders it invaluable as a fungicide, either in elementary form or combined in organic compounds. In the vulcanization of rubber, it increases the strength by establishing cross-linkages between adjacent rubber fibres. Its tendency towards exothermic oxidation results in its use as the element in gunpowder and as phosphorus sulphide in matches. Large quantities are also consumed in the manufacture of carbon disulphide and sulphuric acid.

Selenium

Selenium (ionic radius, 200 pm) can replace sulphur (ionic radius, 180 pm) in many sulphide lattices and is often found associated with sulphides. In a concentrated state it is contained in the 'anode mud' obtained during the extraction of copper from copper pyrites. If the anode mud is roasted, selenium dioxide is formed; this volatilizes and the vapour is dissolved in water to give selenious acid, which is then reduced to the element by adding hydrochloric acid and passing sulphur dioxide:

$$\text{Se (impure)} \rightarrow \text{SeO}_2 \xrightarrow{\text{H}_2\text{O}} \text{SeO}_3^{2-} \xrightarrow{\text{SO}_2} \text{Se} \downarrow \text{(pure)}$$

The last reaction is essentially between the selenite and sulphite ions. As the reducing tendency declines with increase in atomic number, it is the sulphite which is oxidized and the selenite which is reduced.

The electrical conductivity of grey selenium is increased in a remarkable manner by exposure to light, and one of its chief uses is in photoelectric cells.

Tellurium

Not much is known of the geochemistry of tellurium, but it is the only element found so far in chemical combination with gold. It is also associated with lead and bismuth ores, but the main commercial source is the anode mud mentioned above. Fusion of this with sodium cyanide converts the sulphur and selenium present into sodium hexacyanosulphate(IV) and hexacyanoselenate(IV), respectively, whereas tellurium is converted into sodium telluride from which tellurium is precipitated by the passage of air through the aqueous solution:

$$\text{Anode mud} \xrightarrow{\text{NaCN}} \text{Na}_2\text{Te} \xrightarrow{\text{air}} \text{Te} \downarrow$$

Tellurium is required for the colouring of glasses and for alloying with lead to improve its tensile strength and resistance to corrosion.

223

Polonium occurs in pitchblende as a product of radioactive decay and can be prepared synthetically by the neutron irradiation of bismuth:

$$^{209}_{83}\text{Bi} + ^{1}_{0}n \rightarrow {}^{210}_{83}\text{Bi} \qquad {}^{210}_{83}\text{Bi} \xrightarrow{-\beta} {}^{210}_{84}\text{Po}$$

HYDRIDES

All the elements of this Group form hydrides of formula H_2X which, with the exception of H_2O, are poisonous, pungent and volatile. The relative lack of volatility of water is the result of association of molecules through hydrogen bonding, a factor which also explains its very low acidity.

	$K = \dfrac{[H^+][HX^-]}{[H_2X]}$	m. pt. °C	b. pt. °C	
H_2O	$\sim 10^{-15}$ M	0	100	
H_2S	$\sim 10^{-7}$ M	-85	-60	Thermal stability decreases
H_2Se	$\sim 10^{-4}$ M	-66	-41	
H_2Te	$\sim 10^{-3}$ M	-51	$+2$	↓

Oxygen also forms a peroxide, H_2O_2, whilst sulphur reveals its ability to form chains by continuing the sequence as far as H_2S_6

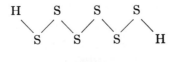

OXIDES

Oxygen combines directly with most elements to give oxides of low energy, so that the reactions are exothermic; e.g.

$$Al_2O_3, \quad \Delta H_f = -1\,590 \text{ kJ mol}^{-1} \qquad SO_2, \quad \Delta H_f = -297 \text{ kJ mol}^{-1}$$

Higher valency states are usually formed, and several elements even give peroxy compounds, e.g. sodium:

$$2Na + O_2 \rightarrow Na_2O_2$$

The ease of combustion depends on the state of subdivision, the partial pressure of oxygen and the energy of activation. Thus some finely divided metals are pyrophoric, i.e. burst into flame on exposure to air, but most elements require heating to effect complete oxidation, and even then actual burning is usually necessary. Otherwise, surface oxidation or tarnishing of the solid is the rule, although the nature of the surface film is important. If the film is very porous, then further oxidation can take place by diffusion of either oxygen or solid through the film, the rate of diffusion (and hence of reaction) being increased with increase in temperature (*Figure 11.4*).

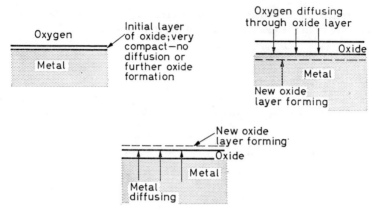

Figure 11.4. Oxide formation

Oxides can also often be prepared by the action of heat on oxycompounds, for example peroxides, carbonates, hydroxides, nitrates and sulphates. Generally, the less electropositive the metal, the greater the ease of decomposition of these compounds:

$$2BaO_2 \rightarrow 2BaO + O_2 \uparrow$$

$$CaCO_3 \rightarrow CaO + CO_2 \uparrow$$

$$Cu(OH)_2 \rightarrow CuO + H_2O$$

$$2Pb(NO_3)_2 \rightarrow 2PbO + 4NO_2 \uparrow + O_2 \uparrow$$

$$2FeSO_4 \rightarrow Fe_2O_3 + SO_2 \uparrow + SO_3 \uparrow$$

Dehydration of weak acids also often gives an oxide. Boron oxide, for instance, is produced by heating boric acid:

$$2H_3BO_3 \rightarrow B_2O_3 + 3H_2O$$

(Many of these so-called acids are, however, possibly no more than hydrated oxides in the first place, so this may not always be a genuine method of preparation.)

Some oxides are precipitated by adding acid or alkali to an appropriate solution. The acidification of a silicate produces silica, whilst alkali converts a solution of a silver salt into a precipitate of silver oxide

$$SiO_3^{2-} + 2H^+ \rightarrow SiO_2 \downarrow + H_2O$$

$$2Ag^+ + 2OH^- \rightarrow Ag_2O \downarrow + H_2O$$

225

Iron(III) oxide is precipitated when an aqueous solution of an iron(III) salt is boiled. This illustrates the preparation of an oxide by hydrolysis:

$$2Fe^{3+} + 3H_2O \rightarrow Fe_2O_3 \downarrow + 6H^+$$

Concentrated nitric acid is a powerful oxidizing agent and often produces an oxide directly from the elements, as with tin and carbon:

$$Sn \xrightarrow{\text{HNO}_3} SnO_2 \qquad\qquad C \xrightarrow{\text{HNO}_3} CO_2$$

Structure of Oxides

Oxides of the more electropositive metals are chiefly ionic and belong to the more common crystal systems, the actual system depending on the ratio of the radius

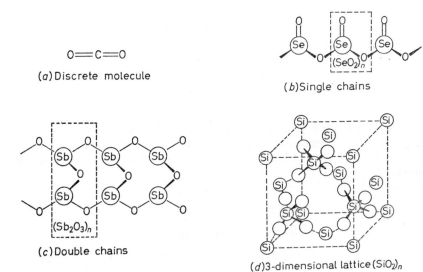

(a) Discrete molecule

(b) Single chains

(c) Double chains

(d) 3-dimensional lattice $(SiO_2)_n$

Figure 11.5. Covalent oxides

of the cation to that of the anion, and upon the empirical formula. For example, oxides of type M_2O_3 have the corundum structure whilst type MO crystallizes either in the rock salt (6:6 coordination) or zinc blende (4:4 coordination) system (see p. 75).

Covalent oxides can exist in the form of

(a) discrete molecules, e.g. carbon dioxide
(b) single chains, e.g. selenium dioxide
(c) double chains, e.g. antimony(III) oxide
(d) 3-dimensional lattices, e.g. β-cristobalite (*Figure 11.5*).

Oxides are usually classified as (*i*) acidic, neutral, basic, amphoteric or (*ii*) simple, higher and sub-oxides. These distinctions are often of degree rather than kind and are not always adequate. It is more rewarding to search for reasons behind differences in behaviour.

Ionic oxides are basic because the oxide ion can react with water to give hydroxide, or with acid to produce water:

$$O^{2-} + H_2O \rightarrow 2OH^- \qquad O^{2-} + 2H^+ \rightarrow H_2O$$

Some metals, with a tendency to complex-ion formation, give reactions with both acids and alkalis and are said to be amphoteric, e.g. zinc oxide

$$ZnO + 2H^+ \rightarrow Zn^{2+} + H_2O \qquad ZnO + 2OH^- + H_2O \rightarrow Zn(OH)_4^{2-}$$

Oxides of non-metals tend to be acidic by virtue of the large electronegativity of oxygen; the inductive effect results in a shift of electrons towards the oxygen, leaving the other element deficient in electrons and able to coordinate with oxide or hydroxide ions:

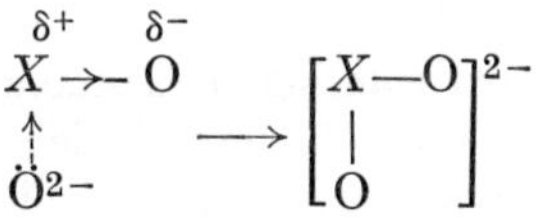

$$\overset{\delta+}{X} \rightarrow \overset{\delta-}{O} \atop \underset{O^{2-}}{\uparrow} \longrightarrow \begin{bmatrix} X-O \\ | \\ O \end{bmatrix}^{2-}$$

Consequently, the more oxygen in the oxide, the more acidic it is, e.g.

CrO	Cr_2O_3	CrO_3
basic	amphoteric	acidic

There is a tendency for oxides of metals and non-metals to react together to form salts, the extent depending on the readiness with which the metallic oxide provides the oxide ion required by the non-metal, e.g.

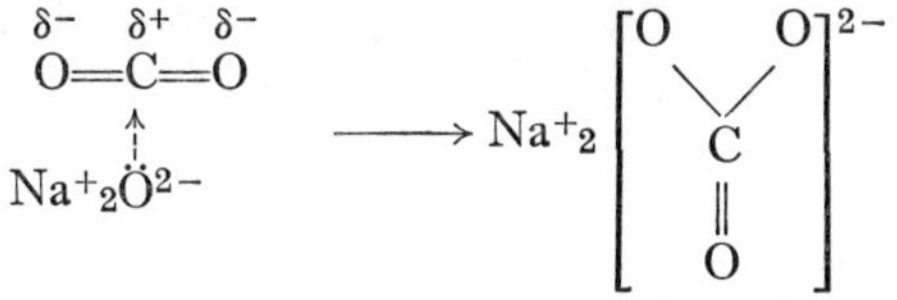

$$\overset{\delta-}{O}=\overset{\delta+}{C}=\overset{\delta-}{O} \atop \underset{Na^+_2 O^{2-}}{\uparrow} \longrightarrow Na^+_2 \begin{bmatrix} O \quad O \\ \diagdown\diagup \\ C \\ \| \\ O \end{bmatrix}^{2-}$$

This point is illustrated by the free energy changes accompanying the reaction between various oxides:

$$\Delta G \atop kJ\ mol^{-1}$$

$K_2O + H_2O$	-193	Increase in acidity of
$K_2O + CO_2$	-352	acidic oxides increases
$K_2O + SO_3$	-659	prospect of reaction
$BaO + CO_2$	-218	Decrease in basicity of
$CaO + CO_2$	-130	basic oxides reduces pros-
$MgO + CO_2$	-67	pect of reaction

PEROXIDES

All true peroxides contain the linkage —O—O— and consequently give hydrogen peroxide when treated with cold acids

$$[-O-O-]^{2-} + 2H^+ \rightarrow H_2O_2$$

Hydrogen peroxide, H_2O_2

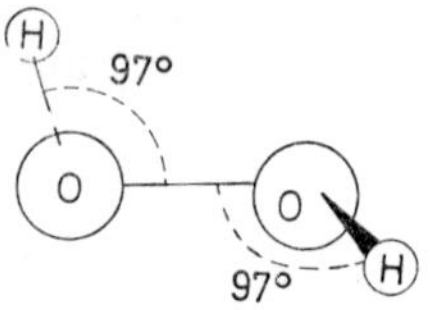

Hydrogen peroxide is required commercially as a propellant (either alone or with substances such as hydrazine) and as a bleaching agent. It is manufactured by two methods:

(*a*) by electrolysis of a cool solution of ammonium hydrogen sulphate, using a platinum anode and a carbon cathode and high current density. Discharge of the hydrogen sulphate ion gives rise to peroxydisulphuric acid

$$HSO_4^- - e \rightarrow HSO_4 \qquad 2HSO_4 \rightarrow H_2S_2O_8$$

This solution is then distilled and hydrogen peroxide formed by hydrolysis

$$S_2O_8^{2-} + 2H_2O \rightarrow 2HSO_4^- + H_2O_2$$

(*b*) by passage of hydrogen into 2-ethyl anthraquinone dissolved in a mixture of benzene and cyclohexanol, in the presence of, for example, nickel as catalyst. Reduction to the quinol takes place. If now a current of oxygen replaces the hydrogen, the quinol is oxidized back to the quinone as the hydrogen is removed from it in the form of hydrogen peroxide:

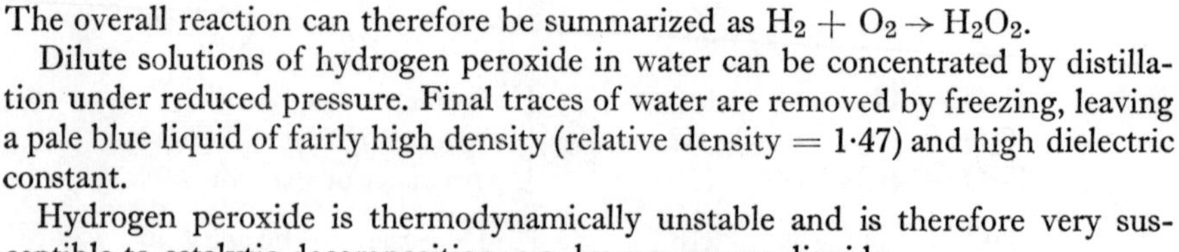

The overall reaction can therefore be summarized as $H_2 + O_2 \rightarrow H_2O_2$.

Dilute solutions of hydrogen peroxide in water can be concentrated by distillation under reduced pressure. Final traces of water are removed by freezing, leaving a pale blue liquid of fairly high density (relative density $= 1\cdot47$) and high dielectric constant.

Hydrogen peroxide is thermodynamically unstable and is therefore very susceptible to catalytic decomposition, e.g. by manganese dioxide

$$H_2O_{2(l)} \rightarrow H_2O_{(l)} + \tfrac{1}{2}O_{2(g)} \qquad \Delta G = -122 \text{ kJ}$$

Hydrogen peroxide is thus a powerful oxidizing agent, especially in acid solution. Its standard redox potential exceeds that for iodine/iodide and for cyanoferrate(III) /cyanoferrate(II) so that both iodides and cyanoferrates(II) are readily oxidized in acid solution

$$2I^- + H_2O_2 \rightarrow 2OH^- + I_2$$
$$2[Fe(CN)_6]^{4-} + H_2O_2 + 2H^+ \rightarrow 2[Fe(CN)_6]^{3-} + 2H_2O$$

In alkaline solution, though, its redox potential is less than that of the latter system and so, under these conditions, cyanoferrate(III) can oxidize hydrogen peroxide

$$2[Fe(CN)_6]^{3-} + H_2O_2 + 2OH^- \rightarrow 2[Fe(CN)_6]^{4-} + 2H_2O + O_2 \uparrow$$

Even in acid solution, the redox potential for the oxidation of hydrogen peroxide is less than that of tetraoxomanganate(VII) (permanganate)/manganese(II), and so acidified potassium permanganate oxidizes it:

$$5H_2O_2 - 10e \rightarrow 10H^+ + 5O_2 \uparrow$$
$$2MnO_4^- + 16H^+ + 10e \rightarrow 2Mn^{2+} + 8H_2O$$

This last reaction is used to estimate the concentrations of hydrogen peroxide solutions

$$2MnO_4^- + 5H_2O_2 + 6H^+ \rightarrow 2Mn^{2+} + 8H_2O + 5O_2 \uparrow$$

that is, one mole (34 g) of hydrogen peroxide requires 2 dm³ of M/5 $KMnO_4$.

Hydrogen peroxide is marketed as 'x volumes' in terms of the reaction

$$2H_2O_2 \rightarrow 2H_2O + O_2 \uparrow$$
$$2 \times 34 \text{ g} \qquad 22\cdot4 \text{ dm}^3$$

so a hydrogen peroxide solution containing 68 g dm^{-3} is a '22·4 volumes' solution.

HYDROXIDES AND OXYACIDS

Hydroxides of the very electropositive metals are formed simply by dissolving their oxides in water. This is essentially a reaction of the oxide ion:

$$O^{2-} + H_2O \rightarrow 2OH^-$$

Those hydroxides that are insoluble in water can be conveniently prepared by precipitation, by treating a solution of a salt of a metal with an alkaline solution, e.g.

$$M^{2+} + 2OH^- \rightarrow M(OH)_2 \downarrow$$

This reaction is of importance in qualitative analysis and is made selective by using ammonium hydroxide as the alkali and suppressing its ionization with ammonium chloride (the common ion effect, p. 158), so that only the very insoluble hydroxides of iron(III), chromium(III) and aluminium are precipitated.

Salts of strong acids and weak bases always hydrolyse to some extent unless prevented by lowering the pH value; in fact, many hydroxides can be regarded as formed from the hydrated cations: e.g.

$$[Al(H_2O)_6]^{3+} \underset{+H^+}{\overset{-H^+}{\rightleftharpoons}} [Al(H_2O)_5(OH)]^{2+} \underset{+H^+}{\overset{-H^+}{\rightleftharpoons}} [Al(H_2O)_4(OH)_2]^+ \rightleftharpoons \text{etc.}$$

Hydroxides of the more electropositive metals are ionic and give hydroxide ions in solution (that is, they are alkalis). As the electronegativity of the element increases, so also does the covalent character; consequently, as the Periodic Table is crossed from left to right, there is a progressive decrease in alkalinity and an increase in acidity:

Period 2

NaOH	Mg(OH)$_2$	Al(OH)$_3$	(HO)$_2$SiO	(HO)$_3$PO	(HO)$_2$SO$_2$	HOClO$_3$
alkaline	basic	amphoteric	weak acid		strong acids→	

Consideration of Fajans' rules leads to the same general conclusions: if a cation of a non-metal were present, it would have such a large polarizing power on account of its small size that the structure would instantly revert to the covalent form. If there *is* any ionization, it is the result of the weakening of the O—H link by the polarising power of the central atom so that hydrogen ions are released:

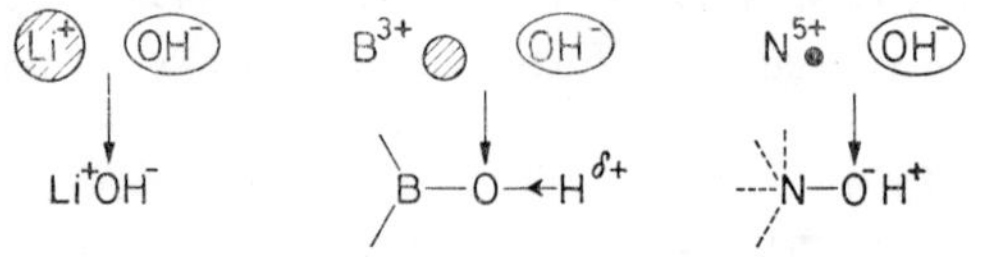

It follows that the acidity of a substance will be increased by an increase in the number of oxygen atoms in the molecule; the effect is particularly marked if some of the oxygen atoms, instead of being attached to hydrogen as well as the element in question, are linked solely to the latter:

	K_1		K_1
B(OH)$_3$	$\sim 4 \times 10^{-10}$M	O$_2$N(OH)	$\sim 10^1$M
Cl(OH)	$\sim 3 \times 10^{-8}$M	O$_2$S(OH)$_2$	$\sim 10^1$M
O.C(OH)$_2$	$\sim 4 \times 10^{-7}$M	O$_3$Cl(OH)	$\sim 10^8$M
O.N(OH)	$\sim 4 \times 10^{-4}$M		
O.S(OH)$_2$	$\sim 2 \times 10^{-2}$M		
O$_2$.I(OH)	$\sim 2 \times 10^{-1}$M		

The more hydroxy groups there are in an acid, the closer are they together and the greater the prospect of elimination

Sometimes, one molecule of water is eliminated from two molecules of acid, and an oxygen 'bridge' results:

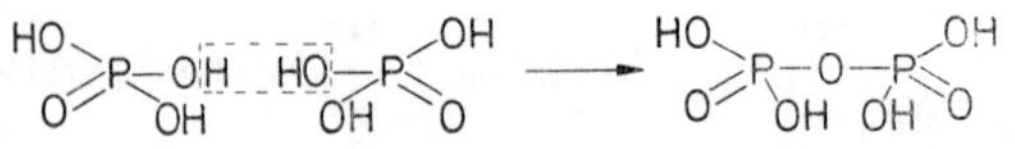

OXY SALTS

The stability of oxy salts is a function of both the polarizability of the oxyanion and the polarizing power of the cation. The cations of Groups 1, for instance, with their low polarizing power, exist in solid hydrogen carbonates, sulphites and nitrites.

Basic salts are very common and, in the case of salts of strong acids, arise through the presence of polarizing cations

In the case of salts of weak oxyanions, hydrolysis leads to a rise in pH and subsequent reaction between liberated hydroxide ions and the cations

$$MX + 2H_2O \rightleftharpoons M^{2+} + H_2X + 2OH^-$$

If the acid produced by hydrolysis is volatile and is lost from the system, the equilibrium shifts in favour of further hydrolysis and basic salt formation. (The situation is often made more complicated by water of crystallization which may shield the anion from polarization.)

It can be seen, then, that there is a wide range of stability of oxy salts towards water. There is similar variation in their stability to heat. The more electropositive the metal and the lower its polarizing power, the more ionic the oxy salts and the greater the resistance to heat although, of course, provided the temperature is high enough, all compounds will decompose.

Nitrates are, on the whole, vulnerable to heat; those of the heavy metals decompose to oxide; e.g.

$$2Zn(NO_3)_2 \rightarrow 2ZnO + 4NO_2 \uparrow + O_2 \uparrow$$

and, in the case of metals of very low electropositivity, even the oxide decomposes

$$Hg_2(NO_3)_2 \rightarrow 2Hg + 2NO_2 \uparrow + O_2 \uparrow$$

Nitrates of the alkali metals decompose only with some difficulty and give chiefly the nitrite

$$2\,NaNO_3 \rightarrow 2NaNO_2 + O_2 \uparrow$$

Carbonates are also generally decomposed by moderate heat, although it is customary to regard the carbonates of the alkali metals as thermally stable. The following decomposition temperatures illustrate the correlation of thermal instability with the extent of covalent character:

	Decomposition temperature, °C
Potassium carbonate	810
Magnesium carbonate	350
Zinc carbonate	300
Mercury(II) carbonate	130

(Significantly, pure aluminium carbonate has not yet even been isolated.)

Sulphates decompose to the oxide only at high temperatures. Sulphur dioxide and trioxide are usually evolved

$$2FeSO_4 \rightarrow Fe_2O_3 + SO_2 \uparrow + SO_3 \uparrow$$

SULPHIDES

Because sulphur is considerably less electronegative than oxygen and the sulphide ion has a very large ionic radius and is easily polarized, most sulphides are covalent. Only those of the alkali and alkaline earth metals are ionized to any extent and, because the sulphide ion is attacked by water, they are readily hydrolysed in aqueous solution:

$$2H_2O + S^{2-} \rightarrow H_2S + 2OH^-$$

(cf. the hydrolysis of ionic oxides).

Most of the other sulphides of metals are insoluble in water and have a metallic lustre (iron pyrite is known as 'fool's gold') and electrical conductivities comparable to those of metals. They are usually prepared by passing hydrogen sulphide through aqueous solutions of their salts. Selective precipitation is achieved to some extent in qualitative analysis by controlling the pH of the solution and therefore the extent of the ionization of the hydrogen sulphide (p. 157).

The very high electronegativity of oxygen produces compounds in which the highest valency states are stabilized. This is not the case with sulphur; for example, whilst Bi_2O_5 and PbO_2 exist, Bi_2S_5 and PbS_2 are unknown.

Hydrogen Sulphide,   H   H
$$\underset{S}{\overset{92°}{\diagdown\diagup}} \quad b.p. = -61°C$$

This colourless gas, of offensive odour, is usually prepared by the action of hydrochloric acid on iron(II) sulphide, but a purer product results from antimony sulphide

$$Sb_2S_3 + 6H^+ \rightarrow 2Sb^{3+} + 3H_2S \uparrow$$

The hydrogen sulphide given off can be washed with water and dried over anhydrous magnesium sulphate. It is a reducing agent and, in the course of reduction, is usually oxidized to sulphur, $H_2S \rightarrow 2H^+ + S \downarrow + 2e$:

$$H_2S + 2Fe^{3+} \rightarrow 2Fe^{2+} + 2H^+ + S \downarrow$$

$$H_2S + Cl_2 \rightarrow 2HCl + S \downarrow$$

$$H_2S + H_2SO_4 \rightarrow 2H_2O + SO_2 \uparrow + S \downarrow$$

The sulphur dioxide in the last example can be further reduced

$$2H_2S + SO_2 \rightarrow 2H_2O + 3S \downarrow$$

Hydrogen sulphide burns in air to form sulphur dioxide and water, provided there is sufficient oxygen to give complete combustion; usually some sulphur is formed as well

$$H_2S \xrightarrow{\;O_2\;} H_2O + S \downarrow + SO_2 \uparrow$$

Hydrogen sulphide is acidic in aqueous solution

$$H_2S \rightleftharpoons H^+_{(aq)} + HS^- \qquad\qquad K_1 = 9 \times 10^{-8}M$$

$$HS^- \rightleftharpoons H^+_{(aq)} + S^{2-} \qquad\qquad K_2 = 1 \times 10^{-15}M$$

The extent of ionization is influenced by the acidity of the solution. In strong acid, the ionization is suppressed by the common ion effect (p. 157) and the sulphide ion concentration is then only sufficient to precipitate the very insoluble sulphides. In alkaline solution, on the other hand, further ionization is encouraged because of the removal of hydrogen ions (to form water), and more soluble sulphides can then be precipitated.

Sulphides of most non-metals exist. As with oxygen, sulphides of the non-metals of Period 2 are present as discrete molecules, e.g. carbon disulphide and sulphur nitride

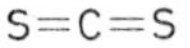

S=C=S
Carbon disulphide

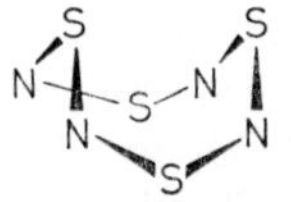

Sulphur nitride

Non-metals of the remaining periods usually form sulphides of macromolecular structure

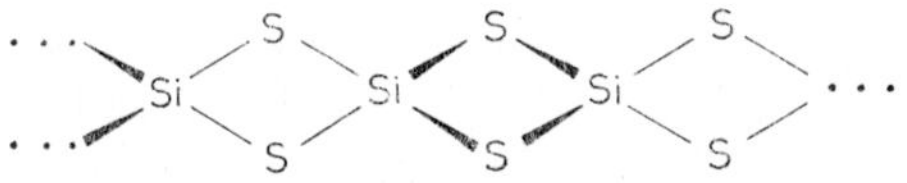

Polysulphides

If sulphur is heated with solutions of sulphides of the alkaline earth and alkali metals, a mixture of polysulphides is formed. Addition of the cold product to hydrochloric acid at 10° or less results in the formation of a yellow oil which on distillation yields hydrides up to H_2S_6. In all of these compounds, sulphur reveals its capacity for catenation:

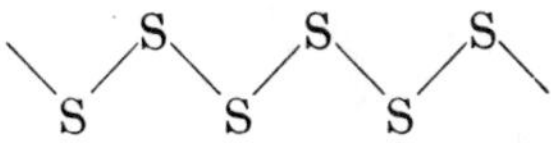

233

SULPHUR DIOXIDE AND SULPHITES

Sulphur Dioxide, SO_2

is obtained as a colourless, pungent gas (b.p. $-10°C$) when sulphur-containing materials are burnt in air. In the case of coal, sulphur dioxide is formed incidentally, and it is the escape of this gas into the atmosphere which plays such a great part in its pollution. Commercially, sulphur dioxide is produced when pyrites and 'spent oxide' are burnt, e.g.

$$4 FeS_2 + 11 O_2 \rightarrow 2 Fe_2O_3 + 8 SO_2 \uparrow$$

Sulphur dioxide is also a by-product of the extraction of metals from sulphide ores. Large quantities result, for example, from the roasting of zinc blende

$$2 ZnS + 3 O_2 \rightarrow 2 ZnO + 2 SO_2 \uparrow$$

The chief uses of sulphur dioxide are in the manufacture of sulphuric acid, as a bleaching agent, disinfectant and preservative and, together with calcium hydroxide (that is, as calcium hydrogen sulphite), for removing lignin from wood and leaving the cellulose in a condition suitable for paper manufacture.

In the laboratory, sulphur dioxide is conveniently prepared by the action of either hot, concentrated sulphuric acid on copper (or of cold dilute acid on sulphites), and collected by downward delivery

$$Cu + 2 H_2SO_4 \rightarrow CuSO_4 + 2 H_2O + SO_2 \uparrow$$

Sulphur dioxide dissolves readily in water to give a solution commonly called 'sulphurous acid' but which is now believed to be made up largely of a clathrate (p. 353), of composition $SO_2.7H_2O$, with sulphur dioxide imprisoned in a water lattice. There is, however, a certain amount of the dibasic acid, H_2SO_3, present as well and the corresponding acid and normal salts (hydrogen or bisulphites, and sulphites) are well characterized.

Sulphites contain the pyramidal ion

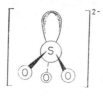

and, because sulphurous acid is not very strong, their aqueous solutions are extensively hydrolysed

$$SO_3^{2-} + 2H_2O \rightleftharpoons H_2SO_3 + 2OH^-$$

Sulphites also react with acids to liberate sulphur dioxide

$$SO_3^{2-} + 2H^+ \rightarrow H_2SO_3 \rightarrow H_2O + SO_2 \uparrow$$

In solution, sulphur dioxide can be both a reducing and an oxidizing agent: when it reduces, it is often oxidized itself to sulphate, and when it is reduced itself by stronger reducing agents, the end product is often sulphur.

Reducing Reactions

$$H_2SO_3 + H_2O - 2e \rightarrow SO_4^{2-} + 4H^+ \qquad E° = -0.20 \text{ V}$$

The electrons provided by this system can be accepted by an oxidizing system such as

$$MnO_4^- + 8H^+ + 5e \rightarrow Mn^{2+} + 4H_2O \qquad E° = 1.52 \text{ V}$$

Combination of the two systems gives

$$2MnO_4^- + 5H_2SO_3 \rightarrow 2Mn^{2+} + 4H^+ + 5SO_4^{2-} + 3H_2O \qquad E° = 1.32 \text{ V}$$

Similarly, sulphur dioxide reduces an acidified solution of potassium dichromate, the colour changing from orange to green

$$\underset{\text{orange}}{Cr_2O_7^{2-}} + 3H_2SO_3 + 2H^+ \rightarrow \underset{\text{green}}{2Cr^{3+}} + 3SO_4^{2-} + 4H_2O$$

The colour change accompanying this reaction is often used as a test for sulphur dioxide.

Oxidizing Reactions

$$H_2SO_3 + 4H^+ + 4e \rightarrow S \downarrow + 3H_2O \qquad E° = 0.45 \text{ V}$$

The system is capable of oxidizing iron to iron(II)

$$Fe - 2e \rightarrow Fe^{2+} \qquad E° = 0.44 \text{ V}$$

that is

$$2Fe + H_2SO_3 + 4H^+ \rightarrow 2Fe^{2+} + S \downarrow + 3H_2O$$

Zinc is also capable of being oxidized by a solution of sulphur dioxide; this time, the sulphite is reduced to dithionate:

$$Zn + 2SO_3^{2-} + 4H^+ \rightarrow Zn^{2+} + S_2O_4^{2-} + 2H_2O$$

As the critical temperature of sulphur dioxide is 158°, it can be liquefied merely by the application of pressure. The liquid formed is of high dielectric constant, and it has been suggested that it may self-ionize slightly:

$$2SO_2 \rightleftharpoons SO^{2+} + SO_3^{2-}$$

(compare with water and liquid ammonia)

Sodium Thiosulphate [sodium trioxothiosulphate(VI)], $Na_2S_2O_3 \cdot 5H_2O$
If sodium sulphite solution is boiled with sulphur, sodium thiosulphate is formed

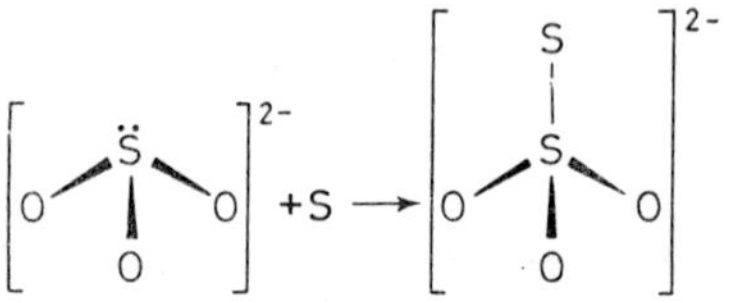

The reaction corresponds to the oxidation of sulphite to sulphate

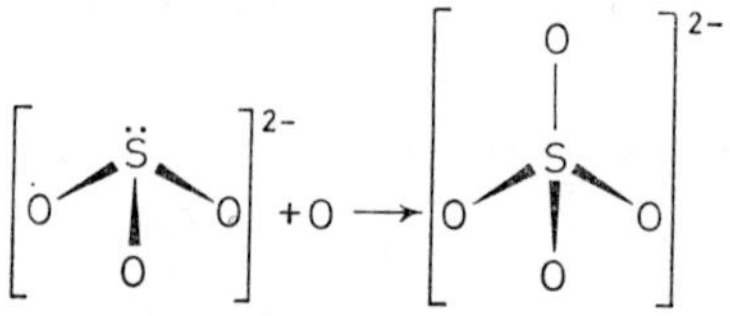

Sodium thiosulphate is very important in the laboratory as a means of reducing iodine in volumetric analysis

$$2S_2O_3^{2-} + \quad I_2 \quad \rightarrow \quad S_4O_6^{2-} \quad + 2I^-$$
$$\text{Tetrathionate}$$

Commercially it is of value (as 'hypo') for the 'fixing' of photographic film: it reacts with the silver ions which have not been exposed to light (and which have therefore not been reduced during development) to form complex ions which can be washed out of the gelatine emulsion on the film, e.g.

$$2S_2O_3^{2-} + Ag^+ \rightarrow [Ag(S_2O_3)_2]^{3-}$$

Thiosulphates react with acids to give sulphur dioxide and a deposit of sulphur; they are therefore readily recognized:

$$S_2O_3^{2-} + 2H^+ \rightarrow H_2O + SO_2 \uparrow + S \downarrow$$

SULPHUR TRIOXIDE

Sulphur trioxide can be made on the small scale by heating sulphates, for example iron(III) sulphate

$$Fe_2(SO_4)_3 \rightarrow Fe_2O_3 + 3SO_3 \uparrow$$

On the large scale, it is manufactured by the *contact process* (*Figure 11.6*). Sulphur dioxide and purified air are passed over a catalyst of vanadium(V) oxide at a temperature of about 450°:

$$2SO_2 + O_2 \rightleftharpoons 2SO_3 \qquad \Delta H = -(2 \times 96)\ kJ$$

Application of Le Chatelier's theorem leads to the conclusion that the forward reaction will be favoured by high pressure and low temperature. In practice, the slight increase in yield resulting from the application of excessive pressure is not thought to warrant the increased capital cost, whilst low temperature means that the equilibrium is reached only very slowly. A compromise temperature of 450° ensures a fairly high reaction velocity and a not-too-unfavourable equilibrium. Excess air is also used, as this brings about a more complete conversion of the sulphur dioxide:

$$K = \frac{[SO_3]^2}{[SO_2]^2[O_2]}$$

Sulphur trioxide is a very hygroscopic solid. It is polymorphic and the more common form melts at 18° and resembles asbestos in appearance.

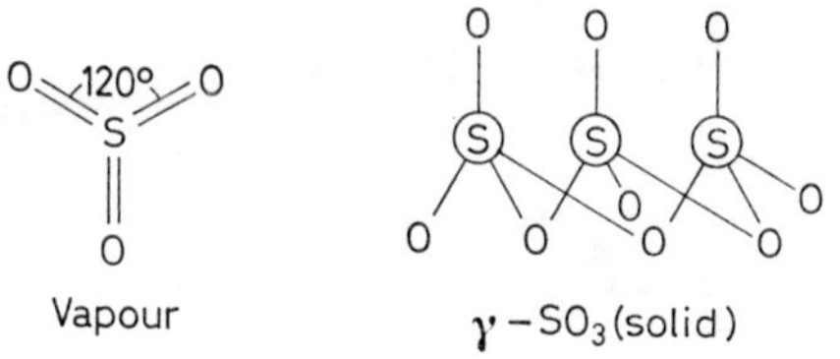

Vapour $\gamma - SO_3$ (solid)

SULPHURIC ACID

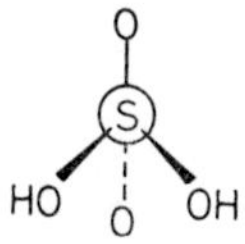

Sulphur trioxide is 'sulphuric anhydride' and when, after being manufactured by the above process, it is passed into water, great heat is evolved and sulphuric acid formed:

$$SO_3 + H_2O \rightarrow H_2SO_4 \qquad \Delta H = -88\ kJ$$

The heat evolved reduces the solubility of the sulphur trioxide and produces an acid mist. Therefore, sulphur trioxide is passed into cold, concentrated sulphuric acid to form disulphuric acid: $H_2SO_4 + SO_3 \rightarrow H_2S_2O_7$. Addition of the requisite amount of water then yields more sulphuric acid

$$H_2S_2O_7 + H_2O \rightarrow 2H_2SO_4$$

237

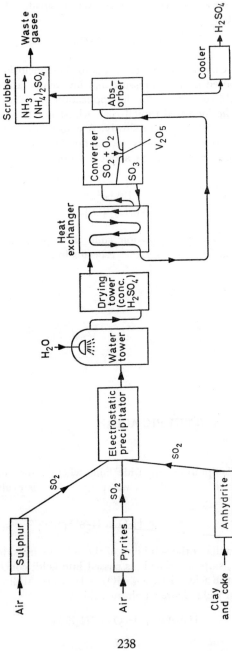

Figure 11.6. Manufacture of sulphuric acid by contact process

The contact process allows an acid of very high purity and concentration to be obtained, particularly if the raw material is sulphur.

Production of Sulphur (1967)		Sources of Sulphuric Acid	
	10^6 tonnes	(U.K. 1968)	per cent
U.S.A.	8·7	Sulphur	58
U.S.S.R.	3·2	Anhydrite	22
Japan	2·2	Pyrites	9
Canada	1·9	Zinc concentrates	6
Mexico	1·7	Spent oxide	5
France	1·5		
Spain	1·1	Uses of Sulphuric Acid	
		(U.K. 1968)	per cent
		Fertilizers	34
		Paints	15
Production of Sulphuric Acid (1962)		Fibres	13
		Detergents	9
U.S.A.	17 555	Plastics	4
U.S.S.R.	6 133	Iron pickling	4
Japan	4 910	Dyestuffs	3
U.K. (1968)	3 282	Manufacture of other acids	3
West Germany	3 101	Petroleum industry	2
France	2 214	Miscellaneous	11

Whilst natural sulphur is easily the major source of sulphur dioxide for the manufacture of sulphuric acid, sulphides such as pyrites and zinc blende are also utilized—sulphur dioxide being produced by roasting the mineral in air:

$$2ZnS + 3O_2 \rightarrow 2ZnO + 2 SO_2$$

In Britain, the *anhydrite process* is important. Anhydrite, $CaSO_4$, clay and coke are finely ground together and heated strongly; the calcium sulphate is reduced by the coke to calcium oxide which combines with silica and alumina to give cement clinker as a by-product. The sulphur dioxide produced simultaneously is then converted to sulphuric acid by the contact process

$$CaSO_4 + C \rightarrow CaO + CO_2 \uparrow + SO_2 \uparrow$$
$$\downarrow$$
$$CaSiO_3, \text{ etc.}$$

Sulphuric acid, when pure, is a colourless, oily liquid of high density and dielectric constant. It undergoes some self-ionization

$$2H_2SO_4 \rightleftharpoons H_3SO_4^+ + HSO_4^-$$

K

The addition of water results in the evolution of considerable heat as oxonium ions are formed

$$H_2SO_4 + H_2O \rightarrow H_3O^+ + HSO_4^-$$

This reaction takes place so readily that the concentrated acid is very hygroscopic and is therefore useful as a drying agent. It can also, by removing the elements of water from compounds, act as a dehydrating agent. For example, glucose can be dehydrated to carbon, and methanoic (formic) acid to carbon monoxide

$$C_6H_{12}O_6 - 6H_2O \rightarrow 6C$$

$$HCOOH - H_2O \rightarrow CO \uparrow$$

When hot and concentrated, sulphuric acid is a strong oxidizing agent. For instance, carbon is oxidized to carbon dioxide and copper to copper(II) ions, the acid in each case being reduced to sulphur dioxide

$$C + 2H_2SO_4 \rightarrow CO_2 \uparrow + 2SO_2 \uparrow + 2H_2O$$

$$Cu + 2H_2SO_4 \rightarrow CuSO_4 + SO_2 \uparrow + 2H_2O$$

Being dibasic, sulphuric acid gives rise to two series of salts, the normal sulphates and the acid hydrogen sulphates (bisulphates).

Most sulphates and hydrogen sulphates are soluble in water and are therefore prepared usually by the neutralization of sulphuric acid by base, followed by crystallization.

Sulphates are, on the whole, fairly stable to heat, especially in the case of the more electropositive metals. Strong heating of sulphates of the transitional metals results in the oxide being formed, e.g. nickel(II) sulphate

$$2NiSO_4 \rightarrow Ni_2O_3 + SO_2 + SO_3$$

The sulphate ion is tetrahedral in shape and often associated with a molecule of water. As the cation is also frequently hydrated, sulphates may contain appreciable water of crystallization, e.g. $FeSO_4 . 7H_2O$.

Isomorphous groups of double sulphates exist. The 'alums' have the general formula $M^I N^{III}(SO_4)_2 . 12H_2O$ where M^I is the ammonium, thallium(I) or alkali metal iron and N^{III} an ion such as aluminium, chromium(III), iron(III) or gallium(III). There is also a series $M_2^I N^{II} (SO_4)_2 . 6H_2O$, typified by ammonium iron(II) sulphate.

One of the few insoluble sulphates is barium sulphate. This is therefore precipitated in the identification and estimation of soluble sulphates

$$Ba^{2+} + SO_4^{2-} \rightarrow BaSO_4 \downarrow$$

Thionic acids, of general formula $H_2S_xO_6$, exist up to $x = 6$ and illustrate further the ability of sulphur to form chains. The sodium salt of tetrathionic acid, $HO.SO_2.S.S.SO_2.OH$, is obtained as a by-product of the reaction between thiosulphate and iodine.

The sulphur cycle

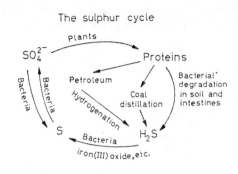

Reactions of sulphur

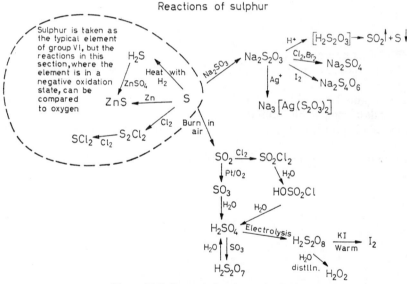

Figure 11.7. General chemistry of sulphur

QUESTIONS

1. Why is sulphur chosen as the typical member of Group **6** rather than oxygen? By comparing the salient features of the elements of this Group, discuss how far this choice is justified.

2. Compare the hydrides of this Group with those of the elements of adjacent groups.

3. Discuss the merits and limitations of the various methods of classification of oxides.

4. In the past, the consumption of sulphuric acid by a country was held to be a reliable yardstick for assessing the standard of living of that country. How far do you think this is true today? Can you think of some more reliable criterion?

5. In the course of this book, the terms Avogadro's hypothesis, Avogadro's law and Avogadro's principle have been used. Discuss their relative merits.

6. Comment on the molar heats of formation of the oxides shown below

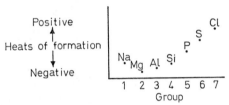

Figure 11.8

7. Write an essay on polymorphism.

8. How would you account for the marked difference in acidity of these two compounds of comparable formula?

Ethanedioic acid $H_2C_2O_4$ $K_a = 6 \times 10^{-3}$ M Sulphuric acid H_2SO_4 $K_a = 10^1$ M

9. Compare the iso-electronic ions, ClO_4^-, SO_4^{2-}, PO_4^{3-} and SiO_4^{4-}.

CHAPTER 12

GROUP 5 ELEMENTS

Group 4	ns	np		Group 6
C	**Nitrogen, N**	2.5		O
Si	**Phosphorus, P**	2.8.5		S
Ge	**Arsenic, As**	2.8.18.5		Se
Sn	**Antimony, Sb**	2.8.18.18.5		Te
Pb	**Bismuth, Bi**	2.8.18.32.18.5		Po

Introduction

A common characteristic of the elements of Group 5 is the existence of five electrons in the outer shell. In the ground state, these consist of two of opposite spin in the

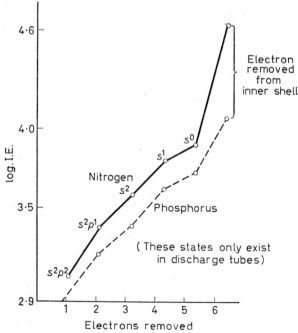

Figure 12.1. Successive ionization energies of nitrogen and phosphorus

243

s orbital and one in each of the *p* orbitals. The energy levels of the outermost electrons of the nitrogen and phosphorus atoms are indicated by the relevant ionization energies shown in *Figure 12.1*. It can be seen that the ease with which an electron can be removed becomes much less with each successive one and that there is a very large increase in the energy required when an electron is removed from the penultimate shell.

An increase in atomic weight and molecular complexity on descending the Group results in nitrogen, with a simple diatomic molecule, being its only gaseous member (cf. Group 6). Another consequence of increasing atomic number is an increase in atomic volume: there is thus less attraction between the nucleus and the outermost electrons and therefore a general decrease in electronegativity in passing from nitrogen to bismuth, particularly between nitrogen and phosphorus. Nitrogen is one of the most electronegative of all elements and can gain three electrons to form the nitride ion, N^{3-}, whereas bismuth is sufficiently electropositive to show metallic properties, sometimes losing three electrons to give Bi^{3+} (*Table 12.1*).

Table 12.1

	Electronic configuration	*Atomic radius* pm	*Ionic radius* (M^{3+}), pm	*m.p.* °C	*b.p.*
N	$2s^2 2p^3$	74		−210	−196
P	$3s^2 3p^3$	110		44 (white)	280
As	$4s^2 4p^3$	120	70	610 (sublimes)	
Sb	$5s^2 5p^3$	140	90	630	1380
Bi	$6s^2 6p^3$	150	120	271	1451

STRUCTURES

Only atoms of small size are capable of forming multiple bonds, and therefore nitrogen is unlike the rest of the Group in that it forms a diatomic molecule, $N \equiv N$, of considerable stability ($\Delta H_f = -921$ kJ mol^{-1}). Doubtless, this fact is chiefly responsible for the general non-reactivity of the element and the comparative instability of many of its compounds, e.g. the oxides.

Phosphorus, arsenic and antimony are all polymorphic, the less dense form in each case being translucent, tetrahedral in molecular shape and soluble in organic solvents, and the denser form opaque, metallic in form, insoluble in organic liquids and an electrical conductor.

As$_4$

'Metallic' arsenic

244

The tetrahedral molecules, having very small inter-bond angles of 60°, are strained and unstable; thus white phosphorus readily ignites in air to form the stable compound, phosphorus(V) oxide, in which the smallest inter-bond angles are 101° (p. 248).

GENERAL CHEMISTRY

Members of this Group have oxidation states ranging from -3 to $+5$, nitrogen showing the greatest versatility in this respect

-3	-2	-1	0	$+1$	$+2$	$+3$	$+4$	$+5$
NH_3	N_2H_4	NH_2OH	N_2	N_2O	NO	N_2O_3	N_2O_4	N_2O_5

Nitrogen, although with little of the facility of its neighbour carbon towards catenation, does form multiple bonds such as

$C=N, \Delta H_f - 616$ kJ
imine

$C\equiv N, \Delta H_f - 891$ kJ
nitrile

$N=N, \Delta H_f - 418$ kJ
diazo

$N\equiv N, \Delta H_f - 920$ kJ

It should be noted that imines and diazo compounds are more readily reduced than nitriles and nitrogen, respectively.

Molecules containing double-bonded nitrogen are angular, and there are possibilities of geometric isomerism (p. 362). For example, azobenzene exists in two forms

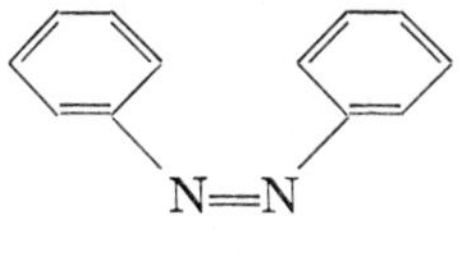

cis 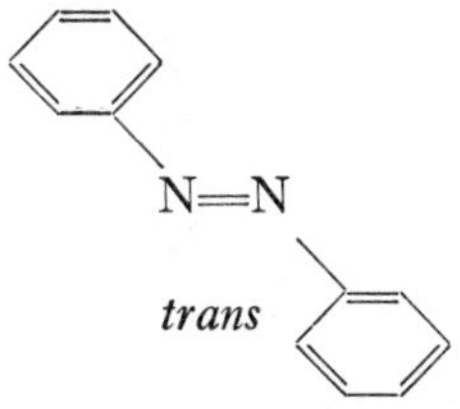   trans

The maximum covalency of nitrogen is four (because of the absence of d orbitals in the second quantum shell) and the resultant 4-covalent molecules, like those for the rest of the Group, are based on a tetrahedral structure. In the other elements, 6-covalent compounds are common and have an octahedral structure, e.g. $Sb(OH)_6^-$ and PCl_6^-

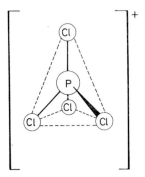

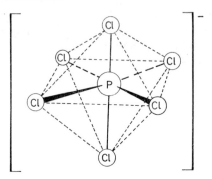

4-covalent phosphorus   6-covalent phosphorus
(tetrahedral)   (octahedral)

The decrease in electronegativity on descending the Group results in the appearance of some cationic chemistry with antimony and bismuth, even though it normally means the antimonyl ion, SbO^+, in the case of the former. Bismuth, on the other hand, gives rise to the bismuth ion, Bi^{3+}, and to a fairly stable series of salts with oxyacids.

The elements of this group, like those of Groups 4 and 6, feature prominently in organic chemistry. Nitrogen is an essential component of protein, and phosphorus, in the form of phosphoric acid, is instrumental in the manufacture of protein in the living cell and in respiration and photosynthesis (p. 476). Nitrogen, phosphorus, arsenic and antimony form compounds of the type MR_3 (where $R =$ alkyl or aryl radical), whereas bismuth participates chiefly in compounds of the type OMR_3.

Hydrides

All the elements form hydrides of the type MH_3, the stability decreasing markedly from nitrogen to bismuth. These are volatile compounds, although ammonia, NH_3, is sufficiently polar to give rise to hydrogen bonding (p. 52), with a consequently lower volatility than would otherwise be expected

	NH_3	PH_3	AsH_3	SbH_3	BiH_3
b.p. (°C)	−34	−87	−55	−18	+20

The effect of the lone pair on the central atom is to give a pyramidal molecule, the angles between the bonds decreasing as the electronegativity of the element decreases. Thus, the H$\widehat{N}$H angle in ammonia is almost tetrahedral, whilst the H$\widehat{P}$H angle in phosphine is only 93°. As nitrogen is much more electronegative than phosphorus, the electron pairs of the covalent bonds are much nearer nitrogen than is the case with phosphorus. There are consequently much stronger mutually repulsive forces between the bonds in ammonia than in phosphine, and so the influence of the lone pair is less pronounced.

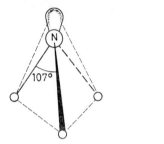 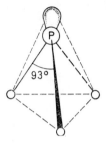

Nitrogen also forms N_2H_4, hydrazine, and N_3H, hydrogen azide, whilst phosphorus forms P_2H_4, diphosphorus tetrahydride.

Halides

Nitrogen forms only trihalides, NX_3. Of these, only NF_3 is stable. NCl_3 is explosive ($2NCl_3 \rightarrow N_2 \uparrow + 3Cl_2 \uparrow; \Delta H = -250$ kJ mol^{-1}), the existence of NBr_3 is doubtful and the tri-iodide is always associated to some extent with ammonia. Trihalides of the remaining elements have been characterized; like the hydrides they are all pyramidal. Although there is an increase in ionic character on descending the Group, they are all hydrolysed to a greater or lesser extent by water

$$PCl_3 + 3H_2O \rightarrow 3HCl + P(OH)_3 \text{ (i.e. } H_3PO_3)$$

$$4AsCl_3 + 6H_2O \rightarrow 12HCl + As_4O_6$$

$$SbCl_3 + H_2O \rightleftharpoons 2HCl + SbOCl$$

$$BiCl_3 + H_2O \rightleftharpoons 2HCl + BiOCl$$

Phosphorus and antimony form penta- as well as tri-halides, the pentachlorides having trigonal bipyramidal structures in the vapour state

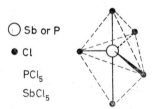

Solid phosphorus pentachloride consists of tetrahedral PCl_4^+ and octahedral PCl_6^- ions (see diagram p. 246) whilst solid phosphorus pentabromide exists as $PBr_4^+Br^-$ These halides are also hydrolysed by water, e.g.

$$PCl_5 + H_2O \rightarrow 2HCl + POCl_3$$

$$POCl_3 + 3H_2O \rightarrow 3HCl + H_3PO_4 \text{ (i.e. } PO(OH)_3)$$

Oxides and Oxyacids

The characteristic oxides have the empirical formula M_2O_3. With the exception of bismuth(III) oxide they can be regarded as acid anhydrides

$$N_2O_3 + H_2O \rightarrow 2HNO_2$$
Nitrous acid
[hydrogen dioxonitrate(III)]

$$P_4O_6 \rightarrow H_3PO_3, H_4P_2O_5, HPO_2$$
Phosphorous acids

$$As_4O_6 \rightarrow X_3AsO_3$$
Arsenites
[trioxoarsenates(III)]

$$Sb_4O_6 \rightarrow XSbO_2$$
Antimonites
[dioxoantimonates(III)]

$$Bi_2O_3 \rightarrow Bi(OH)_3$$
Bismuth hydroxide

N_2O_3 ionizes when in the liquid state

$$N_2O_3 \rightleftharpoons NO^+ + NO_2^-$$

The dimeric molecules of the phosphorus, arsenic and antimony compounds are tetrahedral in shape

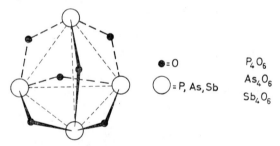

$\bullet = O$

$\bigcirc = P, As, Sb$

P_4O_6
As_4O_6
Sb_4O_6

There is a general increase in basic character on descending the Group, As_4O_6 reacting with concentrated sulphuric acid to give a basic sulphate, whilst Sb_4O_6 yields a normal sulphate under the same conditions and bismuth trioxide even reacts with phosphoric acid.

All give oxides of empirical formula M_2O_5, although in the case of bismuth, the compound has apparently never been prepared in the pure state. With the exception of the latter, these oxides too can be regarded as acid anhydrides, e.g.

$$P_4O_{10} + 6H_2O \rightarrow 4H_3PO_4$$

although the antimony 'acids' are probably no more than hydrated forms of the oxide.

Like the trioxides, the pentoxides of phosphorus, arsenic and antimony are dimeric molecules

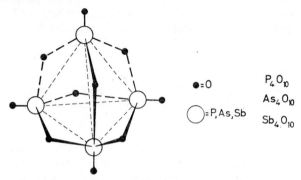

$\bullet = O$

$\bigcirc = P, As, Sb$

P_4O_{10}

As_4O_{10}

Sb_4O_{10}

Both the trioxides and the pentoxides can be obtained by direct combination of the relevant elements under suitable conditions.

Sulphides

The sulphides are formed in an analogous manner to the oxides. As would be expected from simple valency considerations, arsenic, antimony and bismuth give trisulphides, the first two reacting with alkali and sulphides to form trithioarsenate (III) and trithioantimonate(III), respectively, e.g.

$$As_2S_3 + 3Na_2S \rightarrow 2Na_3AsS_3$$

(cf. $As_2O_3 + 3Na_2O \rightarrow 2Na_3 AsO_3$).

In addition, arsenic, like nitrogen, forms a tetrasulphide. Both have the structure

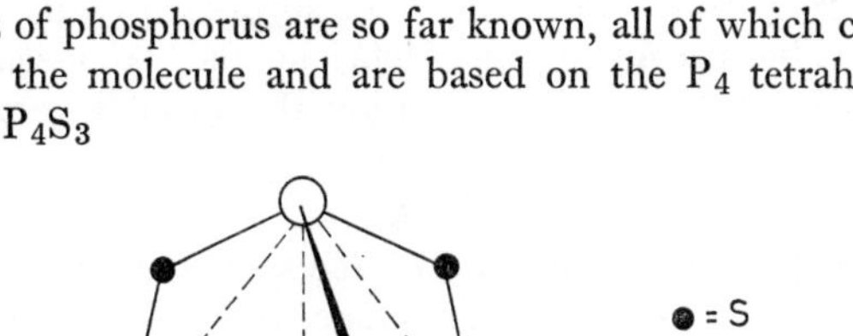

$\bullet = S$

$\bigcirc = N, As$

N_4S_4

As_4S_4

Four sulphides of phosphorus are so far known, all of which contain four phosphorus atoms in the molecule and are based on the P_4 tetrahedron. The most stable of these is P_4S_3

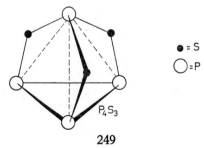

$\bullet = S$

$\bigcirc = P$

P_4S_3

249

Nitrides and Phosphides

Binary compounds of nitrogen and phosphorus with elements of lower electronegativity are known as nitrides and phosphides, respectively, whilst similar compounds with the elements of Groups 6 and 7 are normally classified as chalkogenides or halides.

(*a*) *Nitrides* (cf. carbides)—There are three possibilities:
(*i*) *Covalent* nitrides are prepared by direct combination, by the action of nitrogen on a mixture of an oxide and carbon or of ammonia on a halide, e.g.

$$3SiCl_4 + 4NH_3 \rightarrow Si_3N_4 + 12\ HCl \uparrow$$

Of particular interest is boron nitride, a solid of remarkable stability, existing in two forms having the structures of the isoelectronic graphite and diamond (p. 277). It is manufactured by heating boron oxide and carbon in an atmosphere of nitrogen at 1 400°C

$$B_2O_3 + 3C + N_2 \rightarrow 2BN + 3CO \uparrow$$

(*ii*) *Electrovalent* nitrides contain the nitride ion, N^{3-}. Lithium, sodium, the metals of Group 2, and possibly also potassium, combine directly with nitrogen on heating, to form compounds of high melting point which are readily hydrolysed by water with the evolution of ammonia, e.g.

$$N^{3-} + 3H_2O \rightarrow 3OH^- + NH_3 \uparrow$$

The metals of Group 1, excluding lithium, are also capable of replacing part of the hydrogen of ammonia, on heating in the gas, to give so-called amides, e.g.

$$2K + 2NH_3 \rightarrow 2KNH_2 + H_2 \uparrow$$

(*iii*) *Interstitial* nitrides also exist, in which the small nitrogen atom occupies holes in the metallic lattice: the metallic structure persists in most cases, so that the product is hard, with high melting point and electrical conductivity. For example, vanadium nitride has a m.p. of 2 570°C and a hardness of 9–10 on the Moh scale (a hardness scale ranging from 1 up to 10). Not surprisingly, these compounds are rarely stoichiometric but usually nitrogen-deficient; they are essentially derivatives of the transitional metals, formed by heating the metal in ammonia to above 1 000°C.

(*b*) *Phosphides* can be roughly divided into the same classes as nitrides, although (*i*) there are fewer volatile covalent compounds, complex polymers usually being formed instead; (*ii*) the electrovalent phosphides, although hydrolysed to give phosphine, e.g. $P^{3-} + 3H_2O \rightarrow 3OH^- + PH_3 \uparrow$, are undoubtedly less ionic than the corresponding nitrides; and (*iii*) the interstitial products are better regarded as alloys (since there is less prospect of the larger phosphorus atom being accommodated in the 'holes' of the lattice).

Ammonolysis of phosphorus pentachloride gives an interesting series of fairly inert polymers, the *phosphonitriles*, of empirical formula $PNCl_2$, such as $(PNCl_2)_3$

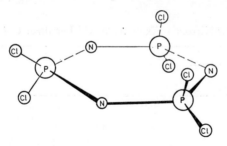

OCCURRENCE AND EXTRACTION

Nitrogen

The extreme stability of the nitrogen molecule, mentioned already in this Chapter, leads one to the expectation that nitrogen compounds would be readily converted to the element. Indeed, molecular nitrogen comprises about 78 per cent (4×10^{15} tons) of the atmosphere and represents the balance of several biological processes. Bacteria play a vital part in the 'nitrogen cycle' (*Figure 12.2*). In the course of their

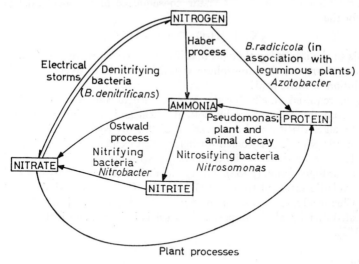

Figure 12.2. The nitrogen cycle

various metabolic processes, some remove nitrogen from the atmosphere whilst others reverse the procedure. The presence of nitrate in the soil is essential for plant growth, as plants require nitrogen in this form for the synthesis of proteins.

Man, having realized this, now manufactures enormous quantities of nitrogenous fertilizers for addition to the soil (*Table 12.2*).

Table 12.2 Nitrogen Fixation: World Fertilizer Consumption

Consumption of nitrogenous Fertilizers 1967	10^6 tons	per cent	Types of Fertilizer	approx. percentage of total
World total	19·8		Ammonium sulphate	38
U.S.A.	5·5	28	'Nitro-chalk'	17
U.S.S.R.	2·7	14	Ammonia	14
			Ammonium nitrate	13
			Calcium nitrate	5
			Calcium cyanamide	5
			Urea	5
			Sodium nitrate	5

Little nitrogen is found in either the hydrosphere or the lithosphere. In the latter, the most important commercial source is Chile saltpetre, 'caliche', $NaNO_3$. The arid climate of Chile is no doubt responsible for the fact that this very soluble material has not been leached out, but there are many theories as to its origin. One of the most interesting is that it is the consequence of the local existence of a catalyst for the reaction

$$\tfrac{1}{2}N_2 + \tfrac{1}{2}H_2O + \tfrac{5}{4}O_2 \rightarrow HNO_3$$

This leads to the intriguing thought that if this, so far undiscovered, catalyst were of general occurrence, the atmosphere would be devoid of oxygen and the sea a dilute solution of nitric acid!

Nitrogen, required chiefly for the manufacture of ammonia (p. 255), can be extracted from the atmosphere by both physical and chemical methods. In the former, the air is dried, freed from carbon dioxide by cooling, and compressed to $200-300 \times 10^5$ N m^{-2}. It is cooled further by heat exchange and suddenly allowed to expand: the work done by the expanding gas results in cooling intense enough to bring about liquefaction (p. 82). The liquid air is then fractionally distilled, when the more volatile nitrogen (b.p. $-195°$) distils off, leaving behind oxygen (b.p. $-182°$). Chemically, oxygen is removed from the air on a small scale by passing it over a heated metal such as copper, and industrially by passing it through red-hot coke (p. 189).

Phosphorus

Phosphorus is found as phosphate in the lithosphere, the most common form being apatite, $CaF_2 . 3Ca_3(PO_4)_2$. It plays a vital part in many processes of the living cell (p. 511) and is a fundamental constituent, in the form of calcium phosphate, of the bones of vertebrates. It can, in fact, be extracted from bones, but it is more usual, on the large scale, to use rock phosphate, which is well mixed with sand

and carbon, and heated to about 1 500°C in an electric furnace: white phosphorus distils over and is solidified under water (*Figure 12.3*).

Extraction of rock phosphate (1966)	10^6 tons	per cent
World	58	
U.S.A.	35	62
Morocco	9	15
Tunisia	3	5

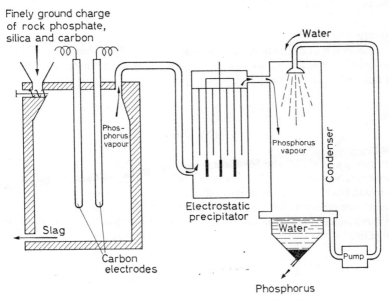

Figure 12.3. Electrothermal extraction of phosphorus

The reaction can be represented by the equations

$$2Ca_3(PO_4)_2 + 6SiO_2 \rightarrow 6CaSiO_3 + P_4O_{10}$$

$$P_4O_{10} + 10C \rightarrow P_4 \uparrow + 10CO \uparrow$$

The silicon dioxide (sand), although only weakly acidic, can displace phosphorus(V) oxide because of the non-volatility of the former compared with the latter at the high temperature of the furnace.

Phosphorus, in the elemental red form, is used in safety matches and in rat poison. As the sulphide, it comprises part of the head of the 'non-safety' match, and in the form of phosphate and 'superphosphate' it is used as a fertilizer.

253

Arsenic

Arsenic is closely associated with sulphide ores; consequently it is found in the flue-dust deposits from the extraction of metals from these ores, particularly of nickel, copper and tin. The arsenic(III) oxide present can be reduced to arsenic by heating with carbon

$$As_4O_6 + 6C \rightarrow As_4 + 6CO \uparrow$$

Arsenic is added in very small quantities to molten lead, especially in the manufacture of lead shot, to increase the tensile strength. Use is made of its toxic character in the preparation of drugs and insecticides.

Antimony

Antimony is found chiefly as the sulphide, stibnite, Sb_2S_3, but it sometimes partly replaces sulphur in galena. Roasting of stibnite gives a sublimate of oxide which is then reduced to the element by heating with carbon

$$Sb_4S_6 + 9O_2 \rightarrow Sb_4O_6 + 6SO_2 \uparrow$$

$$Sb_4O_6 + 6CO \rightarrow 4Sb + 6CO \uparrow$$

Antimony confers greater hardness on Group 4 metals. It is alloyed with lead in the manufacture of accumulators and piping, and with lead and tin in pewter and type metal. [In the latter, an important aspect is the extension of the range of temperature throughout which it is plastic (*Figure 12.4*)]. Antimony has been used for adornment since prehistoric times, and today lead antimonate is a constituent of some paints.

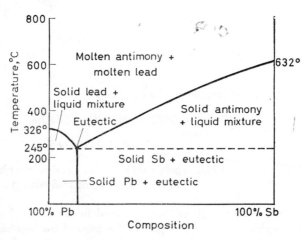

Figure 12.4. Effect of antimony on melting point of lead

254

Bismuth

In the upper lithosphere this element is found chiefly as sulphide or as the trioxide formed by weathering. Galena acts as host for it and the flue dusts from the smelting of galena can be 'worked up' for bismuth: the oxide present is reduced by heating with iron or carbon. Bismuth is particularly useful for making readily fusible alloys; e.g. 'Wood's fusible metal', consisting of 4 parts bismuth, 2 parts lead and 1 part each of tin and cadmium, melts at 71°. Bismuth telluride shows the rare combination of low thermal but high electrical conductivity and is therefore used in thermoelectric materials.

HYDRIDES OF NITROGEN

Ammonia, NH_3, b.p. $= -33\cdot4°$

This colourless gas of characteristic odour is prepared on the small scale by hydrolysing nitrides with acid (sometimes, water) or ammonium compounds with alkali

$$N^{3-} + 4H^+ \rightarrow NH_4^+ \xrightarrow{OH^-} NH_3 \uparrow$$

$$NH_4^+ + OH^- \rightarrow H_2O + NH_3 \uparrow$$

On the large scale, it is made by the commercially more attractive method of direct synthesis

$$N_2 + 3H_2 \rightleftharpoons 2NH_3 \qquad \varDelta H = -46 \text{ kJ mol}^{-1}$$

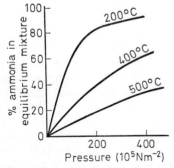

Figure 12.5. Effect of pressure and temperature in ammonia manufacture

Uses of ammonia (U.S.A.)	
	per cent
Fertilizers	76
Explosives	5
Fibres	3
Plastics	3

Application of Le Chatelier's principle (p. 133) leads to the conclusion that the yield of ammonia will be increased by the use of low temperature and high pressure. Unfortunately, reaction velocities are low at low temperature, and therefore equilibrium would be approached only very slowly; in practice, a compromise of about 500°C is used. A very high pressure of 200×10^5 N m^{-2} is also employed, together with a catalyst of iron, the action of which may be promoted by the addition of molybdenum. Under these conditions the yield is about 12 per cent (*Figures 12.5* and *12.6*).

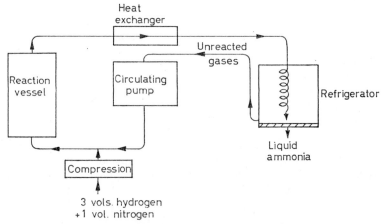

Figure 12.6. Flow sheet for ammonia manufacture

Ammonia is extremely soluble in water, giving a solution often known as ammonium hydroxide but which appears to consist mainly of physically dissolved ammonia

$$NH_3 + H_2O \rightleftharpoons NH_3.H_2O \rightleftharpoons NH_4OH \rightleftharpoons NH_4^+ + OH^- \qquad pK = 4.7$$

The nitrogen of the ammonia molecule is a proton acceptor

$$H^+ + :NH_3 \rightleftharpoons NH_4^+$$

and thus ammonia is basic and gives rise to a series of ammonium salts, of which the sulphate and nitrate are used in large quantities as fertilizers. The overwhelming use of ammonia is in the manufacture of such products, including urea (p. 509). It is also important as a means of manufacturing diamines prior to the synthesis of nylon (p. 512) and as a feedstock in nitric acid manufacture. In the laboratory, ammonia solution is very useful as a source of hydroxide ions in qualitative analysis, being preferred to, e.g. sodium hydroxide because strong heating removes all the ammonium ion and prevents any possible interference at later stages of the analysis. The common ion effect can also be used to control the pH of the solution.

256

The pyramidal molecule of ammonia is capable of inversion, i.e. the nitrogen atom can pass through the base of the pyramid to form an apex at the other side.

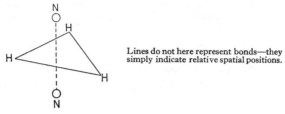

Lines do not here represent bonds—they simply indicate relative spatial positions.

If ammonia is exposed to heterogeneous radio waves, radio energy of a certain fixed frequency is absorbed as the nitrogen atoms oscillate to and fro; this constant frequency can be used as a means of stabilizing the frequency of a microwave oscillator and thus of increasing the accuracy of what is known as the 'atomic clock'.

The fact that ammonia is readily liquefied and has a high heat of vaporization ($1 \cdot 4$ kJ g^{-1} at the boiling point) results in its being used as a refrigerant.

Ammonia will not burn in the air but pure oxygen supports its combustion to nitrogen and water

$$4NH_3 + 3O_2 \rightarrow 2N_2 \uparrow + 6H_2O$$

A platinum catalyst modifies the reaction in favour of the far more valuable nitric oxide (p. 263). Oxidation to nitrogen also takes place if ammonia is passed over the heated oxides of metals such as copper

$$3CuO + 2NH_3 \rightarrow 3Cu + 3H_2O + N_2 \uparrow$$

Liquid ammonia is a useful solvent, having a high dielectric constant and being to some extent ionized (p. 158)

$$2NH_3 \rightleftharpoons NH_4^+ + NH_2^-$$

Ammonia can easily be recognized by its smell and by the fact that it will turn red litmus blue; since all ammonium compounds, on heating with alkali, liberate ammonia, these too can be similarly identified. Many nitrogenous compounds, such as proteins, can be estimated by the methods of Kjeldahl and Dumas. In the former, the nitrogen is converted into ammonium sulphate by reaction with concentrated sulphuric acid

$$\text{nitrogenous compound} \xrightarrow{H_2SO_4} (NH_4)_2SO_4$$

The salt is then treated with excess alkali

$$(NH_4)_2SO_4 + 2OH^- \rightarrow 2NH_3 \uparrow + SO_4^{2-} + 2H_2O$$

and the ammonia generated is absorbed in standard acid. The excess acid is determined volumetrically. In the latter method, the ammonia (produced as

257

before) is passed over heated copper oxide and the volume of nitrogen liberated is measured

$$2NH_3 \longrightarrow N_2$$

$$34 \text{ g} \qquad 22\text{·}4 \text{ dm}^3 \text{ at s.t.p.}$$

The composition of ammonia can be established by reacting excess ammonia with a known volume of chlorine in the apparatus shown (*Figure 12.7*). Nitrogen

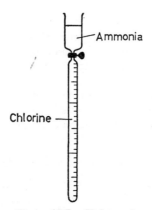

Figure 12.7. Volumetric composition of ammonia

and hydrogen chloride are formed and the latter is absorbed in water. The remaining volume of nitrogen is found to have one-third that of the original chlorine when measured under the same conditions of temperature and pressure, i.e. 3 vols. chlorine → 1 vol. nitrogen. But 3 vols. chlorine combine with 3 vols. hydrogen and therefore 1 vol. nitrogen is associated with 3 vols. hydrogen.

By Avogadro's hypothesis, 1 molecule (2 atoms) of nitrogen is associated with 3 molecules (6 atoms) of hydrogen. Thus the empirical formula of ammonia is N_1H_3, and the molecular formula is $(NH_3)_x$. The vapour density is 8·5 and the molecular weight therefore 17. Thus, $x = 1$, and the empirical formula, NH_3, is also the molecular formula.

Hydrazine,
$$\begin{array}{cc} H & H \\ \diagdown & \diagup \\ N\!-\!N \\ \diagup & \diagdown \\ H & H \end{array}$$
m.p. 2°C, b.p. 113°C

Hydrazine is nowadays in much demand, and great interest centres on the possible methods of manufacture.

Direct synthesis $\qquad$ $N_2 + 2H_2 \rightarrow N_2H_4$ $\qquad$ $\Delta G = + 159 \text{ kJ}$

The large positive free energy change in this reaction means that it is most unlikely to be a feasible method of preparation, especially as the free energy change for the formation of ammonia is negative

$$\tfrac{1}{2}N_2 + \tfrac{3}{2}H_2 \rightarrow NH_3 \qquad \Delta G = - 17 \text{ kJ}$$

This latter reaction is accordingly far more likely to occur.

Oxidation of ammonia

$$2NH_3 + \tfrac{1}{2}O_2 \rightarrow N_2H_4 + H_2O_{(g)} \qquad \Delta G = - 38 \text{ kJ}$$

But

$$2NH_3 + \tfrac{3}{2}O_2 \rightarrow N_2 + 3H_2O_{(g)} \qquad \Delta G = - 653 \text{ kJ}$$

An increase in pressure would improve the chance of the first reaction occurring but nevertheless it must be expected that most of the ammonia would still be oxidized to nitrogen, as in the second equation, because of the considerable decrease in free energy.

Nitrogen (I) reduction of oxide

$$N_2O + 3H_2 \rightarrow N_2H_4 + H_2O_{(g)} \qquad \Delta G = - 172 \text{ kJ}$$

Also $\qquad N_2O + H_2 \rightarrow N_2 \quad + H_2O_{(g)} \qquad \Delta G = - 331 \text{ kJ}$

$$N_2O + 4H_2 \rightarrow 2NH_3 + H_2O_{(g)} \qquad \Delta G = - 364 \text{ kJ}$$

This time, the energy picture is more encouraging; furthermore, increase in pressure would particularly favour the first reaction. Indeed, high yields have been claimed for this method, using an iron catalyst.

Despite the research undertaken on the various possibilities described, the usual method of manufacture remains the oxidation of ammonia with hypochlorite in the presence of a little gelatin or ethylene diamine tetra-acetate, which probably catalyses the relevant reaction and inactivates, by chelation (p. 324), various metallic ions which would otherwise catalyse decomposition reactions

$$OCl^- + NH_3 \rightarrow OH^- + NH_2Cl$$
$$NH_2Cl + OH^- + NH_3 \rightarrow N_2H_4 + Cl^- + H_2O$$

This reaction produces a weakly alkaline solution of hydrazine, mainly in the form of its monohydrate (cf. NH_3), from which the anhydrous compound can be obtained by distillation. It is a diacid base but occurs largely in the form of the monovalent hydrazinium cation when in acid solution:

$$N_2H_4 + H^+ \rightleftharpoons N_2H_5^+ \qquad (\text{cf. } NH_3 + H^+ \rightleftharpoons NH_4^+)$$

Hydrazine is a powerful reducing agent, especially in alkaline solution

$$N_2H_{4(aq)} + 4OH^- \rightarrow N_2 \uparrow + 4H_2O + 4e \qquad E^\circ = + 1 \cdot 16 \text{ V}$$

Under alkaline conditions it reacts vigorously with hydrogen peroxide

$$N_2H_4 + 2H_2O_2 \rightarrow N_2 \uparrow + 4H_2O$$

This reaction, as well as that with oxygen, has been utilized in rocket propulsion and in 'fuel cells'. Such cells convert the chemical energy of a conventional fuel directly into electrical energy rather than into the intermediate forms of heat and

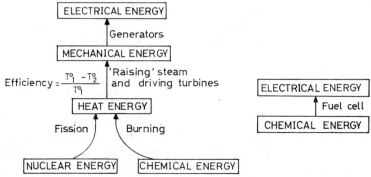

Figure 12.8. Energy transformation in power stations and fuel cells

mechanical energy, when limitations of efficiency, expressed in the Second Law of Thermodynamics, are inevitably present (*Figure 12.8*). In one type of fuel cell, hydrazine hydrate is present in alkaline solution. At the fuel electrode the hydrazine decomposes

$$N_2H_4 \rightarrow N_2 \uparrow + 4H^+ + 4e$$

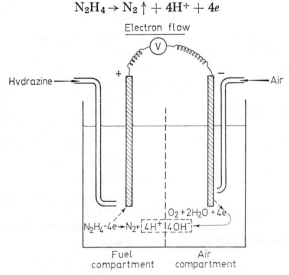

Figure 12.9. Diagram of a fuel cell

The electrons travel externally to the oxygen electrode and form hydroxide ions which react with hydrogen ions from the fuel electrode (*Figure 12.9*)

$$O_2 + 2H_2O + 4e \rightarrow 4OH^- \qquad 4H^+ + 4OH^- \rightarrow 4H_2O$$

The reactions taking place can therefore be summarized as

$$N_2H_4 + O_2 \rightarrow N_2 \uparrow + 2H_2O$$

The affinity of hydrazine for oxygen also results in its being used for removing the latter from boiler water and for stabilizing various oxygen-susceptible liquids such as aniline.

Hydrazine, like its derivatives such as 2,4-dinitrophenylhydrazine, is also of great importance in organic chemistry (p. 460). It condenses with carbonyl compounds to give hydrazones and then azines, e.g.

$$R . CHO + H_2N.NH_2 \rightarrow R.CH:N.NH_2 \rightarrow R.CH:N.N:CH.R$$

and it also reacts with acid chlorides

$$R . COCl + H_2N.NH_2 \rightarrow R.CO.HN.NH_2 + HCl$$

The resultant hydrazides have various applications. For instance, maleic hydrazide is used as a plant-growth regulator and benzenesulphonyl hydrazide as a blowing agent in the manufacture of foam plastics because of the nitrogen gas released on heating.

Hydrogen Azide, HN_3,

$$\begin{array}{c} H \\ \backslash\text{-}111° \\ N \doteq N \doteq N \end{array}$$

b.p. 35·7°C

Hydrogen azide is a colourless liquid of unpleasant smell, prepared by heating sodium in dry ammonia gas and treating the sodamide so formed with nitrogen (I) oxide

$$2Na + 2NH_3 \rightarrow 2NaNH_2 + H_2 \uparrow$$

$$NaNH_2 + N_2O \rightarrow NaN_3 + H_2O$$

Distillation of the product with sulphuric acid gives hydrogen azide

$$N_3^- + H^+ \rightarrow HN_3 \uparrow$$

Derivatives of this acid, the *azides*, are either ionic and fairly stable, e.g. $Na^+N_3^-$, or covalent and explosive, e.g. lead azide, $Pb(N_3)_2$, which is used as a detonator.

Hydroxylamine, NH_2OH,

$$\begin{array}{c} H\text{---}\bigcirc N \text{---} O \diagup^H \\ | \\ H \end{array}$$

m.p. 33°C

Hydroxylamine can be regarded as being derived from ammonia by replacement of one hydrogen atom by a hydroxyl group. It is a weaker base than ammonia, presumably owing to the lone pair of electrons on the nitrogen atom not being so available because of its attraction towards the more electronegative oxygen atom. However, hydroxylamine can react with a proton to give hydroxyammonium compounds, e.g. $[NH_3OH]^+Cl^-$, which, being more stable than the parent base, are the usual sources of the latter. They are normally prepared by cathodic reduction of nitrates or nitrites, e.g.

$$CH_3ONO \rightarrow NH_2OH + CO \uparrow$$

$$HONO_2 + 6H^+ + 6e \rightarrow NH_2OH + 2H_2O$$
Nitric acid

but increasing amounts come from the petrochemicals industry, nitration of hydrocarbons being followed by reduction and distillation with sulphuric acid, e.g.

$$RCH_3 \rightarrow RCH_2NO_2 \rightarrow RCH_2NHOH \rightarrow NH_2OH$$

Hydroxylamine and its compounds are reducing agents, e.g.

$$4Fe^{3+} + 2NH_2OH \rightarrow 4Fe^{2+} + N_2O \uparrow + 4H^+ + H_2O$$

Under certain circumstances, however, they can act as oxidizing agents; for instance, the above reaction takes place in acid solution, but in alkaline solution the base can oxidize iron(II) to iron(III)

$$2Fe^{2+} + NH_2OH + H_2O \rightarrow 2Fe^{3+} + NH_3 \uparrow + 2OH^-$$

In organic chemistry, hydroxylamine is of interest because of its ability to condense with carbonyl compounds to form oximes (p. 460), e.g.

$$R . CHO + NH_2OH \rightarrow R . CH{:}NOH + H_2O$$

Phosphine, PH_3, b.p. $= -87 \cdot 4°C$

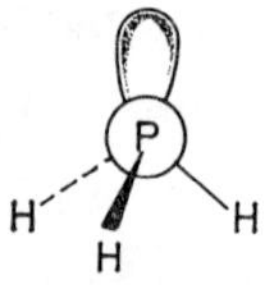

This gas, which has a smell of rotting fish, is normally obtained by the action of strong alkali on white phosphorus

$$P_4 + 3OH^- + 3H_2O \rightarrow PH_3 \uparrow + 3H_2PO_2^-$$

The gas is usually contaminated with hydrogen and with diphosphorus tetrahydride, P_2H_4, formed in side-reactions. A purer product can be made by the action of dilute acid on calcium phosphide: $P^{3-} + 3H^+ \rightarrow PH_3 \uparrow$.

The phosphorus atom, being considerably less electronegative than that of nitrogen, cannot partake in hydrogen bonding. As a result of this, phosphine (in contrast to ammonia) has a very low solubility in water. For the same reason it is also much less basic than ammonia: phosphonium salts have been prepared, but they are rapidly hydrolysed by water.

The P—H bond is very weak, and phosphine is easily oxidized either to phosphorus or one of its oxides or oxyacids.

OXIDES OF NITROGEN

The most common oxides are nitrogen(I) oxide (nitrous oxide, N_2O), nitrogen (II) oxide (nitric oxide, NO) and nitrogen(IV) oxide (nitrogen dioxide, NO_2, which is usually in equilibrium with its dimer, N_2O_4). All these can be obtained by the reduction of nitric acid under differing conditions. Nitrogen(III) and nitrogen(V) oxides are also known but are of little importance.

The great stability of the nitrogen molecule means that these oxides are formed endothermically from their elements and thus dissociate fairly readily on heating. (With nitrous oxide, the dissociation into nitrogen and oxygen is so marked that it can easily be mistaken for oxygen.) Consequently, all three oxides will support combustion to some extent, although nitric oxide, the most stable of them in this respect, does so only if the substance is burning at a temperature of at least $1\,000°$. There is thus the apparent paradox of burning sulphur being extinguished by this gas whilst strongly-burning phosphorus burns even more brightly than in air

$$P_4 + 10NO \rightarrow P_4O_{10} + 5N_2 \uparrow$$

Advantage is taken of their thermal instability in establishing the molecular formulae. A measured volume of the gas confined by mercury is decomposed into its elements by means of an electrically heated iron wire. The oxygen formed is then removed from the gaseous phase by combination with the iron. When the reaction is complete, the volume of the remaining nitrogen (at the original temperature and pressure) is measured. A knowledge of the vapour density is then sufficient to establish the molecular formula (see p. 63).

Nitrogen(I) Oxide, N_2O, b.p. $-88·5°C$

Nitrous oxide is obtained by the reduction of nitric acid with zinc

$$4Zn + 10HNO_3 \rightarrow 4Zn^{2+} + 8NO_3^- + 5H_2O + N_2O \uparrow$$

or by the action of heat on ammonium nitrate

$$NH_4^+NO_3^- \rightarrow N_2O \uparrow + 2H_2O \qquad \Delta H = -21 \text{ kJ}$$

The molecule is linear, $N{=}N{=}O$, and is isoelectronic with that of carbon dioxide.

Nitrogen(II) Oxide, NO, b.p. $-151·7°C$

This colourless, insoluble gas can be conveniently made in small quantities by the action of 50 per cent nitric acid on copper, and collected over water

$$3Cu + 8HNO_3 \rightarrow 3Cu^{2+} + 6NO_3^- + 4H_2O + 2NO \uparrow$$

On the large scale, it is made as an intermediate in the manufacture of nitric acid, by the catalytic oxidation of ammonia (p. 257). Small quantities are formed in the atmosphere by direct synthesis in the vicinity of lightning flashes.

Nitric oxide is paramagnetic, and therefore an unpaired electron is present in the molecule. The structure appears to be intermediate between the two formulations

$$:\dot{N}{=}\dot{\underline{O}}. \quad \text{and} \quad :\overset{+}{N}{=}\bar{\underline{O}}:$$

Dimerization, involving the unpaired electron, takes place increasingly at very low temperatures, especially in the solid.

Nitric oxide is a reactive compound and shows considerable versatility:

(*a*) *loss of one electron*, forming the nitrosyl ion, NO^+. This is isoelectronic with the molecule of carbon monoxide and can react similarly with transitional metals to form nitrosyls, corresponding to carbonyls. An example of this type of reaction is afforded by the formation of the brown complex $[Fe^I(NO^+)(H_2O)_5]^{2+}$ when nitric oxide is passed into a solution of iron(II) ions, as in the 'brown ring' test for a nitrate. Nitrosyl hydrogen sulphate, $NO^+HSO_4^-$, is formed in the course of the manufacture of sulphuric acid by the lead chamber process.

(*b*) *gain of one electron*, to give NO^-. When nitric oxide is passed into a solution of sodium in liquid ammonia, the compound Na^+NO^- is formed.

(*c*) *sharing of electrons*, as for example when nitric oxide and chlorine are passed over charcoal to form nitrosyl chloride

$$2NO + Cl_2 \rightarrow 2Cl{-}N{=}O$$

Reaction takes place instantly when nitric oxide comes into contact with air or oxygen at room temperature, brown fumes of nitrogen dioxide being formed

$$2NO + O_2 \rightarrow 2NO_2 \rightleftharpoons N_2O_4$$

Dinitrogen(IV) Oxide, N_2O_4, b.p. 21·2°C

Commonly called nitrogen dioxide, because there is endothermic dissociation into the paramagnetic monomer, NO_2, at room temperatures and above (*Figure 12.10*), it is prepared in the laboratory by the action of heat on lead nitrate

$$2Pb(NO_3)_2 \rightarrow 2PbO + 4NO_2 \uparrow + O_2 \uparrow$$

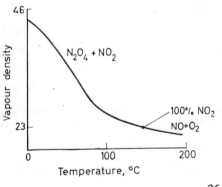

Figure 12.10. Equilibrium between NO_2 and N_2O_4

Cooling of the gas evolved results in the condensation of a diamagnetic liquid, which is green because of traces of blue N_2O_3 with the pale yellow N_2O_4. Present evidence suggests that the structure of the N_2O_4 is chiefly

This liquid boils at 21°C, giving a pale yellow vapour made up almost entirely of N_2O_4 molecules. As the temperature is increased so is dissociation, accompanied by a steady darkening of colour until, at 150°, dissociation is just about complete and the vapour is black. The structure of the monomer which, like nitric oxide, contains an unpaired electron, is perhaps best represented as

$$:\overset{..}{O}\diagdown^{\overset{\centerdot}{N}}\diagup\overset{..}{O}:$$

Decomposition into nitric oxide and oxygen begins to take place if the temperature is increased further

$$N_2O_4 \underset{cool}{\overset{heat}{\rightleftharpoons}} 2NO_2 \underset{cool}{\overset{heat}{\rightleftharpoons}} 2NO + O_2$$

The electrical conductivity of liquid dinitrogen tetroxide is very low, suggesting that any self-ionization is very slight. Dissolution in liquids of high dielectric constant, however, encourages ionization into nitrosyl and nitrate ions, especially if subsequent reaction takes place

$$N_2O_4 \rightleftharpoons NO^+ + NO_3^-$$

(cf. the isoelectronic oxalate ion: $C_2O_4^{2-} \rightarrow CO + CO_3^{2-}$)

Liquid dinitrogen tetroxide will dissolve various metals and salts, the course of reaction indicating self-ionization of the above type. In the case of zinc, the zinc nitrate initially formed gives rise to an interesting complex

$$Zn + 2N_2O_4 \rightarrow Zn(NO_3)_2 + 2NO \uparrow$$

$$Zn(NO_3)_2 + 2N_2O_4 \rightarrow (NO^+)_2[Zn(NO_3)_4]^{2-}$$

This should be compared with the reaction between zinc and sodium hydroxide which yields $Na_2^+[Zn(OH)_4]^{2-}$.

Several previously unknown anhydrous nitrates of the transition metals have also been isolated from systems involving anhydrous dinitrogen tetroxide as solvent.

The compound is also a fairly strong oxidizing agent; for example, it will oxidize hydrogen sulphide to sulphur and carbon monoxide to carbon dioxide

$$NO_2 + H_2S \rightarrow NO \uparrow + H_2O + S \downarrow$$

$$NO_2 + CO \rightarrow NO \uparrow + CO_2 \uparrow$$

OXYACIDS OF NITROGEN

Nitrous Acid, $\begin{array}{c} H \\ \diagdown \\ O-N \\ \diagdown\diagdown \\ O \end{array}$

Nitrous acid is known only in dilute aqueous solution, in which state it behaves as a weak acid ($K \sim 10^{-5}$M). Attempts to concentrate the solution result in its decomposition

$$3HNO_2 \rightleftharpoons HNO_3 + H_2O + 2NO \uparrow$$

The back reaction affords a method of preparing a dilute solution. An equivalent method is the addition of water to a mixture of nitric oxide and dinitrogen tetroxide.

A 'pure' dilute solution of nitrous acid can be obtained by carefully treating barium nitrite with sulphuric acid

$$Ba(NO_2)_2 + H_2SO_4 \rightarrow BaSO_4 \downarrow + 2HNO_2$$

For most practical purposes, however, the presence of extraneous stable ions is no limitation, so that by 'nitrous acid' is often meant the product of the mixing of cold, dilute solutions of sodium nitrite and hydrochloric acid

$$NaNO_2 + HCl \rightarrow HNO_2 + NaCl$$

The nitrites are made by heating the corresponding nitrates, either alone or with a reducing agent such as lead; addition of the latter allows a lower temperature for reaction and hence prevents decomposition of heavy metal nitrites to oxides, e.g.

$$Ba(NO_3)_2 + 2Pb \rightarrow Ba(NO_2)_2 + 2PbO$$

Nitrites of the electropositive metals are ionic and fairly stable. Several covalent nitrites also exist, including organic derivatives.

Nitrous acid can be oxidized to nitric acid and reduced to various oxides and hydrides of nitrogen so that, under different conditions, it will act as a reducing or as an oxidizing agent. For example, in acid solution, iodides are oxidized to iodine

$$2I^- + 4H^+ + 2NO_2^- \rightarrow I_2 + 2H_2O + 2NO \uparrow$$

whereas, in neutral solution, iodine is reduced to iodide

$$I_2 + H_2O + NO_2^- \rightarrow 2H^+ + 2I^- + NO_3^-$$

A very important reaction involving a nitrite is that of 'diazotization', by which dyestuffs can be prepared (p. 414).

It has been said already that nitrous acid is unstable. Evolution of brown fumes of nitrogen dioxide on the addition of dilute mineral acid is indicative of the presence of a nitrite, as is the formation of the characteristic brown complex, $[Fe(NO^+)(H_2O)_5]^{2+}$, upon the addition of iron(II) sulphate solution. These reactions, as well as that of diazotization, can be used in identifying nitrites.

Quantitative estimation can be effected either volumetrically, by oxidation with acidified potassium permanganate,

$$2MnO_4^- + 5NO_2^- + 6H^+ \rightarrow 2Mn^{2+} + 5NO_3^- + 3H_2O$$

or colorimetrically by formation of the red azo dye with 4-aminobenzene sulphonic acid (sulphanilic acid) and 1-aminonaphthalene.

Nitric Acid and Nitrates

Nitric acid is prepared in the laboratory by the action of hot, concentrated sulphuric acid on sodium nitrate, followed by condensation of the vapour evolved

$$NO_3^- + H_2SO_4 \rightarrow HSO_4^- + HNO_3$$

It is manufactured by the catalytic oxidation of ammonia. Dry ammonia, together with excess air, is passed over a platinum catalyst at about 500°C: this causes nitric oxide to be formed

$$4NH_3 + 5O_2 \rightarrow 4NO \uparrow + 6H_2O$$

The reaction mixture is then cooled to favour formation of nitrogen dioxide (see *Figure 12.10*), which is dissolved in water in the presence of air to give nitric acid (*Figure 12.11*)

$$2NO + O_2 \rightarrow 2NO_2$$
$$4NO_2 + 2H_2O + O_2 \rightarrow 4HNO_3$$

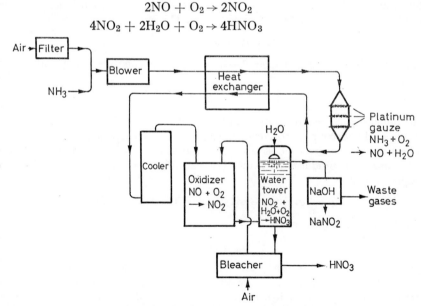

Figure 12.11. Manufacture of nitric acid

267

Nitric acid is required in considerable quantities for the manufacture of fertilizers, e.g. ammonium nitrate, and explosives, e.g. trinitrotoluene. The formation

Production of nitric acid (1966) (excluding Communist bloc)	10^6 metric toness	per cent
World total	15	
U.S.A.	4·8	32
W. Germany	2·8	19
France	2·2	15

of these compounds illustrates various features of the acid: the pure substance, which is a colourless, volatile liquid of boiling point 84°, self-ionizes to some extent

$$2HNO_3 \rightleftharpoons H_2NO_3^+ + NO_3^-$$

One molecule of the acid is here acting as a proton donor and the other as a proton acceptor. Other substances can affect this position: for example, water can assume the role of proton acceptor to give the oxonium ion

$$HNO_3 + H_2O \rightleftharpoons H_3O^+ + NO_3^-$$

The acidity of nitric acid is thus enhanced and becomes the most obvious feature of dilute aqueous solutions, so that, e.g. neutralization of ammonia solution by dilute nitric acid gives ammonium nitrate. Concentrated sulphuric acid, on the other hand, encourages dehydration of the $H_2NO_3^+$ ion

$$H_2NO_3^+ \rightarrow H_2O + NO_2^+$$

Aromatic hydrocarbons and their derivatives are particularly prone to 'nitration' by this system

$$R . H + NO_2^+ \rightarrow R . NO_2 + H^+$$

Thus toluene gives rise to trinitrotoluene (see also p. 389).

Nitric acid is also a powerful oxidizing agent, especially in concentrated solution. On heating alone, it decomposes into water, nitrogen dioxide and oxygen, and with most non-metals produces oxyacids. For example, phosphorus is converted to phosphoric acid

$$P_4 + 20HNO_3 \rightarrow 4H_3PO_4 + 20NO_2 \uparrow + 4H_2O$$

and sulphur to sulphuric acid

$$S + 2HNO_3 \rightarrow H_2SO_4 + 2NO \uparrow$$

This oxidizing capacity is responsible for the almost complete absence of hydrogen in the evolved gases when metals are dissolved in nitric acid, although magnesium gives a reasonable yield of hydrogen with cold, very dilute acid. In the case of metals above hydrogen in the electrochemical series, nascent hydrogen is possibly first formed and then instantly oxidized by the remaining nitric acid, the extent of reduction of the acid depending to a large extent on its concentration. In the case of zinc, nitrous oxide is evolved and might be the result of the sequence

$$4Zn + 8HNO_3 \rightarrow 4Zn(NO_3)_2 + 8H$$

$$8H + 2HNO_3 \rightarrow 5H_2O + N_2O$$

$$4Zn + 10HNO_3 \rightarrow 4Zn(NO_3)_2 + 5H_2O + N_2O \uparrow$$

Formation of hydrogen by reaction with metals below hydrogen in the electrochemical series is highly improbable. Here, initial reaction is probably the formation of the oxide, followed by reaction between this and more acid. Thus, in the case of copper

$$Cu + HNO_3 \rightarrow CuO + HNO_2$$

$$CuO + 2HNO_3 \rightarrow Cu(NO_3)_2 + H_2O$$

$$HNO_2 + HNO_3 \rightarrow 2NO_2 + H_2O$$

$$Cu + 4HNO_3 \rightarrow Cu(NO_3)_2 + 2H_2O + 2NO_2 \uparrow$$

It should be pointed out, however, that there is much conflicting evidence on this subject and the reactions given are certainly oversimplified.

Metal nitrates are very soluble salts, prepared in the usual way by the action of the acid on metal, base or carbonate. They are all unstable to heat, particularly those of the more noble metals, and the course of the reaction depends upon the electronegativity of the metal. Thus, nitrates of the alkali metals decompose only at fairly high temperature to give nitrites, whereas the heavy metal nitrates form the oxides: if the oxide is itself thermally unstable, then the metal is formed as well

$$2NaNO_3 \rightarrow 2NaNO_2 + O_2 \uparrow$$

$$2Pb(NO_3)_2 \rightarrow 2PbO + 4NO_2 \uparrow + O_2 \uparrow$$

$$Hg_2(NO_3)_2 \rightarrow 2Hg + 2NO_2 \uparrow + O_2 \uparrow$$

Ammonium nitrate is unique in that violent shock or heating results in explosion, whilst gentle heating causes it to melt and decompose into nitrous oxide and water

$$NH_4NO_3 \rightarrow N_2O \uparrow + 2H_2O \uparrow$$

Covalent nitrates are less stable than the ionic; like the covalent azides they tend to be explosive, such as glyceryl trinitrate ('nitroglycerine') and 'fluorine nitrate',

NO_3F. The halogen nitrates can, however, be stabilized by pyridine, dative cova-lence from the nitrogen of the latter enhancing the ionic character of the nitrate

$$C_5H_5N: + Cl.NO_3 \longrightarrow C_5H_5N \rightarrow Cl^+NO_3^-$$

Inorganic nitrates can be recognized by means of the 'brown ring test': a solution of the nitrate is treated with iron(II) sulphate solution in dilute sulphuric acid; concentrated sulphuric acid is then carefully introduced to give two liquid phases. A brown ring develops at the junction of the two liquids (cf. test for nitrites, p. 266)

$$NO_3^- + 3Fe^{2+} + 4H^+ \rightarrow 3Fe^{3+} + 2H_2O + NO$$

$$Fe^{2+} + NO \rightarrow [Fe^I(NO^+)]^{2+}$$

This test is invalid in the presence of bromides, since concentrated sulphuric acid oxidizes bromides to bromine, which also gives a brown ring at the interface. In this case, the nitrate is identified by reduction to ammonia by Devarda's alloy (45 per cent Al, 5 per cent Zn, 50 per cent Cu) in the presence of alkali

$$8Al + 5OH^- + 18H_2O + 3NO_3^- \rightarrow 8Al(OH)_4^- + 3NH_3 \uparrow$$

The ammonia evolved can be identified by the usual qualitative methods or ab-sorbed in standard acid if it is desired to estimate the nitrate quantitatively. (If ammonium ions are originally present, it is necessary to boil with excess sodium hydroxide solution until no more ammonia is evolved, before introducing Devarda's alloy.)

PHOSPHORIC ACIDS AND PHOSPHATES

*Ortho*phosphoric acid, H_3PO_4, can be manufactured by digesting rock phosphate with sulphuric acid for several hours

$$Ca_3(PO_4)_2 + 3H_2SO_4 \rightarrow 3CaSO_4 + H_3PO_4$$

When made by this process, it is contaminated with soluble calcium dihydrogen phosphate. (This 'superphosphate', of great importance as a source of soluble phosphorus in fertilizers, can be manufactured by a modification of this procedure.) A purer product is obtained by oxidizing phosphorus to the pentoxide and dissolv-ing this in hot water. In the laboratory, it can be made by oxidizing phosphorus with concentrated nitric acid (p. 268). Evaporation of the resultant solution gives a syrupy liquid, the high viscosity being caused by association through hydrogen bonds (cf. glycerol)

Deliquescent crystals of the compound (m.p. 42°) can be obtained by vacuum desiccation. It is a tribasic acid, and progressive neutralization with sodium hydroxide gives rise in turn to NaH_2PO_4, Na_2HPO_4 (commonly called 'sodium

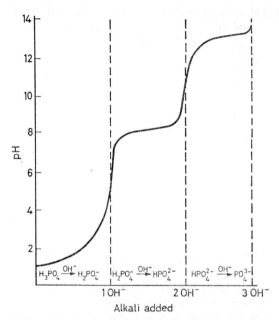

Figure 12.12. Neutralization of phosphoric acid with strong alkali

phosphate') and Na_3PO_4 (*Figure 12.12*). It is a weak acid with the dissociation constants

$$K_1 = \frac{[H^+][H_2PO_4^-]}{[H_3PO_4]} = 7 \times 10^{-3}M \quad K_2 = \frac{[H^+][HPO_4^{2-}]}{[H_2PO_4^-]} = 6 \times 10^{-8}M$$

$$K_3 = \frac{[H^+][PO_4^{3-}]}{[HPO_4^{2-}]} = 5 \times 10^{-13}M$$

so that the salts in aqueous solution are extensively hydrolysed.

Phosphoric acid also forms esters with alcohols, e.g.

$$(HO)_3PO + 3CH_3OH \rightarrow (CH_3O)_3PO + 3H_2O$$
$$\text{Trimethyl phosphate}$$

On heating to about 220°, *ortho*phosphoric acid steadily loses water and forms diphosphoric acid

$$2H_3PO_4 - H_2O \rightarrow H_4P_2O_7$$

271

CUA—L

This colourless, crystalline solid loses more water on heating to about 320°, to form a translucent mass of *meta*phosphoric acid

$$n\text{H}_4\text{P}_2\text{O}_7 - n\text{H}_2\text{O} \rightarrow (\text{HPO}_3)_{2n}$$

All phosphates and phosphoric acids have structures based on the PO_4 tetrahedron, although only *ortho*- and diphosphates give discrete ions

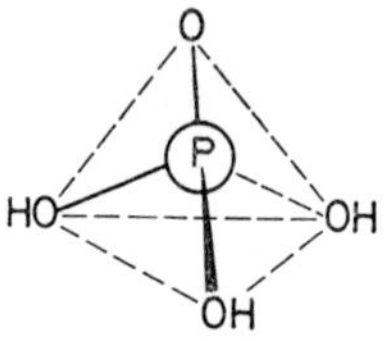

ortho 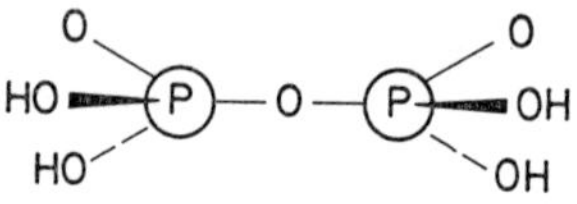   di – (pyro-)

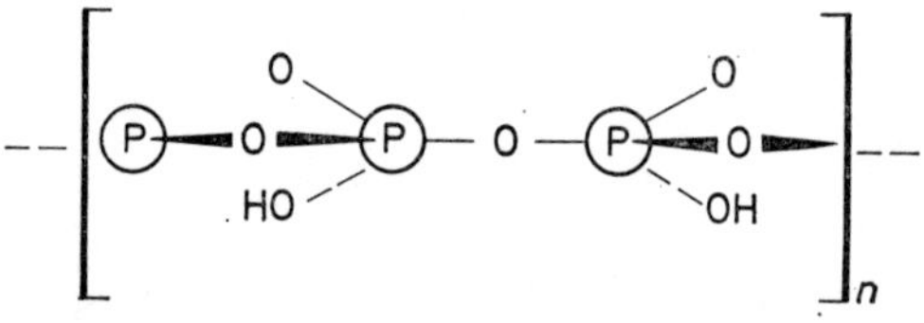

meta –

*Ortho*phosphoric acid is an important constituent of living matter, playing a vital and varied part in many metabolic processes (p. 511).

Phosphates also play a diverse part in industrial life. It is apparent from the occurrence of phosphorus in plant and animal life that it is necessary as a plant fertilizer; it is usually added to the soil in the form of the rather insoluble calcium phosphate (as bone meal) or as the more soluble 'superphosphate'.

*Meta*phosphates of the alkali metals are used in water softening. Calcium and magnesium ions are 'sequestered', i.e. rendered inactive, by reaction with the colloidal metaphosphate polymer, giving a molecule of structure

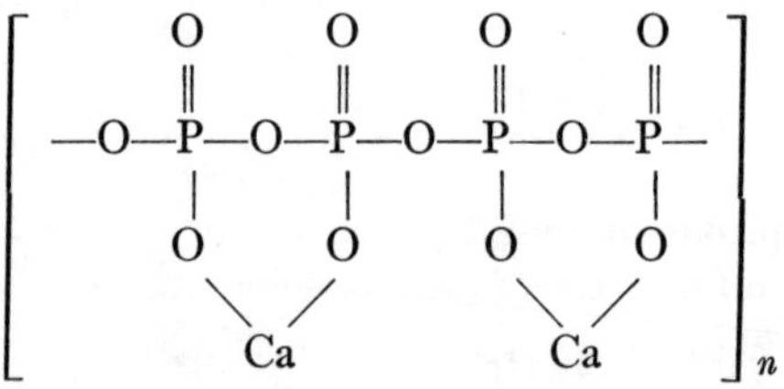

Esters of alcohols and phosphoric acid, such as tributyl phosphate, are often used as plasticizers in the production of rubber and plastic articles, and also for the solvent extraction of metals.

Phosphates can be identified by the slow formation of a yellow precipitate of the complex ammonium phosphomolybdate on mixing a solution of the salt with dilute nitric acid and ammonium molybdate solution. (Arsenates form a similar precipitate but only on warming the mixture.)

Magnesium ammonium phosphate is quantitatively precipitated when an aqueous solution of a magnesium salt is added to an ammoniacal solution of a phosphate; this precipitate can be filtered, washed and ignited to constant weight

$$2MgNH_4PO_4 \rightarrow Mg_2P_2O_7 + 2NH_3 \uparrow + H_2O$$

In this way, phosphates are quantitatively estimated.

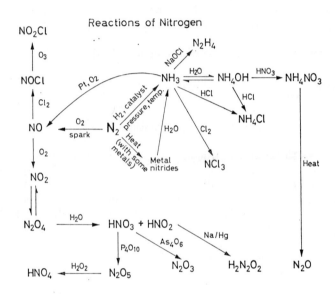

Reactions of Nitrogen

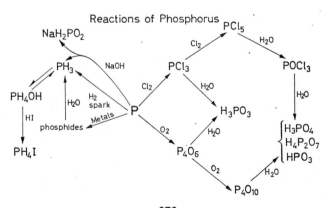

Reactions of Phosphorus

QUESTIONS

1. Suggest why the compound H_3NO_3 is not known, whereas H_3PO_3 is well known.

2. Why do you think phosphine is much less basic than ammonia?

3. What reaction would you expect to occur when bromine is passed into nitrous acid? How could you test your conclusion?

4. Discuss, and as far as possible explain, the changes in the properties of the compounds $X(OH)_3$, where X is N, P, As, Sb, Bi, respectively.

5. How might traces of arsenic in a compound be detected?

6. Hypophosphorous acid, H_3PO_2, behaves as a monobasic acid. What structure does this suggest for the acid?

7. Write a brief account of the polymorphism of phosphorus.

8. Give possible electronic structures for N_4S_4, $(PNCl_2)_3$, P_4S_3.

9. Why are both nitrogen and phosphorus required by plants? Why are plants generally unable to utilize these elements when in their natural condition as N_2 and $Ca_3(PO_4)_2$? How does the chemist make them more freely available?

CHAPTER 13

GROUP 4 ELEMENTS

		ns	np	
Group 3		⊞	⊞	Group 5

B	**Carbon, C**	2.4	N
Al	**Silicon, Si**	2.8.4	P
Ga	**Germanium, Ge**	2.8.18.4	As
In	**Tin, Sn**	2.8.18.18.4	Sb
Tl	**Lead, Pb**	2.8.18.32.18.4	Bi

Introduction

There is *superficially* a closer relationship between the elements of this Group than has been the case hitherto: all the elements are solid at room temperature and all are capable, in one form or another, of conducting electricity. Carbon electrodes are used in dry cells, silicon and germanium in transistors and lead in accumulators. Carbon, in the form of diamond, could be mistaken at a cursory glance for silicon dioxide and various silicates, whilst in the form of graphite it is used in 'lead' pencils.

There is, however, a profound difference between the chemistry of carbon and that of the rest of the group. Carbon is unique in the extent to which it can combine with itself, either with single or multiple bonds. Its chemical versatility is exploited by the living cell, and over a million organic compounds are already known: in this profusion it is surpassed only by hydrogen (which is able to do this only because it is attached to carbon in organic compounds).

The outer shell of all these elements contains four—two s and two p—electrons. The earlier elements are able to hybridize their valency electrons after promotion of an s electron to a p orbital to give a tetrahedral configuration with a valency of four:

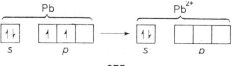

This ability declines, however, as the series is descended: the main valency state of lead is two. This is attributed to an 'inert pair' effect, the two p electrons being lost, leaving behind the s electrons of lower energy

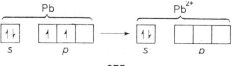

Although tin also gives an ion, Sn^{2+}, it is easily hydrolysed and is a powerful reducing agent, e.g.

$$Hg^{2+} + Sn^{2+} \rightarrow Hg \downarrow + Sn^{4+}$$

The carbon atom readily forms π-bonds with other carbon atoms (p. 50). The fact that its outer shell can hold only eight electrons means that the compounds it forms are quite stable. Silicon not only shows great reluctance to π-bonding, but its outer shell contains available d orbitals, so that its compounds are readily hydrolysed (p. 195). Because of the possibility of expansion of the outer shells of silicon, germanium, tin and lead, complex ions are numerous, e.g. SiF_6^{2-}, SnO_3^{2-}, $Pb(OH)_6^{2-}$.

The progressive increase in atomic volume on passing from carbon to lead, together with an increase in shielding power from the inner shells, produces a fall in electronegativity and a decrease in ionization energy (*Table 13.1*).

Table 13.1

Outer electronic configuration		Atomic radii, pm	Ionic radii, M^{2+} M^{4+} pm		Ionization energies 1st 2nd kJ mol⁻¹	
C	$2s^2 2p^2$	77			1 090	2 360
Si	$3s^2 3p^2$	117		39	788	1 580
Ge	$4s^2 4p^2$	122		53	762	1 540
Sn	$5s^2 5p^2$	141		71	706	1 418
Pb	$6s^2 6p^2$	154	132	84	715	1 458

The bond energies for single bonds between atoms of the same element from carbon down to lead fall sharply, and permit some prediction to be made about the likely reactivity of the elements. The bond energy for the carbon–oxygen bond is only slightly larger than that for carbon–carbon, whereas the silicon–oxygen bond has twice the energy of silicon–silicon. It is to be expected then that hydrocarbons will be far more resistant to oxidation than silanes (especially as the carbon–hydrogen bond is much stronger than the silicon–hydrogen bond), and this is found to be the case in practice. The decline in bond energies from carbon to lead for the M–M bonds also suggests that the ability to catenate will become less pronounced (*Table 13.2*).

Table 13.2

Bond energies, kJ mol⁻¹	
C—C 348	Si—Si 209
C—O 356	Si—O 418
C—H 415	Si—H 315

There is a marked increase in density on descending the Group, in accordance with an increase in metallic character. On the other hand, the volatility increases, due to a weakening of the $M–M$ bond as the Group is descended (*Table 13.3*); both density and volatility are to some extent dependent on the structures assumed by the elements.

Table 13.3

	Density, g cm^{-3}	Melting point °C	Boiling point °C
Carbon	2·2; 3·5	3 575	4 200
Silicon	2·3	1 414	2 300
Germanium	5·4	958	2 700
Tin	5·7; 7·3	232	2 360
Lead	11·3	327	1 755

STRUCTURES

Carbon

Carbon exists in the crystalline condition as both diamond and graphite (*Figure 13.1*). In diamond, the four covalent bonds are directed tetrahedrally, giving a three-dimensional framework of high melting point and hardness.

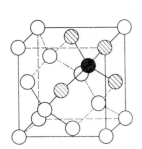

(*a*) Diamond, showing tetrahedral distribution of bonds around each carbon atom

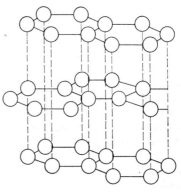

(*b*) The layer structure of graphite

Figure 13.1

Graphite has a layer structure with sheets of carbon atoms held together by van der Waals forces at a distance of 340 pm. It is therefore less dense than diamond and also very soft: application of stress causes the layers to slide over each other, which helps explain its usefulness as a lubricating agent. The C—C distances in the hexagonal sheets are uniform, and it is believed that the remaining p orbitals participate in π-bonding; the increased electron availability results in graphite being a good conductor of heat and electricity. The distance between the carbon layers is large enough to permit the entry of various atoms; for example, alkali metals can be directly absorbed to give substances of increased conductivity, the π-bonds are removed and the product is non-conducting.

Graphite is more stable than diamond at ordinary temperatures and pressures; this type of allotropy (polymorphism) is *monotropy* (p. 84). Conversion of diamond into graphite is a very slow process though. 'Amorphous' carbon, such as lampblack and charcoal, contains minute crystals of graphite.

Silicon and Germanium

The only known forms of silicon and germanium correspond to the diamond structure, with sp^3 hybridized orbitals directed tetrahedrally.

Tin

Tin, like carbon, is polymorphic. Grey (α-)tin has the diamond structure. At $13°$ it is in equilibrium with white (β-)tin (which has a metallic structure), i.e. the allotropy is *enantiotropy* (p. 86).

Lead

As might be expected from the increase in metallic character on descending the Group, lead exists only in the metallic form, with cubic close packing.

OCCURRENCE AND EXTRACTION

Carbon

Carbon occurs in the atmosphere to the extent of about 0·03 per cent as carbon dioxide. This is formed as a result of respiration and other combustion processes and removed by photosynthesis; recent evidence from investigations of the upper atmosphere suggests that the proportion of carbon dioxide is increasing.

In the lithosphere, carbon is found both combined as carbonates and elementally as diamond and graphite. It is a vital constituent of living tissue and is consequently present in their products of decay, for instance as coal and petroleum.

Carbon is extracted in the form of coke by the carbonization of coal. Coal is heated in the absence of air and the volatile material is removed. Coke remains behind in the retorts, whilst condensation and washing of the gases evolved removes coal tar (p. 376), ammonia, naphthalene, etc. (*Figure 13.2*). Hydrogen sulphide is removed from the gas by passing over iron(III) oxide, which itself is converted into sulphide and from which sulphur dioxide can later be obtained by roasting when the oxide has become 'spent'. Any benzole that is present can be dissolved in gas oil which is subsequently distilled. The remaining gas (coal gas) has the approximate composition: hydrogen $= 50\%$, methane $= 32\%$, carbon monoxide $= 8\%$, nitrogen $= 6\%$ and ethene $= 4\%$, and is used as a gaseous fuel although this traditional route has more recently been supplemented by the reforming of hydrocarbons from the petroleum industry (where 'rich' gas is converted to larger quantities of 'lean' gas) and by the increasing supply of natural gas. See (*Figure 13·3*.) Some of the many products obtained from coal are shown in *Figure 13.4* and *Table 13.4*.

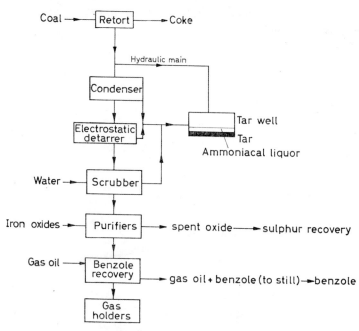

Figure 13.2. Flow sheet for gas works

Low-temperature carbonization gives a less pure product than coke but one which burns more readily (smokeless fuel).

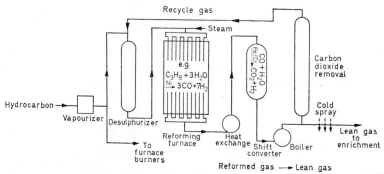

Figure 13.3. *Reforming of hydrocarbons*

279

Silicon

Silicon comprises about a quarter of the Earth's crust in the form of silica and silicates. It is extracted from silica by reduction with carbon in the electric furnace

$$SiO_2 + 2C \rightarrow Si + 2CO \uparrow$$

If silica is mixed with some iron(III) oxide, ferro-silicon, a very important reducing agent, is formed.

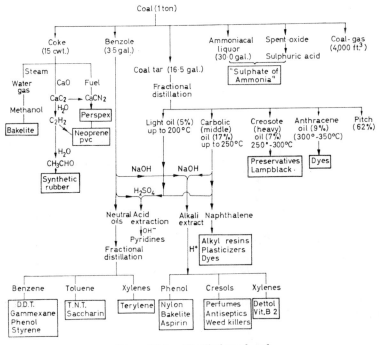

Figure 13.4. Distillation of coal

Germanium is associated with the sulphides of copper, silver, lead, tin and zinc. The element is extracted by reduction of the dioxide with hydrogen or carbon

$$GeS_2 \xrightarrow[H_2SO_4(impure)]{HNO_3} GeO_2 \xrightarrow{HCl} GeCl_4 \xrightarrow{H_2O} GeO_2 \xrightarrow{H_2} Ge$$
$$\text{(pure)}$$

Both silicon and germanium are required in an exceedingly pure form for use as semi-conductors in transistors. In each case, the element is heated with halogen to convert it into the tetrahalide. The silicon tetrahalide is fractionally distilled and converted back to the element by reduction with hydrogen

$$Si \xrightarrow{X_2} SiX_4 \xrightarrow{H_2} Si$$

Table 13.4

Coal production (1966) 10^6 tonnes		Per cent
World total	2 080	
U.S.A.	490	24
U.S.S.R.	440	21
U.K.	180	9
West Germany	130	6
Poland	120	6

About 30 million tons of coal are converted to coke each year in this country. From this, 3 million tons of coal tar are obtained, which yields

	10^3 tonnes		10^3 tonnes
Benzene	120	Phenol	15
Toluene	35	Cresols	70
Xylene	19	Anthracene	3·5
Naphthalene	49	Pyridines	1·5

The germanium tetrahalide is hydrolysed to germanium dioxide which is also reduced to the element with hydrogen. The final stage of purification for both elements involves zone refining (p. 520).

Tin

Tin is found native in small quantities, but its chief source is the dioxide, cassiterite. After the ore has been purified by washing, roasting (to eliminate arsenic and sulphur as volatile oxides) and by removing any iron magnetically, the metal is obtained by smelting with carbon (together with some calcium oxide, which removes silica as slag) in a reverberatory furnace

$$SnO_2 + 2C \rightarrow Sn + 2CO$$

As tin is not attacked to any extent by organic acids, it is used for coating iron in making tin plate for food preservation, even though this material is very susceptible to corrosion. Its low melting point is utilized and enhanced by alloying with lead to make solder. From the phase diagram

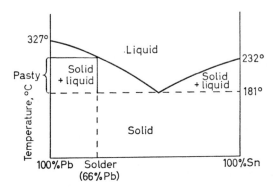

it is seen that not only does tin depress the melting point of lead but gives a range of temperature through which the solder has a desirable 'pasty' quality, with solid and liquid present together.

Table 13.5. Production and Consumption of Tin (1966)

Production			Consumption		
	ore (Sn content) 10^3 tonnes	per cent		10^3 tonnes	per cent
World total (excl. Communist bloc)	159		World total	169	
			U.S.A.	61	36
Malaysia	72	45	U.K.	19	11
U.K.	18	11	Japan	19	11
Thailand	17	11	West Germany	11	7
Netherlands	13	8	France	10	6
Nigeria	10	6			

Table 13.6. Production of Lead (1966)

	Primary metal (10^3 tonnes)	per cent
World total (excl. Communist bloc)	2 140	
U.S.A.	410	19
Australia	196	9
Canada	168	8
Mexico	164	8
West Germany	110	5

Lead

Although lead occurs as sulphate, carbonate and chromate, its main ore is galena, PbS. This is roasted in air to convert it partially to oxide and sulphate

$$2PbS + 3O_2 \rightarrow 2PbO + 2SO_2 \qquad PbS + 2O_2 \rightarrow PbSO_4$$

The oxidizing atmosphere is then replaced by one of reduction by cutting off the air supply; more lead sulphide or carbon (and calcium oxide if the ore is contaminated with silica) is added and the temperature is raised. The following reactions take place:

$$2PbO + PbS \rightarrow 3Pb + SO_2 \uparrow$$

$$PbSO_4 + PbS \rightarrow 2Pb + 2SO_2 \uparrow$$

$$PbO + C \rightarrow Pb + CO \uparrow$$

If the quantity of silver in the lead so obtained warrants it, it is then 'desilvered' (p. 99). The product may be finally purified by electrolysis, with the impure lead as anode of the cell and lead(II) hexafluorosilicate as electrolyte.

Its softness, low melting point and resistance to corrosion make lead very popular with the plumber and it is used widely in piping and roofing. It is also used for lining tea chests and lead chambers. The invention of the internal combustion engine has had a profound effect on the demand for lead; it forms the electrodes of the accumulator and is converted into lead tetraethyl for improving the burning qualities of petrol. The element is a very efficient absorber of radiation and is used in x-ray shields and as a container for radioactive isotopes. Its oxides and carbonate find application in paints, e.g. as red and white lead.

REACTIONS OF THE GROUP 4 ELEMENTS

Tin and lead reveal their metallic character by tarnishing to some extent in air at ordinary temperatures. If all the elements of Group 4 are heated sufficiently strongly in air or oxygen, they burn to give various oxides.

Lead also reacts with water at room temperature, especially if the water is soft and oxygenated: under these conditions lead(II) hydroxide is formed, and as this is fairly soluble, the reaction continues. However, in hard waters, the presence of carbonate or sulphate ions produces protective coatings of carbonate or sulphate. Carbon and silicon will react on heating strongly in steam

$$C + H_2O \rightarrow CO \uparrow + H_2 \uparrow$$

$$Si + 2H_2O \rightarrow SiO_2 + 2H_2 \uparrow$$

Concentrated nitric acid attacks all the elements except silicon; in the cases of carbon, germanium and tin, the dioxide is formed, e.g.

$$Sn + 4HNO_3 \rightarrow SnO_2 + 2H_2O + 4NO_2 \uparrow$$

Lead reveals its more metallic nature by forming the nitrate

$$3Pb + 8HNO_3 \rightarrow 3Pb(NO_3)_2 + 4H_2O + 2NO \uparrow$$

Neither tin nor lead are as reactive to acids as their electrode potentials might indicate, probably owing to overpotential and surface protection; the only acid to which silicon is vulnerable is hydrofluoric

$$Si + 6HF \rightarrow H_2SiF_6 + 2H_2 \uparrow$$
$$\text{hexafluorosilicic acid}$$

Fused sodium hydroxide attacks all except carbon to give oxyanions, e.g.

$$Si + 2NaOH + H_2O \rightarrow Na_2SiO_3 + 2H_2 \uparrow$$

HYDRIDES

Carbon is exceptional in the way in which it can form compounds with hydrogen; it gives straight-chain and cyclic compounds in which there may be single bonds and localized or non-localized π orbitals (p. 49). The profusion of hydrides of the remaining elements falls off as the Group is descended, and the only type so far prepared with certainty are straight-chain compounds corresponding to the alkanes (p. 354). In the case of silicon and germanium, compounds up to M_6H_{14} have been obtained by treating magnesium silicide or germanide with dilute hydrochloric acid, e.g.

$$Mg_2Si + H^+ \rightarrow Mg^{2+} + SiH_4 \uparrow, \text{ etc.}$$

The lower members of the *silane* and *germane* series are gases at room temperature; as with hydrocarbons, their volatility decreases with increase in molecular weight. Where they differ sharply from alkanes, though, is in their ease of hydrolysis by water or alkali, particularly with silanes: this reactivity can be ascribed to the availability of d orbitals able to accept electrons from the oxygen atom (p. 195).

The silanes react with hydrogen halides to give halogen derivatives, e.g.

$$SiH_4 + HCl \rightarrow SiH_3Cl \uparrow + H_2 \uparrow$$

and they are also powerful reducing agents, converting hydroxide ions to hydrogen

$$Si_2H_6 + 8OH^- \rightarrow 2SiO_4^{4-} + 7H_2 \uparrow$$

Only one hydride of tin, stannane, SnH_4, has so far been identified, as the result

284

of treating tin(IV) chloride with lithium tetrahydridoaluminate in ether at low temperature. This hydride is unstable at room temperature, although it is not readily hydrolysed.

Plumbane, PbH_4, is believed to be formed transiently by the action of dilute acid on magnesium–lead alloys.

HALIDES

All the possible tetrahalides are known except $PbBr_4$ and PbI_4, the bond energies of these presumably being too small to allow of their existence.

As with the hydrides, the halides of carbon are on the whole comparatively inert and, in marked contrast with the rest, resist hydrolysis. (The tetrachloride is prepared commercially by chlorinating carbon disulphide in the presence of an iron catalyst, $CS_2 + 3Cl_2 \rightarrow CCl_4 \uparrow + S_2Cl_2 \uparrow$.)

Catenation of carbon atoms is also much in evidence. Polytetrafluoroethylene (PTFE), for example, is made up of macromolecules with the repeating unit

$$\left[\begin{array}{c} \overset{\displaystyle F}{\underset{\displaystyle |}{|}} \quad \overset{\displaystyle F}{\underset{\displaystyle |}{|}} \\ -C-C- \\ \underset{\displaystyle F}{|} \quad \underset{\displaystyle F}{|} \end{array} \right]$$

All the tetrahalides of the remaining elements are very susceptible to hydrolysis and are therefore made by reactions under anhydrous conditions, usually by direct combination of the two elements.

Both carbon and silicon form oxohalides; characteristically, carbon gives discrete molecules, e.g. $COCl_2$, carbonyl chloride, obtained by direct union of carbon monoxide and chlorine, whilst silicon gives more complex molecules with union through Si—O bonds; for instance, treatment of silicon halides with moist ether gives compounds containing the structure

$$\left[\begin{array}{c} X \quad\quad X \\ | \quad\quad\quad | \\ -Si-O-Si- \\ | \quad\quad\quad | \\ X \quad\quad X \end{array} \right]_n$$

Germanium, tin and lead form dihalides with some ionic character, as befits the increase in metallic nature, although methods of preparation differ greatly: germanium dihalides can be made by disproportionation reactions (i.e. by heating germanium(IV) halides with germanium) whilst, because they are fairly insoluble, lead dihalides are obtained by double decomposition

$$GeCl_4 + Ge \rightleftharpoons 2GeCl_2$$

$$Pb^{2+} + 2Cl^- \rightarrow PbCl_2 \downarrow$$

285

OXIDES

All the elements give monoxides and dioxides. There is great variation in the stability of these compounds, and only carbon oxides are present as discrete molecules corresponding to the empirical formulae.

Carbon monoxide has a structure intermediate between the two forms

$$\overset{\times}{\underset{\times}{\times}}C \overset{\times}{\underset{\times}{}} O \colon \quad (C\equiv O) \quad \text{and} \quad \overset{\times}{\times}C \overset{\times\,\bullet\bullet}{\underset{\bullet\bullet}{\colon}}O \quad (C=O)$$

Although it is obtained by dehydration of methanoic (formic) acid, it cannot be regarded as the true anhydride of this acid, as the reverse reaction does not occur, presumably because the electronic structures of the two compounds are not compatible

$$\underset{\overset{\|}{O}}{H-C-OH} \xrightarrow[\text{H}_2\text{SO}_4]{\text{conc.}} CO$$

Carbon monoxide is formed when carbon dioxide is reduced by carbon at high temperature, and so it tends to predominate when carbon burns in a limited amount of oxygen

$$C + O_2 \rightarrow CO_2 \uparrow \qquad\qquad CO_2 + C \rightarrow 2CO \uparrow$$

It is also obtained on the large scale by passing steam over white-hot coke: the mixture of hydrogen and carbon monoxide so formed is a useful fuel and is often used unchanged as 'water gas'

$$C + H_2O \rightarrow CO \uparrow + H_2 \uparrow \qquad\qquad \Delta H = +126 \text{ kJ}$$

Carbon monoxide is a colourless, odourless gas which burns in air with a characteristic blue flame. It is extremely poisonous, as it combines with haemoglobin in the blood to form a compound more stable than oxyhaemoglobin, so that the normal process of respiration is impaired.

It is a powerful reducing agent and much of its large-scale use can be attributed to this, e.g. the reduction of iron ores

$$Fe_2O_3 + 3CO \rightarrow 2Fe + 3CO_2 \uparrow$$

It combines with many non-metals; for example, with sulphur it gives carbonyl sulphide, COS, and with chlorine, in the presence of light, carbonyl chloride (phosgene) is formed. With hydrogen, in the presence of zinc oxide and chromium (III) oxide, the multiple bond is saturated and methanol formed

$$CO + 2H_2 \rightarrow CH_3OH$$

Under different conditions, the carbon monoxide can be reduced completely to methane

$$CO + 3H_2 \xrightarrow{\text{Ni}} CH_4 \uparrow + H_2O$$

286

With certain of the transitional metals, carbon monoxide serves as a *ligand* and forms carbonyls which are covalent and volatile. For example, when it is passed over heated nickel, nickel carbonyl, $Ni(CO)_4$, is formed by coordinating to the metallic atom

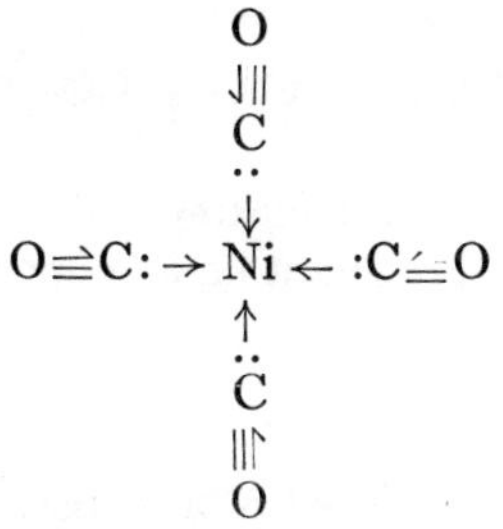

$$
\begin{array}{c}
O \\
\downarrow\!\parallel \\
C \\
\cdots \\
\downarrow \\
O\equiv C:\ \rightarrow\ \overset{\uparrow}{Ni}\ \leftarrow\ :C\!\equiv\!O \\
\overset{\cdots}{C} \\
\parallel \\
O
\end{array}
$$

This can be removed as vapour (leaving impurities behind) and decomposed to the pure metal at a higher temperature

$$Ni + 4CO \xrightarrow{90°} Ni(CO)_4 \xrightarrow{180°} Ni \downarrow + 4CO \uparrow$$

When carbon monoxide is passed under pressure into fused sodium hydroxide, it reacts to form sodium methanoate (formate)

$$Na^+OH^- + CO \rightarrow H.COO^-Na^+$$

The gas is insoluble in water but it dissolves in ammoniacal copper(I) chloride to form complexes with the copper, e.g.

$$Cu(NH_3)_2Cl \xrightarrow{CO} Cu(CO)Cl$$

Carbon Dioxide, $O\!=\!C\!=\!O$

Carbon dioxide is the true anhydride of carbonic acid

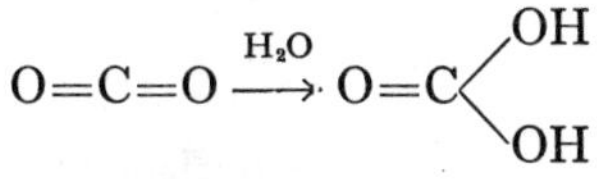

$$O\!=\!C\!=\!O \xrightarrow{H_2O} O\!=\!C\!\!\begin{array}{c} \diagup OH \\ \diagdown OH \end{array}$$

Because the equilibrium lies well over to the left, carbon dioxide is evolved when carbonates are treated with acids stronger than carbonic acid

$$CO_3^{2-} + 2H^+ \rightarrow H_2CO_3 \rightarrow CO_2 \uparrow + H_2O$$

It also results from heating hydrogen carbonates and most carbonates

$$HCO_3^- \rightarrow CO_2 \uparrow + OH^- \qquad\qquad CO_3^{2-} \rightarrow CO_2 \uparrow + O^{2-}$$

The colourless, heavy gas can be dried by passing through concentrated sulphuric acid and collected by upward displacement of air.

Industrially, it is a by-product of fermentation processes and of the manufacture of calcium oxide

$$C_6H_{12}O_6 \xrightarrow{\text{enzymes}} 2C_2H_5OH + 2CO_2 \uparrow$$

$$CaCO_3 \rightarrow CaO + CO_2 \uparrow$$

It is used in the Solvay process for making sodium hydrogen carbonate (p. 308) and, when solid, as a mobile refrigerant on account of its large heat of sublimation

$$CO_{2(s)} \rightarrow CO_{2(g)} \qquad \Delta H = +25 \text{ kJ at} - 56°C$$

The evolution of carbon dioxide when acid reacts with carbonate is utilized in baking powders (the acid used is solid and so only reacts to produce carbon dioxide when there is moisture present to dissolve it), in sherbets and in fire extinguishers.

Silicon monoxide is claimed to be formed by the reduction of silica by silicon at high temperature. *Silica* itself, silicon dioxide, occurs widely and is formed when silicon burns in oxygen. It exists in three crystal forms, each with a high- and low-temperature modification:

$$\text{quartz} \rightleftharpoons \text{tridymite} \rightleftharpoons \text{cristobalite}$$

In all these forms, silicon is surrounded tetrahedrally by four oxygen atoms in a macromolecular lattice.

Although silica is not hydrated to form silicic acid, it does react on fusion with alkalis to give silicates, and in this sense is an acid anhydride.

Germanium monoxide is unstable and, like all the divalent compounds of germanium, tends to disproportionate on heating

$$2Ge^{II} \rightarrow Ge^{IV} + Ge$$

Tin(II) oxide is prepared by heating the oxalate or by carefully dehydrating the compound formed when tin(II) chloride solution is treated with alkali. It is amphoteric, dissolving in caustic alkalis to form stannates(II), e.g.

$$SnO + H_2O + OH^- \rightarrow Sn(OH)_3^-$$

The stable oxide of tin is *tin(IV) oxide*, which is commonly made by treating tin with concentrated nitric acid and strongly heating the white residue obtained (see above), but is better prepared by the hydrolysis of tin(IV) compounds, e.g.

$$SnCl_4 + 2H_2O \rightarrow SnO_2 + 4HCl$$

This, too, is amphoteric, although one of the few acids in which it will dissolve is concentrated sulphuric. With caustic alkalis it dissolves, giving the stannate(IV) ion

$$SnO_2 + 2H_2O + 2OH^- \rightarrow Sn(OH)_6^{2-}$$

Lead gives three oxides. *Lead(II) oxide* is obtained by the general methods for metallic oxides, such as the action of heat on the metal (in air), the carbonate, hydroxide or nitrate.

This oxide is also amphoteric and dissolves both in acids (giving lead(II) salts) and in caustic alkalis (giving plumbates(II))

$$PbO + 2H^+ \rightarrow Pb^{2+} + H_2O$$

$$PbO + 2OH^- \rightarrow PbO_2^{2-} + H_2O$$

Lead(II)(IV) tetraoxoplumbate, red lead, Pb_3O_4, is usually obtained by heating lead(II) oxide in air at about 400°

$$6PbO + O_2 \rightarrow 2Pb_3O_4$$

If this is treated with nitric acid, which produces a soluble salt, *lead(IV) oxide* is left behind

$$Pb_3O_4 + 4H^+ \rightarrow 2Pb^{2+} + 2H_2O + PbO_2\downarrow$$

Lead dioxide is also precipitated by the action of oxidizing agents on solutions of lead(II) compounds. It is a strong oxidizing agent, converting, e.g. hydrochloric acid to chlorine, and it forms the positive plate of the lead accumulator; as an accumulator discharges, the lead(IV) oxide gains electrons and the lead electrode loses them

$$Pb^{4+} + Pb \rightarrow 2Pb^{2+}$$

If lead(IV) oxide is treated with, for example, cold, concentrated hydrochloric acid or glacial ethanoic (acetic) acid, it dissolves to form essentially covalent compounds

$$PbO_2 + 4HCl \rightarrow PbCl_4 + 2H_2O$$

$$PbO_2 + 4CH_3COOH \rightarrow (CH_3COO)_4Pb + 2H_2O$$

With alkalis it forms plumbates(IV), e.g.

$$PbO_2 + 2H_2O + 2OH^- \rightarrow Pb(OH)_6^{2-}$$

Tin(IV) oxide and lead(IV) oxide both have the rutile structure (p. 74).

289

OXYACIDS AND SALTS

Carbonates

Carbonates contain the discrete ion

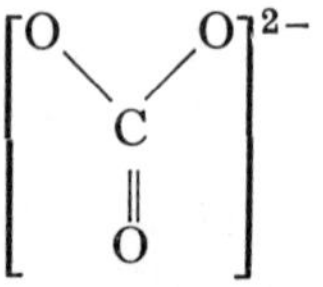

$$\left[\begin{array}{c} O \diagdown_{}{}^{O} \\ C \\ \| \\ O \end{array}\right]^{2-}$$

and are largely insoluble; they are therefore often prepared by double decomposition between a soluble salt of the metal and a solution of a soluble carbonate, although this procedure can result in a basic carbonate being formed

$$5Zn^{2+} + 6OH^- + 2CO_3^{2-} \rightarrow 2ZnCO_3 \cdot 3Zn(OH)_2 \downarrow$$

(The hydroxide ions result from the hydrolysis of the carbonate ions in aqueous solution.)

To prevent the basic carbonate being precipitated, a solution of a hydrogen carbonate is used, the concentration of hydroxide ions produced by hydrolysis of this ion being much less than from the carbonate

$$Zn^{2+} + HCO_3^- \rightarrow ZnCO_3 \downarrow + H^+$$

Although most metals (excluding those of Group 1) can form normal and basic carbonates, there are some, such as copper, which appear to be capable of forming the basic variety only.

The carbonates of the Group 1 metals are soluble and are prepared by first passing carbon dioxide through a solution of the alkali until saturated, and then adding a second quantity of alkali equal in volume and in concentration to the first

$$OH^- + CO_2 \rightarrow HCO_3^-$$

$$HCO_3^- + OH^- \rightarrow CO_3^{2-} + H_2O$$

Since carbonic acid is a weak acid, it is readily displaced from carbonates by the addition of acids stronger than itself, the more so because it breaks up into water and carbon dioxide and disturbs the equilibrium

$$CO_3^{2-} + 2H^+ \rightleftharpoons H_2CO_3 \rightleftharpoons H_2O + CO_2$$

Carbonates, with the exception of those of Group 1, are also fairly easily decomposed by heat, e.g.

$$CaCO_3 \rightleftharpoons CaO + CO_2 \uparrow$$

Only the hydrogen carbonates (bicarbonates) of the alkali metals (and of ammonium) with low polarizing power exist in the solid state; the rest are known

290

only in solution and are decomposed during attempts to isolate them

$$Mg(HCO_3)_2 \rightleftharpoons MgCO_3 \downarrow + H_2O + CO_2 \uparrow$$

The reversibility of this reaction means that they can be obtained in solution by treatment of a suspension of the carbonate (or of a solution of the hydroxide) with carbon dioxide until the solution is saturated or the suspension has disappeared. Like carbonates, the hydrogen carbonates are decomposed on heating and by the action of most acids.

Silicates

Although silicic acid, H_2SiO_3 (probably better represented as $SiO_2 . 2H_2O$), is weaker than carbonic acid, silicates are obtained by heating silica with carbonates because of the non-volatility of silica

$$CO_3^{2-} + SiO_2 \rightarrow SiO_3^{2-} + CO_2 \uparrow$$

With the exception of those of Group 1, silicates, like the acid itself, are insoluble in water. Whilst there are some silicates with discrete ions, e.g. olivine, $Mg_2^{2+} SiO_4^{4-}$, the strength of the Si—O bond is such that most of the naturally occurring silicates have condensed systems (*Figure 13.5*)

Continuation of this trend results in silica, with its three-dimensional lattice, being formed. Replacement of Si^{IV} by Al^{III} can take place, however, because of its similar size and coordination number, giving one negative charge per replacement. The resulting complex oxy-anion is then associated with the requisite number of cations, usually Na^+, K^+, Ca^{2+}, Mg^{2+}, Al^{3+}. The choice of resulting structures is very great, accounting for the wide variety of silicates in nature, such as felspars, e.g. orthoclase, $KAlSi_3O_8$, and zeolites, e.g. natrolite, $Na_2Al_2Si_3O_{10} . 2H_2O$. In the case of the latter, the alkali cation is capable of ion exchange—an effect now much employed for the softening of water. The lattice is also often capable of allowing the passage of molecules below a certain critical size; this property is employed in 'molecular sieves'.

The fusion of sand with alkali carbonates and basic oxides results in the formation of glasses of fairly variable composition. These materials have a structure similar to that of a liquid, that is, only short-range ordered regions can be detected. They are in fact sometimes loosely described as supercooled liquids.

Sulphides

All the elements of Group 4, except lead, give a disulphide, MS_2, which can be obtained by direct combination. Characteristically, carbon disulphide exists as small molecules, $S{=}C{=}S$, whilst the rest are macromolecular; for example, SiS_2 consists of the repeating unit

$$\left[\begin{array}{c} S \quad\quad S \\ \diagdown \diagup \diagdown \diagup \diagdown \\ Si \quad Si \\ \diagup \diagdown \diagup \diagdown \diagup \\ S \quad\quad S \end{array} \right]$$

and typically is hydrolysed by water to the oxide.

Tin gives a monosulphide as well as the disulphide, although this is easily oxidized; for instance, if it is dissolved by ammonium polysulphide to form ammonium trithiostannate(IV), it is the disulphide that is precipitated on acidification

$$\text{SnS} \xrightarrow{\text{(NH}_4)_2\text{S}_x} \text{(NH}_4)_2\text{SnS}_3 \xrightarrow{\text{H}^+} \text{SnS}_2 \downarrow$$

Lead(II) sulphide, like tin(II) sulphide, is precipitated by passing hydrogen sulphide through an alkaline or only weakly acidic solution of a suitable salt, but it is not dissolved by ammonium sulphide nor by alkaline solutions, and so the two can easily be distinguished.

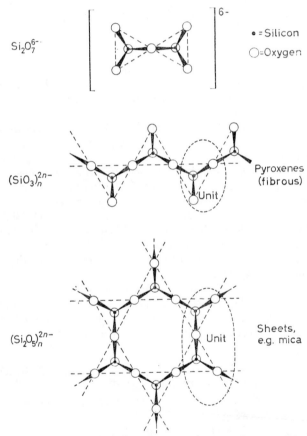

$Si_2O_7^{6-}$

$(SiO_3)_n^{2n-}$ Pyroxenes (fibrous)

$(Si_2O_5)_n^{2n-}$ Sheets, e.g. mica

$\bullet$ = Silicon
$\bigcirc$ = Oxygen

Figure 13.5. Examples of condensed systems in naturally occurring silicates

Organic Compounds

Organic carbon compounds can be made to yield derivatives with the other elements of the Group. Thus, if an alkyl halide is heated to about 300° with silicon, in the presence of a copper catalyst, a series of compounds with silicon attached to carbon and halogen is obtained, e.g.

$$CH_3Cl + Si \overset{Cu}{\to} (CH_3)_3SiCl + (CH_3)_2SiCl_2 + CH_3SiCl_3$$

The mechanism of this reaction involves free radicals

$$2Cu + CH_3Cl \to CuCl + CuCH_3$$

$$CuCH_3 \to Cu + \cdot CH_3$$

$$(4-n)CuCl + Si + nCH_3\cdot \to (CH_3)_nSiCl_{4-n} + (4-n)Cu \quad (n \not> 3)$$

Hydrolysis of these intermediates gives fibrous or resinous polymers called *silicones* which, because of their electrical resistance, water repellancy, inertness and elasticity or viscosity, have a wide variety of uses.

$$(CH_3)_2SiCl_2 \to (CH_3)_2Si(OH)_2 \to \begin{bmatrix} CH_3 & CH_3 & CH_3 \\ | & | & | \\ -Si-O-Si-O-Si- \\ | & | & | \\ CH_3 & CH_3 & CH_3 \end{bmatrix}_n$$

Because silicon is less electronegative than carbon, the silicon–carbon bond is polar, with silicon electrophilic and carbon, unusually, nucleophilic.

The strength of the bond with carbon falls off as the Group is descended, but lead tetraethyl, $(C_2H_5)_4Pb$, obtained by treating lead(II) chloride with ethyl magnesium chloride, is comparatively stable

$$2PbCl_2 + 4C_2H_5MgCl \to (C_2H_5)_4Pb \uparrow + 4MgCl_2 + Pb \downarrow$$

At elevated temperatures, however, it decomposes to give free ethyl radicals, which promote a more regular oxidation of hydrocarbon; it thus finds use as an 'anti-knock' agent in petrol.

Cyanogen, $(CN)_2$

Cyanogen is evolved when certain metallic cyanides are heated. Significantly, these metals are usually not very electropositive and have variable oxidation states: the reaction involves the reduction of the metal by the cyanide ion, e.g.

$$2Cu(CN)_2 \to 2CuCN \downarrow + (CN)_2 \uparrow$$

It is also formed from copper(II) sulphate and potassium cyanide solutions

$$2Cu^{2+} + 4CN^- \to 2CuCN \downarrow + (CN)_2 \uparrow$$

Both these reactions can be compared with the reduction of copper(II) by the iodide ion

$$2Cu^{2+} + 4I^- \to 2CuI \downarrow + I_2$$

Cyanogen is a very poisonous, colourless gas, smelling of almonds. It readily polymerizes to *para*cyanogen $(CN)_x$. Many of its reactions earn it the name of a pseudohalogen, e.g.

$$(CN)_2 + 2K \to 2KCN$$

$$(CN)_2 + 2OH^- \to CN^- + CNO^- + H_2O$$
$$\text{cyanide} \quad \text{cyanate}$$

Cyanides

Potassium cyanide is made by heating potassium carbonate and carbon in the presence of ammonia

$$K_2CO_3 + 2C + 2NH_3 \rightarrow 2KCN + CO \uparrow + 2H_2O \uparrow + H_2 \uparrow$$

Distillation of potassium cyanide with sulphuric acid results in the evolution of hydrogen cyanide, which can be condensed on cooling to a colourless, very poisonous liquid which behaves as a weak, monobasic acid.

Hydrogen cyanide has a high dielectric constant, and there is evidence for self-ionization of the form

$$HCN \rightleftharpoons H_2CN^+ + CN^-$$

The cyanide ion is a powerful ligand and readily forms complexes with metals. This is the basis for its identification: the cyanide is treated with iron(II) sulphate in acid solution—as oxidation of some of the iron to iron(III) takes place, so a precipitate of Prussian blue forms

$$Fe^{2+} + 2CN^- \rightarrow Fe(CN)_2 \xrightarrow{CN^-} Fe(CN)_6^{4-} \xrightarrow{Fe^{3+}} [Fe^{II} Fe^{III} (CN)_6]^-$$

Cyanides occur naturally as glucosides (p. 472), e.g., amygdalin, and can be used in the manufacture of Perspex and acrylonitrile (p. 385).

Carbides

Carbides are often made by heating the element or its oxide with carbon

$$Si + C \rightarrow SiC \qquad CaO + C \rightarrow CaC_2 + CO \uparrow$$

They can be divided broadly into three classes:

(a) *Saline*—Most of these contain the C_2^{2-} ion and are consequently decomposed by water with the evolution of ethyne (acetylene). They are thus known as acetylides

$$C_2^{2-} + 2H^+ \rightarrow C_2H_2 \uparrow$$

A few saline carbides, however, contain discrete carbon atoms or C^{4-} ions; these evolve methane when treated with water and are called methanides, e.g.

$$Al_4C_3 + 12H_2O \rightarrow 3CH_4 \uparrow + 4Al(OH)_3 \downarrow$$

The more covalent acetylides are insoluble in water and can be prepared by passing acetylene through a suitable solution of the metal, e.g. ammoniacal copper(I) chloride.

(b) *Covalent*—The covalent carbides are not attacked by acids and are normally very hard. Silicon carbide, for example, is a powerful abrasive, known as carborundum, which approaches diamond in hardness.

(c) *Interstitial*—The covalent carbides are stoichiometric (or Daltonide). Non-stoichiometric carbides also exist, particularly those of the transition elements. If the radius of the metallic atoms is at least 130 pm, the octahedral holes in the lattice are sufficiently large for carbon atoms to fit in. These carbides are very hard and refractory, inert and able to conduct electricity.

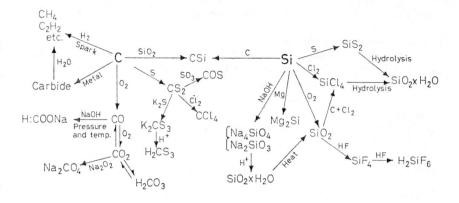

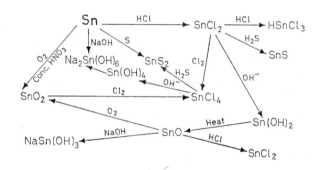

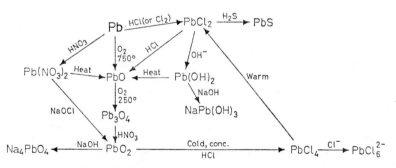

Figure 13.6. Reactions of carbon, silicon, tin and lead

QUESTIONS

1. Comment on the following:

(a) When hydrogen sulphide is passed through an acidified solution of a salt of a metal, a brown precipitate is produced. This dissolves in ammonium sulphide solution, and on acidification a yellow precipitate is formed.

(b) A dilute solution of sodium hydrogen carbonate is alkaline to methyl orange but acidic to phenolphthalein.

(c) Red lead gives a brown deposit with nitric acid; if the filtrate is treated with potassium iodide solution, a yellow precipitate is obtained.

(d) If sand and fluorspar are heated with concentrated sulphuric acid, a gas is evolved which gives a white deposit on coming into contact with water.

2. How far do you think cyanogen can be called a pseudo-halogen and hydrogen cyanide a pseudo-hydrogen halide?

3. A light yellow solid heated with dilute sulphuric acid produced a gas smelling of almonds, and with concentrated sulphuric acid, a gas which burnt with a pale blue flame. After this latter reaction, the residue was diluted with water and a dark blue precipitate developed. Identify the nature of the original substance and suggest equations for the reactions.

4. It has been said that the acidity of an oxide depends upon the relative amounts of metal and of oxygen contained in it. Discuss this by reference to the oxides of tin and lead.

5. Write an essay on silicate minerals.

6. 'As one descends a group of the Periodic Table, one encounters elements of increasing metallic character.' Discuss this statement from the point of view of the elements of Group 4.

7. Carbon, because of the fundamental role which its atoms play in cellular processes, is often regarded as the element of life. Yet several of its compounds are very toxic, carbon monoxide and cyanides being particularly notorious. Indicate how these two poisons function and whether their efficacy reveals a weakness in evolutionary processes.

8. The gas industry is in a state of flux. Indicate why this is and summarize the steps being taken by the industry to solve its problems.

9. 'The commercial future of lead is inextricably bound up with that of the car.' Indicate if and why this should be so.

CHAPTER 14

THE ELEMENTS OF GROUPS 1, 2 AND 3

ns	np	ns	np	ns	np
↑	☐☐☐	↑↓	☐☐☐	↑↓	↑ ☐☐

Group 0				Group 4
He	Lithium, Li	Beryllium, Be	Boron, B	C
Ne	Sodium, Na	Magnesium, Mg	Aluminium, Al	Si
Ar	Potassium, K	Calcium, Ca	Gallium, Ga	Ge
Kr	Rubidium, Rb	Strontium, Sr	Indium, In	Sn
Xe	Caesium, Cs	Barium, Ba	Thallium, Tl	Pb
Rn	Francium, Fr	Radium, Ra		

Introduction

The increase in non-metallic character in going along each Period means that of the three Groups, Group 3 is the least metallic. Also, the increase in metallic character on descending each Group results in boron being the least metallic of this selection of elements. However, the ionization energies (*Table 14.1*) of the first element of each Group are sufficiently high for several compounds of lithium, beryllium and boron to be mainly covalent. Furthermore, the increase in the charge on the ions in proceeding from Group 1 to 3 adds to the effect of the high ionization energy of boron to give this element the properties of a non-metal. For these reasons, much of the chemistry of aqueous solutions of beryllium, aluminium, and particularly boron, is concerned with the element in the form of an anion, such as BeO_2^{2-}, $Al(OH)_4^-$, $H_2BO_3^-$, the relevant cations having such considerable polarizing power that under normal conditions they would not be expected to exist.

Table 14.1. Ionization energies of the elements of Groups 1–3

Group 1	1st I.E. kJ mol^{-1}	Group 2	2nd I.E. kJ mol^{-1}	Group 3	3rd I.E. kJ mol^{-1}
Li	519	Be	1 759	B	3 660
Na	494	Mg	1 445	Al	2 740
K	418	Ca	1 145	Ga	2 950
Rb	406	Sr	1 061	In	2 690
Cs	377	Ba	965	Tl	2 860

Although regular gradations in properties may reasonably be anticipated for the elements of Groups 1 and 2 and their compounds, a less regular change must be expected for Group 3. This is the result of the transition elements producing an irregular increase in size after the second element in the Group (*Table 14.2*).

Table 14.2. Atomic radii of the elements of Groups 1–3

	Group 1 Atomic		Group 2 Atomic		Group 3 Atomic
	Element	radius, pm	Element	radius, pm	Element	radius, pm
Progressive increase	Li	123	Be	89	B	80
in atomic size on	Na	157	Mg	136	Al	125
descending Group	K	203	Ca	174	Ga	125*
gives steady de-	Rb	216	Sr	191	In	150
crease in electro-	Cs	235	Ba	198	Tl	155
negativity						

* Intervening transitional metals, with their contraction in atomic volume, arrest increase in atomic size in Group 3.

Valency States

The s^1 electron of the Group 1 elements allows only the monovalent state. The elements of Group 2 similarly can only be divalent because of the s^2 electrons in the outer electron shell. Elements of Group 3 have three electrons in their outer, valency, shell, but these are not all of the same type, being designated s^2p^1 (p.23). The maximum valency of the Group 3 elements is produced either by ionization of these electrons, giving a trivalent cation or, as always occurs for boron, by excitation of the form

giving three unpaired electrons and, hence, a covalency of three. Because the s and p electrons can behave independently of each other, it is possible for the p electron to be involved in bonding whilst the pair of s electrons remains inert, so that both monovalent and trivalent forms of the Group 3 elements can occur. Indeed, for thallium, because of this 'inert pair' effect, the monovalent state is more stable than the trivalent. However, the remaining elements of the Group are essentially trivalent. In covalent bond formation, the maximum number of electrons which can be accommodated in the second quantum shell is eight (s^2p^6), giving beryllium and boron a possibility of tetrahedral coordination by the further acceptance of two pairs or one pair of electrons, respectively, i.e. by the formation of dative bonds. As a result, beryllium and boron often occur in complexes in which they have acquired the necessary number of electrons, e.g.

$$BeF_4^{2-} \qquad [Be(H_2O)_4]^{2+} \qquad BF_3 . NH_3 \qquad BF_4^-$$

298

OCCURRENCE AND EXTRACTION

Because of their high electropositivity, none of these elements are found native; those of Group 1 occur as their halides, of Group 2 as sulphates and carbonates and of Group 3 as oxides or silicates (often accompanied by the elements of Groups 1 and 2). The general solubility of the salts results in many being found in sea water and salt beds; for example, the approximate concentrations of the more common metallic ions in sea water are Na^+ 1·1 per cent, Mg^{2+} 0·13 per cent, Ca^{2+} and K^+, 0·04 per cent (*Table 14.3*).

Table 14.3. Some common naturally-occurring compounds of the elements of Groups 1–3

	Group 1	*Group 2*	*Group 3*
Halides	NaCl, rock salt KCl . MgCl₂ . 6H₂O carnallite	CaF₂, fluorspar	Na₃AlF₆, cryolite
Sulphates		CaSO₄, anhydrite CaSO₄ . 2H₂O, gypsum BaSO₄, barytes MgSO₄ . 7H₂O, Epsom salts	
Oxides			AlO(OH), bauxite Na₂B₄O₇ . 10H₂O, borax
Silicates	NaAlSi₃O₈, felspar LiAl(SiO₃)₂, spodumene	Be₃Al₂(SiO₃)₆, beryl CaMg₃(SiO₃)₄, asbestos Mg₃Si₄O₁₀(OH)₂, talc	Al₂Si₂O₅(OH)₄, kaolin
Carbonates		MgCO₃, magnesite CaCO₃ . MgCO₃, dolomite CaCO₃ chalk, limestone, marble BaCO₃, witherite	

Although all the oxides can be reduced to the corresponding element by carbon at suitable temperatures, this method is often not convenient, chiefly because of carbide formation between the element produced and excess carbon, and on account of the high temperatures sometimes needed (*Table 14.4*).

Table 14.4. Extraction and uses of the elements of Groups 1–3

Element	Extraction	Uses
Lithium ⎱ Sodium ⎰	electrolysis of fused chloride	scavenger; alloys lamps, coolant (atomic reactors); manufacture of 'anti-knock', Na_2O_2 and NaCN; reducing agent, e.g. for titanium extraction
Potassium	electrolysis of fused chloride or hydroxide	reducing agent
Rubidium ⎱ Caesium ⎰	electrolysis of fused cyanide	photoelectric cells
Beryllium	electrolysis of fused chloride	x-ray 'windows'; scavenger; alloys; moderator in fission reactions
Magnesium	electrolysis of fused chloride and thermal reduction of oxide by Fe/Si or C	light alloys; reducing agent, e.g. extraction of uranium; container for nuclear fuel; 'Grignards'
Calcium	electrolysis of fused chloride or reduction of oxide by Al	scavenger, e.g. removal of water from ethanol, air from radio valves
Strontium Barium	electrolysis of fused chloride reduction of oxide with Al	⎱scavengers⎰
Boron	reduction of the oxide with Mg or Al	borohydrides
Aluminium	electrolysis of fused Al_2O_3 + Na_3AlF_6	light alloys; electrical conductor; reducing agent (Thermit process
Gallium ⎱ Indium ⎰ Thallium	electrolysis of aqueous solutions	thermometry—low m.p. (30°C) and high boiling point (2 070°C)

Electrolytic methods—All these elements except boron, barium and, more recently, magnesium are obtained by the electrolytic reduction of the fused salts, with graphite serving as the anode, where halogen and oxygen are evolved, and iron generally as the cathode. Mixed salts are normally employed to lower the melting point of the system (p. 103) and for increasing the conductivity. Examples of the electrolytes used are

(a) Group 1 chlorides fused in the presence of calcium chloride (as impurity)

(b) fused beryllium chloride made from beryl (with sodium chloride as impurity)

(c) fused carnallite, the magnesium chloride often being replenished by producing it from magnesite

(d) fused calcium chloride, with calcium fluoride or potassium chloride as impurity

(e) aluminium oxide dissolved in molten cryolite, Na_3AlF_6; the latter acts as a conducting solvent for the oxide, and is itself significantly consumed.

Thermal reduction—The high temperature required for the electrolysis of fused barium chloride causes too many difficulties, and barium is therefore usually isolated by the reduction of barium oxide (from calcined witherite) with aluminium

$$3BaO + 2Al \rightarrow 3Ba + Al_2O_3$$

300

An alternative method for the extraction of magnesium is by the reduction of magnesium oxide, either with ferrosilicon if the oxide is MgO.CaO (from calcined dolomite) or with coke, as in the Pidgeon process (p. 180). In the former case, the ferrosilicon removes the calcium as a slag of calcium silicate

$$\text{MgO.CaO} \xrightarrow{\text{Fe/Si}} \text{Mg} + \text{CaSiO}_3$$

Boron is prepared by reducing boron oxide (itself obtained from borax) with magnesium or aluminium

$$\text{B}_2\text{O}_3 + 3\text{Mg} \rightarrow 2\text{B} + 3\text{MgO}$$

PROPERTIES OF THE ELEMENTS

(a) Physical Properties
The metals, in contrast to boron, have low melting points (*Table 14.5*) and are good conductors of electricity.

Table 14.5

Element	Melting point °C	Element	Melting point °C	Element	Melting point °C
Li	178	Be	1 285	B	2 300
Na	97	Mg	650	Al	658
K	63	Ca	803	Ga	30
Rb	39	Sr	800	In	156
Cs	29	Ba	850	Tl	449

Table 14.6. Structures of the elements

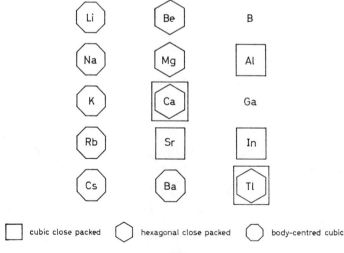

☐ cubic close packed   ⬡ hexagonal close packed   ⬯ body-centred cubic

301

With the exception of boron and gallium, the elements crystallize in the typical metallic cubic close-packed, hexagonal close-packed and body-centred cubic structures (*Table 14.6*). Gallium forms complex orthorhombic crystals, whilst boron has a complex non-metallic lattice of uncertain structure, the atoms being too short of electrons to form an orthodox three-dimensional covalent lattice.

The elements have low densities (*Table 14.7*), except gallium, indium and thallium, and for this reason and because of their tensile strength, magnesium and aluminium find uses in light alloys, e.g. Magnalium (Al 90 per cent, Mg 10 per cent) and Duralumin (Al 95 per cent, Mg 0·5 per cent, Cu 4 per cent, Mn 0·5 per cent); the low electron density of beryllium makes it suitable for 'windows' in x-ray work, whilst radium and its salts find applications in radiochemistry.

Table 14.7. Densities of the elements

Group 1		Group 2		Group 3	
Element	Density g cm^{-3}	Element	Density g cm^{-3}	Element	Density g cm^{-3}
Li	0·53	Be	1·86	B	2·4
Na	0·97	Mg	1·75	Al	2·7
K	0·86	Ca	1·55	Ga	5·9
Rb	1·53	Sr	2·6	In	7·3
Cs	1·90	Ba	3·6	Tl	11·9

(b) Chemical Properties

The elements of Groups 1 and 2, and aluminium, have the largest negative electrode potentials (*Table 14.8*). Accordingly, they generally displace hydrogen from suitable compounds such as water and acids, their reactivity increasing with increase in atomic number.

Table 14.8. Electrode potentials (25°C)

V	V	V
Li − 3·02	Be − 1·70	
Na − 2·71	Mg − 2·38	Al − 1·67
K − 2·92	Ca − 2·87	
Rb − 2·99	Sr − 2·89	
Cs − 3·02	Ba − 2·90	

Since the hydroxides of the Group 1 elements are freely soluble in water, these elements react most violently with water, e.g.

$$Na \rightarrow Na^+ + e \qquad H_2O \rightleftharpoons H^+ + OH^-$$

$$H^+ + e \rightarrow \tfrac{1}{2}H_2 \uparrow$$

$$(2Na + 2H_2O \rightarrow 2NaOH + H_2 \uparrow)$$

whilst the hydroxides of the Group 2 elements are not so readily soluble and therefore their initial formation by the above type of reaction hinders the dissolution of the metal in water. Elements of these two Groups also react with alcohols, but with less vigour than with water; such systems, particularly that of sodium and ethanol, are useful as reducing agents. The elements of Group 3, except boron, also react with alcohols, provided that a catalyst is available, e.g.

$$Al + 3ROH \xrightarrow{\text{HgCl}_2} (RO)_3Al + \tfrac{3}{2}H_2 \uparrow$$

Other reactions of the free elements can be summarized as follows:
(*i*) beryllium, aluminium and gallium dissolve in caustic alkalis, although the anions produced are more complex than the usual equations suggest

$$Be + 2OH^- \rightarrow BeO_2^{2-} + H_2 \uparrow$$

$$2Al + 2OH^- + 2H_2O \rightarrow 2AlO_2^- + 3H_2 \uparrow$$

The beryllate and aluminate ions are usually hydrated, aluminate occurring mainly as $[Al(OH)_4]^-$.
(*ii*) Heating with hydrogen gives the hydrides of the Group 1 metals as well as those of calcium, strontium and barium in Group 2.
(*iii*) Anhydrous halides are readily obtained by direct union with the free halogen, e.g.

$$2Al + 3Cl_2 \xrightarrow{\text{heat}} Al_2Cl_6$$

(*iv*) Oxides, peroxides and superoxides are obtained when the elements are heated to only a moderate temperature with oxygen

$$Li \xrightarrow{O_2} Li_2O \text{ (oxide)}$$

$$Na \xrightarrow{O_2} Na_2O_2 \text{ (peroxide)}$$

$$K \xrightarrow{O_2} KO_2 \text{ (superoxide)}$$

When heated in the air, some nitride is also formed from magnesium. The nitrides of lithium, the remaining Group 2 elements, boron and aluminium are obtained if the relevant element is heated in nitrogen itself. With the exception of the latter two products, they contain the nitride ion, N^{3-}, and are readily hydrolysed, e.g.

$$Mg_3N_2 + 3H_2O \rightarrow 3MgO + 2NH_3 \uparrow$$

303

(Aluminium nitride is also readily hydrolysed, but boron nitride, one form of which has a covalent structure of the graphite type

reacts only slowly with water.)

Sulphides are obtained by direct combination in a similar manner to the oxides.

(*v*) Heating with ammonia or dissolution in liquid ammonia results in the formation of amides of the Group 1 metals and those of calcium, strontium and barium. Like the nitrides these are readily hydrolysed

$$Na + NH_3 \rightarrow Na^+NH_2^- + \tfrac{1}{2}H_2 \uparrow$$
$$\text{sodamide}$$

$$Na^+NH_2^- + H_2O \rightarrow Na^+OH^- + NH_3 \uparrow$$

(The name 'amide' is unfortunate, since in organic chemistry an amide has the characteristic group —CONH₂.)

COMPOUNDS OF GROUPS 1-3

Hydrides

The very electropositive metals of Groups 1 and 2 form ionic hydrides on heating in dry hydrogen. As the hydride ion, H^-, is unstable in the presence of protons, H^+, these solids are decomposed by water and by acids, e.g.

$$Ca^{2+}H_2^- + 2H_2O \rightarrow Ca^{2+}(OH^-)_2 + 2H_2 \uparrow$$

The hydrides of Group 3 are less well characterized; boron gives a series of hydrides, all of which are electron-deficient. The parent member, diborane, B_2H_6, is prepared by adding a boron trifluoride–ether complex slowly to an ethereal suspension of lithium hydride and refluxing

$$6H^- + 8(C_2H_5)_2O.BF_3 \rightarrow 6BF_4^- + 8(C_2H_5)_2O + B_2H_6 \uparrow$$

Diborane is a gas which is rapidly hydrolysed by water to boric acid and hydrogen. The shape of the molecule is known to be

but the method of bonding in the bridge between the boron atoms is uncertain.

304

Boron, aluminium and gallium form complex hydrides containing the ions BH_4^-, AlH_4^- and GaH_4^-, respectively. Of these, the tetrahydridoaluminate, as a lithium salt, is much used as a reducing agent, especially in organic chemistry. Aluminium also forms a polymeric hydride, $(AlH_3)_n$.

Halides

The halides can be obtained by the action of halogen acids on the metals, oxides, hydroxides or carbonates

$$Mg \xrightarrow{HCl} MgCl_2$$

$$KOH \xrightarrow{HI} KI$$

$$Na_2CO_3 \xrightarrow{HBr} NaBr$$

The halides of most of the elements of the three Groups readily form hydrates, and it is these that this method will produce. An alternative method, for preparing the anhydrous salts, is to pass anhydrous hydrogen halide or halogen over the heated metal (or over a mixture of the metal oxide and carbon), e.g.

$$2Al + 3Cl_2 \rightarrow Al_2Cl_6$$

$$MgO + C + Cl_2 \rightarrow MgCl_2 + CO$$

In each Group, the degree of hydration decreases as the cationic size increases. The most marked difference is between the first and second members, owing to the very small ions of lithium, beryllium and boron. The lithium ion is so small and its charge density so great that it exerts a powerful attraction towards the lone pairs of electrons on the oxygen atom in water or alcohol molecules, with the result that lithium chloride is deliquescent and soluble in alcohol as well as in water, whereas the other halides in this Group are soluble only in water. This characteristic behaviour is more pronounced in Group 2. For example, beryllium halides hydrolyse when their aqueous solutions are heated

$$BeCl_2 + H_2O \rightarrow BeO \downarrow + 2HCl$$

Magnesium chloride produces oxychlorides on evaporating its aqueous solution

$$MgCl_2 + H_2O \rightarrow Mg(OH)Cl \downarrow + HCl$$

On the other hand, calcium chloride is used as a drying agent because of the low vapour pressure of its several hydrates, the polarizing power of the cation being insufficient to cause hydrolysis of the salt.

The boron halides have low boiling points and undergo complete hydrolysis in water, e.g.

$$BCl_3 + 3H_2O \rightarrow B(OH)_3 + 3HCl$$

They are electron-deficient and therefore behave as strong Lewis acids in attempting to complete their octet of electrons, e.g. $BF_3 + F^- \rightarrow [BF_4]^-$. This is in contrast to aluminium chloride which completes the octet of electrons around the aluminium atom by forming a dimer. Like many other covalent halides, aluminium chloride fumes in moist air and is hydrolysed by small amounts of water

$$Al_2Cl_6 + 6H_2O \rightarrow 2Al(OH)_3 \downarrow + 6HCl$$

Oxides and Hydroxides

The hydroxides, nitrates and carbonates, especially those of the Group 1 metals, are so stable that they do not readily decompose into the oxides on heating. The latter (or peroxides) can be prepared by heating the elements in oxygen or in some cases with other, less stable, oxides; for example, in the Thermit process, aluminium oxide is formed by heating aluminium with the oxide of a transitional metal such as iron or chromium

$$Fe_2O_3 + 2Al \rightarrow 2Fe + Al_2O_3$$

The oxides of the Group 1 elements are alkaline; being strongly ionic, they react with water to produce hydroxides

$$O^{2-} + H_2O \rightarrow 2OH^-$$

There is an increase in alkalinity of all the oxides and hydroxides as the cation increases in size, so that, whereas magnesium hydroxide is only a weak base, barium hydroxide is sufficiently strong to function as a volumetric alkali. Whereas boron oxide dissolves in hot water to give a weak *acid*

$$B_2O_3 + 3H_2O \rightleftharpoons 2B(OH)_3 \text{ (i.e. } H_3BO_3)$$

$$B(OH)_3 \rightleftharpoons H^+ + H_2BO_3^-$$

aluminium hydroxide is amphoteric (as also is the oxide)

$$Al_{(aq)}^{3+} \underset{H+}{\overset{OH-}{\rightleftharpoons}} Al(OH)_3 \underset{H+}{\overset{OH-}{\rightleftharpoons}} Al(OH)_4^-$$

The effect of the intervening transition metals before gallium, indium and thallium is to produce comparatively small cations (see Chapter 15) for these metals, with the result that their hydroxides, when in oxidation state $+3$, are not very alkaline. However, thallium, because of the inert-pair effect, exists also in the monovalent condition, with a correspondingly larger cation; as a result, Tl(I)OH rivals the hydroxides of the Group 1 metals in its alkalinity.

The increase in alkalinity of the hydroxides as the cation increases in size can be explained by the fact that, as the cationic size increases, so the positive charge upon it becomes further away from the negative charge on the hydroxide ion. There will thus be a decrease in the coulombic attraction between the two ions, and the hydroxide ion is more free to act as such.

COMPOUNDS OF GROUPS 1–3

Of the hydroxides, that of sodium is prepared in large quantities, mainly by the electrolysis of sodium chloride solution. One such process, the Kellner–Solvay, has been described earlier (p. 201) when dealing with the preparation of chlorine. To obtain sodium hydroxide, the sodium amalgam obtained is run into large tanks and then decomposed by water in the presence of iron or graphite

$$2Na/Hg + 2H_2O \rightarrow 2NaOH + H_2 \uparrow + 2Hg \downarrow$$

The resultant solution of the hydroxide can then be evaporated to dryness.

An alternative method of preparation is by electrolysis of brine in a diaphragm cell, so called because of the porous partition which separates the anode from the cathode. An example of such is the Nelson cell, shown diagrammatically in *Figure 14.1*. The asbestos diaphragm ensures that the chlorine evolved at the anode is not able to react with the solution of sodium hydroxide produced at the cathode.

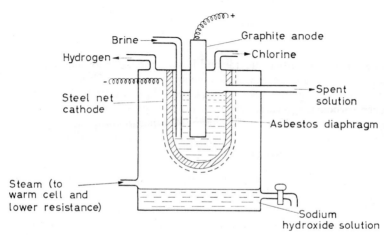

Figure 14.1. Preparation of sodium hydroxide in a diaphragm cell

$$NaCl \rightarrow Na^+ + Cl^- \qquad H_2O \rightleftharpoons H^+ + OH^-$$

Anode Cl⁻ and OH⁻ attracted   *Cathode* Na⁺ and H⁺ attracted

$$Cl^- - e \rightarrow Cl \rightarrow \tfrac{1}{2}Cl_2 \uparrow \qquad H^+ + e \rightarrow H \rightarrow \tfrac{1}{2}H_2 \uparrow$$

Table 14.9. World production and uses of sodium hydroxide (1966)

Production	10^6 tonnes	per cent	Uses	per cent
World	15		Artificial silk	25
U.S.A.	6·6	44	Chemicals,	
Japan	1·4	9	e.g. NaCN, Na₂O₂	25
West Germany	1·3	9	Soap	13
U.S.S.R.	1·3	9	Paper	10
			Textiles	7
			Oil refining	5

Carbonates

The soluble hydrogen carbonates and carbonates may be obtained by passing carbon dioxide through an aqueous solution of the alkali, the compound produced depending upon the relative amounts of alkali and carbon dioxide used

$$OH^- + CO_2 \rightarrow HCO_3^-$$

$$2OH^- + CO_2 \rightarrow CO_3^{2-} + H_2O$$

The insoluble carbonates are obtained by double decomposition reactions, e.g.

$$Ca^{2+} + CO_3^{2-} \rightarrow CaCO_3 \downarrow$$

The carbonate ion is easily polarized by small cations of high charge

breaking down into carbon dioxide and leaving the oxide ion on heating. For this reason, the carbonates of the weakly polarizing ions of Group 1 (i.e. excluding lithium) are the only ones stable to heat and soluble in water. Furthermore, their hydrogen carbonates (bicarbonates) are the only ones which can be isolated from aqueous solution; hydrogen carbonates of other metals, although existing in solution, cannot be obtained in the solid state because the equilibrium

$$2HCO_3^- \rightleftharpoons CO_3^{2-} + H_2O + CO_2 \uparrow$$

favours formation of the carbonate, particularly if water is lost as vapour and the carbonate is insoluble. In other words, if a solution of such a hydrogen carbonate is heated, or even allowed to evaporate at low temperature, the carbonate is precipitated (e.g. as stalactites). Boiling is, of course, a means of removing the temporary hardness of water caused by the presence of the hydrogen carbonates of calcium and magnesium; since the calcium and magnesium cations are removed from solution as insoluble carbonates, they are no longer able to react with the soap anions, and the water is thus softened.

Aluminium ions have a very large polarizing power, and attempts to prepare aluminium carbonate result in the precipitation of aluminium hydroxide. This result could also be predicted from consideration of the hydrolysis of aluminium salts in solution. Aluminium hydroxide is only weakly basic, and carbonic acid is a very weak acid: accordingly, complete hydrolysis of aluminium carbonate in water would be expected.

Thallium(I) carbonate *does* exist, showing that the polarizing effect of the monovalent cation is much less than that of trivalent ones.

Sodium hydrogen carbonate and sodium carbonate are of commercial value and are manufactured by the Ammonia–Solvay process in which ammonia, carbon dioxide and sodium chloride solution produce an equilibrium which is displaced in favour of the formation of sodium hydrogen carbonate, since this is the least soluble component of the system and is precipitated from the solution

$$Na^+ + H_2O + NH_3 + CO_2 \rightleftharpoons Na^+HCO_3^- \downarrow + NH_4^+$$

The ammonia also serves to remove hydrogen ions which would otherwise polarize the hydrogen carbonate ions. The flow sheet (*Figure 14.2*) emphasizes the overall economy of the process, the net result of which can be summarized by the equation

$$2NaCl + CaCO_3 \rightarrow CaCl_2 + Na_2CO_3$$

Table 14.10. Production of Sodium Carbonate (1966)

	10^6 tonnes	per cent
World total	15	
U.S.A.	4·6	31
U.S.S.R.	2·8	19
West Germany	1·2	8
France	1·1	7

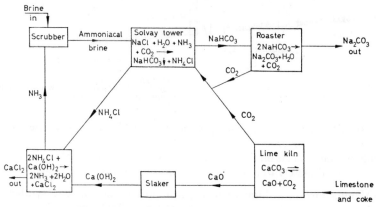

Figure 14.2. Flow sheet for the Solvay process

Sodium hydrogen carbonate is used in baking powders, 'health' salts and fire extinguishers, carbon dioxide being produced when hydrogen ions are made available from some suitable source (such as tartaric acid in 'health' salts). Sodium carbonate finds use as a water softener in washing powders

$$CO_3^{2-} + Ca^{2+} \rightarrow CaCO_3 \downarrow$$

and as a high-temperature alkali, particularly in glass manufacture

$$Na_2CO_3 + SiO_2 \rightarrow Na_2SiO_3 + CO_2 \uparrow$$

Calcium carbonate is also of considerable industrial importance, mainly for the manufacture of cement and calcium oxide (quicklime). For the former, a fine slurry of calcium carbonate and clay (which can be regarded as a hydrated double oxide of aluminium and silicon) are reacted together at high temperatures to form an intimate mixture of the oxides of calcium, aluminium and silicon. It is believed that when water is added to this mixture, complex silicates are formed.

Calcium oxide is produced by the decomposition of calcium carbonate in a lime kiln

$$CaCO_3 \rightleftharpoons CaO + CO_2$$

Sodium

Manufacture

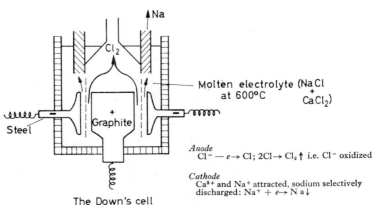

The Down's cell

Anode
$Cl^- - e \rightarrow Cl; 2Cl \rightarrow Cl_2 \uparrow$ i.e. Cl^- oxidized

Cathode
Ca^{2+} and Na^+ attracted, sodium selectively discharged: $Na^+ + e \rightarrow Na \downarrow$

About 200 000 tons of sodium are produced annually throughout the world, of which 98 per cent comes from the Downs Process.

The molten sodium floats on the denser electrolyte and is drawn off at will. (Calcium chloride is added merely to lower the melting point of the electrolyte see p. 103.)

Uses—Manufacture of sodium peroxide, sodium cyanide; organic synthesis and as a reducing agent; sodium vapour discharge lamps; lead tetraethyl.

Reactions

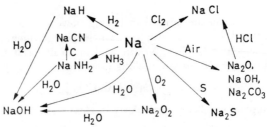

Figure 14.3. Reactions of sodium

The equilibrium is shifted towards the formation of calcium oxide by providing a current of air to remove the carbon dioxide as it is evolved. The calcium oxide is used for improving the texture and reducing the acidity of soils and for the production of calcium silicate in glass making.

Salts of Oxyanions of Strong Acids

Although boron trioxide reacts with certain oxyacids, the compounds formed are not true salts; for example, the so-called sulphate and phosphate are $B_2O_3.3SO_3$

Magnesium

Extraction

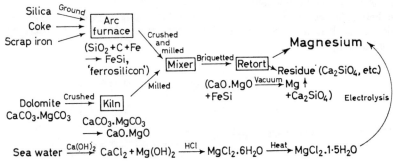

Reactions

Figure 14.4. Extraction and reactions of magnesium

Table 14.11. Primary production (1966) and uses of magnesium

Production	10^3 tonnes	per cent	Uses
World total (except Communist bloc)	124		light alloys, e.g. Magnalium
U.S.A.	72	58	reducing agent for titanium, uranium, etc.
Norway	28	21	flashlight photography; pyrotechnics
			sacrificial protection, e.g. of marine vessels
			organic synthesis (as Grignards)

and $2B_2O_3.P_4O_{10}$, respectively. All the other elements of the three Groups are sufficiently metallic to form salts with nitric, sulphuric and phosphoric acids.

(a) *Sulphates*—The sulphates of the Group 1 metals are all soluble in water, whilst the solubility of the sulphates of the Group 2 metals decreases in passing from magnesium to barium, presumably owing to a progressive increase in the lattice energies—barium sulphate is one of the least soluble of all sulphates. Typically, the thermal stability depends on the electronegativity of the metal, so that the sulphates of Groups 1 and 2 are very resistant to decomposition by heat; aluminium sulphate, on the other hand, is decomposed to the oxide.

Calcium sulphate occurs as an economic mineral in the form of gypsum, $CaSO_4.2H_2O$, and anhydrite, $CaSO_4$. The former is a source of 'plaster of Paris', which is made by partial dehydration to give $(CaSO_4)_2H_2O$, the setting of the paste being caused by hydration back to interlocking crystals of gypsum. Anhydrite is used as a source of sulphur compounds such as sulphuric acid and ammonium sulphate

$$CaSO_4 + 4C \rightarrow CaS + 4CO \uparrow$$

$$CaS + 2H_2O \rightarrow Ca(OH)_2 + H_2S \uparrow$$

$$2H_2S + 3O_2 \rightarrow 2H_2O + 2SO_2 \uparrow$$

$$2SO_2 + O_2 + 2H_2O \rightarrow 2H_2SO_4$$

$$CaSO_4 + NH_3 + CO_2 \rightarrow CaCO_3 \downarrow + (NH_4)_2SO_4$$

The *alums* are a well known series of double sulphates of general formula $M^IM^{III}(SO_4)_2 12H_2O$, e.g. $KAl(SO_4)_2 . 12H_2O$, which crystallize as octahedra. They are acidic in solution, because of the hydrolysis which invariably occurs with salts of strong acids and polarizing cations. Aluminium sulphate and sodium alum ($M^I = Na$, $M^{III} = Al$) are used as flocculating agents in sewage disposal because of the increased efficiency of a multivalent ion ($Al^{3+}_{(aq)}$) in removing the stabilizing charge on colloidal particles (p. 91).

(b) *Nitrates*—As all the nitrates are soluble in water, they are prepared by reaction of nitric acid with the oxides, hydroxides or carbonates of the metals (although aluminium oxide, like the free metal, becomes passive in concentrated nitric acid), e.g.

$$NaOH + HNO_3 \rightarrow NaNO_3 + H_2O$$

$$BaO + 2HNO_3 \rightarrow Ba(NO_3)_2 + H_2O$$

The nitrates of the alkali metals yield the corresponding nitrite and oxygen when heated, but the remainder decompose to the oxide, with the evolution of nitrogen dioxide and oxygen

$$2KNO_3 \rightarrow 2KNO_2 + O_2 \uparrow$$

$$2Sr(NO_3)_2 \rightarrow 2SrO + 4NO_2 \uparrow + O_2 \uparrow$$

312

Some of the nitrates find commercial application. Potassium nitrate is used in gun powder, in preference to the cheaper, naturally-occurring sodium nitrate because the latter is hygroscopic. Both sodium and potassium nitrates are important nitrogenous fertilizers. The nitrates of the Group 2 metals are used in pyrotechnics to produce characteristic coloured fires.

Aluminium

Extraction

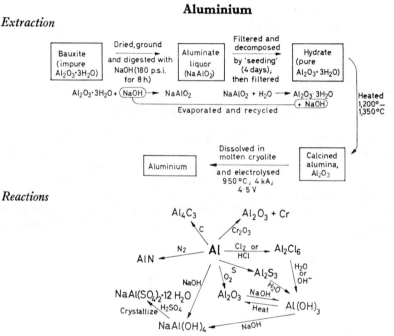

Reactions

Figure 14.5. Extraction and reactions of aluminium

Table 14.12. Output of bauxite and production and uses of aluminium (1966)

	Bauxite		Aluminium primary production		Uses
	10^6 tonnes	per cent	10^6 tonnes	per cent	
World total	37·1		5·6		electrical cables; domestic
U.S.A.	2·2	7	2·7*	49	articles; building material
Canada	—	—	0·8	15	(after anodizing and dye-
Jamaica	9·2	25	—	—	ing); paints; light alloys,
Surinam	5·6	14	—	—	e.g. Duralumin; ex-
Guyana	4·3	12	—	—	plosives, e.g. Ammonal;
			* incl. secondary sources		foil

Organo-metallic Compounds

Mention has been made of the reaction between these metals and alcohols (p. 303); many other organo-metallic compounds can also be made. Thus magnesium dissolves in alkyl halides to produce Grignard reagents (p. 304) which react with metals or metallic halides to give *metal alkyls*, e.g.

$$Na^+CH_3^-$$

$$Mg \xrightarrow{CH_3I} CH_3MgI$$

with Na going to $Na^+CH_3^-$ and $BeCl_2$ going to

$$Be(CH_3)_2$$

Such compounds hydrolyse in water, e.g.

$$Be(CH_3)_2 + 2H_2O \rightarrow Be(OH)_2 \downarrow + 2CH_4 \uparrow$$

The simple covalent compounds of the elements beryllium, boron and aluminium are usually unstable because of electron deficiencies, and accordingly they often achieve stability by the acquisition of extra electrons by coordination. Organo-metallic derivatives of this type, involving two or more bonds from the same organic group, are called *chelate* (Greek chēlē = claw) compounds. Examples of these are 'basic' beryllium ethanoate (acetate), $Be_4O(CH_3COO)_6$, a covalent compound possessing the characteristic properties of high volatility and solubility in organic solvents

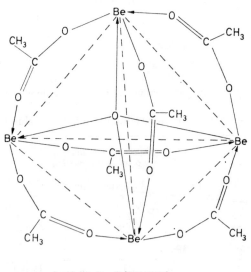

$$Be_4O(CH_3COO)_6$$

and the compound formed between aluminium ions and 8-hydroxyquinoline

ANALYSIS

The characteristic visible spectra of the elements sodium (yellow), potassium (lilac), calcium (red), strontium (crimson), barium (green), indium (indigo) and thallium (green) can be used both as qualitative and quantitative methods of analysis. Very few insoluble compounds of the alkali metals are known; those which are relatively insoluble and hence afford a means of identification when all other metallic ions have been removed, are potassium perchlorate, $KClO_4$, potassium hexanitrocobaltate(III), $K_3[Co(NO_2)_6]$, and sodium dizinc diuranyl(VI) acetate, $NaZn_2(UO_2)_2(CH_3COO)_9$.

The general solubility of the salts of the Group 2 metals results in their being detected towards the end of the conventional scheme of qualitative analysis. The solubility products of the carbonates of calcium, strontium and barium are exceeded, however, when ammonium carbonate is added to solutions containing these cations, even in the presence of ammonium chloride (common ion effect, p. 157), which does still prevent the precipitation of magnesium carbonate. Magnesium is precipitated as magnesium ammonium phosphate, $MgNH_4PO_4$, after removal of all other elements except the alkali metals by previous precipitations. The phosphate is quantitatively precipitated and can be filtered, washed and ignited to convert it into magnesium diphosphate, $Mg_2P_2O_7$; this affords a convenient gravimetric estimation of the element.

Aluminium hydroxide is precipitated as a colourless, gelatinous precipitate by ammonium hydroxide. This, too, is quantitative, and ignition to form aluminium oxide provides a suitable gravimetric estimation. Other quantitative precipitations which form the bases for gravimetric estimations of the relevant metals, are calcium as oxalate and strontium and barium as sulphate.

Boron can be estimated as boric acid by titration in the presence of glycerol.

QUESTIONS

1. Show that the oxidation numbers of the elements of Groups 1, 2 and 3 in the following compounds are consistent with their valencies: $Na_2B_4O_7$, $Mg_2P_2O_7$, $Na_3Co(NO_2)_6$, $K_4Fe(CN)_6$ and $NaZn_2(UO_2)_2(CH_3COO)_9$.

2. Discuss the changes that would occur to a piece of sodium left exposed to the air.

315

3. Burning the alkali metals in air gives the following compounds: Li_2O, Na_2O_2, KO_2, RbO_2 and CsO_2. Discuss the structures of these compounds and suggest reasons for their formation.

4. Explain under what conditions the following substances react together and say why the reactions take place:
calcium hydroxide and sodium carbonate; aluminium chloride and lithium hydride; calcium nitride and water; calcium sulphate and ammonium carbonate.

5. Discuss the structures of diborane, B_2H_6, and its addition compound with ammonia, $B_2H_6.2NH_3$. On heating the latter, borazole, $B_3N_3H_6$, and hydrogen are formed. Borazole reacts with hydrogen chloride, first giving $B_3N_3H_9Cl_3$ and then $B_3N_3H_6Cl_3$. Because of the similarity in structure, borazole has been called 'inorganic benzene'. On this basis give the structures of the compounds mentioned.

6. The radii of Rb^+, Ca^{2+}, I^- and O^{2-} are 148, 99, 216 and 140 pm, respectively. What would be the expected crystal structures of rubidium iodide and calcium oxide? (Refer to Chapter 4.)

7. Why is it possible to use sodium carbonate as an alkali? Why are heavy-metal carbonates precipitated by sodium hydrogen carbonate in preference to the former?

8. How would you titrate (a) borax against acid, (b) boric acid against alkali?

9. How would you analyse calcium, magnesium, sodium and lithium in the presence of each other?

10. How far can the ammonium ion be regarded as a member of Group 1?

11. 'There exists a diagonal relationship between lithium and magnesium, beryllium and aluminium, and boron and silicon.' Provide evidence for this statement and suggest a reason for there being such a relationship.

12. Explain the following:
A little methyl orange was added to some borax solution. The resultant colour was yellow, but it turned pink on the addition of 20 cm³ of 0·1 M hydrochloric acid. Some neutral glycerol was then added, followed by phenolphthalein. 0·1 M sodium hydroxide was subsequently run into the solution until a permanent red colour was obtained; the volume of NaOH required was 10 cm³.

CHAPTER 15

THE TRANSITIONAL ELEMENTS—'d BLOCK ELEMENTS'

Group $(n-1)d$ ns $(n-1)d$ ns Group
2 [↑| | | |] [↑↓] ⟶ [↑↓|↑|↑|↑|↑|↑|↑|↑|↑] [↑↓] 3

Ca	Sc	Ti	V	Cr	Mn	Fe	Co	Ni	Cu	Zn	Ga
Sr	Y	Zr	Nb	Mo	Tc	Ru	Rh	Pd	Ag	Cd	In
Ba	La ↑	Hf	Ta	W	Re	Os	Ir	Pt	Au	Hg	Tl

Lanthanides

There are three series of so-called 'transitional' elements in which the $3d$, $4d$ and $5d$ orbitals, respectively, are in the process of being filled with electrons. (From the point of view of the Periodic Table, they can be regarded as effecting a *transition* from the electropositive elements of the s block, on the one hand, to the more electronegative elements of the p block on the other.)

There are ten elements in each series, if one includes zinc, cadmium and mercury in which the final electron needed to complete the d orbitals has been added. These latter three elements will therefore show departures from those characteristic properties of transitional metals which are the consequence of the d orbitals being incomplete.

Table 15.1

	$3d$	$4s$	Oxidation states (common ones boxed)
Scandium, Sc	[↑| | | |]	[↑↓]	[3]
Titanium, Ti	[↑|↑| | |]	[↑↓]	−1, +2, 3, [4]
Vanadium, V	[↑|↑|↑| |]	[↑↓]	−1, +1, 2, [3], 4, 5
Chromium, Cr	[↑|↑|↑|↑|↑]	[↑]	−2, −1, +1, 2, [3], 4, 5, 6
Manganese, Mn	[↑|↑|↑|↑|↑]	[↑↓]	−3, −1, +1, [2], 3, 4, 5, 6, 7
Iron, Fe	[↑↓|↑|↑|↑|↑]	[↑↓]	−2, +1, [2], [3], 4, 5, 6
Cobalt, Co	[↑↓|↑↓|↑|↑|↑]	[↑↓]	−1, +1, [2], 3, 4
Nickel, Ni	[↑↓|↑↓|↑↓|↑|↑]	[↑↓]	−1, +1, [2], 3, 4
Copper, Cu	[↑↓|↑↓|↑↓|↑↓|↑↓]	[↑]	+[1], [2], 3
Zinc, Zn	[↑↓|↑↓|↑↓|↑↓|↑↓]	[↑↓]	2

Table 15.1 shows how the orbitals of the d block are filled for the first transition series. In accordance with Hund's rule, electrons occupy the orbitals singly first, before there is any pairing. The increased stability when all five d orbitals possess either one or two electrons is indicated by the way in which an electron is transferred from the $4s$ orbital to a d orbital in the case of chromium and copper. One thing that quickly emerges from consideration of *Table 15.1* is the large electron availability of most of these metals. They are therefore very good conductors, and both copper and silver find extensive commercial application on this account. Furthermore, there is powerful metallic bonding, and the elements are, on the whole, strong and with high melting points. These are, in fact, the substances readily recognizable to the non-scientist as metals. They are widely used where great strength and hardness are required; they are largely miscible with each other, and the resultant alloys often have enhanced properties (*Table 15.2*).

Table 15.2. Hardness and tensile strength of some transitional metals and alloys

Relative hardness (Brinell)		Tensile strength kg cm^{-2}
Copper, wrought	50	2 200
Brass, 70/30	160	5 500
Iron, wrought	100	3 140
Mild steel	130	4 710
Nickel–chromium steel	400	13 500
For comparison: wrought aluminium	27	944

Melting points tend to become higher on descending a series and reach a peak with tungsten, which is therefore used in the filament of electric light bulbs. Zinc, cadmium and mercury are exceptional: their melting points not only are very low but decrease from zinc to mercury, which is the only liquid metal at room temperature—at least in temperate climates—and so is used in thermometers, barometers, etc. (*Table 15.3*).

Table 15.3

Element	Melting point °C	Element	Melting point °C	Element	Melting point °C
Ti	1 725	Zr	1 860	Hf	2 200
V	1 700	Nb	2 410	Ta	2 850
Cr	1 800	Mo	2 625	W	3 380
⋮		⋮		⋮	
Zn	419	Cd	321	Hg	−39

It is also found that, as a particular series is traversed horizontally, the size of the atoms steadily contracts (*Figure 15.1, Table 15.4*). The explanation lies in the fact that the diffuse d orbitals do not appreciably affect the overall size of the atom, whilst the steadily increasing number of protons exerts a progressively greater attractive force on the outer s orbitals, causing them to contract in size. In the case of the third series, where the $5d$ orbitals are being filled, the further effect of the lanthanide contraction must be added. One immediate consequence of this is that the atom of hafnium, coming directly after the lanthanides, is no bigger than that of zirconium: not only does hafnium replace zirconium atoms in minerals, but their chemistry is virtually identical, and their separation is a matter of great difficulty.

With the atomic volume actually decreasing as nucleons are added, the density must inevitably increase: metals on the right of a series are very dense, those in later series particularly so (*Table 15.4*). Thus osmium, with a relative density of 22·5, is the densest element, closely followed by iridium.

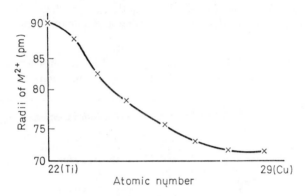

Figure 15.1. Size of atoms

Table 15.4. Atomic size and density

Element	Atomic size, pm	Density, g cm⁻³	Element	Atomic size, pm	Density, g cm⁻³	Element	Atomic size, pm	Density, g cm⁻³
Ti	132	4·5	Zr	145	6·53	Hf	144	13·07
V	122	6·0	⋮			⋮		
Cr	117	7·1	⋮			⋮		
Mn	117	7·4	⋮			⋮		
Fe	116	7·9	Ru	124	12·2	Os	126	22·5
Co	116	8·7	⋮			⋮		
Ni	115	8·9	⋮			⋮		
Cu	117	8·9	⋮			⋮		
Zn	125	7·1	Cd	141	8·6	Hg	144	13·6

Note: very high density of osmium; atomic radii of zirconium and hafnium; atypical behaviour of zinc, cadmium and mercury.

319

Another consequence of shrinking atomic size and increased nuclear charge is a stronger attraction between nucleus and electrons; that is, the electronegativity increases, together with ionization energies (*Table 15.5*). Therefore, metals towards the end of a series are inactive and resistant to corrosion; in the last two series are found the noble metals which are so valuable because of their resistance to atmospheric oxidation and which are used extensively in jewellery and coinage.

Whilst there is no great change in atomic dimensions throughout the transitional series, nonetheless the differences are sufficient to bring about 'locking' of the crystal planes during alloying, with increase in strength. For instance, alloys of iron and manganese are tough enough to be used in railway lines, whilst alloys of iron with chromium and tungsten have the hardness required in high-speed tools. The transitional metals are also able to accommodate small atoms such as boron, carbon and nitrogen in holes in the lattice, the resultant, non-stoichiometric borides, carbides and nitrides often possessing desirable qualities.

Table 15.5. Electrode potentials and ionization energies
(electrode potential for most common ion)

Element	Electrode potentials, V	Ionization energies, kJ mol^{-1}				
		1st	2nd	3rd	4th	5th
Ti	−0·95	659	1 313	2 720	4 180	9 610*
V	−1·5	650				
Cr	−0·71	654				
Mn	−1·55	717	1 517	3 090†		
Fe	−0·44	762	1 561	2 960		
Co	−0·28	758				
Ni	−0·25	738	1 754	3 440		
Cu	+0·34	747	1 960†	2 850		
Zn	−0·76	910	1 735	3 870†		

* Electron removed from inner shell.
† High energy required to remove a d electron when all the orbitals are full or half full.

The energy levels of the penultimate d orbitals are so close to each other and to the energy level of the outer s orbital that it is relatively easy to remove or add varying numbers of electrons and to give different oxidation states (see *Table 15.1*). As is to be expected, this feature is particularly pronounced in the middle of each row (where seven electrons are available for valency purposes) and declines towards the end of the series, so that zinc, cadmium and mercury are always divalent. Many of the different oxidation states are able to absorb light of different energy (and wavelength) from the visible spectrum and hence appear coloured. Because of the small energy differences involved, conversion of a transition metal from one oxidation state to another is usually a fairly simple operation:

Oxidation Number

0	+1	+2	+3	+4	+5	+6	+7

V V^{2+} $\overset{Mg}{\rightleftharpoons}$ V^{3+} $\overset{Zn}{\rightleftharpoons}$ $[VO]^{2+}$ $\overset{SO_2}{\rightleftharpoons}$ $[VO_4]^{3-}$
         violet   green   blue   yellow-red

Cr $Cr^{3+}\overset{C_2H_5OH}{\underset{H^+}{\longleftarrow\!\!\!\!\!-\!\!\!\!\!-\!\!\!\!\!-}}[CrO_4]^{2-}$
             green   yellow

Mn Mn^{2+} $MnO_2 \longleftarrow [MnO_4]^{2-}\overset{Alkene}{\underset{OH^-}{\longleftarrow}}[MnO_4]^-$
         pink   brown   green   purple
                              Fe^{2+}, H^+

Cu $Cu_2O\overset{tartrate}{\underset{OH^-}{\longleftarrow\!\!\!-\!\!\!-}}Cu^{2+}$
     red   blue

One use of unstable oxidation states is afforded by the NIFE cell in which a higher oxide of nickel (of approximate composition Ni_2O_3) occurs

$$Fe \text{ (negative electrode)} + 2OH^- \text{ (from electrolyte, KOH)} \longrightarrow Fe(OH)_2 + 2e$$

$$Ni_2O_3 \text{ (on nickel electrode)} + H_2O + 2H^+ + 2e \longrightarrow 2Ni(OH)_2$$

Those oxidation states in which the d orbitals are either completely empty or full, or in which each orbital contains one electron, are characterized by greater stability relative to the others, e.g.

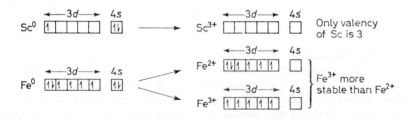

Paramagnetism

Because of its charge, a moving electron, whether the movement is translational or rotational, is associated with a magnetic field. If an atom contains an unpaired electron, then it gives positive interaction with an external magnetic field; that is, there is attraction and the lines of force become more concentrated. Such a substance is said to be *paramagnetic*. In the case of a *diamagnetic* substance, there are no unpaired electrons, and repulsion occurs when the substance is placed in a magnetic field (*Figure 15.2*). Measurements of magnetic moments can therefore provide significant information about the electronic condition of substances.

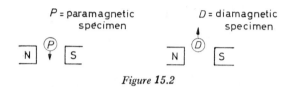

Figure 15.2

Figure 15.3 shows how the paramagnetic moments of the ions of the first transition series vary regularly with the number of unpaired electrons.

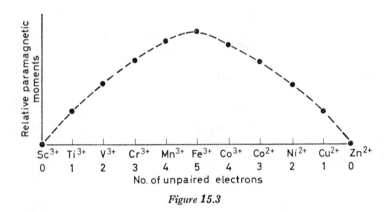

Figure 15.3

Coordination Compounds

Coordination compounds are formed when ions or molecules (*ligands*) possessing one or more lone pairs of electrons donate these electrons to a central metallic atom or ion to form a complex ion or molecule. Transitional metals, by virtue of the vacant orbitals which they possess and because of the high polarizing power of the small ions of high charge which attract lone pairs towards them, are outstanding in their ability to form coordination compounds. As the water molecule is itself a ligand, it would be naive to expect transitional metal ions to remain unhydrated in aqueous solution. They usually coordinate to four or six molecules of water and it is this water which, persisting in the crystal lattice of the solid, is known as water of crystallization. Not all the water of crystallization, however, need be coordinated to the metallic atom. In the case of copper(II) sulphate pentahydrate, $CuSO_4.5H_2O$, there are four ligands attached to the copper ion: the remaining water molecule is linked by hydrogen bonding to the sulphate ion. (That one molecule of water is attached differently to the other four is shown by the fact that the last molecule of water is removed only by prolonged heating at 200°)

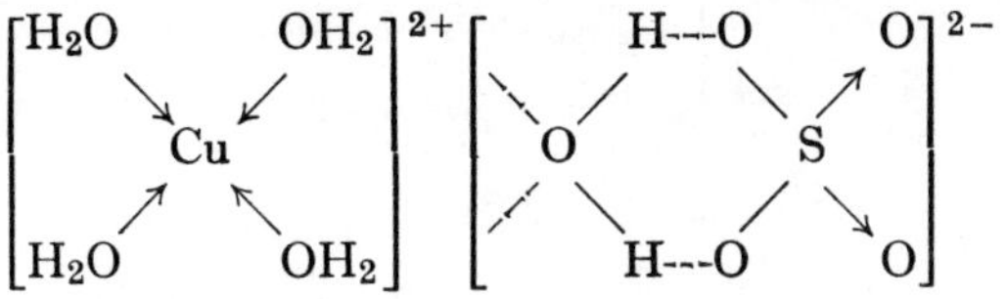

$$\begin{bmatrix} H_2O & & OH_2 \\ & Cu & \\ H_2O & & OH_2 \end{bmatrix}^{2+} \begin{bmatrix} H\text{---}O & & O \\ & O & S \\ H\text{---}O & & O \end{bmatrix}^{2-}$$

Other atoms containing lone pairs capable of being donated to a metallic ion are carbon (especially in the form of the cyanide ion, CN^-), halogen, when present as anion X^-, and nitrogen, particularly as ammonia (when it forms *ammines*) and its organic derivatives such as amines. If the ligand is neutral, then the complex carries the positive charge of the central metallic ion and is known as a cationic complex; if it carries a sufficiently negative charge, then the overall charge of the complex will be negative and it will be anionic. Some examples are

Cationic complexes

$[Cr^{II}(H_2O)_6]^{2+}$ hexaaquachromium(II)

$[Cu^{II}(H_2O)_4]^{2+}$ tetraaquacopper(II)

$[Cr^{III}(NH_3)_6]^{3+}$ hexaamminachromium(III)

$[Cu^{II}(NH_3)_4]^{2+}$ tetraamminecopper(II)

$[Cu^{I}(CO)_2]^+$ dicarbonyl copper(I)

Anionic complexes

$[Fe^{II}(CN)_6]^{4-}$ hexacyanoferrate(II)

$[Fe^{III}(CNS)_6]^{3-}$ hexathiocyanatoferrate(III)

$[Ni^{II}(CN)_4]^{2-}$ tetracyanoniccolate(II)

$[Co^{II}Cl_4]^{2-}$ tetrachlorocobaltate(II)

$[Ta^{V}F_8]^{3-}$ octafluorotantalate(V)

$[Mn^{VII}O_4]^-$ tetraoxomanganate(VII)

In some cases, a ligand can donate more than one lone pair and is then known as e.g. a, tri-dentate ligand. Two well-known and useful examples of ligands of this type are 1,2-diaminoethane (ethylenediamine, *en*) and its tetra-acetate derivative (E.D.T.A.)

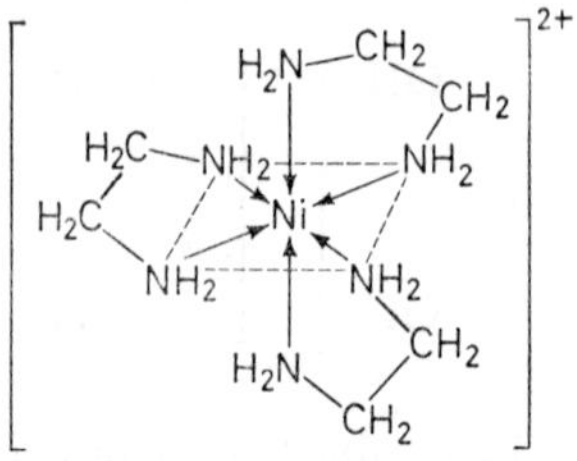

Nickel complex of ethylenediamine

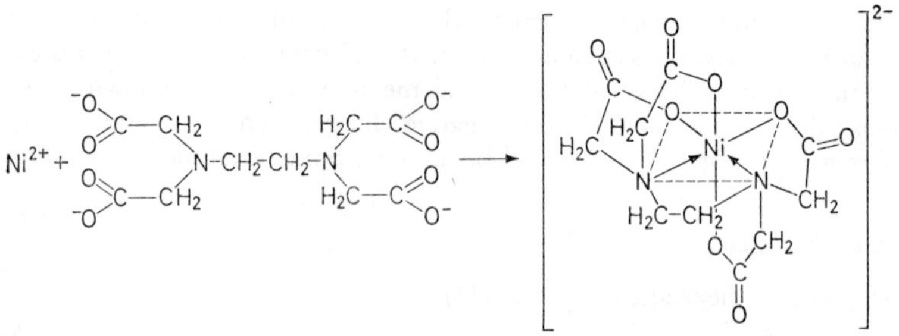

Nickel complex of
ethylenediamine tetra-acetate

Ligands which give ring formation are often described as *chelating* agents and the ion is said to be *sequestered*.

Complete or partial replacement of one ligand by another can take place, e.g.

$$[Co(NH_3)_6]^{2+} + 6H_3O^+ \rightarrow [Co(H_2O)_6]^{2+} + 6NH_4^+$$

$$[Cu(H_2O)_6]^{2+} + 4NH_3 \rightarrow [Cu(NH_3)_4(H_2O)_2]^{2+} + 4H_2O$$
$$\text{Diaquatetraammine}$$
$$\text{copper(II)}$$

The formation of complexes can stabilize a transitional metal ion. For example, simple cobalt(III) ions do not exist, and yet its complex with the nitrite ion is formed readily from the cobalt(II) ion by aerial oxidation

$$2Co^{2+} + 12NO_2^- + 2H^+ + \tfrac{1}{2}O_2 \rightarrow 2\,[Co^{III}(NO_2)_6]^{3-} + H_2O$$
$$\text{Hexanitrocobaltate(III)}$$

Those complexes tend to be most stable which contain completed orbitals, and this explains the oxidation of cobalt(II) in the above example and also the stability of hexacyanoferrate(II)

324

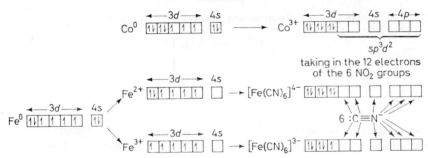

Whereas Fe^{3+} is more stable than Fe^{2+}, $[Fe(CN)_6]^{3-}$ is less stable than $[Fe(CN)_6]^{4-}$ and needs one electron to complete the d orbitals, i.e. it is an oxidizing agent.

It can be said very roughly that ligands are so disposed in space about the central atom that there is the maximum distance between the lone pairs. A more reliable and precise guide is given in terms of hybridization of the relevant orbitals:

sp hybridization gives a linear arrangement, e.g. $[H_3N \rightarrow Ag \leftarrow NH_3]^+$

sp^3 hybridization gives a tetrahedral arrangement, e.g.

sp^3d hybridization gives a trigonal bipyramid, e.g.

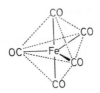

sp^3d^2 gives an octahedral configuration, e.g.

325

sp^2d hybridization gives a planar arrangement, e.g.

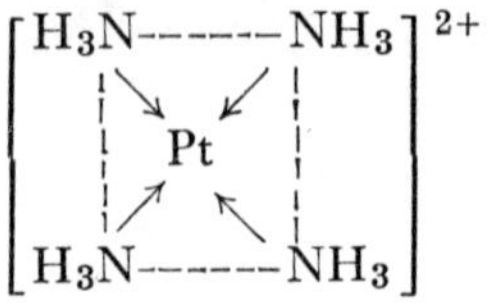

$$\left[\begin{array}{c} H_3N\text{------}NH_3 \\ Pt \\ H_3N\text{------}NH_3 \end{array}\right]^{2+}$$

Stereoisomerism (p. 361) is possible with many coordinated compounds. Thus, two geometric isomers of the planar diamminedichloroplatinum(II) exist:

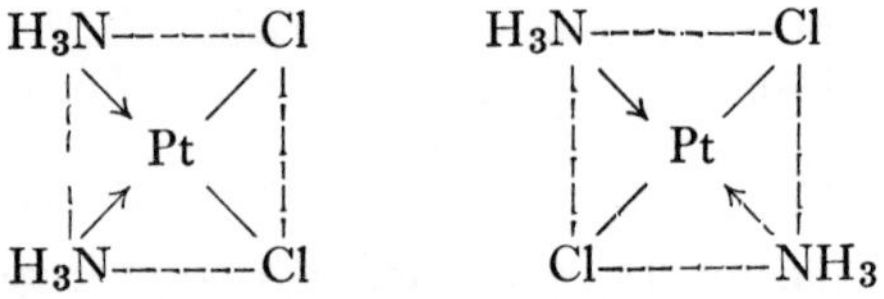

cis-   trans-

For the ion, bis(ethylenediamine)dichlorocobalt(III), three structures based on an octahedral configuration are possible

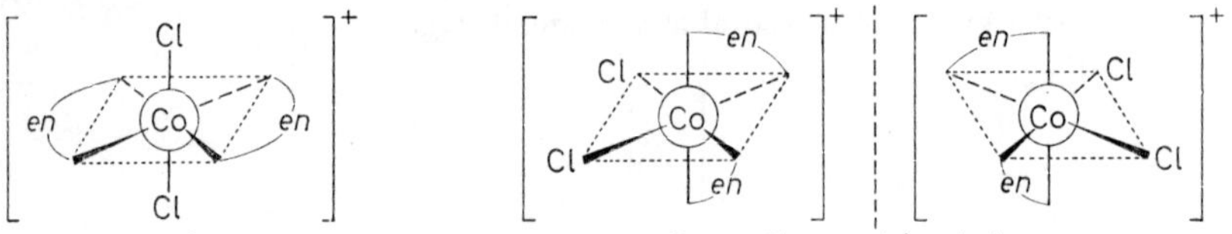

trans form   2 enantiomorphic *cis* forms

en = ethylenediamine

Here the two *cis* forms, although having two ligands disposed in the same order around the central atom, cannot be superimposed and, in fact, are mirror images of each other; that is, they are optical isomers (enantiomorphs) and rotate the plane of polarized light in different directions.

Occurrence

A large number of the transitional elements occur either as their oxides or as the central atom in complex oxyanions (see *Table 15.6*). Notable exceptions are the sulphide ores of molybdenum, iron, cobalt, nickel, copper and the zinc group.

Extraction

The elements at the beginning of each row are quite electropositive, their affinity for oxygen being such that reduction with vigorous reducing agents like aluminium is necessary for extraction of the metal, unless one has recourse to indirect methods.

Table 15.6. Sources of the transitional elements

Element	Source(s)
Titanium	ilmenite, $FeTiO_3$; rutile, TiO_2
Zirconium	baddeleyite, ZrO_2; zircon, $ZrSiO_4$ (with 1–2 per cent hafnium)
Vanadium	vanadinite, $3Pb_3(VO_4)_2 . PbCl_2$
Niobium	niobite, $Fe(NbO_3)_2$ (containing tantalum)
Tantalum	tantalite, $Fe(TaO_3)_2$ (containing niobium)
Chromium	chromite, $Fe(CrO_2)_2$; crocoisite, $PbCrO_4$
Molybdenum	molybdenite, MoS_2; wolfenite, $PbMoO_4$
Tungsten	wolframite, $FeWO_4/MnWO_4$; scheelite, $CaWO_4$; tungstite, WO_3
Manganese	pyrolusite, MnO_2; hausmannite, Mn_3O_4
Technetium	traces only in nature
Rhenium	molybdenite—up to 20 p.p.m.
Iron	magnetite, Fe_3O_4; haematite, Fe_2O_3; pyrites, FeS_2
Cobalt	in sulphides of nickel and copper; smaltite, $CoAs_2$
Nickel	pentlandite, $NiS . 2FeS$; garnierite, $MgNi$; silicates
Platinum metals	ruthenium, rhodium, palladium, osmium, iridium; native platinum
Copper	native; copper pyrites, $CuFeS_2$; cuprite, Cu_2O; malachite, $CuCO_3 . Cu(OH)_2$
Silver	argentite, Ag_2S; horn silver, $AgCl$; native
Gold	native
Zinc	zinc blende,wurtzite, ZnS; calamine, $ZnCO_3$ (with small amounts of cadmium)
Mercury	cinnabar, HgS

For instance, in the production of titanium, zirconium and hafnium, the oxide is converted to the tetrachloride and this is then reacted with molten magnesium in an inert atmosphere (cf. *Figure 8.8*, p. 182)

$$XCl_4 + 2Mg \rightarrow X + 2MgCl_2$$

Aluminothermal reduction of the oxide is the normal method for the production of vanadium, chromium and manganese

$$3M_2^{n+}O_n + 2nAl \rightarrow 6M + nAl_2O_3$$

Several of the remaining metals can be extracted from their oxides by reduction with carbon; this applies particularly to molybdenum, iron, tungsten, nickel, zinc and cadmium. Sulphides must first be converted to oxide by roasting in air, e.g.

$$2ZnS + 3O_2 \rightarrow 2ZnO + 2SO_2 \uparrow$$

$$ZnO + C \rightarrow Zn + CO \uparrow$$

Nickel obtained in like manner is then usually purified by the Mond process which consists of passing carbon monoxide over the crude metal at about 90°. Reaction occurs, and gaseous nickel tetracarbonyl is formed

327

The carbonyl is then heated to about 180°, when it decomposes and deposits pure nickel

$$Ni + 4CO \underset{180°}{\overset{90°}{\rightleftharpoons}} Ni(CO)_4$$

Reactions

Many of the transition metals to the right of the series (particularly at bottom right) are noble in character and come below hydrogen in the electrochemical series. They are therefore unattacked by non-oxidizing and non-complexing acids. (Iron, cobalt, nickel and chromium, with negative electrode potentials, are rendered surprisingly passive by concentrated nitric acid.) The tendency of the transitional metals to complex formation is revealed by copper dissolving in concentrated hydrochloric acid, and by gold in aqua regia as well as in potassium cyanide solution

$$Cu \xrightarrow{HCl} H_2[CuCl_4] + H_2 \uparrow$$

$$Au \xrightarrow[O_2]{CN^-} [Au(CN)_2]^-$$

$$Au \xrightarrow[HNO_3]{HCl} [AuCl_4]^-$$

The elements react on heating with halogen, and often on heating with sulphur

$$Ni \xrightarrow{Cl_2} NiCl_2$$

$$Fe \xrightarrow{S} FeS$$

Those to the left of the series will also form nitrides on heating in nitrogen and will decompose steam

$$Ti \xrightarrow[800°]{N_2} TiN$$

$$Cr \xrightarrow{steam} Cr_2O_3$$

Many transitional elements are able to accommodate hydrogen atoms in the interstices of the crystal lattice to give non-stoichiometric hydrides. Palladium is particularly prominent in this respect: at 80°C and $10^5 Nm^{-2}$ pressure it can absorb 900 times its own volume of hydrogen.

COMPOUNDS

Halides

As the oxidation number of the metal increases, so also does the covalent nature of

VANADIUM

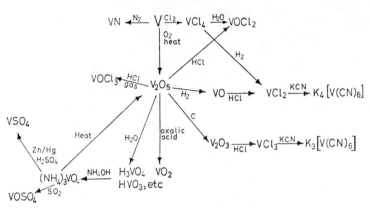

CHROMIUM

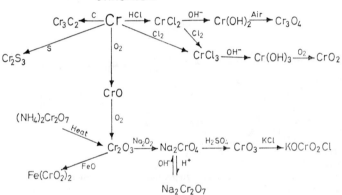

Figure 15.4a. Reactions of vanadium and chromium

the halide. Thus iron(II) chloride is a simple salt, $Fe^{2+}Cl_2^-$, whilst iron(III) chloride can exist as discrete Fe_2Cl_6 molecules

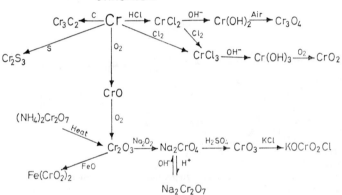

In keeping with its covalent character, iron(III) chloride is volatile and soluble in many organic liquids, when it often appears to exist as $FeCl_3$ molecules.

The consequence of increasing oxidation number is increasing polarizing power. The strongly polarizing Fe^{3+} attracts the oxygen of water, so that its halides are readily hydrated. Furthermore, the oxygen–hydrogen bond in the water molecule

MANGANESE

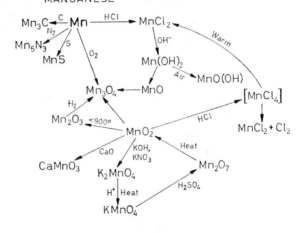

COBALT

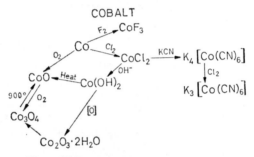

Figure 15.4b. Reactions of manganese and cobalt

is weakened, so that a hydrogen ion can be lost and the system reacts as an acid

$$Fe_2Cl_6 \overset{H_2O}{\rightleftharpoons} [Fe . 6H_2O]^{3+} \rightleftharpoons [Fe . 5H_2O(OH)^-]^{2+} + H^+, \text{ etc.}$$

Iron(II) chloride also forms a hexahydrate, but in this case there is less tendency to hydrolysis.

Similar considerations apply to compounds of the transitional elements with other non-metals and their oxy compounds.

Which particular oxidation state is the most stable depends to some extent on the halogen involved. The very electronegative fluorine tends to reveal the highest oxidation state of the metal. In the case of tantalum, for example, all the complexes TaF_6^-, TaF_7^{2-} and TaF_8^{3-}, with tantalum in its highest oxidation state of five, are known. Sometimes, the fluoride of the metal in its highest oxidizing state is not very stable and decomposes fairly easily to give 'nascent' fluorine, so that it can be used as a fluorinating agent, e.g. silver(II) fluoride, $AgF_2 \rightarrow AgF + [F]$.

IRON

Extraction

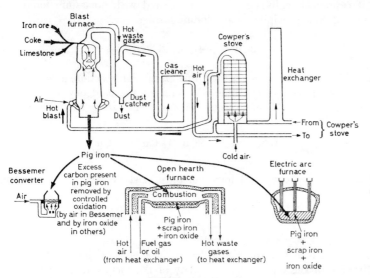

The linings of steel furnaces are either basic or acidic, to remove various impurities; e.g. calcined dolomite, CaO.MgO, removes SiO_2 (as $CaSiO_3$) and P_4O_{10} e.g. as $Mg_3(PO_4)_2$.

Reactions

Figure 15.5

Crude steel production (1966)		*Extraction of iron ore* (1966)		
10^6 tonnes	per cent		(Fe content) 10^6 tonnes	per cent
World (excl. China) 460		World (excl. China)	319	
U.S.A. 122	26	U.S.S.R.	93	29
U.S.S.R. 97	21	U.S.A.	52	17
Japan 48	10			
West Germany 35	8			
U.K. 25	5			

The iodide ion, on the other hand, has a tendency to be associated with the metal in a lower oxidation state (in other words, it is a reducing agent: $I^- \to \frac{1}{2}I_2 + e$). Thus, if copper(II) sulphate solution is treated with potassium iodide solution, copper(I) iodide is precipitated and iodine formed

$$2Cu^{2+} + 4I^- \to 2CuI + I_2$$

It should be added, though, that this particular reaction is favoured by the insolubility of the copper(I) iodide which offsets the tendency of copper(I) to be converted into copper(II) in aqueous systems, indicated by the following data

$$K = \frac{Cu^{2+}}{Cu^+} \sim 10^6: \qquad 2Cu^+ \to Cu^{2+} + Cu \qquad E° = 0.37 \text{ V}$$

It follows that free copper(I) ions can only exist in very low concentration in aqueous solution.

Oxides and Oxy Compounds

The heats of formation of the oxides of the metals of the first transition series are indicative of the high electropositivity of the metals at the beginning of the period declining as the atomic number increases

	TiO_2	$\frac{1}{2}V_2O_5$	CrO_3	MnO	$\frac{1}{2}Fe_2O_3$	CoO	NiO	CuO
ΔH	-946	-775	-580	-387	-412	-240	-244	-156 kJ mol^{-1}

With the exception of scandium, the common oxidation number of two is evident from the formation of oxides and hydroxides of general formula MO and $M(OH)_2$, respectively. These are predominantly basic and many of them, for example VO, CrO and MnO, are reducing because of their tendency to acquire higher oxidation states. The hydroxides are especially reactive in this respect, e.g.

$$Mn(OH)_2 \xrightarrow{\text{air}} MnO(OH) \xrightarrow{\text{air}} MnO_2$$

i.e. $Mn^{+2}[Mn^{IV}O_2(OH)_2]^{2-}$
Manganese(II) dioxodihydroxomanganate(IV)

The preparations of the lower oxides have, in fact, often to be carried out under reducing conditions, e.g.

$$FeC_2O_4 \xrightarrow{\text{heat}} FeO + CO_2 \uparrow + CO \uparrow$$
Iron(II) oxalate

$$Cu^{II} \xrightarrow[\text{complex in alkaline soln.}]{\text{reduction of tartrate}} Cu_2O$$

$$V_2O_5 \xrightarrow[\text{high temperature}]{H_2} VO$$

NICKEL

Extraction (simplified)

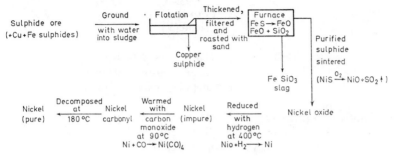

Reactions

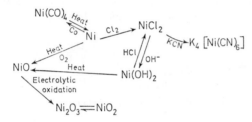

Figure 15.6

Whilst 2/3 of nickel ore comes from Canada, the U.S.A. extracts about half the total amount of metal. The main uses of nickel, as percentages are:

Stainless steel	42	electroplating	16
other alloys	36	chemicals	6

As the oxygen content rises, so also does the acidity of the oxide

Basic	Basic	Amphoteric		Acidic
MnO	Mn_2O_3	MnO_2		Mn_2O_7
$\downarrow H^+$	$\downarrow H^+$	$H^+\nearrow$	$\searrow OH^-$	$H_2O\downarrow$
Mn^{2+}	Mn^{3+}	Mn^{4+}	MnO_3^{2-}	$H^+[Mn^{VII}O_4]^-$

This trend accompanies the decrease in size and increase in cationic charge of the metallic atom, which results in high polarizing power and weakening of the bond between oxygen and hydrogen of any hydroxy group that becomes attached; that is, hydrogen ionizes and an oxyanion of the transitional metal remains

$$V_2O_5 \xrightarrow{\;OH^-\;} VO_3^-$$
Trioxo-
vanadate(V)

$$CrO_3 \xrightarrow{\;OH^-\;} [CrO_4]^{2-} \xrightarrow{\;H^+\;} [Cr_2O_7]^{2-}$$
Tetraoxo- μ-oxobistrioxo-
chromate(VI)   chromate (VI)

$$MnO_2 \xrightarrow[\text{air}]{\;OH^-\;} [Mn^{VI}O_4]^{2-} \xrightarrow{\;H_2O\;} [Mn^{VII}O_4]^- + MnO_2 + OH^-$$
Tetraoxo-   Tetraoxo-
manganate(VI)   manganate(VII)

'*Mixed*' *oxides* consist of metal atoms in different oxidation states; for example, Fe_3O_4, which is prepared by the action of steam on hot iron

$$3Fe + 4H_2O \rightleftharpoons Fe_3O_4 + 4H_2$$

should be represented as

$$Fe^{II}[Fe^{III}O_2]_2$$
Iron(II) bisdioxoferrate(III)

It reacts with acids, in accordance with this, to produce cations of different charge

$$Fe^{II}[Fe^{III}O_2]_2 \xrightarrow{\;H^+\;} Fe^{2+} + 2Fe^{3+} + H_2O$$

Many oxides of the transitional metals are non-stoichiometric. This again is a consequence of variable valency: ions of different charge are present in the crystal lattice, and so electrical neutrality does not require a simple ratio of atoms. For example, the formula of iron(II) oxide approximates to $Fe_{0.95}O$ because the crystal contains some Fe^{3+} ions.

Organometallic Compounds

The ultimate members of the transitional periods, zinc, cadmium and mercury, resemble the Group 2 elements in possessing two s electrons outside completed inner orbitals. But whereas the elements of Group 2 have only s and p orbitals complete in the penultimate shell, zinc, cadmium and mercury have the d orbitals complete as well. The correspondingly higher nuclear charge, the low shielding effect of d orbitals and the contraction of atomic size combine to make the latter three elements far less electropositive than their alkaline earth counterparts (*Table 15.7.*)

COPPER

Extraction

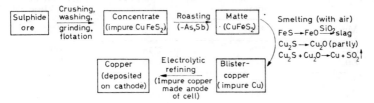

Reactions

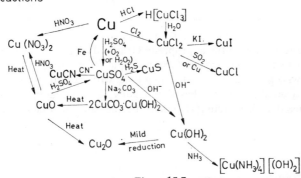

Figure 15.7

Production of refined copper (1966)	10^6 tonnes	per cent	Uses
World total (excl. Communist bloc)	5·3		metal: wire; ornaments; steam pipes; catalysis
U.S.A.	2·0	38	alloys: brass (Cu 75, Zn 25)
Zambia	0·5	9	bronze (Cu 92, Sn 8)
Japan	0·4	8	coinage 'silver' (Ag 50, Cu 40, Ni and Zn 5 each)
			sulphate: fungicide

Bearing in mind that elements high in a group are less electropositive than those below, comparisons with magnesium will be sought rather than with calcium, strontium and barium. Such a resemblance is found in organometallic compounds. Whilst the Grignard reagent, $RMgX$, is now far more popular, the first organometallic compounds to be discovered were the zinc alkyls, obtained by reacting alkyl halides with a zinc–copper couple and decomposing the intermediate, $RZnX$:

$$R\text{—}X \xrightarrow{\text{Zn/Cu}} R\text{—}Zn\text{—}X \xrightarrow{\text{heat}} ZnX_2 + R\text{—}Zn\text{—}R$$

(It is now doubtful whether the simple molecules $RZnX$ and $RMgX$ do in fact exist, see p. 404.)

335

Table 15.7

	Atomic radii, pm	Electrode potential, V	First ionization energy, kJmol^{-1}
Mg	136	−2·38	736
Ca	174	−2·76	590
Zn	125	−0·76	910
Sr	191	−2·89	549
Cd	141	−0·40	868
Ba	198	−2·90	507
Hg	144	+0·80	1 005

Cadmium and mercury alkyls also exist; CdR_2 and HgR_2 are formed when halides of the metal are reacted with the appropriate Grignard reagent. (It also seems that, unlike the other metals, mercury *can* form a simple $RHgX$ molecule.)

Mercury shows a great facility for forming links with carbon; for example, 'mer-curation' of the aromatic nucleus takes place readily, if it is suitably activated, e.g. as phenol, and reacted with mercury(II) ethanoate (acetate) in ethanoic (acetic) acid.

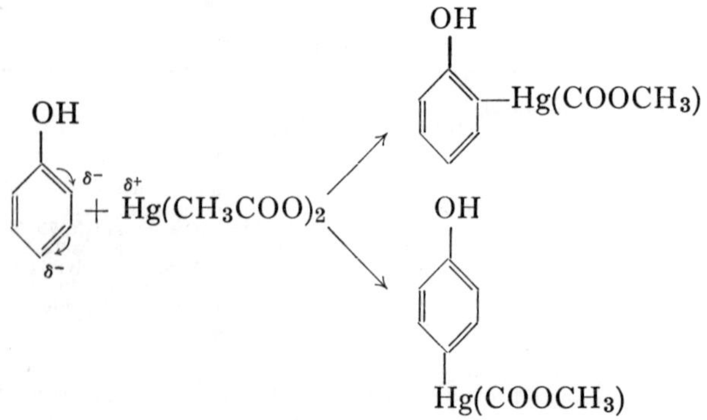

All these organometallic compounds appear to be simple covalent compounds with σ-bonding. There is also the possibility of π-bonding between the available d orbitals of the transitional metals and p electrons of unsaturated organic com-pounds. Great interest has been aroused in this aspect since the discovery of a very stable compound, ferrocene, by combination of Fe^{II} with the cyclopentadiene ion, $C_5H_5^-$. All the $3d$ elements have now been found to give the same type of compound. Other aromatic ring systems, possessing non-localized π-bonding, have also been

employed successfully

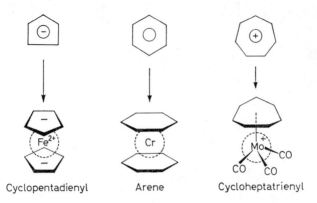

Cyclopentadienyl   Arene   Cycloheptatrienyl

The precipitates of acetylides, obtained when acetylene is passed through ammoniacal solutions of copper(I) and silver, may also involve interaction between the metal and the π-bonding of the acetylene.

ANALYSIS

The variable oxidation states and ability to form coordination compounds are widely utilized in the identification of the transitional metals.

Chromium

The green hydroxide precipitated by ammonium hydroxide, in the presence of ammonium chloride, dissolves in sodium peroxide with formation of yellow chromate

$$Cr(OH)_3 \rightarrow CrO_4^{2-}$$

Manganese

Mn^{2+} can be converted into green manganate by fusion with sodium carbonate and potassium chlorate, and into purple permanganate by oxidation with sodium bismuthate

$$Mn^{2+} \begin{array}{c} \nearrow^{NaBiO_3} MnO_4^- \\ \\ \searrow_{Na_2CO_3 + KClO_3} MnO_4^{2-} \end{array}$$

337

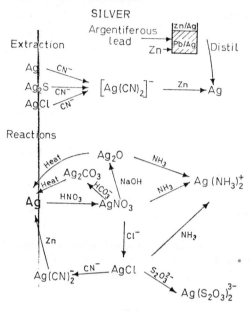

Figure 15.8

Production of silver (1966)	10^3 tonnes	per cent	Uses
World total	7·0		electroplating
Mexico	1·3	19	photography
U.S.A.	1·3	19	jewellery
Canada	1·0	14	coinage
Peru	1·0	14	

Iron

Fe^{3+}, precipitated as $Fe(OH)_3$, is converted into red Fe^{3+} $[Fe^{III}(CNS)_6]^{3-}$ by treatment with potassium thiocyanate.

Cobalt

Co^{2+} forms a similar complex, this time blue in colour, when reacted with the thiocyanate ion

$$Co^{2+} \xrightarrow{CNS^-} [Co(CNS)_4]^{2-}$$

Nickel

Black nickel sulphide is dissolved in acid and then treated with dimethyl glyoxime in alkaline solution; a red precipitate forms

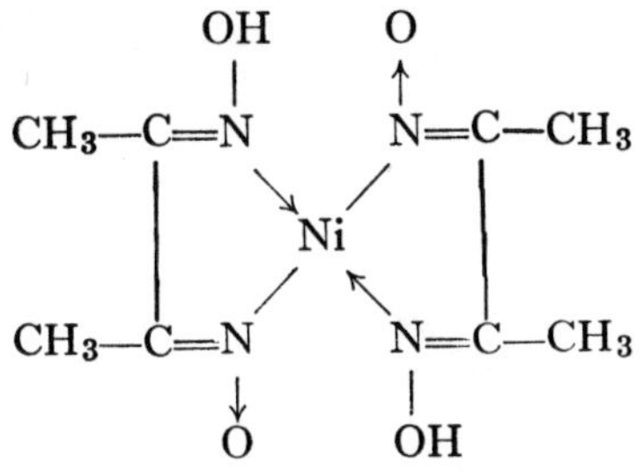

$$
\begin{array}{c}
\underset{\displaystyle CH_3-C=N}{\overset{\displaystyle OH}{|}} \qquad \underset{\displaystyle N=C-CH_3}{\overset{\displaystyle O}{\uparrow}} \\
\\
\text{Ni} \\
\\
\underset{\displaystyle \underset{O}{\downarrow}}{CH_3-C=N} \qquad \underset{\displaystyle OH}{N=C-CH_3}
\end{array}
$$

This precipitation is sufficiently quantitative to provide the basis for the estimation of nickel.

Copper

The presence of the copper(II) ion is indicated by the development of a deep blue complex when excess ammonia is added to an aqueous solution

$$[Cu(H_2O)_4]^{2+} \rightarrow [Cu(NH_3)_4]^{2+}$$

Copper is estimated volumetrically in terms of the iodine liberated when copper(II) is reduced to copper(I) by the addition of potassium iodide

$$2Cu^{2+} + 4I^- \rightarrow 2CuI\downarrow + I_2$$

Both copper and cadmium form complexes with cyanide ions

$$Cd^{2+} + 4CN^- \rightleftharpoons [Cd(CN)_4]^{2-} \qquad Cu^{2+} + 4CN^- \rightarrow [Cu(CN)_4]^{2-}$$

The cadmium complex dissociates sufficiently to allow cadmium to be precipitated by hydrogen sulphide (in contrast to the far more stable copper complex), and thus cadmium can be identified in the presence of copper.

Silver

Silver is precipitated, along with mercury(I) and lead, when an aqueous solution of its ions is treated with hydrochloric acid. It can, however, be distinguished from the other two metals by dissolving in excess ammonia with the formation of the soluble complex $[Ag(NH_3)_2]^+$.

339

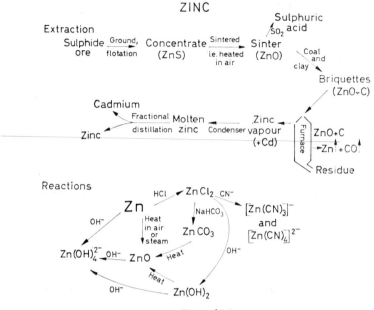

Figure 15.9

Production, consumption and uses of Zinc

Production (1966)	10^6 tonnes	per cent	Consumption (10^3 tonnes)	Uses
World total	3·4		U.K. 338	metal: galvanizing; batteries; roofing
U.S.A.	1·0	30	(incl. brass 114	alloys: brass (Zn 25, Cu 75)
Japan	0·5	15	galvanizing 95	compounds:
			sheet metal 27	oxide, paints and rubber
			oxide 25)	chloride, flux
				sulphate, mordant
				sulphide, paints

Mercury

Dimercury(II), unlike silver, does not dissolve in ammonia; instead, its *s* electron is used to form a covalency with the amino group; at the same time, free mercury is formed and the colour becomes black.

$$Hg_2^{2+} + NH_3 \rightarrow [Hg - NH_2]^+ + Hg + H^+$$

The mercury(II) ion is recognized by being reduced by tin(II), first to Hg_2^{2+} and then to metallic mercury

$$Hg^{2+} + Sn^{2+} \rightarrow Hg\downarrow + Sn^{4+}$$

340

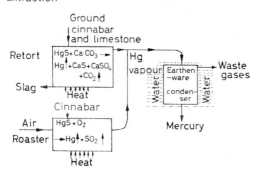

Extraction

Mercury

Reactions

Figure 15.10

Production and uses of Mercury

Production (1966) 10³ tonnes	per cent.	Uses.
World total 7.3		thermometers and barometers
Spain 2.7	38	discharge lamps
Italy 1·8	24	pumps
		amalgams
		Solvay cell
		medicine
		paints

QUESTIONS

1. Starting from the appropriate metal, how could the following compounds be prepared: iron(III) chloride, potassium hexacyanoferrate(III), potassium tetraoxomanganate(VII), potassium tetraoxochromate(VI) and potassium dicyanoargentate(I)?

2. Suggest how the alums of some of the transition elements might be prepared.

3. The formulae of the nitrosyl carbonyl and carbonyl of iron are $Fe(NO)_2(CO)_2$ and $Fe(CO)_5$, respectively. Show that these are in agreement with the attainment of a noble gas-type electronic structure.

4. Rationalize the following information by giving the structures of the molecules

$$KCrO_3Cl \xleftarrow{\text{KCl}} CrO_3 \xrightarrow{\text{KOH}} K_2CrO_4$$
$$\downarrow \text{HCl}$$
$$CrO_2Cl_2$$

5. How many compounds of the molecular formula $Co(NH_3)_xCl_3$ ($x \not> 4$) could there be? How might they be distinguished by using silver nitrate solution?

6. Given the following information:

$$[Fe(CN)_6]^{4-} \xrightarrow{\text{HNO}_3} [Fe(CN)_5(NO)]^{2-} \xrightarrow[\text{H}_2\text{O}]{\text{NH}_3} [Fe(CN)_5(NH_3)]^{3-} + HNO_2$$

describe the changes in the oxidation number of iron.

7. Give an account of the organometallic compounds of some of the transitional metals.

8. Give structures for the following compounds so as to accord with the observed molar conductivities (Ω^{-1} cm^2 mol^{-1})

$PtCl_4 . 6NH_3$	523	$PtCl_4 . 3NH_3$	97
$PtCl_4 . 5NH_3$	404	$PtCl_4 . 2NH_3$	0
$PtCl_4 . 4NH_3$	299		

9. How many isomers with formulae MA_2B_2, MA_2BC and $MABCD$ can exist if M is the central metallic atom and A, B, C and D are the ligands attached (i) tetrahedrally, (ii) in a planar form?

10. Predict what you can of the chemistry of an element with an outer configuration of d^7s^2.

11. In the presence of excess of mercury, the equilibrium constant for the reaction between Hg(I) and Hg(II) is given by [Hg(II)]/[Hg(I)] and not by [Hg(II)]/[Hg(I)]^2. Show that this is in agreement with the existence of mercury in the dimeric form, Hg$_2^{2+}$, rather than as Hg$^+$.

12. Discuss the coordination compounds of the coinage metals, copper, silver and gold, paying special attention to the stabilization of the silver(II) oxidation state.

13. Of what use in analysis are the coordination compounds of mercury?

14. In what respects is the chemistry of mercury unlike that of zinc?

15. Compare the stabilities of the oxides, chlorides and complex cyanides of five of the transition elements in relation to their oxidation numbers.

16. If the effect of heat on zinc oxide is represented by the equation:

$$Zn^{2+} + O^{2-} \rightarrow Zn_i^{2+} + \tfrac{1}{2}O_2 + 2e$$

where Zn_i^{2+} are the interstitial zinc ions formed, show that $[Zn_i^{2+}] \propto [O_2]^{-\frac{1}{6}}$

17. Some authorities consider zinc, cadmium and mercury to be transitional metals whilst others do not. State and justify your own views on this.

18. Explain the following:
(a) black copper oxide dissolves in hot, concentrated hydrochloric acid, to give a greenish solution; this, on boiling with copper is decolorized. When this latter solution is poured into water, a white precipitate is obtained;
(b) an aqueous solution of chromium(III) ions gives a green precipitate with ammonia solution; this precipitate dissolves in sodium peroxide to form a yellow solution, which turns orange on the addition of excess acid;
(c) on passing sulphur dioxide into ammonium vanadate solution, the colour turns from orange to blue; the addition of zinc powder changes the blue to green, whilst addition of magnesium gives a violet coloration;
(d) black manganese(IV) oxide gives a green mass on fusing with potassium chlorate and potassium hydroxide. Boiling a solution of this in the presence of excess carbon dioxide converts the colour to purple. If the carbon dioxide is replaced by sulphur dioxide, however, the solution is decolorized.

CHAPTER 16

THE f BLOCK ELEMENTS—LANTHANIDES AND ACTINIDES

La   Ce Pr Nd Pm Sm Eu Gd Tb Dy Ho Er Tm Yb Lu   Hf
Ac   Th Pa U Np Pu Am Cm Bk Cf Es Fm Md No Lw

In these two series of elements, inner f orbitals are being filled with electrons. As there are seven such orbitals, there are fourteen lanthanides and fourteen actinides in which the $4f$ and $5f$ orbitals, respectively, are being filled. *Table 16.1* shows, however, that these orbitals are not filled in a perfectly regular order.

Table 16.1

Lanthanides		Actinides	
Element [Xe]*	Electronic configuration	Element [Rn]*	Electronic configuration
Cerium	$4f$ ⊡ $\uparrow\uparrow$	Thorium	$5f$ ⊡ $6d^2$
Praseodymium	$\uparrow\uparrow\uparrow$	Protoactinium	$\uparrow\uparrow$ $6d^17s^2$
Neodymium	$\uparrow\uparrow\uparrow\uparrow$	Uranium	$\uparrow\uparrow\uparrow$
Promethium	$\uparrow\uparrow\uparrow\uparrow\uparrow$ $6s^2$	Neptunium	$\uparrow\uparrow\uparrow\uparrow$
Samarium	$\uparrow\uparrow\uparrow\uparrow\uparrow\uparrow$	Plutonium	$\uparrow\uparrow\uparrow\uparrow\uparrow\uparrow$ $7s^2$
Europium	$\uparrow\uparrow\uparrow\uparrow\uparrow\uparrow\uparrow$	Americium	$\uparrow\uparrow\uparrow\uparrow\uparrow\uparrow\uparrow$
Gadolinium	$\uparrow\uparrow\uparrow\uparrow\uparrow\uparrow\uparrow$ $5d^16s^2$	Curium	$\uparrow\uparrow\uparrow\uparrow\uparrow\uparrow\uparrow$ $6d^17s^2$
Terbium	$\uparrow\downarrow\uparrow\downarrow\uparrow\uparrow\uparrow\uparrow\uparrow$	Berkelium	$\uparrow\downarrow\uparrow\downarrow\uparrow\uparrow\uparrow\uparrow\uparrow$
Dysprosium	$\uparrow\downarrow\uparrow\downarrow\uparrow\downarrow\uparrow\uparrow\uparrow$	Californium	$\uparrow\downarrow\uparrow\downarrow\uparrow\downarrow\uparrow\uparrow\uparrow$
Holmium	$\uparrow\downarrow\uparrow\downarrow\uparrow\downarrow\uparrow\downarrow\uparrow\uparrow$ $6s^2$	Einsteinium	$\uparrow\downarrow\uparrow\downarrow\uparrow\downarrow\uparrow\downarrow\uparrow\uparrow$ $7s^2$
Erbium	$\uparrow\downarrow\uparrow\downarrow\uparrow\downarrow\uparrow\downarrow\uparrow\uparrow$	Fermium	$\uparrow\downarrow\uparrow\downarrow\uparrow\downarrow\uparrow\downarrow\uparrow\uparrow$
Thulium	$\uparrow\downarrow\uparrow\downarrow\uparrow\downarrow\uparrow\downarrow\uparrow\downarrow\uparrow$	Mendelevium	$\uparrow\downarrow\uparrow\downarrow\uparrow\downarrow\uparrow\downarrow\uparrow\downarrow\uparrow$
Ytterbium	$\uparrow\downarrow\uparrow\downarrow\uparrow\downarrow\uparrow\downarrow\uparrow\downarrow\uparrow\downarrow$	Nobelium	$\uparrow\downarrow\uparrow\downarrow\uparrow\downarrow\uparrow\downarrow\uparrow\downarrow\uparrow\downarrow$
Lutetium	$\uparrow\downarrow\uparrow\downarrow\uparrow\downarrow\uparrow\downarrow\uparrow\downarrow\uparrow\downarrow\uparrow\downarrow$ $5d^16s^2$	Lawrencium	$\uparrow\downarrow\uparrow\downarrow\uparrow\downarrow\uparrow\downarrow\uparrow\downarrow\uparrow\downarrow\uparrow\downarrow$ $6d^17s^2$

*The elements in each row have an inner core of electrons corresponding to either xenon or radon

Of these elements, promethium (61), americium (95) and all subsequent ones are unknown in nature but have been prepared artificially. The naturally occurring lanthanides (and thorium) are usually found together, e.g. in monazite sands, because of their great chemical similarity, which follows from their two outer electron shells having the same configuration for all except gadolinium, lutetium and thorium.

In Chapter 15 it was stated that the d orbitals are particularly stable when they are either completely empty (d^0), completely full (d^{10}) or half full (d^5). In the same way, the most stable states of f orbitals are f^0, f^7 and f^{14}. *Table 16.1* shows how elements tend to adjust their filling of other orbitals so as to attain this configuration. This tendency is also evident in the oxidation states exhibited by the different elements.

The normal oxidation state is $+3$. The reason for this is not immediately obvious from the electronic configurations, which suggest divalency by the loss of the $6s^2$ or $7s^2$ electrons. However, the energy acquired by the hydration of the tripositive ion of many of the elements is sufficient to compensate for that required to ionize the third electron. An added complication is introduced by the tendency to attain, or at least approach, f^0, f^7 or f^{14} configurations, which leads to cerium, praseodymium, neodymium, terbium and dysprosium of the lanthanide series also having a valency of four, whilst samarium, europium, thulium and ytterbium exhibit divalency. For the same reason, gadolinium and lutetium are particularly stable when in the trivalent state.

Variation of valency is more pronounced among the actinides, although the tripositive oxidation state tends to become more common as the series proceeds. *Table 16.2* shows the known oxidation states and corresponding electronic configurations for the two series of elements.

Table 16.2. Oxidation states and electronic configurations

Oxidation state	2	3	4	5		2	3	4	5	6
Ce		f^1	f^0		Th	f^0d^2	f^0d^1	f^0d^0		
Pr		f^2	f^1	f^0	Pa		f^2	f^1	f^0	
Nd		f^3	f^2		U		f^3	f^2	f^1	f^0
Pm		f^4			Np		f^4	f^3	f^2	f^1
Sm	f^6	f^5			Pu		f^5	f^4	f^3	f^2
Eu	f^7	f^6			Am		f^6	f^5	f^4	f^3
Gd		f^7			Cm		f^7	f^6		
Tb		f^8	f^7		Bk		f^8	7		
Dy		f^9	f^8		Cf		f^9			
Ho		f^{10}			Es		f^{10}			
Er		f^{11}			Fm		f^{11}			
Tm	f^{13}	f^{12}			Md		f^{12}			
Yb	f^{14}	f^{13}			No		f^{13}			
Lu		f^{14}			Lw		f^{14}			

In both series of elements there is a contraction in atomic and ionic sizes as the atomic number increases. This is because, as successive protons are added, their increased attraction for the outer electrons causes the orbits of the latter to be pulled closer to the nucleus, especially as they are not particularly well shielded by diffuse f orbitals. These *lanthanide and actinide contractions* are illustrated graphically in *Figure 16.1*.

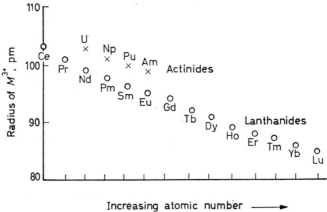

Figure 16.1. Lanthanide and actinide contractions

Like the transitional elements discussed earlier, many ions of these elements exhibit paramagnetism, caused by the presence of unpaired electrons. The magnetic behaviour cannot, however, be as closely correlated with the electronic configurations as was the case with the transition metals, presumably because the f orbitals are effectively shielded from external forces by the s and p orbitals of the outer shells. The greatest paramagnetism in the lanthanide series is shown by dysprosium, and this has been made use of in the production of very low temperatures. Isothermal magnetization of a dysprosium salt, followed by adiabatic

Table 16.3

Number of unpaired electrons	Ions		Colour
0		Lu^{3+}	—
1	Ce^{3+}	Yb^{3+}	—
2	Pr^{3+}	Tm^{3+}	green
3	Nd^{3+}	Er^{3+}	pink
4	Pm^{3+}	Ho^{3+}	rose/yellow
5	Sm^{3+}	Dy^{3+}	yellow
6	Eu^{3+}	Tb^{3+}	—
7	Gd^{3+}		

demagnetization, when repeated several times has permitted a temperature as low as 0·09 K to be attained.

The unpaired electrons give rise to absorption bands in the visible region of the spectrum (as already noted for the d block elements), and so the salts of some of these elements are coloured (*Table 16.3*).

The four elements uranium, neptunium, plutonium, americium, and possibly also protoactinium, form very stable oxycations of formulae MO_2^+ and MO_2^{2+}, and much of their chemistry revolves around these ions.

Production of the Elements

Owing to their very pronounced chemical similarities, the elements of these series usually occur together and are very difficult to separate by common methods of precipitation or crystallization. It is possible to separate some of the elements by treatment of solutions with alkali to precipitate the hydroxides which, owing to their differing solubility products, are precipitated at different pH values, e.g.

	$La(OH)_3$	$Nd(OH)_3$	$Gd(OH)_3$	$Lu(OH)_3$	$Ce(OH)_4$
Precipitation, pH	7·82	7·31	6·83	6·30	2·6

More recently, *solvent extraction* methods have been utilized in the separation of the lanthanides. A solution of the elements (in the form of their compounds) is allowed to flow counter to a second solvent. The elements are extracted into the latter in accordance with their respective extraction coefficients. A very satisfactory system consists of a solution of lanthanides in nitric acid flowing counter to a solution of tributyl phosphate in kerosene.

The most efficient method of separation is *ion exchange*, which depends on differences in the absorbing power of a suitable resin or other absorbent. For example, if a solution of lanthanides is passed through a 15 m zeolite column, the elements separate into different bands. This method has also been applied very successfully to the artificial actinide elements, not only for their separation but for their discovery. The accuracy of the work is such that seventeen atoms of mendelevium were produced *and* characterized, although its half-life is only $1·26 \times 10^4$s.

Prior to separation by such methods, the artificial elements have first to be produced by suitable radiochemical processes. Americium and curium (in the form of their compounds) are obtained by neutron bombardment of plutonium, which is itself first separated from uranium by solvent extraction:

$$^{239}_{94}\text{Pu} \xrightarrow{(n,\gamma)} {}^{240}_{94}\text{Pu} \xrightarrow{(n,\gamma)} {}^{241}_{94}\text{Pu} \xrightarrow[t_{\frac{1}{2}} = 4·2 \times 10^8\text{s}]{\beta-} {}^{241}_{95}\text{Am} \quad (t_{\frac{1}{2}} = 1·4 \times 10^{10}\text{s})$$

where (n,γ) signifies bombardment with a neutron, followed by emission of γ-radiation

$$^{239}_{94}\text{Pu} \xrightarrow[t_{\frac{1}{2}} = 1·8 \times 10^4\text{s}]{(4n,\gamma)} {}^{243}_{94}\text{Pu} \xrightarrow{\beta-} {}^{243}_{95}\text{Am} \xrightarrow{(n,\gamma)} {}^{244}_{95}\text{Am} \xrightarrow[t_{\frac{1}{2}} = 1·6 \times 10^3\text{s}]{\beta-} {}^{244}_{96}\text{Cm} \quad (t_{\frac{1}{2}} = 6·0 \times 10^8\text{s})$$

Micro and sub-micro amounts of heavier elements have been obtained by bombardment with α-particle and stripped-carbon (i.e. carbon nuclei) e.g.

$$^{241}_{95}\text{Am} \xrightarrow{(\alpha,2n)} {}^{243}_{97}\text{Bk}$$

$$^{238}_{92}\text{U} \xrightarrow{(^{13}\text{C},6n)} {}^{244}_{98}\text{Cf}$$

More recently, bombardment has involved other elements than carbon (see *Table 16.4*).

Table 16.4 Production of transuranium elements

$$^{238}_{92}\text{U} + {}^{2}_{1}\text{H} \rightarrow {}^{238}_{93}\text{Np} + 2{}^{1}_{0}n$$
$$^{238}_{92}\text{U} + {}^{4}_{2}\text{He} \rightarrow {}^{240}_{94}\text{Pu} + 2{}^{1}_{0}n$$
$$^{239}_{94}\text{Pu} + {}^{4}_{2}\text{He} \rightarrow {}^{241}_{95}\text{Am} + {}^{1}_{1}\text{H} + {}^{1}_{0}n$$
$$^{239}_{94}\text{Pu} + {}^{4}_{2}\text{He} \rightarrow {}^{240}_{96}\text{Cu} + 3{}^{1}_{0}n$$
$$^{244}_{96}\text{Cu} + {}^{4}_{2}\text{He} \rightarrow {}^{245}_{97}\text{Bk} + {}^{1}_{1}\text{H} + 2{}^{1}_{0}n$$
$$^{238}_{92}\text{U} + {}^{12}_{6}\text{C} \rightarrow {}^{245}_{98}\text{Cf} + 5{}^{1}_{0}n$$
$$^{238}_{92}\text{U} + {}^{14}_{7}\text{N} \rightarrow {}^{247}_{99}\text{Es} + 5{}^{1}_{0}n$$
$$^{238}_{92}\text{U} + {}^{16}_{8}\text{O} \rightarrow {}^{250}_{100}\text{Fm} + 4{}^{1}_{0}n$$
$$^{253}_{99}\text{Es} + {}^{4}_{2}\text{He} \rightarrow {}^{256}_{101}\text{Md} + {}^{1}_{0}n$$
$$^{246}_{96}\text{Cm} + {}^{13}_{6}\text{C} \rightarrow {}^{251}_{102}\text{No} + 8{}^{1}_{0}n$$
$$^{252}_{98}\text{Cf} + {}^{10}_{5}\text{B} \rightarrow {}^{257}_{103}\text{Lw} + 5{}^{1}_{0}n$$

These elements are all appreciably electropositive (see *Table 16.5*), and thus very reactive. They react readily with hydrogen, oxygen, water and the halogens. In consequence of this reactivity, the metals themselves are obtained by vigorous reductive methods, such as electrolysis of the fused halides, or by metallic reduction of the halides, using sodium, magnesium or barium vapour.

Table 16.5 Electrode potentials, $M \rightarrow M^{3+}$ (aq)

	V		V
Ce	−2·52	U	−1·80
Gd	−2·40	Np	−1·83
Lu	−2·25	Pu	−2·03

QUESTIONS

1. Write an essay on the extraction of ^{235}U.

2. Of what value are the compounds of the lanthanides?

3. Where, in the Periodic Table, would you expect the newly discovered element 104 to be placed? Predict the main chemical features of this element.

4. Suggest explanations for the apparent anomalies of atomic sizes in consecutive elements in *Figure 16.2*.

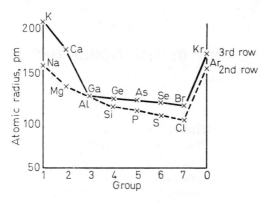

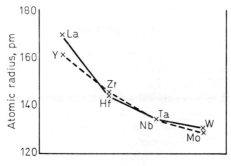

Figure 16·2

5. Write an essay on ion-exchange resins.

CHAPTER 17

GROUP 0: THE NOBLE GASES

Group 7 ns np Group 1

Group 7			Group 1
	Helium, He	2	Li
F	**Neon, Ne**	2.8	Na
Cl	**Argon, Ar**	2.8.8	K
Br	**Krypton, Kr**	2.8.18.8	Rb
I	**Xenon, Xe**	2.8.18.18.8	Cs
At	**Radon, Rn**	2.8.18.32.18.8	Fr

INTRODUCTION

The noble gases, helium, neon, argon, krypton, xenon and radon, constitute Group 0 of the Periodic Table (p. 25). Accordingly, the ultimate electron shell has the s and three p orbitals filled (except helium, which has only an s orbital of suitable energy available), and it requires considerable energy either to add or to remove any electrons. Hence they are very unreactive elements, commonly called *inert gases*. This unreactivity is shown by their high ionization energies and negligible electron affinities. However, the effect of the increase in atomic size on descending the Group is that electrons can be lost more readily, i.e. ionization energies decrease (*Table 17.1*) allowing the larger elements to take part in a limited number of chemical reactions.

Table 17.1   Atomic size and ionization energies

Element	Atomic radius, pm	First ionization energy, kJ mol^{-1}
He	93	2 372
Ne	112	2 080
Ar	154	1 522
Kr	169	1 353
Xe	190	1 163
Rn	—	1 040

The almost total lack of interaction between the atoms, consequent upon the complete pairing of electrons in closed shells, results in their existing as monatomic gases of low boiling point (see *Table 17.2*). Helium has the lowest boiling point of any known substance, 4·2 K. Between 4·2 and 2·2 K liquid helium behaves normally, but below the latter transition temperature it loses its electrical resistance and has such a low viscosity that it is virtually frictionless. It is also able to flow uphill.

Table 17.2. Volatility of the noble gases

	He	*Ne*	*Ar*	*Kr*	*Xe*	*Rn*
m.p. K	0·9 (26×10^5 Nm^{-2})	24·4	83·6	116	161	202
b.p. K	4·2	27·1	87·3	120	166	208

Occurrence, Extraction and Uses

All the gases occur in the atmosphere; their volume percentages are shown in *Table 17.3*. Since helium, like hydrogen, has such a low density, it is lost to space fairly quickly but tends to be replenished from radioactive decay products.

Table 17.3. The composition of the atmosphere

	Volume, per cent		*Volume,* per cent
O_2	20·95	CH_4	$1·5 \times 10^{-4}$
N_2	78·09	Kr	$1·1 \times 10^{-4}$
Ar	$9·3 \times 10^{-1}$	N_2O	5×10^{-5}
CO_2	3×10^{-2}	H_2	5×10^{-5}
Ne	$1·8 \times 10^{-3}$	O_3	4×10^{-5}
He	$5·2 \times 10^{-4}$	Xe	9×10^{-6}

Natural gas acts as a reservoir for α-particles produced from natural radioactivity, and these readily form helium atoms by removing two electrons from (i.e. by oxidizing) elements of variable valency

$$_2^4He^{2+} + 2e^- \rightarrow {}_2^4He$$

Neon, argon, krypton and xenon are extracted from the atmosphere by liquefaction, followed by fractional distillation. The low vapour density of neon results in it being associated with the nitrogen fraction, whilst the remaining gases come off with the oxygen fraction. Final purification is often effected by selective adsorption on activated charcoal.

All isotopes of radon are radioactive and are themselves products of radioactive decay. The most stable of these is produced in the decay of radium

$$_{88}^{226}Ra \xrightarrow{(-\alpha)} {}_{86}^{222}Rn \xrightarrow[t_{\frac{1}{2}}=3·3 \times 10^5 s]{(-\alpha)} {}_{84}^{218}Po \rightarrow \text{etc.}$$

351

The chemically unreactive nature of the noble gases, combined with their physical properties, makes them useful as inert atmospheres in arc welding (especially for aluminium, magnesium, titanium and stainless steel) and in electric light bulbs and discharge tubes. Liquid helium is used in low-temperature research, gaseous helium for filling research and weather balloons and also as a diluent for oxygen in diving systems, since it is far less soluble in blood than nitrogen: divers can then surface fairly quickly without ill effect.

Chemical Properties

Although the ionization energies of these gases are high, the decrease in these values on descending the Group suggests that those of higher atomic number are more likely to enter into chemical combination than those at the beginning of the Group. Indeed, some compounds have been made, including the tetrafluoride, the oxide difluoride and the trioxide of xenon, and the tetrafluorides of krypton and radon. It is understandable that elements of high electron affinity would be the most likely ones to enter into combination with any of these gases; the lattice energies of the solid products must also be high to compensate for the large ionization energies of the gas (see Born-Haber cycle, p. 37).

There is also the possibility of the noble gas atoms being trapped in a crystal lattice of suitable dimensions, to give a substance, called a *clathrate* (*Figure 17.1*), that hardly merits the description of a 'compound' in the classical sense. Such clathrates are obtained when, e.g. 1,4-dihydroxybenzene (quinol) crystallizes in the presence of the noble gas under high pressure. The lattice is such that three quinol molecules linked by hydrogen bonding comprise a cage large enough to hold one atom of the gas, which can be released on dissolving the crystals in water or by heating. Significantly, helium, the smallest member, is not trapped in this way and so does not form a clathrate with quinol.

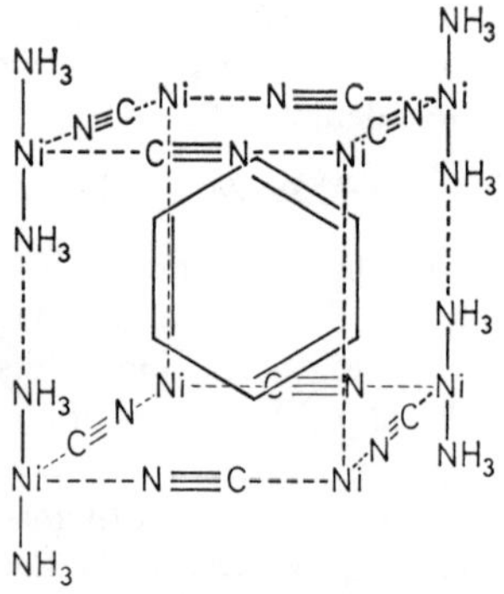

Figure 17.1. A clathrate: Unit cell of benzene ammino-nickel cyanide

The four heaviest elements form crystalline hydrates of variable composition but approximating to hexahydrates. These also appear to be clathrate-type compounds, the atoms of inert gas being trapped in cages of hydrogen-bonded water molecules. In this case, however, there is a possibility that electrons of the gas atoms are attracted towards the positive end of the polar water molecules.

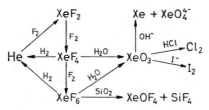

Figure 17.2 Xenon compounds

QUESTIONS

1. In the light of what you have read in this Chapter, discuss the meaning of the word 'compound'.

2. How far do you think it true to say that the noble gases provide the key to the problem of chemical combination?

3. The noble gases have also been called 'rare gases' and 'inert gases'. Discuss the merits of these various names.

4. Do you think the noble gases could form coordination compounds? If you do, what compounds do you think might react in this way and how would you attempt to verify your conclusion?

5. Write an account of the utilization by industry of the noble gases.

CHAPTER 18

ORGANIC CHEMISTRY: AN INTRODUCTION

The chemistry of the living cell revolves around compounds containing carbon and hydrogen, with or without other elements such as oxygen, nitrogen and halogens; compounds that are therefore conventionally known as *organic* compounds. Over a million such compounds are known already, most of them being based structurally upon a framework of carbon atoms. This profusion bears ample testimony to the versatility of carbon and the stability of carbon bonding, which in turn are partly attributable to the following facts:

(*a*) the oxidation number of carbon can be between the limits of -4 and $+4$, presumably because carbon lies in the middle of the Second Period;

(*b*) being in the Second Period, carbon has only one *s* and three *p* orbitals available for bonding, so that the maximum coordination number is four, and four-covalent carbon is therefore in a saturated and stable state, well shielded from attack;

(*c*) multiple bonding is at its most conspicuous in Period Two. Thus silicon, in the same group as carbon but in Period 3, does not form such stable multiple bonds as carbon (p. 276);

(*d*) the power of catenation (Lat. *catena*, chain), i.e. the linking together of identical atoms, reaches a maximum in Group 4 of the Periodic Table. Carbon is pre-eminent in this respect, with silicon exhibiting catenation to a much smaller extent (see Chapter 13).

Hydrocarbons

Even the number of compounds containing carbon and hydrogen alone (hydrocarbons) is very large, but fortunately they can be readily classified as

(*a*) *Aliphatic* hydrocarbons, open-chain structures which may be

    (*i*) *saturated*, when they are called *alkanes* or *paraffins*;

    (*ii*) *unsaturated*, with either double bonds between carbon atoms (the *alkenes* or *olefins*) or triple bonds (*alkynes* or *acetylenes*);

(*b*) *Alicyclic* hydrocarbons, containing closed rings of carbon atoms;

(*c*) *Aromatic* hydrocarbons, also containing closed rings but with non-localized orbitals, giving in effect a completely closed, conjugated double bond system (see Chapter 3).

Table 18.1 depicts some representative compounds of each of these classes.

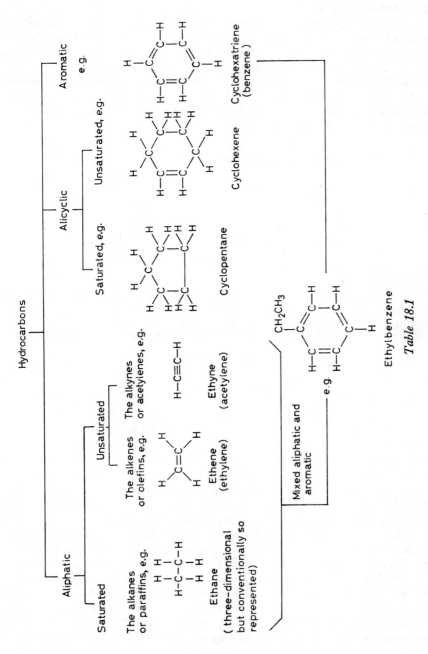

Table 18.1

355

Homologous Series

The carbon chain of a hydrocarbon can be increased by replacing a terminal hydrogen atom by a methyl $(-CH_3)$ group, so that the overall addition is of $(CH_3$ minus $H)$, i.e. CH_2. The series developed in this manner is termed an homologous series and the members, whose formulae differ from each other by $(CH_2)_n$, are called *homologues*. *Table 18.2* shows the general formulae of some of the aliphatic homologous series. As the members of any series usually differ significantly from each other in physical properties only, the *typical* reactions of one homologue can be regarded as being representative of the series as a whole.

Table 18.2. Some homologous series

Name of series	General formula
Alkanes	C_nH_{2n+2}
Alkenes	C_nH_{2n}
Alkynes	C_nH_{2n-2}
Monosubstituted derivatives of alkanes	$C_nH_{2n+1}X$
	$X =$ halogen, $-OH$, etc.

Functional Groups

The replacement of hydrogen atoms in hydrocarbons by other atoms or groups of atoms, either directly or indirectly, gives rise to different homologous series. For example, the replacement of one hydrogen atom in an aliphatic or alicyclic compound by the hydroxy group, $-OH$, produces an alcohol

$$H-\overset{\displaystyle H}{\underset{\displaystyle H}{C}}-\overset{\displaystyle H}{\underset{\displaystyle H}{C}}-\overset{\displaystyle H}{\underset{\displaystyle H}{C}}-H \qquad H-\overset{\displaystyle H}{\underset{\displaystyle H}{C}}-\overset{\displaystyle H}{\underset{\displaystyle H}{C}}-\overset{\displaystyle H}{\underset{\displaystyle H}{C}}-OH$$

An alkane   An alcohol

(The hydrocarbon residue is referred to as the alkyl group (Alk for short) if derived from an aliphatic hydrocarbon and as the aryl group (or Ar) if derived from an aromatic hydrocarbon. When possible, both Alk and Ar are represented by *R*.)

The replacing group is called the *functional group* since, being more labile than the hydrocarbon residue, it confers reactivity upon what would otherwise be fairly inert compounds. In general, all organic compounds containing the same functional group have very similar chemical properties, though these may be modified through interaction between different functional groups in the same molecule. It is due to such interaction between the functional group and the aromatic system to which it is attached that aromatic compounds have modified properties compared with the aliphatic analogues. Nevertheless, it is possible and helpful to consider the typical reactions not only of an homologous series but also of the functional group present, especially if the latter is regarded from the point of view of the nature of the atoms comprising it. This will be the underlying approach in the following Chapters.

356

Table 18.3 lists some of the more common functional groups encountered in organic chemistry.

Table 18.3. Functional groups

Functional group	Formula	Structure	Systematic name	
			prefix	*suffix*
Double bond		$\diagup$C=C$\diagdown$		-ene
Triple bond		$-$C$\equiv$C$-$		-yne
Halide	F, Cl, Br, I		halo	
Amine	$-NH_2$	$-N\diagup^H_{\diagdown H}$	amino	amine
Alcohol and phenol	$-OH$	$\diagdown_O\diagup^H$	hydroxy	-ol
Ether	$-O-$	$\diagdown_O\diagup$	alkoxy	ether
Aldehyde	$-CHO$	$-C\diagup^H_{\diagdown O}$	oxo	-al
Ketone	$\diagup$CO	$\diagup$C=O	oxo	-one
Carboxylic acid	$-COOH$	$-C\diagup^O_{\diagdown O\diagup}{}^H$	carboxy	{ carboxylic (or -oic) acid
Acid chloride	$-COCl$	$-C\diagup^O_{\diagdown Cl}$		carbonyl (or -oyl) chloride
Ester	$-COO-$	$-C\diagup^O_{\diagdown O\diagup}$		-oate
Amide	$-CONH_2$	$-C\diagup^O_{\diagdown N\diagup^H_{\diagdown H}}$	amido	amide
Nitrile and isonitrile	$-CN$, $-NC$	$-C\equiv N$, $-N\equiv C$	cyano, isocyano	nitrile, isonitrile
Nitro	$-NO_2$	$-N\diagup^O_{\diagdown O}$	nitro	

Nomenclature

It is convenient at this juncture to discuss the methods used for naming organic compounds. The systematic naming of molecules should be such that the formula can be constructed unambiguously from the name. However, many common trivial names are so well established that the systematic names are seldom used, and in the course of this book, both trivial and systematic names are used, as appropriate. The systematic name denotes: (a) the length of the chain; (b) the branching of the chain; (c) the position and nature of any functional groups.

Aliphatic Compounds

(1) Saturated aliphatic hydrocarbons are given the termination -ane, which is prefixed by a suitable contraction of the Greek (occasionally Latin) word for the number of carbon atoms contained in the chain (with the exception of the first four members which retain their original names):

CH_4	methane	C_4H_{10}	butane
C_2H_6	ethane	C_5H_{12}	pentane
C_3H_8	propane	C_6H_{14}	hexane

A monovalent radical derived from such a hydrocarbon is given the name of the hydrocarbon, except that -ane is replaced by -yl. For instance, C_2H_5 is called the *ethyl* radical, from the parent hydrocarbon *ethane*.

The chain length of the molecule is obtained by drawing the structure so that there are *the largest number of carbon atoms in a straight chain*, and the compound is then named as a derivative of the hydrocarbon possessing this chain. The position of any substituent is indicated by prefixing with the number of the carbon atom to which it is attached, the carbon atoms being numbered from that end of the chain which results in the smallest number as prefix. Thus

$$\overset{1}{C}H_3\overset{2}{C}H_2\overset{3}{C}H\overset{4}{C}H_2\overset{5}{C}H_2\overset{6}{C}H_3 \quad \text{3-(not 4-)methylhexane}$$
$$|$$
$$CH_3$$

$$CH_3$$
$$|$$
$$CH_3-C-CH_3 \quad \text{2,2-dimethylpropane}$$
$$|$$
$$CH_3$$

(2) Hydrocarbons containing double bonds are named in a similar fashion except that the termination is -ene and the position of the double bond is indicated by the smallest possible number. Similarly, hydrocarbons containing a triple bond end in -yne, for example

$CH_3CH_2CH{=}CH_2$ but-1-ene, from the saturated hydrocarbon butane

$CH_3C{\equiv}CCH_2CH_3$ pent-2-yne, from the saturated hydrocarbon pentane

358

(3) Halogen derivatives are named by prefixing the name of the parent hydrocarbon by fluoro-, chloro-, bromo- or iodo-

CH_3Br, bromomethane

$CH_3CHCH_2CH_2I$ 1-iodo-3-chlorobutane
|
Cl

(4) Simple alcohols are named by replacing the final -e of the corresponding hydrocarbon by -ol

CH_3CH_2OH, ethanol

$CH_3CHCHCH_3$, 3-methylbutan-2-ol
with OH above and CH_3 below

(5) Aldehydes are derived in a similar fashion, the final -e being replaced by -al

$CH_3CH_2CH_2CHO$, butanal

(6) In the case of ketones, the final -e is replaced by -one. If there is any possible ambiguity, then the position of the carbonyl (ketone) group is indicated numerically in the usual way, e.g.

$CH_3CH_2COCH_2CH_3$, pentan-3-one

(7) Carboxylic acids are named by replacing the final -e of the parent hydrocarbon by -oic acid

CH_3CH_2COOH, propanoic acid

In dibasic acids, the final -e is not removed, e.g.

COOH
| ethanedioic acid
COOH

(8) Alicyclic compounds are prefixed by -cyclo, e.g.

H_2C——CH_2
\   /   cyclopropane
CH_2

Aromatic Compounds

(9) Aromatic compounds are also named, where convenient, as derivatives of the corresponding hydrocarbon, e.g.

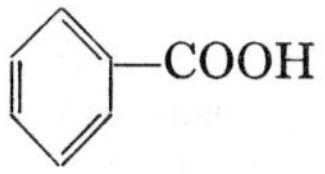

⬡—COOH   benzene carboxylic acid

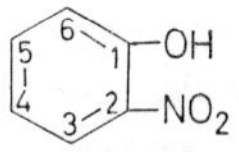

⬡ with OH and NO_2 1-hydroxy-2-nitrobenzene

(numbering of the ring commences at one of the substituted positions)

359

Where this type of nomenclature is difficult to use, it is permissible to refer to the

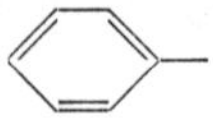

⬡— radical as the phenyl radical. This leads (e.g.) to the name triphenyl-

methanol for the compound having the formula

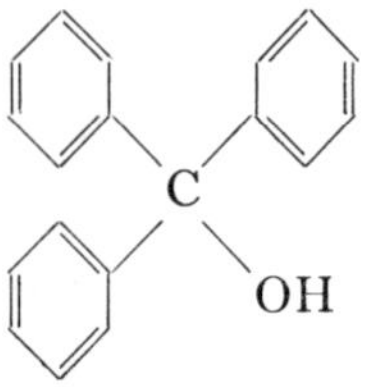

Isomerism

A single molecular formula can represent more than one actual compound. Such substances, having a common molecular formula, are termed *isomers*.

There are two fundamental types of isomerism: structural and stereoisomerism. *Structural isomerism* is that type of isomerism in which different groups are present in the isomers:

(*i*) *Chain isomers* differ only in the arrangement of the carbon atoms relative to each other; for example, the molecular formula C_4H_{10} represents the two compounds

$$CH_3CH_2CH_2CH_3 \quad \text{and} \quad \begin{matrix} CH_3 \\ > CH.CH_3 \\ CH_3 \end{matrix}$$

$$\text{butane, or n-butane} \qquad \text{2-methylpropane (or isobutane)}$$

The prefix n, or normal, implies the presence of a straight chain. The prefix iso- is often used loosely to indicate a branched chain.

(*ii*) *Position isomers* possess the same carbon framework but differ in the relative positions of functional groups; for instance, there are three, and only three, straight-chain position isomers with molecular formula $C_6H_{13}Cl$:

$$\overset{1}{C}H_2\overset{2}{C}H_2\overset{3}{C}H_2\overset{4}{C}H_2\overset{5}{C}H_2\overset{6}{C}H_3 \quad CH_3CHCH_2CH_2CH_2CH_3 \quad CH_3CH_2CHCH_2CH_2CH_3$$
$$|\qquad\qquad\qquad\qquad\qquad | \qquad\qquad\qquad\qquad\qquad |$$
$$Cl \qquad\qquad\qquad\qquad\qquad Cl \qquad\qquad\qquad\qquad\qquad Cl$$
$$\text{1-chlorohexane} \qquad\quad \text{2-chlorohexane} \qquad\quad \text{3-chlorohexane}$$

1- and 6-chlorohexane can be superimposed on each other and therefore represent the same substance; similarly, 2- and 5-chlorohexane are identical, and so also are 3- and 4-chlorohexane. The carbon atom attached to chlorine in 1-chlorohexane is attached to only *one* other carbon atom and the substance is therefore described as a *primary* chloride; if attached to *two* other carbon atoms, as in 2-chlorohexane, it is said to be *secondary*; if to *three*, *tertiary*.

(*iii*) *Functional group isomerism*—Here the isomers possess atoms arranged in the form of different functional groups; for example, the molecular formula C_2H_6O can be represented as

$$CH_3CH_2OH \text{ and as } CH_3OCH_3$$
$$\text{ethanol} \qquad \text{methoxymethane}$$

(*iv*) *Metamerism*—Isomers which differ only in the nature of the alkyl groups present are known as metamers, e.g. $CH_3OC_3H_7$ and $C_2H_5OC_2H_5$, both of which are ethers.

Stereoisomers possess the same groups of atoms, but their relative positions in space are different:

(*i*) *Nuclear isomers* are those compounds which differ only in the relative positions of substituent atoms or groups in cyclic compounds; e.g. there are three possible dichlorobenzene molecules

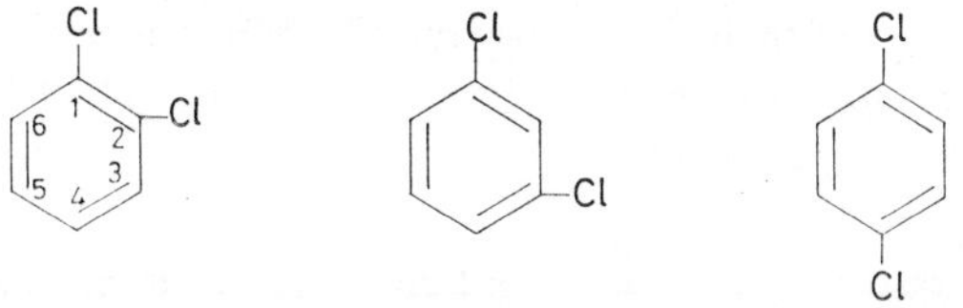

1,2-dichlorobenzene 1,3-dichlorobenzene 1,4-dichlorobenzene
   (*ortho-*)   (*meta-*)   (*para-*)

(*ii*) *Geometric isomerism*—The possibility of non-aromatic compounds differing only in the disposition of the parts of the molecule relative to each other arises from the directional character of the bonds from a carbon atom (p. 46). It is believed that 'free rotation' occurs about a single bond, and it may therefore be thought that compounds could exist in an infinite variety of ways, each having a slightly different spatial arrangement from the others. In a sense this is correct, although the very slight energies required to convert one form into another mean that a substance exists as an 'average' molecule with uniformly characteristic properties; i.e. there is no apparent difference between the structures represented as

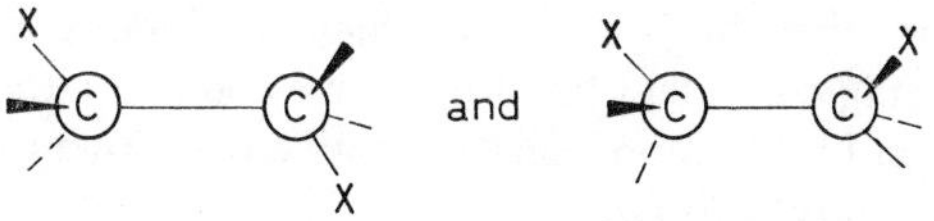

Of course, at lower temperatures the more stable forms of the molecule, with lower energies, predominate. These tend to be such that the repulsive forces present are at a minimum; for example, the more stable form of 1,2-dibromoethane will be

361

(*a*) rather than (*b*), and it will become more predominant as the temperature is decreased

(*a*) (*b*)

The formation of a double bond between carbon atoms prevents free rotation and enables so-called *geometric isomers* to exist, where different atoms or groups occupy permanently different positions about the double bond. An isomer with two identical groups in proximity, i.e. situated on the same side of the double bond, is known as the *cis*-isomer and the other as the *trans*-isomer; for example, the 1,2-dichloroethenes

cis− *trans*−

(*iii*) *Optical isomerism*—If four *different* atoms or groups are attached to a carbon atom, then, by virtue of their tetrahedral distribution, it is possible to represent the resultant structure in two ways, one of which is the mirror image or *enantiomorph* of the other, e.g.

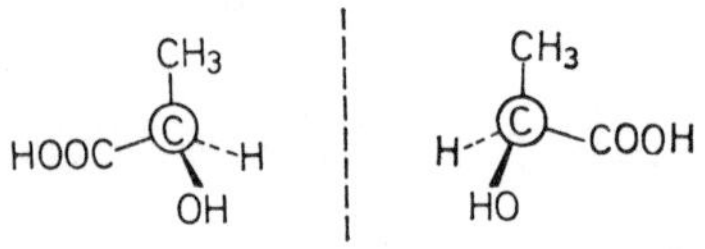

The two enantiomorphs differ only in their ability to rotate the plane of polarized light by equal amounts in opposite directions: that rotating the plane to the right is called dextro- (d- or +) and that to the left laevo- (l- or −). They are therefore known as optical isomers.

N.B. Optical activity does not necessitate the presence of an asymmetric carbon atom, i.e. one with four different groups attached; provided that there is sufficient absence of symmetry in a molecule to enable two enantiomorphic forms to exist, the substance will exist in two optically active forms.

ORGANIC REACTIONS

Organic reactions involve the breaking of chemical bonds, which in general are largely covalent. A single covalent bond may break so that either (*a*) both electrons of the bond are taken by the more electronegative constituent and two ions result. This is known as *heterolytic fission*

$$A{:}B \longrightarrow A^+ + {:}B^-$$

(here and elsewhere, the curved arrow indicates the movement of the electrons)

362

or (*b*) each constituent of the bond takes an electron by *homolytic fission* and neutral free radicals are formed

$$A{-}B \quad \longrightarrow \quad A\cdot + B\cdot$$

The kinetics of the reaction will be determined by the slowest stage (see Chapter 6); if fission is the rate-determining step and involves only the molecule itself, then the reaction is *unimolecular*. If, on the other hand, a second reactant is involved in the breaking of the bond, the reaction is *bimolecular*. The second reactant may be a negative ion or possess a lone pair of electrons capable of donation; it will then seek a positive site and is called a *nucleophilic* reagent. If it is a positive ion or is in any way deficient in electrons, then it is an *electrophilic* reagent. *Table 18.4* lists some of the more common reagents.

Table 18.4. Some common reagents

Nucleophilic	Electrophilic
OH^-	$NO_2{}^+$
halide$^-$	H_3O^+
:O$\big\langle$	
:N$\big\langle$	BF_3
carbanions	carbonium ions
i.e. R^-	i.e. R^+

It is as well to bear in mind that different types of mechanisms can operate under different conditions: ionizing solvents, for instance, favour ionic mechanisms, liquids of low dielectric constant non-ionic mechanisms. Furthermore, the mechanism can change in type on ascending an homologous series or with increased branching of the carbon framework, owing to the electron-releasing tendency (inductive effect) of alkyl groups. Thus, bromoethane is hydrolysed chiefly by a bimolecular reaction involving the halide and hydroxide ions:

$$HO^- + CH_3{-}Br \rightarrow [HO\cdots CH_3 \cdots Br]^- \rightarrow HO{-}CH_3 + Br^-$$
nucleophile

The inductive effects of methyl groups in 2-methyl-2-bromopropane are so marked, however, that the rate-determining step is ionization into a positively charged carbonium ion, which then instantly reacts with a hydroxide ion:

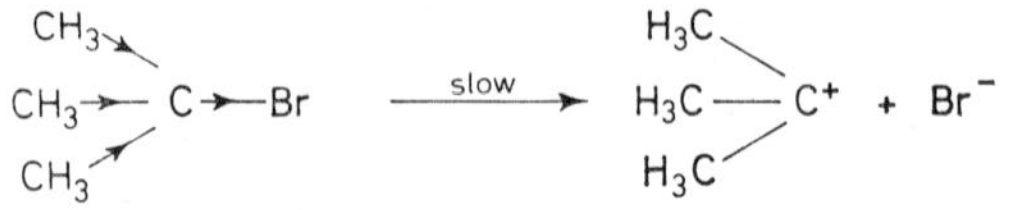

$$CH_3 \rightarrow C \rightarrow Br \xrightarrow{\text{slow}} H_3C - C^+ + Br^-$$

A carbonium ion

($\rightarrow$ = direction in which electrons tend to move)

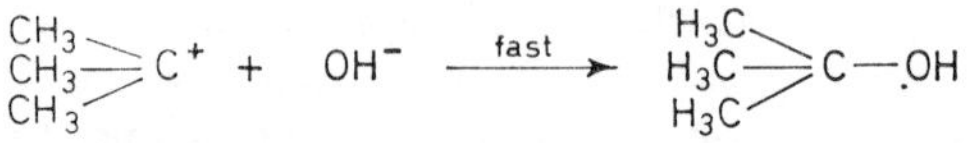

$$CH_3 \rightarrow C^+ + OH^- \xrightarrow{\text{fast}} H_3C \rightarrow C-OH$$

A complete stoichiometric equation often represents more than one reaction mechanism. In the course of this book, we shall give more or less detail of reactions as we think appropriate.

Reactions can often be classified as

1. *Displacement* (or substitution) reactions

$$X^- + AY \longrightarrow XA + Y^-$$

X^- is said to displace Y^- from substance AY.

2. *Elimination* reactions, where an ion or radical is removed from the system, e.g.

$$AH^+ \longrightarrow A + H^+$$

3. *Rearrangement* reactions, which involve the migration of an ion from one atom to another in the same molecule

$$A\text{–}B\text{–}X \longrightarrow X\text{–}A\text{–}B$$

4. *Addition* reactions, where unsaturated compounds become saturated (or less unsaturated) by addition of a reagent to a multiple bond

$$X^+ + A{=}B \longrightarrow [X\text{–}A\text{–}B]^+$$

These reactions do not always give rise to a stable compound; in such cases, the product rapidly undergoes another reaction, and sometimes the overall reaction is given a single title. For example, 'condensation' is the name given to the elimination of water or another small molecule from two reactants; in fact, the elimination is preceded by the addition of one reactant to the other, as in the case of the condensation of a ketone with hydroxylamine

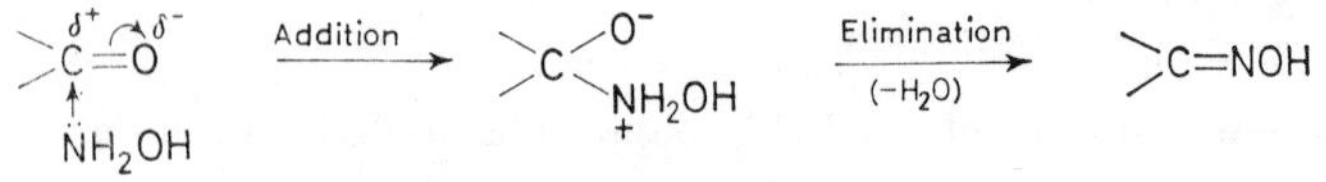

$$\overset{\delta^+ \frown \delta^-}{C{=}O} \xrightarrow{\text{Addition}} C\overset{O^-}{\underset{\overset{+}{NH_2OH}}{}} \xrightarrow[(-H_2O)]{\text{Elimination}} C{=}NOH$$

$$NH_2OH$$

Figure 18.1 consists of a scheme of synthesis which illustrates some of the ways in which functional groups can be changed by the types of reactions listed.

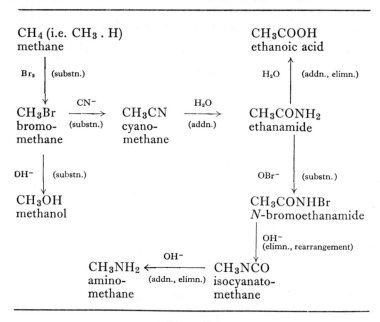

Figure 18.1

QUESTIONS

1 Name the following compounds:

$$CH_3-\underset{\underset{Br}{|}}{\overset{\overset{OH}{|}}{C}}-CH_3 \qquad CH_3CH=CH-C\equiv CH \qquad HO-\langle\ \rangle-NO_2$$

$$CH_3CH_2CO-\underset{\underset{CH_3}{|}}{\overset{\overset{CH_3}{|}}{CH}} \qquad CH_3COCH_2COOH$$

2. Write down the formulae of
2-methylbutanal, buta-1,3-diene, 3-aminobenzene carboxylic acid, cyclopentanone, 1-methoxypropan-2-ol.

3. Draw and name all possible isomers of C_5H_{12} and of $C_5H_{11}X$.

4. Draw the structures of all possible compounds called trihydroxybenzene.

5. Discuss the isomerism of all possible compounds of formula

$$\mathrm{CH_3CH_2CH.Cl.COOH}, \quad \begin{array}{c} \mathrm{CH_3.C{=}CHC_2H_5}, \\ | \quad\quad\quad\quad \\ \mathrm{Cl} \quad\quad\quad\quad \end{array} \quad \begin{array}{l} \mathrm{CH.COOH} \\ \| \\ \mathrm{CH.COOH} \end{array}$$

6. 'Organic chemistry is the chemistry of the functional group.' Discuss.

CHAPTER 19

THE HYDROCARBONS

The structures and methods of classification of the simpler hydrocarbons are shown in *Tables 19.1* and *19.2*.

Table 19.1. Aliphatic and alicyclic hydrocarbons

(a) Aliphatic		
Saturated: *Alkanes (Paraffins)*	*Unsaturated:* *Alkenes (Olefins)*	*Unsaturated:* *Alkynes (Acetylenes)*
General formula C_nH_{2n+2}	C_nH_{2n}	C_nH_{2n-2}
methane, CH_4	[methene (methylene) radical $=CH_2$]	[methyne radical $\equiv CH$]
ethane, C_2H_6 or $CH_3.CH_3$	ethene(ethylene), C_2H_4 or $CH_2:CH_2$	ethyne(acetylene), C_2H_2 or $CH:CH$
propane, C_3H_8 or $CH_3.CH_2.CH_3$	propene(propylene), C_3H_6 or $CH_3.CH:CH_2$	propyne, C_3H_4 or $CH_3.C:CH$
butanes, C_4H_{10} n-butane, $CH_3.CH_2.CH_2.CH_3$ 2-methylpropane (isobutane), $CH_3.CH.CH_3$ $\qquad\ \ |$ $\qquad CH_3$	butenes(butylenes), C_4H_8 but-1-ene, $CH_2:CH.CH_2.CH_3$ but-2-ene, $CH_3.CH:CH.CH_3$ 2-methylpropene, $CH_3.C:CH_2$ $\qquad\ \ |$ $\qquad CH_3$	butynes, C_4H_6 but-1-yne, $CH_3.CH_2.C:CH$ but-2-yne, $CH_3.C:C.CH_3$
(b) Alicyclic		
General formula C_nH_{2n}	C_nH_{2n-2}	C_nH_{2n-4}
cyclopropane	cyclohexene	

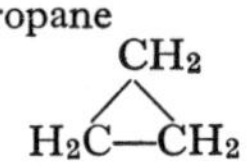

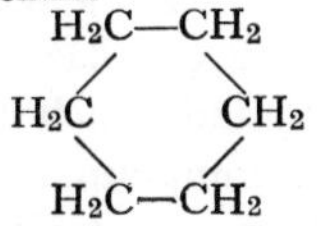

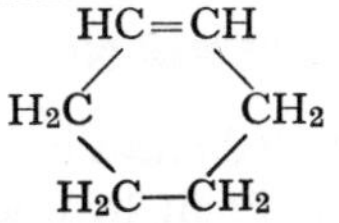

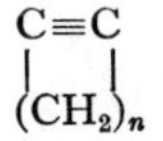

367

Table 19.2. Aromatic hydrocarbons

(1) *Benzene and its homologues*

Benzene, C_6H_6 Methylbenzene, $C_6H_5 . CH_3$ (toluene) 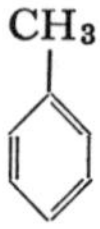   Dimethylbenzenes, $C_6H_4(CH_3)_2$ (xylenes)

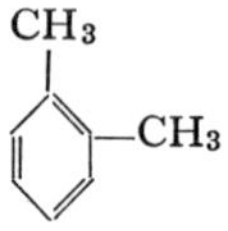

CH_3 CH_3 CH_3—CH_3 CH_3—CH_3 CH_3 CH_3

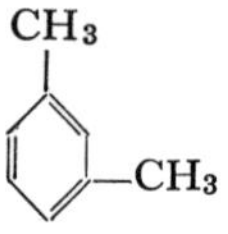

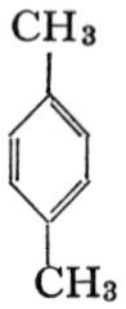

1,2-   1,3-   1,4-
o- (*ortho*)   *m-* (*meta*)   *p-* (*para*)

(2) *Polynuclear*

(*a*) Rings isolated, e.g.

Diphenyl, $C_6H_5 . C_6H_5$ Triphenylmethane, $(C_6H_5)_3CH$

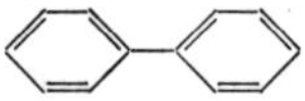

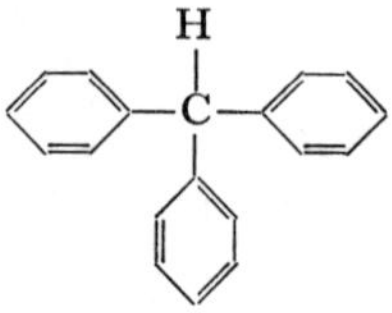

H
—C—

(*b*) 'Condensed' ring systems, e.g.

Naphthalene, $C_{10}H_8$ Anthracene, $C_{14}H_{10}$    Phenanthrene, $C_{14}H_{10}$ 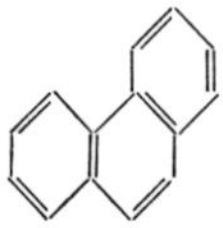   Pyrene, $C_{16}H_{10}$

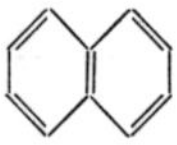

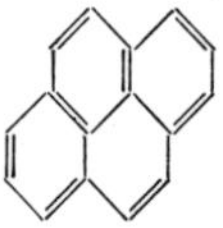

OCCURRENCE

Hydrocarbons are found in concentrated forms in fossilized fuels such as petroleum, peat and coal. A common and important hydrocarbon found in living tissue is 2-methylbut-1,3-diene (*isoprene*), which is synthesized from acetic acid units present from respiratory and photosynthetic processes (p. 476) and which acts as the precursor of a wide variety of hydrocarbons and related compounds. Polymerization of isoprene into long fibres gives natural *rubber* or *gutta-percha*, depending on the geometric isomerism about the double bonds present in the polymer

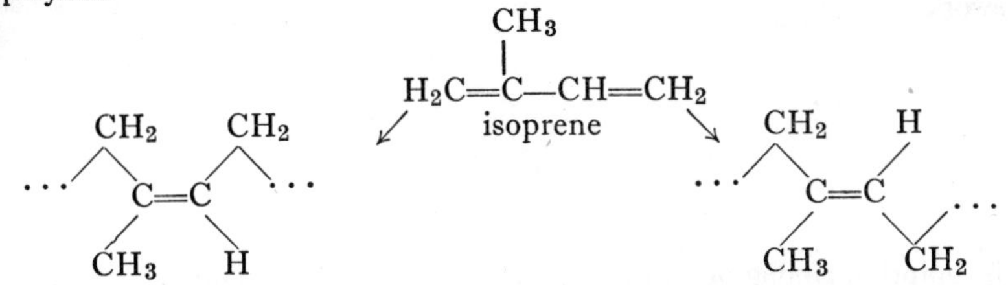

$$CH_3$$
$$H_2C{=}C{-}CH{=}CH_2$$
isoprene

rubber (poly*cis*isoprene)

gutta-percha (poly*trans*isoprene)

Carotenes, the fat-soluble yellow pigments of plants, contain eight isoprene units and are, in some cases, the precursors of Vitamin A:

β-Carotene

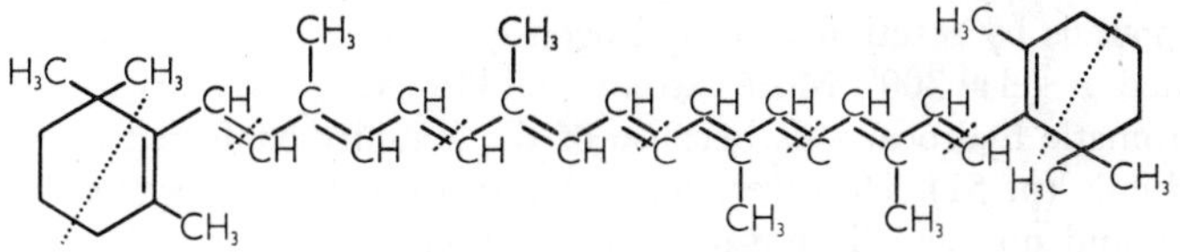

(dotted lines indicate the isoprene links)

Vitamin A

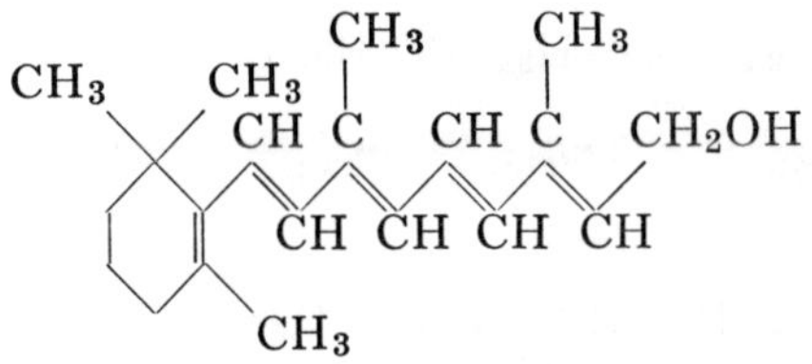

or more shortly written

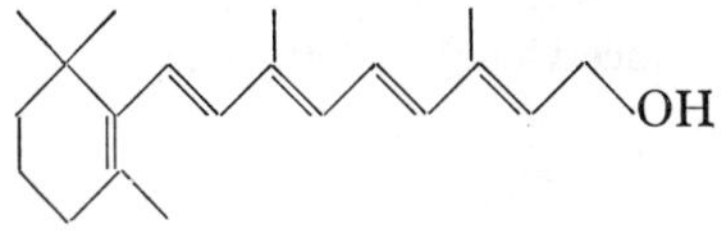

A group of volatile, fragrant oils giving the characteristic odours to many plants are the *terpenes*. These, too, are built up from isoprene units; for example, limonene, $C_{10}H_{16}$, contains two isoprene residues

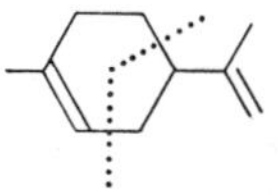

Squalene is a triterpene, i.e. a C_{30} compound, built up from six isoprene units and found particularly in shark liver oil

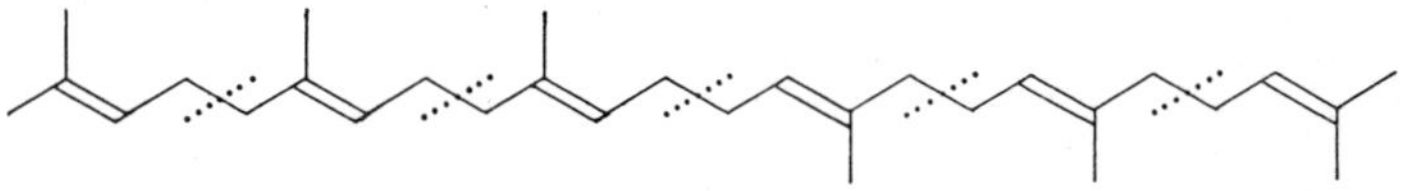

369

It is of some interest because it is known to be the precursor of cholesterol (p. 433) and thus effects a link between isoprene and the *steroids*, compounds which contain the framework

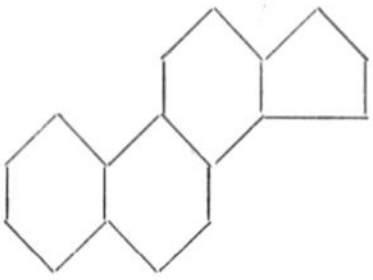

and which comprise, among other things, bile acids and sex hormones.

GENERAL METHODS OF PREPARATION

(i) Hydrogenation

Unsaturated hydrocarbons can be converted into less unsaturated, or fully saturated, compounds by reaction with hydrogen in the presence of a catalyst such as finely divided nickel at 300°. More extreme conditions are necessary for the hydrogenation of aromatic hydrocarbons because of the stability conferred by the non-localized π orbitals (p. 51). However, once this is overcome, the system becomes very reactive and no partially unsaturated compounds can be isolated by this method. In general

$$C_nH_{2n-2} \xrightarrow[\text{Ni}]{H_2} C_nH_{2n} \xrightarrow[\text{Ni}]{H_2} C_nH_{2n+2}$$

$$\text{alkyne} \qquad \text{alkene} \qquad \text{alkane}$$

e.g.

$$HC{\equiv}CH \rightarrow H_2C{=}CH_2 \rightarrow H_3C . CH_3$$

$$\underset{\text{(acetylene)}}{\text{ethyne}} \qquad \underset{\text{(ethylene)}}{\text{ethene}} \qquad \text{ethane}$$

and

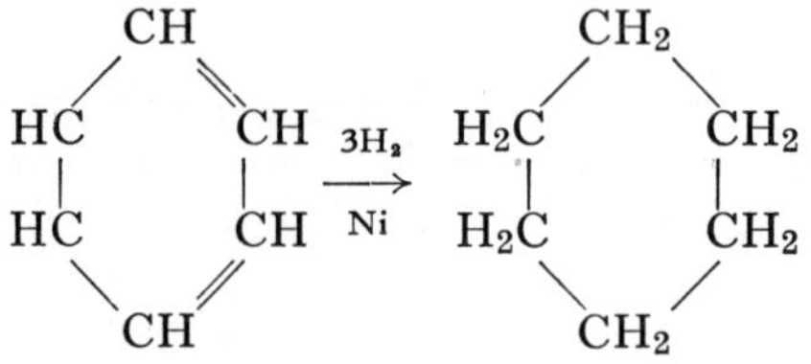

cyclohexatriene (benzene)   cyclohexane

or in abbreviated form

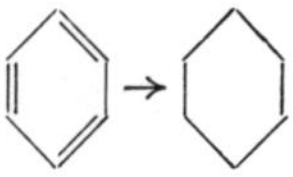

A second method of hydrogenation is by the reduction of halogen derivatives of the hydrocarbons, using reducing agents such as sodium in ethanol:

$$C_2H_5OH + Na \rightarrow C_2H_5O^-Na^+ + [H]$$

$$RI + 2[H] \rightarrow RH + HI$$

(ii) The Kolbé process

Electrolysis of the sodium or potassium salts of carboxylic acids in strong aqueous solution produces hydrocarbons possessing an even number of carbon atoms. Monocarboxylic acids give rise to alkanes, dicarboxylic acids to alkenes and unsaturated dicarboxylic acids to alkynes. The reactions at the anode can be summarized by the equations

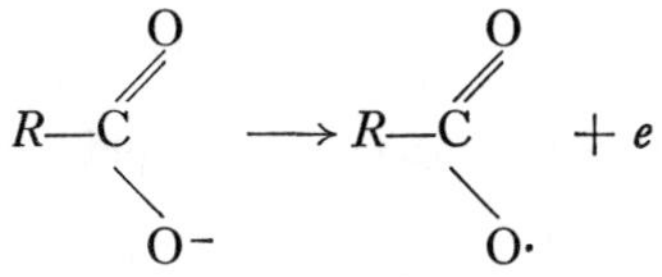

(free radical)

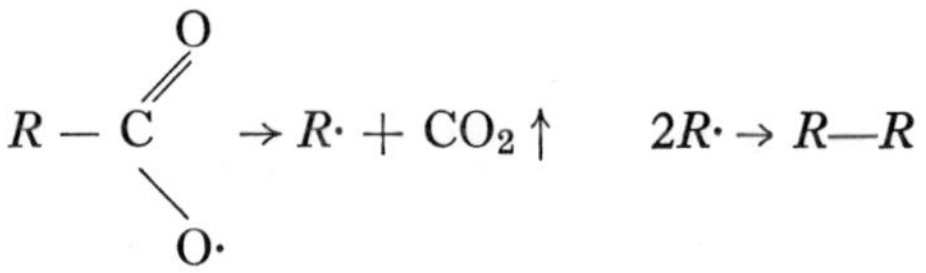

$$\rightarrow R\cdot + CO_2 \uparrow \qquad 2R\cdot \rightarrow R-R$$

e.g.

$$2CH_3COO^- \rightarrow C_2H_6 \uparrow + 2CO_2 \uparrow + 2e$$
ethanoate   ethane
(acetate)

$$\begin{array}{c} CH_2COO^- \\ | \\ CH_2COO^- \end{array} \rightarrow \begin{array}{c} CH_2 \\ || \\ CH_2 \end{array} \uparrow + 2CO_2 \uparrow + 2e$$
butane-1,4-dioate   ethene
(succinate)   (ethylene)

$$\begin{array}{c} CHCOO^- \\ || \\ CHCOO^- \end{array} \rightarrow \begin{array}{c} CH \\ ||| \\ CH \end{array} \uparrow + 2CO_2 \uparrow + 2e$$
cis- or trans-   ethyne
but-2-ene-1,4-dioate   (acetylene)
(maleate or fumarate)

371

METHODS FOR PREPARATION OF ALKANES AND/OR AROMATIC HYDROCARBONS

There are several methods which are applicable only to the preparation of alkanes and aromatic hydrocarbons, of which the following may be mentioned.

(i) Decarboxylation

Carboxylic acids and their sodium salts, when strongly heated with alkali, eliminate a molecule of carbon dioxide. Poor yields are obtained if the hydrocarbon is unstable, and so the method is only of some value for the preparation of saturated and aromatic hydrocarbons. The alkali normally used is soda lime (calcium oxide slaked with sodium hydroxide), but the aromatic acids will react when heated with calcium oxide alone

$$RCOO^- + OH^- \rightarrow CO_3^{2-} + RH$$

By this means sodium ethanoate (acetate) will yield methane

$$CH_3COO^- + OH^- \rightarrow CO_3^{2-} + CH_4 \uparrow$$

and benzene can be obtained from sodium benzene carboxylate (benzoate)

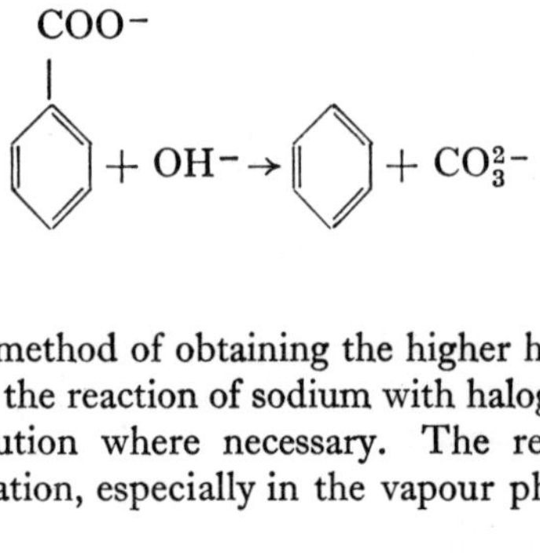

(ii) The Wurtz Reaction

This is a seldom-used method of obtaining the higher homologues of the alkane series, and depends upon the reaction of sodium with halogen derivatives of hydrocarbons, in ethereal solution where necessary. The reaction mechanism may involve free-radical formation, especially in the vapour phase or in liquids of low dielectric constant

$$RX + Na \rightarrow R\cdot + Na^+X^- \qquad\qquad 2R\cdot \rightarrow R\text{---}R$$

Alternatively, in more polar solutions the mechanism may involve

$$RX + 2Na \rightarrow R^-Na^+ + Na^+X^-$$

followed by

$$R^-Na^+ + RX \rightarrow Na^+X^- + R\text{---}R$$

The Wurtz reaction is satisfactory only for the production of hydrocarbons having an even number of carbon atoms, e.g.

$$CH_3Br + 2Na + CH_3Br \rightarrow C_2H_6\uparrow + 2Na^+Br^-$$

and

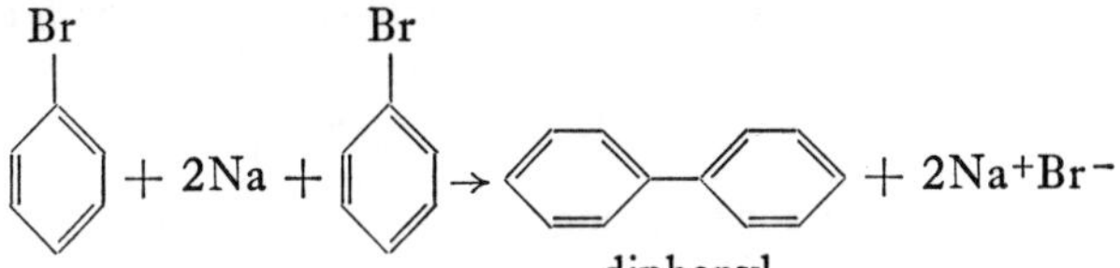

$$\text{Br} \qquad \text{Br}$$

$+ \; 2\text{Na} + $ $\rightarrow$ $+ \; 2\text{Na}^+\text{Br}^-$

diphenyl

The use of mixed halides, e.g. $RX + R'X$, results in the formation of R—R, R—R' and R'—R' which in general are not easy to separate, as there is little difference in their boiling points.

The reaction is more useful as a method of preparing the alkyl derivatives of aromatic hydrocarbons—a modification due to Fittig—e.g.

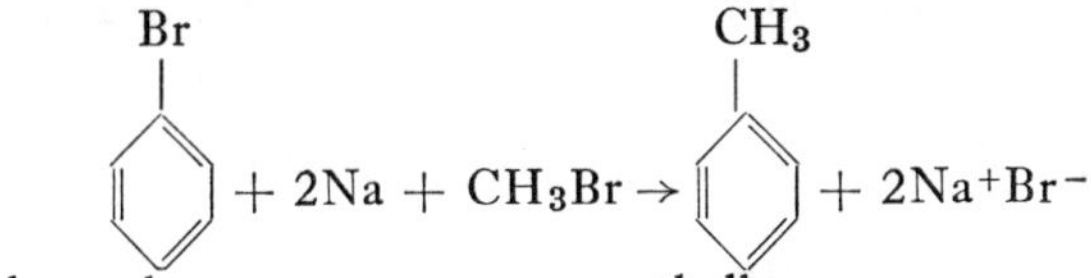

$$\text{Br} \qquad\qquad\qquad \text{CH}_3$$

$+ \; 2\text{Na} + \text{CH}_3\text{Br} \rightarrow$ $+ \; 2\text{Na}^+\text{Br}^-$

bromobenzene   methylbenzene
(toluene)

Although the products obtained in such Wurtz–Fittig reactions may be mixtures, separation is easily effected because of their wide range of boiling points.

(iii) The Friedel–Crafts Reaction

The Friedel–Crafts alkylation reaction entails the reaction between an aromatic hydrocarbon and a halogen derivative of an alkane in the presence of a catalyst which is normally anhydrous aluminium chloride, although in certain cases boron trifluoride or hydrogen fluoride can be used. It is essential for all the reactants to be free of moisture:

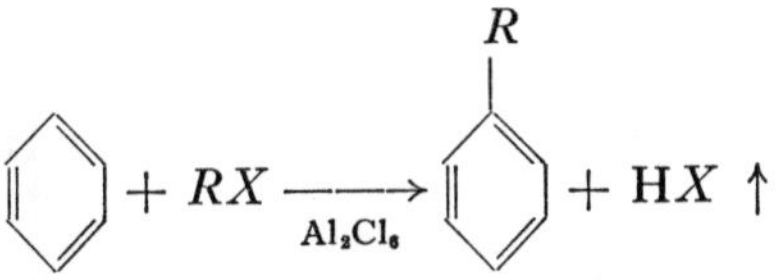

$$R$$

$+ \; RX \xrightarrow[\text{Al}_2\text{Cl}_6]{}$ $+ \; HX \uparrow$

The non-localized π orbitals present in a molecule of benzene (p. 51) will tend to become localized on any part of the molecule which is in proximity to a positive charge. This induced 'polarization' can be represented in terms of the conventional structure by postulating a tendency for the electrons in one of the 'double bonds' to migrate to one end of the bond:

$\longrightarrow$ δ^-

373

The mechanism of the Friedel–Crafts reaction can then be described in terms of this effect as follows. Firstly, the halogen compound reacts with the anhydrous aluminium chloride

$$2RX + Al_2Cl_6 \rightarrow 2R^+[AlCl_3X]^-$$

then the ion R^+ attacks the benzene ring

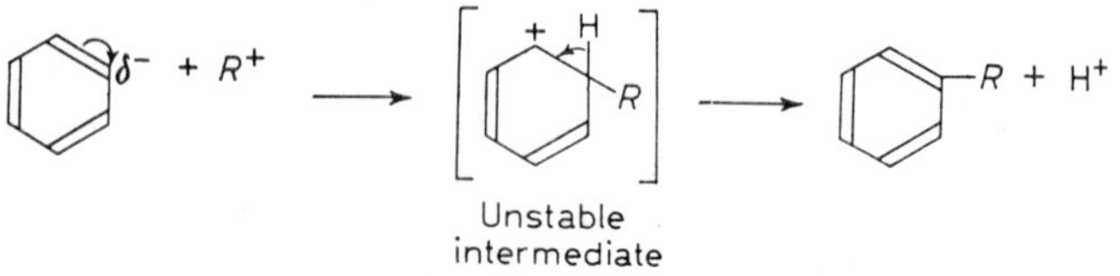

Unstable
intermediate

Methylbenzene and ethylbenzene can be prepared by this method but not propylbenzene because the organic ion isomerizes:

$$2CH_3CH_2CH_2X + Al_2Cl_6 \rightarrow 2CH_3CH_2CH_2^+[AlCl_3X]^-$$

$$CH_3CH_2CH_2^+ \rightarrow CH_3\overset{+}{C}HCH_3$$

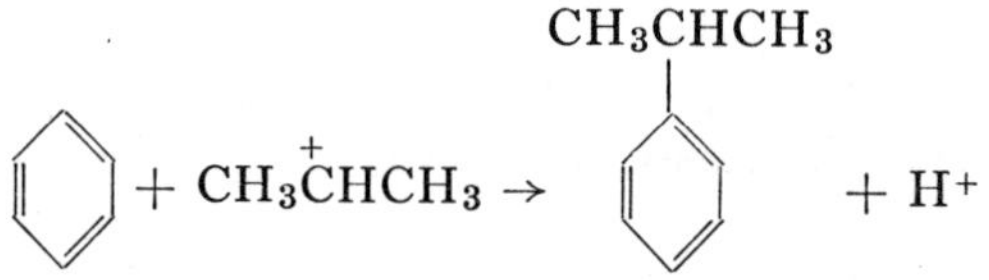

Similar rearrangements occur with all larger carbon chains.

A further difficulty is that the reaction cannot always be stopped at the mono-substituted stage; e.g. the reaction of benzene with bromomethane yields some 1,2- and 1,4-dimethylbenzene as well as monomethylbenzene.

Alcohols and alkenes can also be made to react with aromatic hydrocarbons

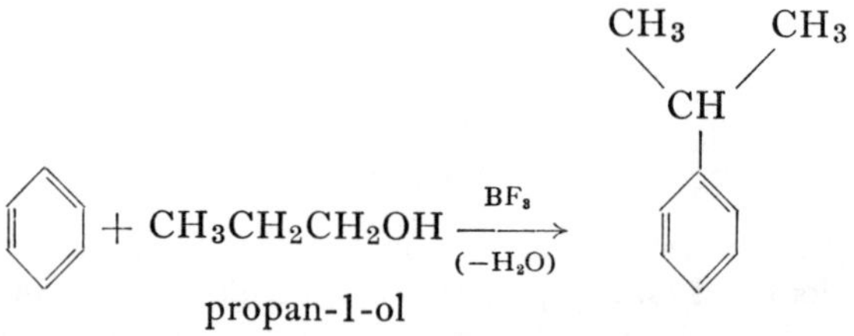

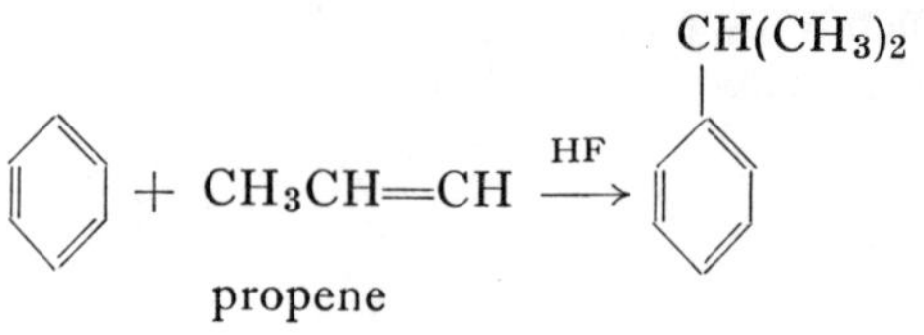

propan-1-ol

propene

METHODS FOR PREPARATION OF ALKENES AND/OR ALKYNES

Besides the methods described so far, there are others suitable only for unsaturated hydrocarbons.

(i) Dehydrohalogenation

The elements of a hydrogen halide can often be removed from a halogen derivative by refluxing with an alcoholic solution of potassium hydroxide. Of the two possible reactions, hydrolysis ($-X \rightarrow -OH$) and dehydrohalogenation, the latter nearly always predominates under these conditions by a first or second order reaction, depending on the concentration of the alkali:

a) 1st order. Rate $\propto$ [halide]

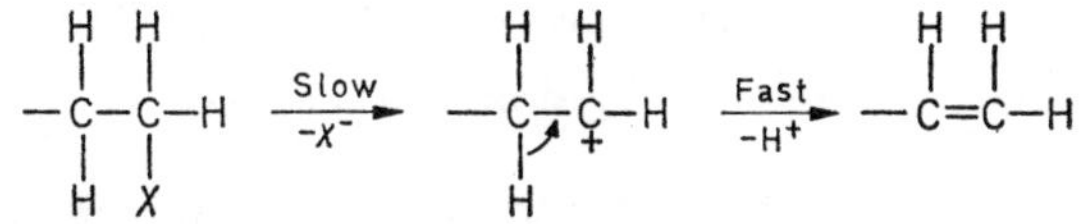

b) 2nd order. Rate $\propto$ [halide] [OH⁻]

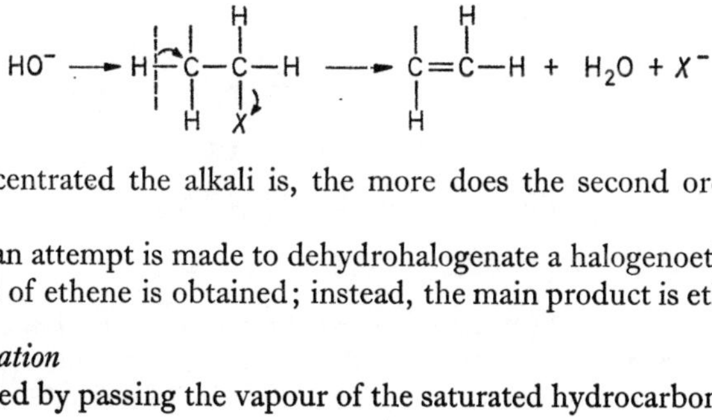

The more concentrated the alkali is, the more does the second order reaction predominate.

However, if an attempt is made to dehydrohalogenate a halogenoethane, no more than 1 per cent of ethene is obtained; instead, the main product is ethanol (p. 435).

(ii) Dehydrogenation

This is effected by passing the vapour of the saturated hydrocarbon over a heated catalyst, e.g.

$$CH_3CH_2CH_2CH_3 \xrightarrow[Cr_2O_3]{Al_2O_3} CH_2{=}CH{-}CH{=}CH_2 + 2H_2 \uparrow$$
butane   buta-1,3-diene

(iii) Dehydration (for alkenes)

The elements of water can be removed from an alcohol, either by heating with excess of concentrated sulphuric or phosphoric acid

$$RCH_2CH_2\overset{..}{O}H \xrightarrow{H^+} RCH_2CH_2\overset{+}{O}{-}H \xrightarrow[(-H_2O)]{} [RCH_2CH_2]^+ \xrightarrow[(-H^+)]{} RCH{=}CH_2$$

375

or by passing the vapour over heated aluminium oxide

$$RCH_2CH_2OH \xrightarrow{Al_2O_3} RCH{=}CH_2 + H_2O$$

The latter reaction proceeds because the C—O bond is weakened by adsorption on to the catalyst:

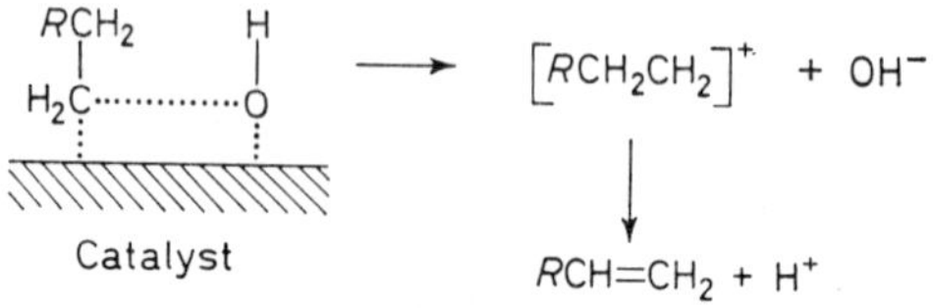

$$[RCH_2CH_2]^+ + OH^-$$

$$RCH{=}CH_2 + H^+$$

Catalyst

INDUSTRIAL SOURCES OF HYDROCARBONS

The two major commercial sources of hydrocarbons are coal and petroleum.

Coal (Figure 19.1)—The destructive distillation of coal (p. 378) yields coal gas, coal tar and coke. The most volatile fraction of coal tar, benzol, is rich in benzene and toluene, which can be either separated by fractional distillation or used together as a means of increasing the calorific and anti-knocking values of petrol. Another constituent of coal tar is the hydrocarbon, naphthalene (see *Table 18.2*) which readily crystallizes out.

The coke remaining in the retort can be heated with calcium oxide in an electric furnace to give calcium carbide

$$CaO + 3C \rightarrow CaC_2 + CO \uparrow$$

Water reacts with this product to yield ethyne (acetylene)

$$C_2^{2-} + 2H_2O \rightarrow 2OH^- + C_2H_2 \uparrow$$

which can be used as a fuel (oxyacetylene welding) or converted into other valuable hydrocarbons. For example, if it is passed into mercury(II) sulphate solution, it is hydrated to ethanal (acetaldehyde) which can then be converted into butadiene

$$CH{\equiv}CH + H_2O \xrightarrow{HgSO_4} CH_3C\overset{H}{\underset{O}{\diagup\diagdown}} \xrightarrow{NaOH} CH_3CH(OH)CH_2CHO$$

$$\Big\downarrow \; {H_2,\,Cu/Cr \atop 300 \times 10^5 Nm^{-2}\; 100°}$$

$$CH_2{=}CH{-}CH{=}CH_2 \xleftarrow[280°]{Na_3PO_4} CH_3CH(OH)CH_2CH_2OH$$

Butadiene, which is also manufactured by dehydrogenation of the C_4 fraction from petroleum 'cracking', is copolymerized with unsaturated compounds such as styrene and acrylonitrile to give various types of synthetic rubbers (p. 407).

376

Alternatively, steam can be passed over white-hot coke to form water gas which, on mixing with more hydrogen and passing, under pressure, over cobalt or nickel catalysts at elevated temperatures (the Fischer–Tropsch reaction), yields various hydrocarbons, e.g.

$$2CO + 4H_2 \rightarrow 2H_2O + C_2H_4 \uparrow$$

The value of coal as a source of hydrocarbons is summarized in *Figure 19.1*.

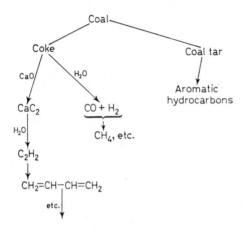

Figure 19.1. Coal as a source of hydrocarbons (see also *Figure 13.4*, p. 280)

Petroleum—Unlike coal tar, petroleum is chiefly composed of aliphatic hydrocarbons, and primary distillation gives a separation of products mainly in terms of molecular weight. Thus hydrocarbons C_1 to C_4 remain in the gaseous condition at the minimum temperature of 30° (*Figure 19.2*).

Petroleum production (1967)

	10^6 tonnes	per cent
World	1 641	
U.S.A.	409	25
U.S.S.R.	265	16
Venezuela	176	11
Kuwait	114	7

To obtain the maximum utility from petroleum it is necessary at present to break down larger hydrocarbons into smaller molecules, a process known as *cracking*. This

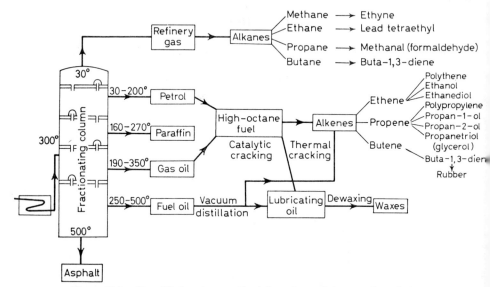

Figure 19.2. Simplified and generalized flow sheet of the petroleum industry (Temperatures in °C)

is accomplished either by the sole application of heat (thermal cracking) or in conjunction with a catalyst (catalytic cracking), e.g. a mixture of silica and alumina

$$CH_3(CH_2)_8CH_3 \longrightarrow CH_3(CH_2)_5CH_3 + CH_3CH{=}CH_2$$
$$\text{decane} \qquad\qquad \text{heptane} \qquad\qquad \text{propene}$$

When decomposition of this type takes place, one of the products at least is unsaturated. Some of the unsaturated hydrocarbons can be used in the petrochemical industry, e.g.

$$CH_2{=}CHCH_3 \xrightarrow[\text{H}_2\text{O}]{\text{H}_2\text{SO}_4} CH_3CHCH_3 \xrightarrow[300°C]{\text{Cu}} CH_3COCH_3 \uparrow + H_2 \uparrow$$
$$\text{propene} \qquad\qquad\quad |\qquad\qquad\qquad \text{propanone}$$
$$\text{OH} \qquad\qquad\qquad \text{(acetone)}$$
$$\text{propan-2-ol}$$

$$CH_2 = CH_{2(g)} + H_2O_{(g)} \xrightarrow[\substack{70 \times 10^5 \text{ Nm}^{-2} \\ 300°C}]{\text{H}_3\text{PO}_4} C_2H_5OH$$

378

or *alkylated, polymerized* or *cyclized* (i.e. *re-formed*) to make hydrocarbons suitable for use in petrol. Cyclic compounds so formed can be *dehydrogenated* to give aromatic hydrocarbons; this *reforming* is carried out in the presence of a platinum catalyst and is usually called *platforming*:

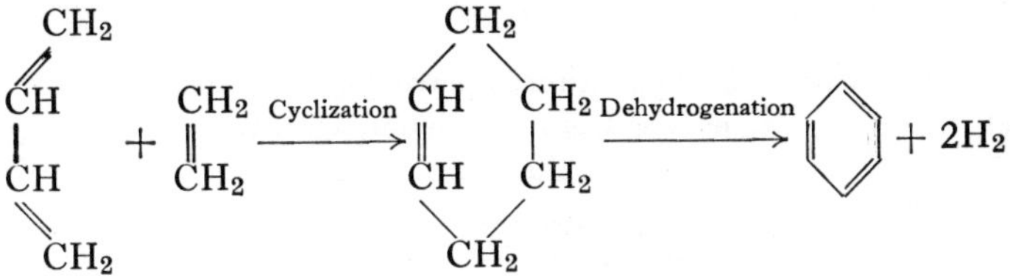

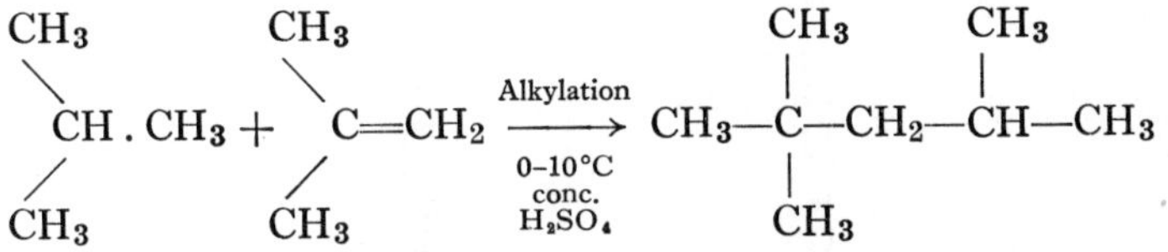

Isomerization also can be used as a means of converting a less valuable into a more valuable substance; by this means, ethyl cyclopentane is changed into methylbenzene (toluene)

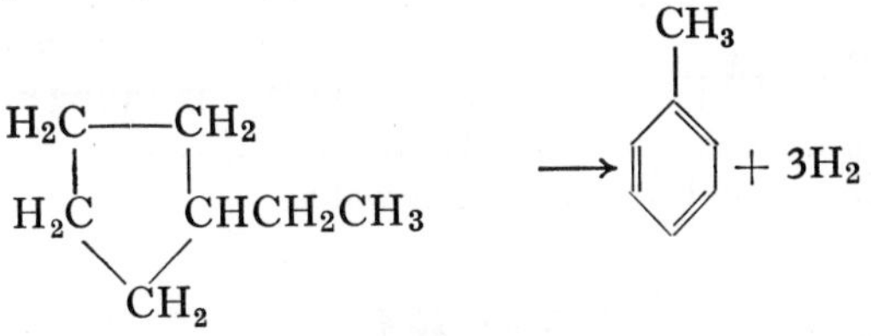

Besides petroleum cracking and reforming, methods based on (*a*) the carbonization and gasification of coal; (*b*) hydrogenation of coal and coal tar; and (*c*) Fischer–Tropsch synthesis, have been used to produce liquid fuels suitable for high-compression engines. The tendency of fuels to detonate prematurely ('knock' or 'pink') sets a limit to the power which can be obtained. It is measured in terms of the 'octane number' which is defined as the volume percentage of 2,2,4-trimethylpentane ('iso-octane', octane number 100) in a mixture of this substance with heptane (octane number 0) which equals the fuel in knock intensity when tested in a standard engine under specified conditions. Branched-chain and aromatic hydrocarbons have low knocking properties, whereas straight-chain hydrocarbons readily produce knocking—hence the value of the above processes.

It is often necessary to remove various objectionable constituents at different stages in petroleum refining. Many of these are sulphur-containing compounds which can be removed by treatment with sodium hydroxide

$$RSH + NaOH \rightarrow RSNa + H_2O$$

379

or by catalytic hydrogenation, when the hydrogen sulphide formed is burnt to sulphur

$$RSH + H_2 \rightarrow RH + H_2S \qquad 2H_2S + O_2 \rightarrow 2H_2O + 2S$$

The importance of petroleum to the chemical industry is indicated in *Table 19.3.*

Table 19.3. Approximate percentage of substance
derived from petroleum, U.S.A.

Substance	Percentage
Propan-2-ol	100
Ethene	100
Methanol	99
Propanone	98
Ethanol	83
Toluene	78
Benzene	50
Glycerol	42

REACTIONS OF THE HYDROCARBONS

Hydrocarbons are generally of zero or low dipole moment (p. 43); there is therefore little or no interaction between similar molecules. They are consequently of relatively high volatility [although this decreases with increasing molecular weight (*Table 19.4*)] and insoluble in polar liquids such as water. The relative densities of the liquid hydrocarbons are less than that of water.

Table 19.4

n-*Alkanes*	b.p. °C	*Alkenes*	b.p. °C	*Alkynes*	b.p. °C
CH_4	−160				
C_2H_6	−88	C_2H_4	−104	C_2H_2	−84
C_3H_8	−44	C_3H_6	−48	C_3H_4	−23
C_4H_{10}	−1	$CH_3CH:CHCH_3$	4	$CH_3C{\equiv}CCH_3$	27
C_5H_{12}	36	$C_2H_5CH:CHCH_3$	37	$C_2H_5C{\equiv}CCH_3$	56
C_6H_{14}	69	$C_2H_5CH:CHC_2H_5$	73	$C_3H_7C{\equiv}CCH_3$	84
cyclohexane, C_6H_{12}	81	cyclohexene, C_6H_{10}	83		

It can be seen from *Table 19.4* that there is a fairly steady increase in boiling point on ascending an homologous series; but the boiling points of isomers differ (*Table 19.5*).

Table 19.5

C$_5$H$_{12}$	CH$_3$(CH$_2$)$_3$CH$_3$	CH$_3$CH$_2$CH(CH$_3$)$_2$	C(CH$_3$)$_4$
	pentane	2-methylbutane	2,2-dimethylpropane
b.p.(°C)	36	28	9
C$_6$H$_6$	CH$_3$C≡CC≡CCH$_3$		
	hexa-2,4-dyne	benzene	
b.p.(°C)	85	80	

Hydrocarbons, when above their respective ignition points, burn in air. If combustion is complete, the flame is non-luminous and the products are carbon dioxide and water. As the molecular complexity increases, progressively more oxygen is required and there is a greater probability of the combustion being incomplete and of resultant free carbon rendering the flame luminous through incandescence. Thus, in the case of methane and naphthalene, for complete combustion

$$CH_4 + 2O_2 \rightarrow 2H_2O \uparrow + CO_2 \uparrow$$

$$C_{10}H_8 + 12O_2 \rightarrow 4H_2O \uparrow + 10CO_2 \uparrow$$

so that, at the same temperature and pressure, six times as much oxygen is required for the complete combustion of a certain volume of naphthalene vapour as for the same volume of methane.

The electron density of the unsaturated hydrocarbons is high (see diagrams on p. 50), and therefore they are most readily attacked by electrophilic reagents, e.g reagents having groups with a partial or complete positive charge. In the case of alkenes and alkynes, the result is usually addition, whereas the aromatic systems are generally so stable that addition of the electrophilic reagent is followed by elimination, leaving the delocalized orbitals intact, so that the reaction is essentially one of substitution and the hydrocarbon is reacting as a saturated compound. However, aromatic hydrocarbons may undergo addition reactions in certain cases, behaving then as unsaturated compounds (see Chapter 3 for electron cloud distribution in hydrocarbons).

1. Addition Reactions of Unsaturated Hydrocarbons

(a) *Hydrogenation* has been described above (p. 371).

(b) *Halogenation*—This involves the addition of a halogen across a multiple bond. As halogens are symmetrical molecules, there can only be one possible product

$$\begin{array}{c} \diagup \\ \diagdown \end{array} C = C \begin{array}{c} \diagdown \\ \diagup \end{array} \rightarrow -\overset{|}{\underset{|}{C}}-\overset{|}{\underset{|}{C}}-$$
$$+$$
$$Br-Br \qquad Br \quad Br$$

Although the addition is shown as a broadside attack, this is less likely to represent the mechanism than an end-on attack. In the latter case, the induced polarization produced as the two reacting molecules approach one another, allows the transition state to be reached with a lower activation energy than that required for a broadside attack

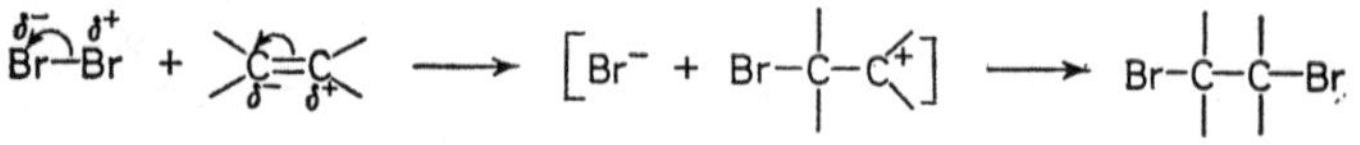

$$\overset{\delta^-}{Br}\overset{\delta^+}{Br} + \underset{\delta^-}{\overset{}{C}}=\underset{\delta^+}{\overset{}{C}} \longrightarrow \left[Br^- + Br-\overset{|}{\underset{|}{C}}-\overset{+}{C}\right] \longrightarrow Br-\overset{|}{\underset{|}{C}}-\overset{|}{\underset{|}{C}}-Br$$

Therefore the addition of the two atoms does not take place simultaneously, as is shown by the fact that bromination of ethene in the presence of chloride or nitrate ions produces some $BrCH_2CH_2Cl$ and $BrCH_2CH_2ONO_2$, respectively, e.g.

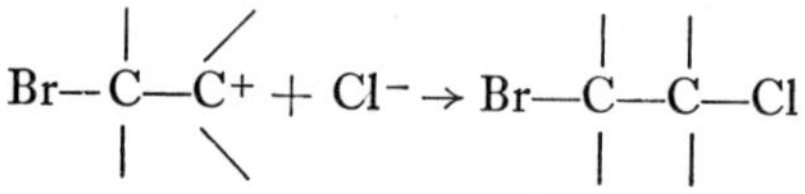

$$Br-\overset{|}{C}-\overset{/}{C^+} + Cl^- \rightarrow Br-\overset{|}{\underset{|}{C}}-\overset{|}{\underset{|}{C}}-Cl$$

The addition of a halogen to an aromatic hydrocarbon takes place in the presence of sunlight or ultra-violet radiation. The light activates some of the halogen molecules and produces free radicals which then attack the aromatic ring, e.g.

$$Cl_2 \xrightarrow[\text{(hv)}]{\text{light}} 2Cl\cdot$$

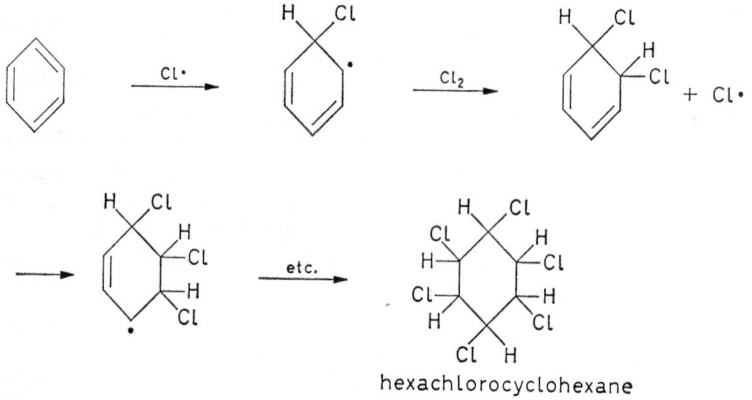

hexachlorocyclohexane

The latter product exists in eight isomeric forms, e.g.

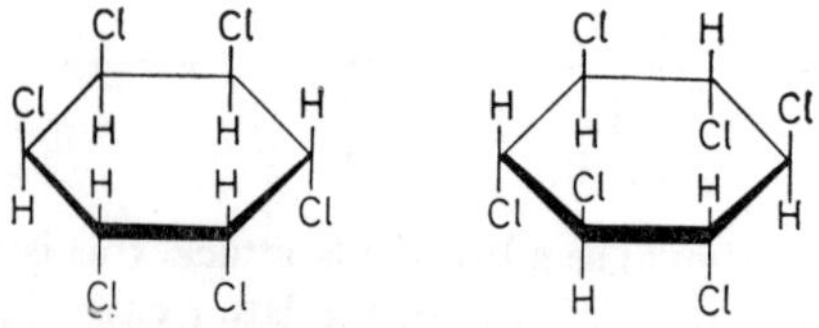

One of the isomers is commonly known as 'gammexane', an insecticide, which is thought to be

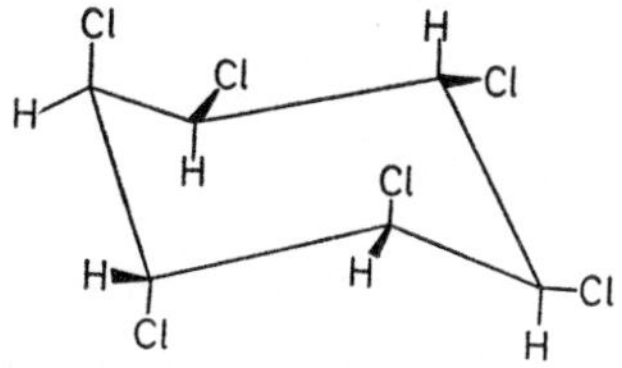

(c) *Hydrohalogenation*—Addition of a hydrogen halide occurs by the initial attack of the electrophilic hydrogen ion on the more electronegative carbon atom to give a *carbonium ion*. Reaction between the halide ion and the carbonium ion completes the process, in accordance with Markownikoff's rule which states that the hydrogen atom of the hydrogen halide attaches to the carbon which already has the greater number of hydrogen atoms attached to it, e.g.

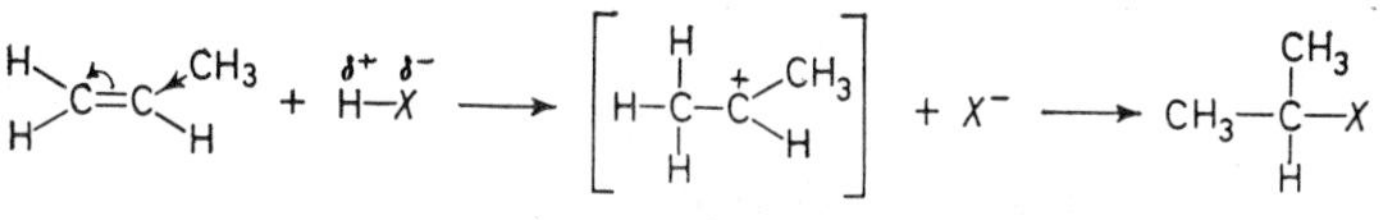

A carbonium ion

(The methyl group tends to be electron-releasing, so that in a suitable environment polarization occurs as shown.)

Two steps are possible in the hydrohalogenation of an alkyne, e.g.

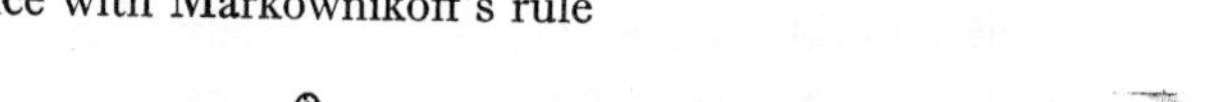

Ethenyl (vinyl) bromide

The ethenyl bromide can then react further, polarization of the molecule occurring by the donation of a lone pair of electrons on the bromine atom to give an extension of the π clouds from the unsaturated ethenyl group. Addition again takes place in accordance with Markownikoff's rule

1,1-Dibromoethane
(ethylidene bromide)

The order of reactivity of the halides is $HI > HBr > HCl > HF$. This is the same order as that of the respective acid strengths and would be expected from a consideration of the bond energies of the four compounds (p. 198).

(d) *Hypohalogenation*—Hypohalogen acids, usually hypochlorous acid, add across multiple bonds. This is a very useful reaction, as two highly reactive groups are introduced by the same reagent into a carbon chain:

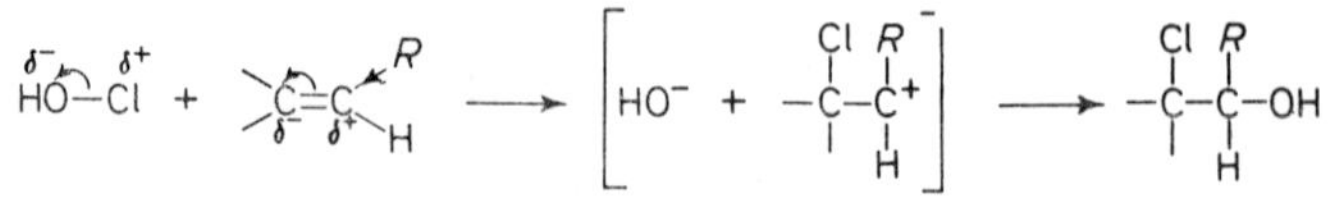

$$\overset{\delta-}{HO}\!\!-\!\!\overset{\delta+}{Cl} + \overset{R}{\underset{H}{>\!C\!=\!C<}} \longrightarrow \left[HO^- + \overset{Cl}{\underset{H}{-\overset{R}{\underset{|}{C}}-\overset{+}{C}}} \right] \longrightarrow \overset{Cl}{\underset{H}{-\overset{R}{\underset{|}{C}}-\overset{|}{C}-OH}}$$

Again the initial attack is by an electrophilic halogen cation, e.g.

$$HOCl + H_2C\!=\!CH_2 \longrightarrow ClCH_2CH_2OH$$

(e) Miscellaneous Addition Reactions

(*i*) Reaction between cold concentrated sulphuric acid and an alkene produces an alkyl hydrogen sulphate:

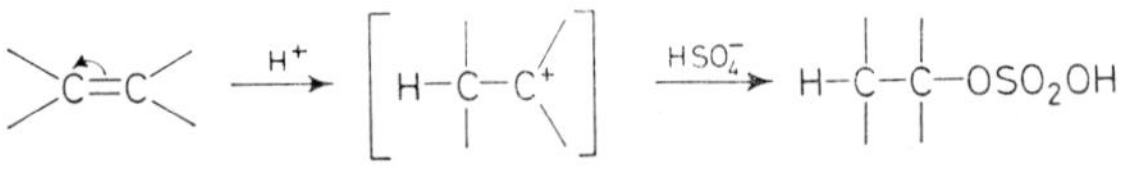

$$>\!\overset{\blacktriangle}{C}\!=\!C< \xrightarrow{H^+} \left[H\!-\!\overset{|}{C}\!-\!\overset{+}{C} \right] \xrightarrow{HSO_4^-} H\!-\!\overset{|}{C}\!-\!\overset{|}{C}\!-\!OSO_2OH$$

This can be readily hydrolysed to an alcohol by diluting and distilling:

$$-\overset{|}{\underset{|}{C}}\!-\!\overset{|}{\underset{|}{C}}\!-\!OSO_2OH + H_2O \longrightarrow -\overset{|}{\underset{|}{C}}\!-\!\overset{|}{\underset{|}{C}}\!-\!OH + H_2SO_4$$

These reactions provide an important route from alkenes to alcohols, e.g.

$$CH_3CH\!=\!CH_2 \xrightarrow{H_2SO_4} \underset{OSO_2OH}{CH_3CH.CH_3} \xrightarrow{H_2O} \underset{\underset{propan\text{-}2\text{-}ol}{OH}}{CH_3CH.CH_3} + H_2SO_4$$
propene

(*ii*) Alkynes react with dilute acids in the presence of a catalyst, the net result being the addition of the elements of water or some other simple molecule:

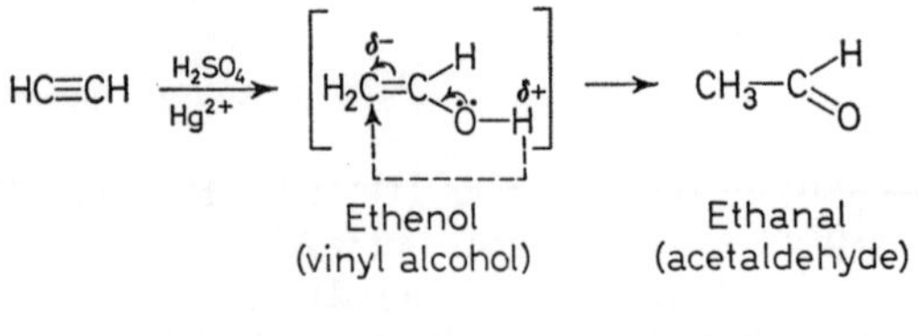

$$HC\!\equiv\!CH \xrightarrow[Hg^{2+}]{H_2SO_4} \left[\overset{\delta-}{H_2C}\!=\!\overset{H}{C} \overset{\delta+}{\underset{O-H}{\diagup}} \right] \longrightarrow CH_3\!-\!C\overset{H}{\underset{O}{\diagup}}$$

Ethenol   Ethanal
(vinyl alcohol)   (acetaldehyde)

(Vinyl alcohol does not exist, as the hydrogen atom of the hydroxyl group is labile because of the polarization shown above.)

Similarly

$$HC\equiv CH \xrightarrow[HCl]{HgCl_2} H_2C\!\!=\!\!CHCl$$
chloroethene
(vinyl chloride)

$$HC\equiv CH \xrightarrow[HCl]{Ba(CN)_2} H_2C\!\!=\!\!CHCN$$
cyanoethene
(vinyl cyanide or acrylonitrile)

The vinyl compounds are important in the manufacture of plastics and paints (p. 406).

(*iii*) Oxidation with alkaline permanganate involves the addition of the elements of hydrogen peroxide, i.e. two hydroxyl groups, e.g.

$$\begin{array}{ccc} CH_2 & & CH_2OH \\ \| & + 2OH \longrightarrow & | \\ CH_2 & & CH_2OH \end{array}$$
ethanediol
(ethylene glycol)

This is a useful diagnostic test for unsaturation because of the appearance of green manganate, but other compounds also reduce alkaline permanganate.

The use of ^{18}O in the permanganate has indicated that both oxygen atoms come from this. The following mechanism has been suggested.

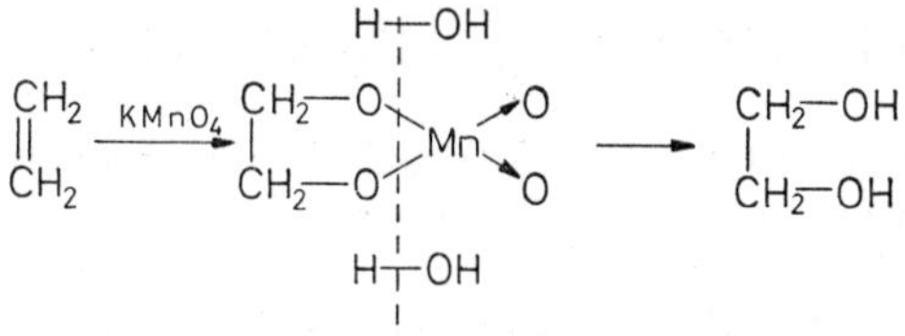

(*iv*) Oxidation with ozone produces ozonides which, on hydrolysis with water, yield aldehydes, ketones, carboxylic acids and hydrogen peroxide. The whole process is known as *ozonolysis*. It is useful for determining the position of multiple bonds between carbon atoms by identification of the products:

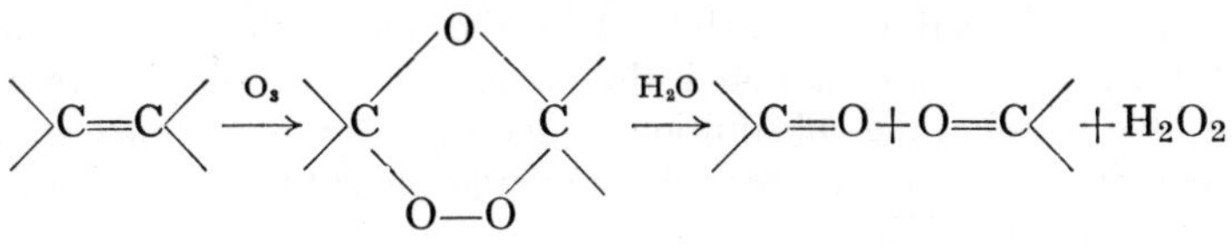

385

and

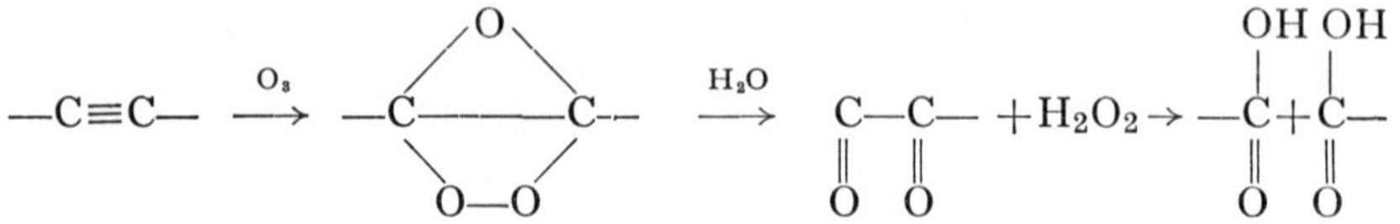

$$-C \equiv C- \xrightarrow{O_3} \underset{\substack{| \quad\quad |\\O—O}}{-C \overset{O}{\diamond} C-} \xrightarrow{H_2O} \underset{\substack{|| \quad ||\\O \quad O}}{C—C-} + H_2O_2 \rightarrow \underset{\substack{|| \quad ||\\O \quad O}}{\overset{OH\ OH}{-C+C-}}$$

Ozonolysis also occurs with aromatic systems.

(v) The alkene link, when ruptured, gives a molecule with two reactive carbon atoms. Thus *polymerization* can occur, giving a long carbon chain system. Ethene is polymerized on a large scale, by using a high pressure and temperature, to give *polythene*. Propene similarly gives polypropylene while styrene forms polystyrene:

$$2n(CH_2{=}CH_2) \rightarrow (-CH_2-CH_2-CH_2-CH_2-)_n$$

$$2n(\underset{\substack{|\\CH_3}}{CH}{=}CH_2) \rightarrow (-\underset{\substack{|\\CH_3}}{CH}-CH_2-\underset{\substack{|\\CH_3}}{CH}-CH_2-)_n$$

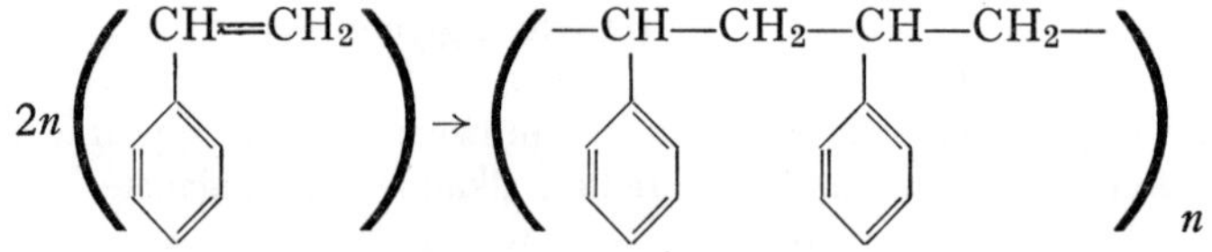

By a similar process, butadiene yields synthetic rubber:

$$nCH_2{=}CH-CH{=}CH_2 \rightarrow (-CH_2-CH{=}CH-CH_2-)_n$$

The chains are terminated by stray free radicals. The properties of the product depend upon the degree of polymerization; polymers with molecular weights of 10000–40000 have desirable properties, e.g. they are thermoplastic, so that by heating them to below their decomposition points they become plastic and can then be moulded.

2. Substitution Reactions of Alkanes and Aromatic Hydrocarbons

Alkanes are much less reactive than unsaturated hydrocarbons, and the unsaturation of aromatic hydrocarbons is stabilized by the non-localized nature of the bonding. Because alkanes are saturated, they cannot undergo addition reactions, but it is possible to effect the substitution of hydrogen atoms by other atoms or groups. Aromatic hydrocarbons normally undergo substitution reactions after preliminary addition (p. 364).

(*a*) *Halogenation*—The halogens can react with alkanes with elimination of hydrogen halide:

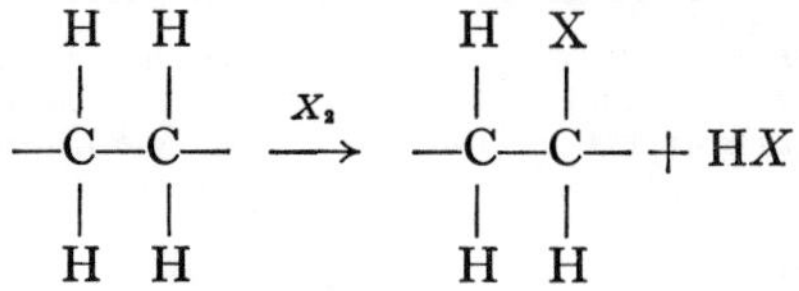

$$\begin{array}{c} \text{H}\ \text{H} \\ |\ \ | \\ -\text{C}-\text{C}- \\ |\ \ | \\ \text{H}\ \text{H} \end{array} \xrightarrow{X_2} \begin{array}{c} \text{H}\ \text{X} \\ |\ \ | \\ -\text{C}-\text{C}- \\ |\ \ | \\ \text{H}\ \text{H} \end{array} + \text{H}X$$

A free-radical mechanism operates, e.g.

$$\text{Cl}_2 \xrightarrow[\text{(}h\nu\text{)}]{\text{light}} 2\text{Cl·}$$

$$\text{CH}_4 + \text{Cl·} \rightarrow \text{CH}_3\text{·} + \text{HCl}$$

$$\text{CH}_3\text{·} + \text{Cl}_2 \rightarrow \text{CH}_3\text{Cl} + \text{Cl·, etc.}$$

The order of reactivity is $\text{F}_2\rangle\text{Cl}_2\rangle\text{Br}_2\rangle\text{I}_2$. Fluorine is so reactive that unless the reaction is moderated, carbon tetrafluoride is the chief product, irrespective of the hydrocarbon used.

Substitution of hydrogen in the aromatic nucleus is usually effected by an electrophile. Consequently, a substituent group already present in the nucleus assists further substitution if it supplies electrons to the nucleus, and hinders it if it withdraws electrons. The *o*- and *p*-positions are most vulnerable to this movement, and so *o*- and *p*-substitution predominates if the nucleus is activated. If the nucleus is deactivated, on the other hand, the effect is most marked in the *o*- and *p*-positions, so that substitution takes place very slowly and then chiefly in the *m*-position. When aromatic hydrocarbons are halogenated at room temperature in the presence of a 'halogen carrier', such as anhydrous aluminium chloride, antimony trichloride or iron filings, substitution occurs in the ring, e.g.

$$2\text{Fe} + 5\text{Br}_2 \rightarrow 2\text{FeBr}_4^- + 2\text{Br}^+$$

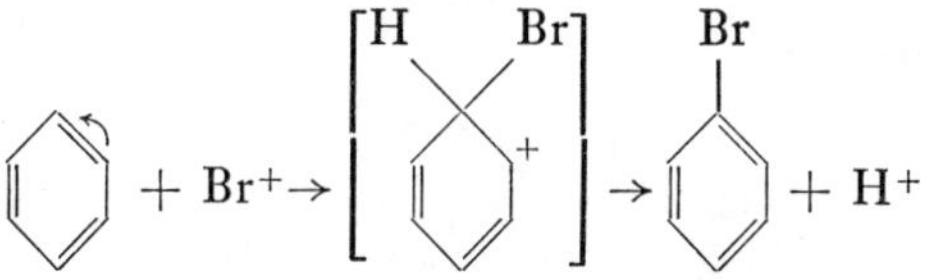

The bromine atom introduced into the benzene nucleus further activates the ring to electrophilic attack (see *Table 19.6*)

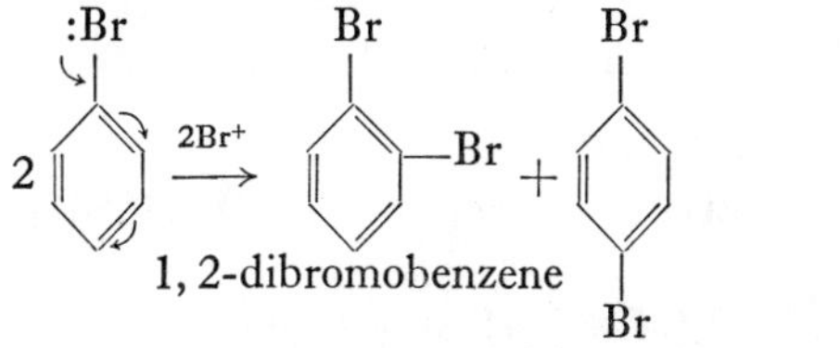

1, 2-dibromobenzene

1, 4-dibromobenzene

Table 19.6

Meta-directing (deactivating) substituents	o- and p-directing (activating) substituents
$-NO_2$	$-CH_3$
$-COOH$	$-Halogen$
$-CHO$	$-NH_2$
$-CN$	$-OH$
$-SO_2OH$	

When the reaction is carried out in the absence of a catalyst and at the boiling point of the hydrocarbon, appreciable substitution occurs in any side chain which is present:

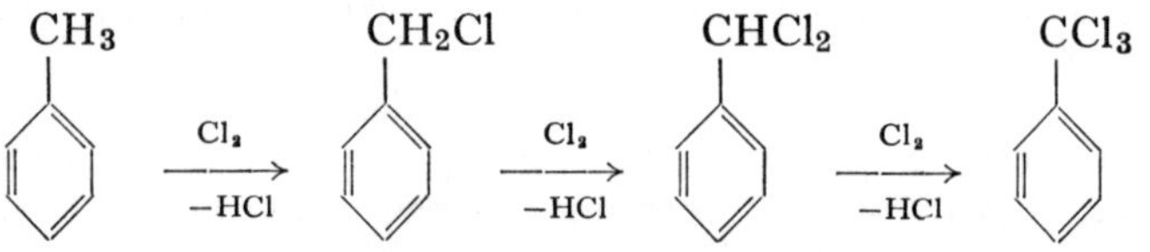

| methylbenzene (toluene) | chloromethyl- benzene (benzyl chloride) | dichloromethyl- benzene (benzal chloride) | trichloromethyl- benzene (benzotrichloride) |

The mechanism involves a chain reaction (see above).

(b) *Sulphonation*—Concentrated or fuming sulphuric acid fairly readily produces sulphonic acids with aromatic hydrocarbons but only with difficulty with alkanes, e.g.

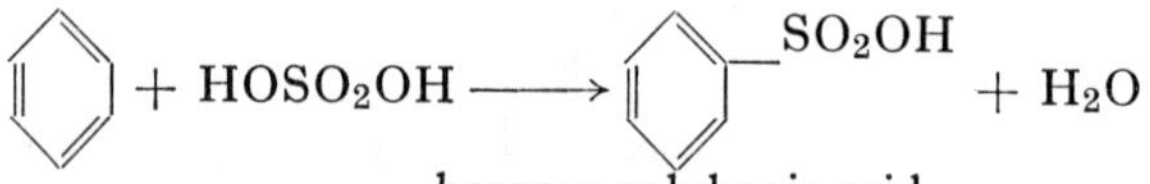

benzene sulphonic acid

It is thought that the attacking electrophile is sulphur trioxide, in which the greater electronegativity of the oxygen atoms produces a positive charge on the sulphur atom.

$$2H_2SO_4 \rightleftharpoons H_3O^+ + HSO_4^- + SO_3$$

or

$$H_2S_2O_7 \rightleftharpoons H_2SO_4 + SO_3$$

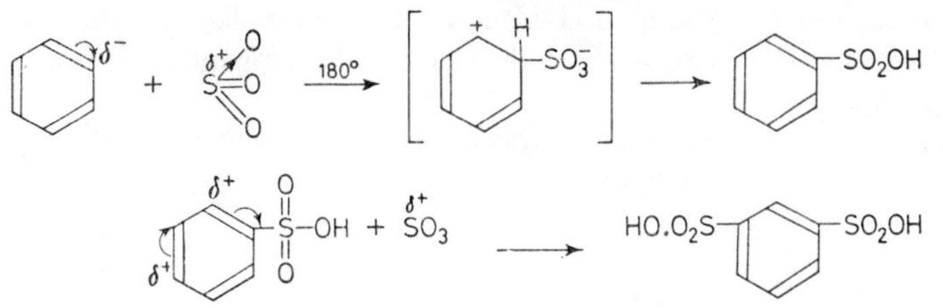

Benzene-1,3-disulphonic
acid

Because the sulphonic acid group is deactivating, substitution virtually ends at this stage.

The sulphonic acids obtained from long-chain alkanes (derived from petroleum) are commercially important as detergents

$$RCH_3 \xrightarrow{\text{H}_2\text{SO}_4/\text{SO}_3} \overset{\delta-\;\delta+}{RCH_2SO_2OH}$$
hydrophobic   hydrophilic

(c) *Nitration*—Nitrating acid consists of a mixture of concentrated or fuming nitric acid and concentrated sulphuric acid. The composition of the mixture has been studied by measurements of the freezing-point depression; the value of the van't Hoff factor (p. 112) is four, in accordance with the ionization

$$2H_2SO_4 + HNO_3 \rightleftharpoons 2HSO_4^- + H_3O^+ + NO_2^+$$
nitryl (nitronium)ion

The electrophilic nitryl ion is the attacking species and, when aromatic hydrocarbons are treated with the above mixture, nitro derivatives are formed (with activated compounds, this can be a very hazardous procedure):

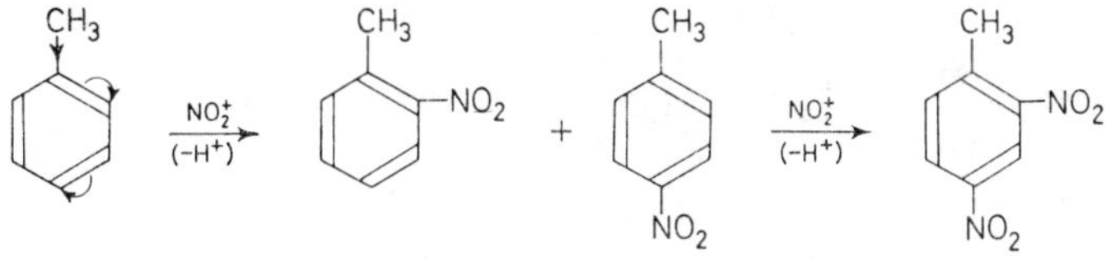

1-Methyl-2-   1-Methyl-4-
nitrobenzene   nitrobenzene

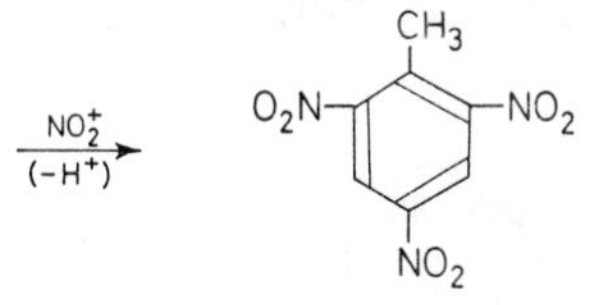

1-Methyl-2,4,6-
trinitrobenzene
(trinitrotoluene
or T.N.T.)

389

It will be noticed that the methyl group (or any other alkyl group) activates the *o*- and *p*-positions of the ring towards the electrophilic reagents, as it is an electron-repelling group.

The nitro group, by electron withdrawal, exerts a deactivating influence on the ring towards the usual electrophilic reagents, e.g.

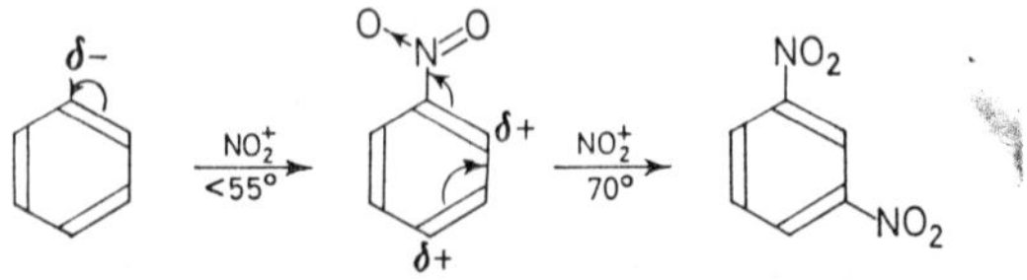

1,3-Dinitrobenzene

The nitro group attacks the ring where it is least deactivated, i.e. where the least positive charge resides (at the *m*-position). It should be noted that no part of the ring has been activated, so that substitution of more than one nitro group into a benzene ring occurs only at higher temperatures.

In the absence of sulphuric acid, nitric acid normally oxidizes any side chain attached to the ring, producing the corresponding carboxylic acid, e.g.

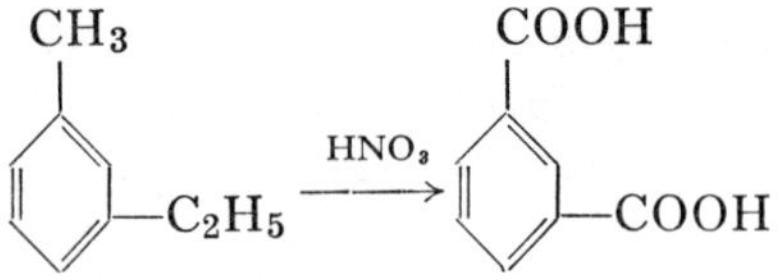

In contrast to aromatic hydrocarbons, the nitration of aliphatic hydrocarbons is slow even at high temperatures, and the process results in much oxidation as well as in the formation of a number of nitro-compounds, e.g.

$$CH_3CH_2CH_3 \xrightarrow[400°]{HNO_3} CH_3CH_2CH_2NO_2, \underset{\underset{NO_2}{|}}{CH_3CH.CH_3}, CH_3CH_2NO_2, CH_3NO_2$$

(vapour-phase nitration)

(*d*) *Friedel–Crafts reaction with aromatic hydrocarbons*—The alkylation reaction has already been discussed (p. 373). A useful modification is the acylation reaction, in which an acyl group (*RCO*—) is introduced into an aromatic ring

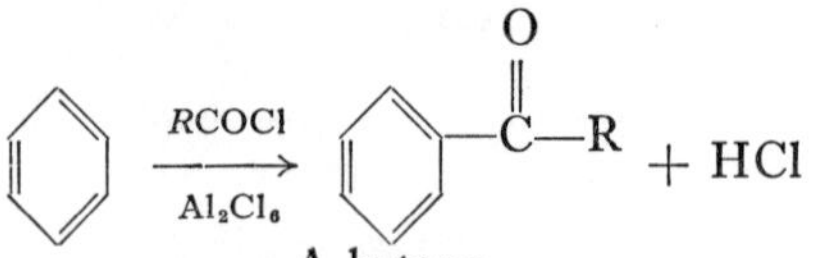

A ketone

This reaction is of value for the preparation of aromatic ketones and also, by

390

subsequent reduction of the product, for the introduction of long straight-chain alkyl groups into an aromatic ring

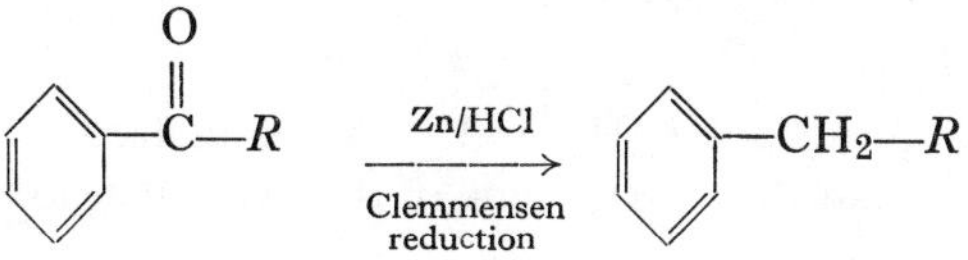

3. Metallic Derivatives of Hydrocarbons

The hydrogen atoms attached to the unsaturated carbon atoms in alkynes show acidic properties, and with ammoniacal solutions of copper(I) or silver chlorides, covalent metallic derivatives are obtained as precipitates

$$HC \equiv CH \xrightarrow{\quad M(NH_3)_2{}^+Cl^- \quad} MC \equiv CM$$

Unlike the ionic carbides (cf. calcium carbide), they detonate easily when dry (cf. covalent and ionic azides and nitrates). With metallic sodium dissolved in liquid ammonia, acetylene reacts to give $NaC \equiv CH$ and ultimately $NaC \equiv CNa$, compounds which are of great value in organic syntheses.

Several structurally interesting metallic derivatives of aromatic systems have been prepared. For example, when cyclopentadiene vapour is passed over heated iron(II) oxide, an orange solid (*ferrocene*) can be condensed

$$2C_5H_6 + FeO \rightarrow (C_5H_5)_2Fe + H_2O$$

By the loss of a proton from the methene group of cyclopentadiene, the resulting cyclopentadienyl anion has six p electrons and can exhibit aromatic properties. These electrons are believed to combine with the d electrons of the iron atom (see p. 337).

ANALYSIS

Water and carbon dioxide are liberated when a hydrocarbon is heated with dry copper(II) oxide and can be identified qualitatively in the usual ways. This procedure is also the basis for the quantitative estimation; when a weighed amount of a hydrocarbon is oxidized in this way, the water produced can be absorbed in weighed calcium chloride tubes and the carbon dioxide in weighed potassium hydroxide bulbs

$$C_xH_y + (x + y/4)O_2 \rightarrow xCO_2 + y/2\ H_2O$$
$$12x + y \qquad\qquad 44x \qquad\quad 9y$$

Example—A hydrocarbon of vapour density 29 gave 0·88 g of carbon dioxide and 0·45 g of water on oxidation. The weights of carbon and hydrogen in the compound are thus $0.88 \times 12/44$ g and $0.45 \times 1/9$ g, respectively. Therefore, the atom ratio of carbon and hydrogen is

$$\frac{0.88 \times 12}{12 \times 44} : \frac{0.45 \times 2}{1 \times 18} \quad \text{i.e. } 2 : 5$$

so that the empirical formula is C_2H_5, and since the molecular weight is 29×2, the molecular formula must be C_4H_{10} (one of the two butanes).

The molecular formula can also be determined volumetrically by the process known as *eudiometry*. A known volume of the hydrocarbon vapour is exploded with a known volume of oxygen (in excess). Cooling results in the elimination of water from the vapour state, and treatment with caustic potash absorbs the carbon dioxide. The final volume represents the volume of unused oxygen. Provided all volumes have been measured at the same temperature and pressure, the molecular formula of the hydrocarbon can be evaluated from the previous equation, i.e. one volume of hydrocarbon vapour requires $(x + y/4)$ volumes of oxygen for complete combustion and yields x volumes of carbon dioxide under the same conditions of temperature and pressure.

Example—After exploding 5 cm³ of a hydrocarbon vapour with 50 cm³ of oxygen and cooling, the volume was 40 cm³, of which 15 cm³ were absorbed by potassium hydroxide solution, i.e.

$$\text{volume of carbon dioxide produced} = 15 \text{ cm}^3$$

$$\text{volume of oxygen used} = [50-(40-15)] = 25 \text{ cm}^3$$

Therefore

$$5 \text{ cm}^3 \text{ C}_x\text{H}_y \text{ requires } 25 \text{ cm}^3 \text{ O}_2 \text{ and produces } 15 \text{ cm}^3 \text{ CO}_2$$

or   1 mole $C_x H_y$ requires 5 mole O_2 and produces 3 mole CO_2

i.e. $x = 3 \qquad x + y/4 = 5 \qquad \text{and} \therefore \quad y = 4(5-3) = 8$

Thus the molecular formula of the hydrocarbon is C_3H_8 (propane).

To confirm the structural formula of a simple organic compound, after identification of the functional groups present, solid derivatives are usually prepared and their melting points compared with the values of known compounds.

No suitable derivatives of aliphatic hydrocarbons can be prepared for the purpose of identification, either because of the unreactivity of the hydrocarbon or because a number of products result from which a solid substance cannot be readily isolated. Therefore, having classified the hydrocarbon as an alkane, alkene or alkyne, further identification is made by measuring its density and, if a liquid, refractive index. These values are then compared with those of known hydrocarbons.

Useful derivatives of aromatic hydrocarbons are the 'picrates', which are generally readily produced by mixing strong alcoholic solutions of picric acid (p. 437) and the hydrocarbon. 'Picrates' are molecular complexes of general formula (hydrocarbon) . n(picric acid), where n is a small integer. Other derivatives of aromatic hydrocarbons suitable for structure determination are the nitro compounds and also the carboxylic acids produced by oxidation of side chains (p. 390).

SUMMARY

The characteristic reaction of saturated hydrocarbons where all the chemical bonds are of the single σ type, is substitution of hydrogen by another element or radical. This can equally be regarded as substitution of alkyl, as for example in the photochemical chlorination of methane:

$$Cl—Cl \xrightarrow{\text{light}} Cl\cdot + \cdot Cl \qquad\qquad \text{dissociation}$$
$$R—H + \cdot Cl \longrightarrow R\cdot + H—Cl \qquad\qquad \text{substitution}$$
$$R\cdot + Cl—Cl \longrightarrow R—Cl + \cdot Cl \qquad\qquad \text{substitution, etc.}$$

In the case of unsaturated hydrocarbons, the bond is a source of attraction to electrophilic reagents and the reaction is of the addition type:

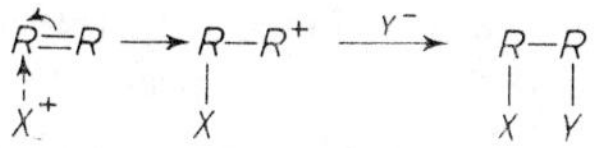

$$R\overset{\curvearrowright}{=}R \longrightarrow R—R^+ \xrightarrow{\;Y^-\;} R—R$$

The stability of the aromatic nucleus is such that, whilst the initial attack is by an electrophilic reagent, instead of the reaction proceeding as an orthodox addition

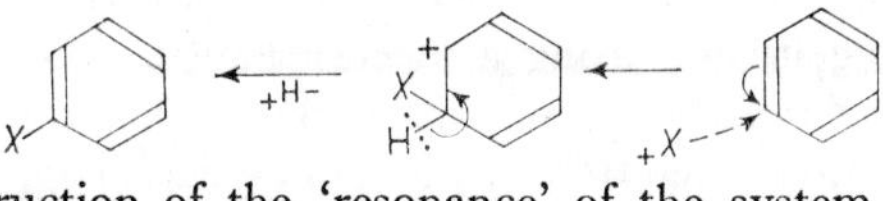

type with consequent destruction of the 'resonance' of the system, a proton is expelled and the non-localized system restored.

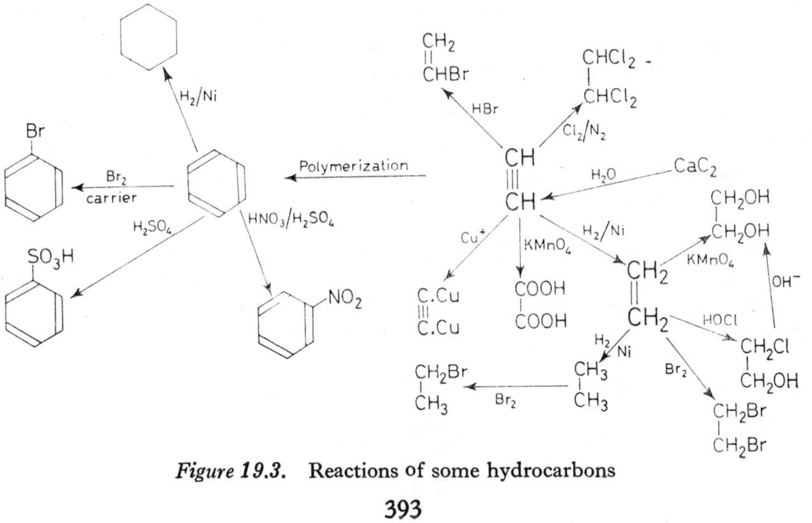

Figure 19.3. Reactions of some hydrocarbons

393

QUESTIONS

1. What are the molecular formulae of rubber, gutta-percha, β-carotene and Vitamin A? What advantages might some synthetic rubbers have over natural rubber?

2. Why is propan-2-ol, rather than propan-1-ol, obtained from propene via sulphuric acid treatment? Using this reaction, how might propan-1-ol be changed to propan-2-ol?

3. Devise a scheme whereby you could distinguish between an alkane, alkene, alkyne and an aromatic hydrocarbon.

4. What forces do you think are operative in 'picrate' compounds?

5. Give the structures and names of the isomers of molecular formula C_5H_{10}.

6. What hydrocarbons could be prepared from butan-2-ol?

7. 10 cm³ of a hydrocarbon were exploded with 130 cm³ of oxygen. After cooling to the original temperature and adjusting the pressure to that originally obtaining, the volume of gas was 95 cm³; 70 cm³ of this was absorbed in potassium hydroxide solution. Suggest possible structural formulae for the hydrocarbon. How would you distinguish between the different possibilities?

8. A hydrocarbon contains 88·9 per cent of carbon and has a vapour density of 27. On hydrogenation, using a nickel catalyst, two molecules of hydrogen were taken up for every molecule of hydrocarbon. The original compound did not give a metallic derivative when treated with ammoniacal copper(I) chloride, but on careful ozonolysis two molecules of methanal and one molecule of a dialdehyde of molecular formula $C_2H_2O_2$ were formed from every molecule of hydrocarbon. Deduce the structure of the hydrocarbon.

9. Suggest a method of preparing 2-bromobenzyl bromide from toluene. What products are possible when this compound reacts with sodium?

10. Given that the bond energies for C—C, C=C, and C—H are 348, 620 and 413 kJ respectively, and that the heat of formation of $C_{(g)}$, $H_{(g)}$ and $C_6H_{6(g)}$ are 716, 218 and 83 kJ respectively, determine if these values are in agreement with an alternate single/double bond structure for benzene.

11. An aromatic hydrocarbon, on oxidation with potassium permanganate gave a monobasic organic acid. This, with soda-lime was decarboxylated to a hydrocarbon having a vapour density of 39. A Friedel-Craft reaction using bromomethane gave the original hydrocarbon. Explain these reactions.

12. Write an essay dealing with the commercial availability of hydrocarbons.

13. Write an essay on polynuclear hydrocarbons..

CHAPTER 20

ORGANIC HALOGEN COMPOUNDS

The halogenation of hydrocarbons was introduced in Chapter 19, where it was seen that, either by addition to unsaturated hydrocarbons or by substitution of hydrogen in saturated hydrocarbons, one or more carbon–halogen bonds are formed. For more than one such bond in a molecule, two possibilities exist, i.e. polyhalogenated hydrocarbons may be divided into (a) those with only one halogen atom attached to any one carbon atom, e.g.

$$ClCH_2 . CH_2Cl$$
1,2-dichloroethane

and (b) compounds in which more than one halogen atom is attached to the same carbon atom, e.g.

$$CH_3 . CHCl_2$$
1,1-dichloroethane

The preparation and properties of the former type are similar to those of the mono-halogenated hydrocarbons, whereas the latter differ in several respects from compounds containing only one halogen atom.

PREPARATION

(i) *Addition to Unsaturated Hydrocarbons* (see also p. 386)
Halogen compounds can be prepared by treating unsaturated hydrocarbons with hydrogen halides or, with the exception of the comparatively unreactive iodine, with the free halogen. The action of fluorine has to be moderated by dilution with nitrogen for the reaction to proceed smoothly, as does also that of chlorine when reacting with an alkyne. Thus

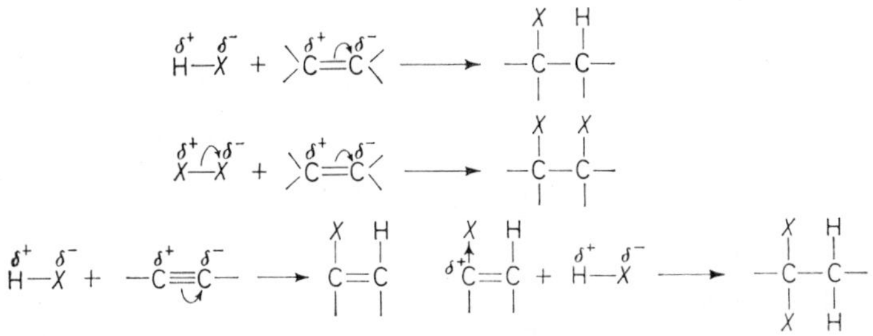

Similarly

$$2X_2 + -C{\equiv}C- \rightarrow \begin{array}{c} X \;\; X \\ | \;\;\; | \\ -C-C- \\ | \;\;\; | \\ X \;\; X \end{array}$$

(ii) Substitution of Hydrogen (see also p. 381)

Polyhalogen derivatives of the alkanes can usually be obtained, in varying yields, by the direct halogenation of the relevant hydrocarbon. For example, methane when treated with chlorine in diffuse daylight, yields a mixture of four compounds produced by a series of free-radical reactions:

$$CH_4 \xrightarrow[(-HCl)]{Cl_2} CH_3Cl \xrightarrow[(-HCl)]{Cl_2} CH_2Cl_2 \xrightarrow[(-HCl)]{Cl_2} CHCl_3 \xrightarrow[(-HCl)]{Cl_2} CCl_4$$

methane   chloro-   dichloro-   trichloro-   tetra-
             methane   methane   methane   chloro-
             (methyl   (methylene   (chloro-   methane
             chloride)   chloride)   form)   (carbon
                                                        tetrachloride)

(The intermediate compounds can also be prepared by reduction of the fully halogenated alkane.)

The fluoro-compounds of the aliphatic series can be conveniently obtained by passing the hydrocarbon vapour and fluorine alternately over cobalt(II) fluoride at 300°C the fluorinating agent being cobalt(III) fluoride

$$2CoF_2 + F_2 \rightleftharpoons 2CoF_3$$

Hydrogen atoms directly attached to a carbon atom in a benzenoid ring can be substituted in the presence of a halogen carrier (see p. 387), e.g.

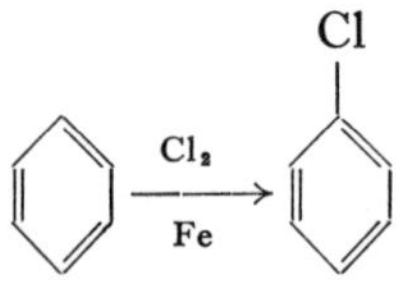

Alternatively, the Sandmeyer reaction can be used (p. 416).

(iii) Substitution of Hydroxyl Groups

Several methods are used for the preparation of compounds of formula RX from the corresponding alcohols ROH. These methods also apply to diols, which give rise to dihalides, but phenols do not react to any significant extent. The re-agents available are

(a) *Hydrogen halides*—The mechanism of attack by hydrogen halide involves initial protonation of the alcohol, followed by reaction with halide ion

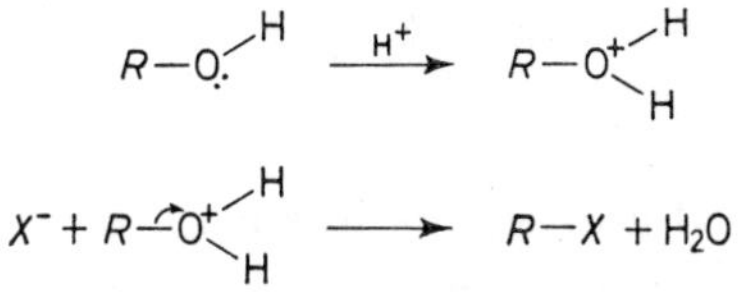

$$R{-}O\overset{H}{\underset{..}{}} \xrightarrow{\ H^+\ } R{-}O^+\overset{H}{\underset{H}{}}$$

$$X^- + R{-}O^+\overset{H}{\underset{H}{}} \longrightarrow R{-}X + H_2O$$

Any factor which can promote ionization of the hydrogen halide will assist the reaction; for example, chloro-compounds can be made by treating an alcohol with dry hydrogen chloride, in the presence of anhydrous zinc chloride, which reacts with hydrogen chloride in accordance with the equation

$$HCl + ZnCl_2 \rightarrow H^+[ZnCl_3]^-$$

The overall reaction can be summarized as e.g.

$$\underset{\text{ethanol}}{C_2H_5OH} + HCl \xrightarrow{\ ZnCl_2\ } \underset{\text{chloroethane}}{C_2H_5Cl} + H_2O$$

Bromo- and iodo-derivatives are usually obtained by preparing the appropriate hydrogen halide *in situ*. The alcohol is distilled with a mixture of the potassium halide and concentrated sulphuric acid and the distillate collected under water, which serves to remove halogen acids, free halogen and sulphur dioxide formed in side reactions, e.g.

$$Br^- + H_2SO_4 \rightarrow HBr + HSO_4^- \ (+ Br_2 + SO_2)$$

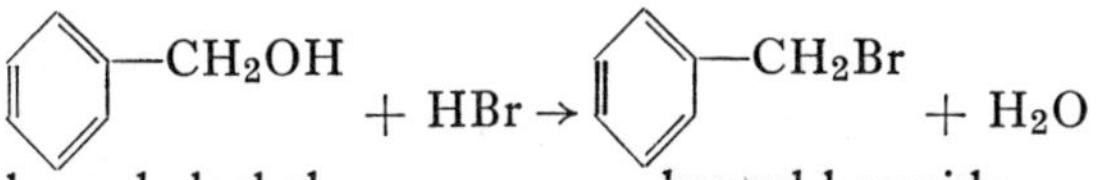

$$\text{benzyl alcohol} \quad {-}CH_2OH + HBr \rightarrow \text{benzyl bromide} \quad {-}CH_2Br + H_2O$$

This method is not very satisfactory for the preparation of derivatives of general formulae

$$\overset{R'}{\underset{R''}{\diagdown{}\diagup}}CHX \quad \text{and} \quad R''{-}\overset{R'}{\underset{R'''}{\diagdown{}\diagup}}CX$$

because the corresponding alcohols so readily react with concentrated sulphuric acid to yield alkenes.

The ease of formation of the halogeno-alkanes is in the order tertiary⟩secondary ⟩primary alcohols, i.e.

$$
\begin{array}{ccc}
R' & & R' \\
\diagdown & & \diagdown \\
R'' - \text{C.OH} & > & \text{CHOH} > R.\text{CH}_2\text{OH} \\
\diagup & & \diagup \\
R''' & & R''
\end{array}
$$

This is the basis of the Lucas test to distinguish between the different alcohol groups. Tertiary alcohols, when shaken with concentrated hydrochloric acid and anhydrous zinc chloride, immediately produce the chloro-derivative, which separates out as an oil; secondary alcohols react only slowly and primary alcohols generally show no change.

(b) *Phosphorus halides*—These react rapidly with alcohols. The chlorides of phosphorus are readily available but the bromides and iodides are made *in situ* by direct combination of red phosphorus and the halogen. Typical of their reaction with alcohols are

$$3R\text{OH} + P\text{Cl}_3 \rightarrow 3R\text{Cl} + P(\text{OH})_3$$
$$R\text{OH} + P\text{Cl}_5 \rightarrow R\text{Cl} + P\text{OCl}_3 + H\text{Cl} \uparrow$$

(c) *Thionyl chloride*—Thionyl chloride, preferably in the presence of pyridine, reacts with alcohols:

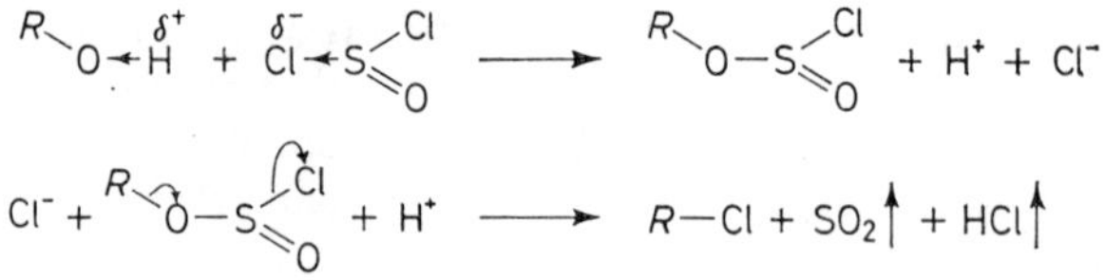

When pyridine is present, its basic character ensures the ready removal of the acidic by-products.

(*iv*) *Substitution of oxygen in a carbonyl group*—Halides containing two halogen atoms attached to the same carbon atom are produced by the reaction of phosphorus pentahalide on aldehydes or ketones, e.g.

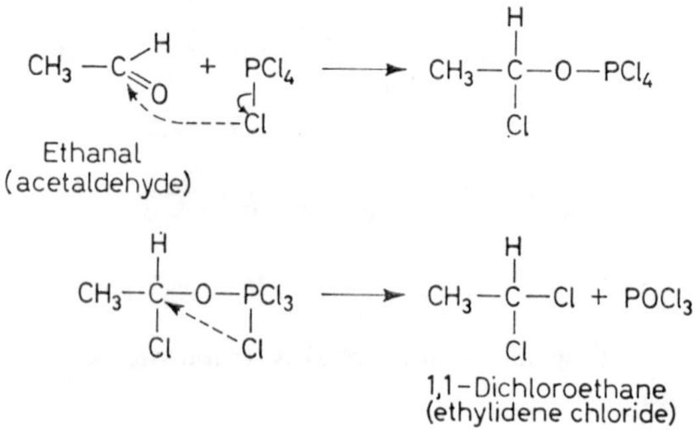

Ethanal
(acetaldehyde)

1,1–Dichloroethane
(ethylidene chloride)

398

Similarly

$$CH_3\overset{\displaystyle O}{\overset{\|}{C}}CH_3 + PCl_5 \rightarrow CH_3-\overset{\displaystyle Cl}{\underset{\displaystyle Cl}{\overset{|}{\underset{|}{C}}}}-CH_3 + POCl_3$$

propanone
(acetone)

2,2-dichloropropane

(v) *The haloform reaction*—The compounds CHX_3 can be prepared by the haloform reaction, which involves a compound of structure

$$CH_3-\overset{|}{\underset{|}{C}}-OH \quad or \quad CH_3-\overset{|}{C}=O$$

(e.g. ethanol, CH_3CH_2OH, or acetone, CH_3COCH_3) undergoing halogenation, followed by hydrolysis. It is usual to perform the reaction with bleaching powder or sodium hypohalite, either of which can be regarded both as a source of halogen and alkali, e.g.

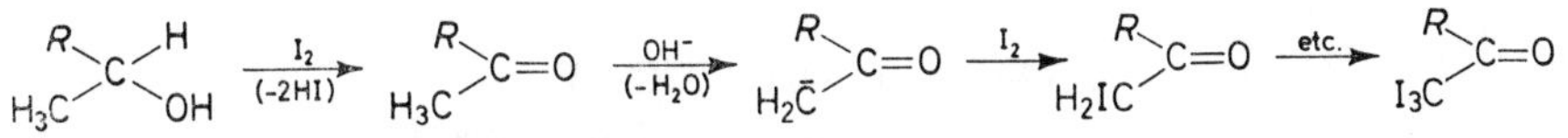

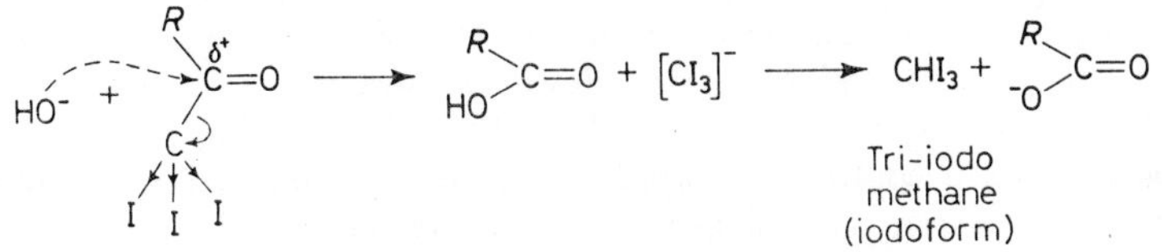

Tri-iodo
methane
(iodoform)

The methylene halides can be prepared by refluxing the appropriate haloform with iron filings and water

$$CHX_3 + 2[H] \rightarrow CH_2X_2 + HX$$

PROPERTIES

Like the hydrocarbons, the halogen derivatives are insoluble in water. Those that are liquid, especially the iodides, are of relatively high density and therefore give a lower, immiscible, layer with water. The increase in molecular weight (which results from replacing a hydrogen atom in a hydrocarbon by a halogen atom), and the increase in interaction of dipole-type forces between the resulting molecules combine to produce a fall in volatility (*Table 20.1*).

CUA—P

Table 20.1. Boiling points of the halogenated hydrocarbons

Compound	b.p. °C	Compound	b.p. °C
CH_4	−164	C_6H_6	80
CH_3Cl	−23	C_6H_5Cl	132
CH_3Br	3·5	C_6H_5Br	156
CH_3I	44	C_6H_5I	186
CH_2Cl_2	42		
$CHCl_3$	61		
CCl_4	76		

The characteristic reaction of halides is the substitution of the halogen atom by another atom or radical. Alkyl halides, and those aromatic halides in which the halogen atom is in a side chain, undergo fairly easy replacement of the halogen atom. The ease of replacement is roughly in the order I>Br>Cl>F, i.e. of increasing bond strengths:

$$\begin{array}{cccc} \text{C—I} & \text{C—Br} & \text{C—Cl} & \text{C—F} \\ 188 & 226 & 278 & 408 \text{ kJ mol}^{-1} \end{array}$$

The order is also dependent upon the proximity of multiple bonds, the number of halogen atoms attached to the carbon atom and the size of the groups surrounding the halogen atom, i.e. on *steric hindrance*. However, more than one mechanism of replacement is possible, so that experimental conditions are also important.

Aryl halides, in marked contrast to alkyl halides, are very unreactive, because the lone pairs of electrons on the halogen atom can interact with the π bonding of the ring system and so extend the delocalized π clouds, with a resulting increase in stability, e.g.

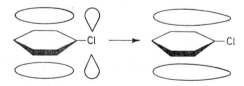

The same considerations apply to unsaturated aliphatic compounds in which the carbon atom to which the halogen is attached is also unsaturated. This change in reactivity between the different halogenated hydrocarbons can be illustrated by considering the necessary conditions required to hydrolyse different compounds:

400

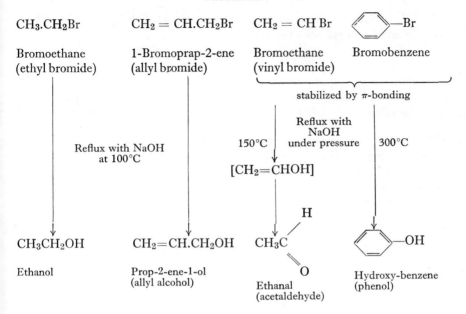

With the above reservations, the reactions given by the halogenated hydrocarbons are:

(*i*) *Reduction to the parent hydrocarbon* by hydrogen iodide (which is an excellent reducing agent). Normally, moist red phosphorus is also added, as by this means hydrogen iodide is regenerated from the iodine produced in the reduction, e.g.

$$RI + HI \longrightarrow RH + I_2$$

$$P + I_2 \rightarrow PI_3 \xrightarrow{\ H_2O\ } P(OH)_3 + HI$$

(*ii*) *The formation of amines with ammonia*—When a halogen compound is treated with ammonia under pressure, a mixture of salts of amines results from continued replacement of halogen

$$H_3N: + R\!-\!X \longrightarrow H_3N^{\pm}\!-\!R + X^- \xrightarrow{\ NH_3\ } NH_4^+ + \underset{\text{Primary amine}}{H_2NR}$$

$$H_2\dot{N}R + R\!-\!X \longrightarrow H_2N^+R_2 + X^- \xrightarrow{\ NH_3\ } NH_4^+ + \underset{\text{Secondary amine}}{HNR_2}$$

$$H\dot{N}R_2 + R\!-\!X \longrightarrow HN^+R_3 + X^- \xrightarrow{\ NH_3\ } NH_4^+ + \underset{\text{Tertiary amine}}{NR_3}$$

$$\dot{N}R_3 + R\!-\!X \longrightarrow \underset{\substack{\text{Quaternary}\\\text{compound}}}{NR_4^+X^-}$$

(*iii*) *Elimination of* HX to yield an alkene, promoted by boiling a halide with an alcoholic solution of potassium hydroxide

$$RCH_2CH_2X \xrightarrow[-HX]{alc.KOH} RCH{=}CH_2$$

This reaction occurs by a bimolecular mechanism involving attack on the halide by the nucleophilic ethoxide ions, $C_2H_5O^-$, present in the alcoholic hydroxide solution.

Two important exceptions to this reaction are provided by the methyl and ethyl halides, which yield the corresponding alcohols (see (*v*) below).

(*iv*) *Formation of ethers*—This often accompanies the previous reaction between a halide and alcoholic potassium hydroxide because of the possibility of the alternative mode of attack by the ethoxide ion.

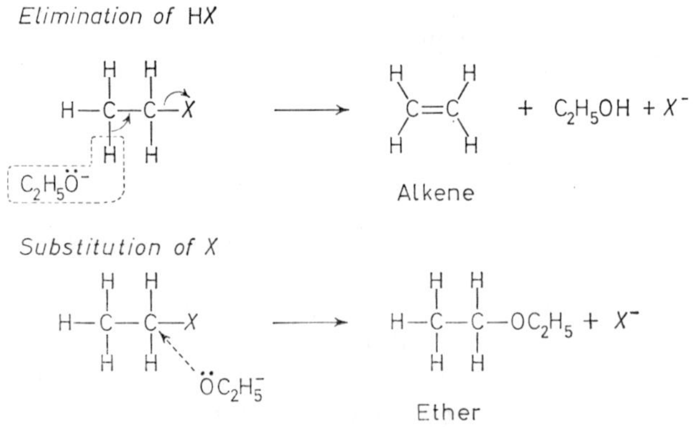

Elimination of HX

Substitution of X

(*v*) *Hydrolysis to an alcohol*—The main reaction when a halide is refluxed with an aqueous solution of an alkali is

$$RX + OH^- \rightarrow ROH + X^-$$

The two possible mechanisms for this reaction have already been described (p. 363). Small amounts of alkene are usually also formed by side reactions.

Hydrolysis of polyhalides initially follows a similar course, but where two or more halogen atoms are attached to the same carbon atom, elimination of water follows upon the first reaction, e.g.

$$CH_3CHCl_2 \rightarrow [CH_3 . CH(OH)_2] \rightarrow CH_3C\overset{\displaystyle H}{\underset{\displaystyle O}{\Big\langle}} + H_2O$$

ethanal (acetaldehyde)

402

$$\text{CHCl}_3 \rightarrow [\text{CH(OH)}_3] \rightarrow \text{H}\cdot\text{C}\overset{\displaystyle O}{\underset{\displaystyle\text{OH}}{\diagup\!\!\diagdown}} + \text{H}_2\text{O}$$

methanoic (formic) acid

(*vi*) *Replacement by the cyano, nitro and nitrite groups*—If refluxed with an aqueous/ alcoholic solution of potassium cyanide, the cyanide ion replaces the halogen, e.g·

$$R\text{—}X + \overset{\delta-}{\text{CN}^-} \rightarrow [\overset{\delta+}{X}\ldots\ldots R\ldots\ldots \overset{\delta-}{\text{CN}}] \rightarrow R\text{—}C\equiv N + X^-$$

A cyanoalkane or nitrile

However, if silver cyanide, which is mainly covalent, is used, the isonitrile is formed

$$X\text{—}R + N\equiv C\text{—}Ag \rightarrow [\overset{\delta-}{X}\ldots R\ldots N\equiv C\overset{\delta+}{\ldots}Ag] \rightarrow X^- + R\text{—}N\equiv C + Ag^+$$

Isonitrile

Similar reactions occur with solutions of potassium and silver nitrites, nitrites and nitro compounds, respectively, being formed. There is, however, some evidence for the view that both of these reagents yield a 2:1 mixture of nitro-compound and nitrite.

$$R\text{—}O\text{—}N\!\!=\!\!O \quad \text{A nitrite}$$

$$RX \overset{\text{KNO}_2}{\underset{\text{AgNO}_2}{\diagup\!\!\diagdown}}$$

$$R\text{—}N\overset{\displaystyle O}{\underset{\displaystyle O}{\diagup\!\!\diagdown}} \quad \text{A nitro-compound}$$

(*vii*) *With metals*—It has been mentioned earlier (p. 372) that sodium reacts with alkyl halides to produce hydrocarbons, e.g.

$$\bigcirc\!\!-\!\!\overbrace{\text{Br} + 2\text{Na} + \text{Br}}\,\text{C}_2\text{H}_5 \longrightarrow \bigcirc\!\!-\!\!\text{C}_2\text{H}_5 + 2\,\text{Na}^+\text{Br}^-$$

Chloroethane reacts with a lead–sodium alloy to form the important compound lead tetraethyl, used as an 'anti-knock' additive in petrol

$$4\text{C}_2\text{H}_5\text{Cl} + 4\text{Na/Pb} \rightarrow (\text{C}_2\text{H}_5)_4\text{Pb} + 3\text{Pb} + 4\text{Na}^+\text{Cl}^-$$

A particularly important reaction of halogen derivatives, including those with the halogen attached directly to an aromatic nucleus, is that with magnesium under anhydrous conditions in ether as solvent. The reaction is normally represented as

$$RX + Mg \rightarrow \overset{\delta-}{R}\!-\!\overset{\delta+}{Mg}\!-\!\overset{\delta-}{X}$$

The product behaves as R—Mg—X although it appears that it might be an equimolecular mixture of R_2Mg and MgX_2. These so-called *Grignard reagents* are very reactive towards many oxygen-containing compounds:

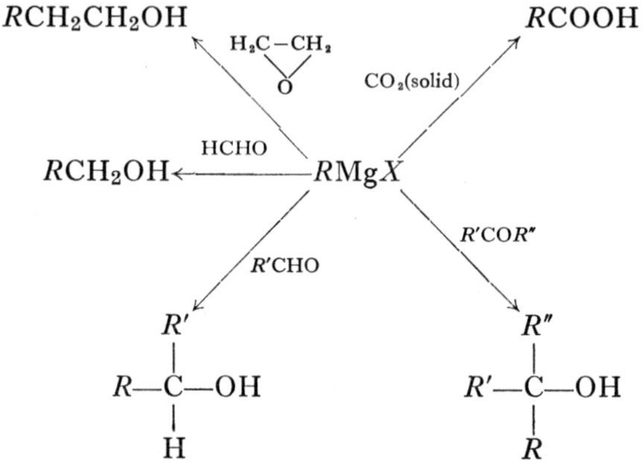

In the examples given, an intermediate complex magnesium compound has to be hydrolysed to produce the required product, e.g.

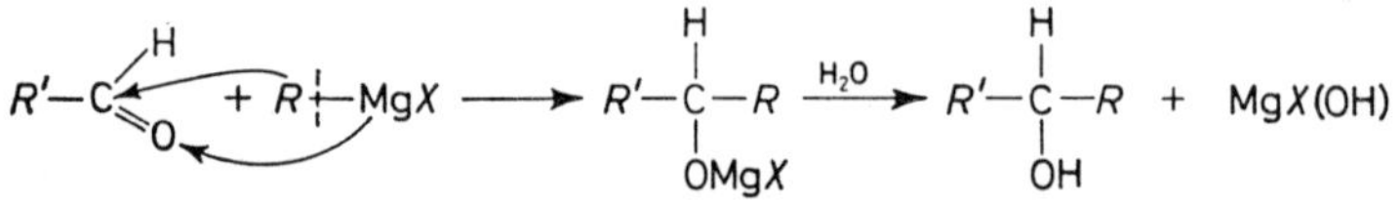

(*viii*) *Formation of esters*—Reaction of a halide with the silver salts of carboxylic acids leads to esters, e.g. silver ethanoate (acetate):

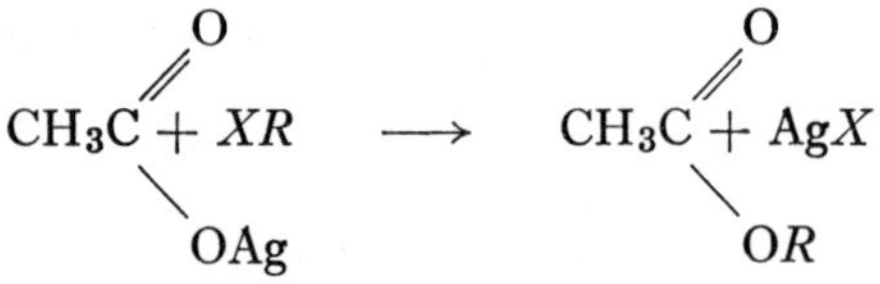

(*ix*) *Addition to aromatic systems*—Alkyl halides take part in the Friedel–Crafts reaction (p. 373), of which a typical example is

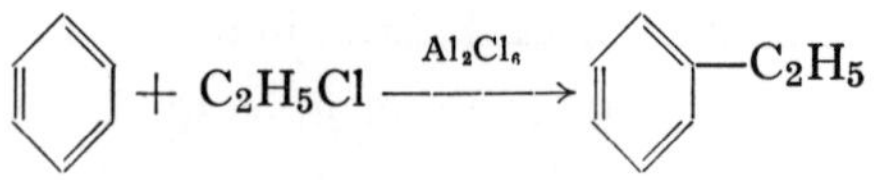

The Directive Influence of Halogens in Aromatic Compounds

A substituent in an aromatic compound directs further attack of the benzene ring. For example, of the three mononitro-derivatives which might be obtained from the nitration of bromobenzene, the *ortho* and *para* compounds predominate:

o-nitrobromobenzene   p-nitrobromobenzene

The reason for this lies in the fact that a lone pair of electrons on the halogen atom becomes incorporated into an extended π electron system of the benzene ring (p. 400). This can be shown using conventional structures such as

so that an electrophilic group, e.g. NO_2^+, will preferentially attack the *o*- and *p*-positions. For aryl halides, the ease of substitution is lower than for aromatic hydrocarbons, because of the high electronegativity of the halogen tending to pull electrons from the aromatic system.

INDUSTRIAL ASPECTS

1. *Polymers*

Chloroethene (vinyl chloride) and similar compounds are very important industrial materials. They are readily polymerized, either separately or together. For example, vinyl chloride when suspended in water and heated under pressure with potassium persulphate or hydrogen peroxide as initiator, can be made to yield polymers of molecular weight between 50 000 and 80 000:

$$n\text{CH}_2\!\!=\!\!\text{CHCl} \rightarrow \left(\!\!\begin{array}{c} -\text{CH}_2-\text{CH}- \\ | \\ \text{Cl} \end{array}\!\!\right)_n$$

The product, commonly referred to as *polyvinyl chloride* or PVC, is thermoplastic (i.e. can be moulded when hot), inert and rigid. As vinyl chloride can be made from acetylene by the addition of hydrogen chloride in the presence of mercury(II) ions as catalyst, the main raw materials are coal, salt and limestone.

If the mercury ion catalyst is replaced by copper(I) chloride, in the presence of ammonium chloride, acetylene dimerizes to *vinylacetylene* which can then react with hydrogen chloride to produce *chloroprene*, in accordance with Markownikoff's rule (p. 383)

$$2HC\equiv CH \rightarrow CH_2{=}CH{-}C\equiv CH$$

$$CH_2{=}CH{-}C\equiv CH + HCl \rightarrow CH_2{=}CH{-}\underset{\underset{Cl}{|}}{C}{=}CH_2$$

Chloroprene readily polymerizes to *neoprene*, which is an important *elastomer* (synthetic rubber):

$$n CH_2{=}CH{-}\underset{\underset{Cl}{|}}{C}{=}CH_2 \rightarrow ({-}CH_2{-}CH{=}\underset{\underset{Cl}{|}}{C}{-}CH_2{-})_n$$

This structure should be compared with that of natural rubber (p. 369).

Treatment of 1,2-dichloroethane with a solution of sodium polysulphides results in a condensation polymerization to give an elastomer, Thiokol, of comparatively high chemical resistance

$$n ClCH_2CH_2Cl \xrightarrow{Na_2S_x} Cl{-}(CH_2{-}CH_2{-}\overset{\overset{S}{\|}}{S}{-}\overset{\overset{S}{\|}}{S})_n{-}Cl$$

A polymer remarkable in view of its stability to a wide range of temperature, its considerable electrical and chemical resistance and also its property of 'self-lubrication', is *polytetrafluoroethylene* (PTFE). It is made by the pyrolysis of chlorodifluoromethane at about 700°, followed by polymerization

$$2CHClF_2 \rightarrow F_2C{=}CF_2 + 2HCl$$

$$n F_2C{=}CF_2 \rightarrow ({-}F_2C{-}CF_2{-})_n$$

2. *Insecticides*

Gammexane, of prime importance in the control of locusts, has been mentioned previously (p. 383).

D.D.T. is made by the reaction between chlorobenzene and chloral hydrate (p. 463) in the presence of concentrated sulphuric acid

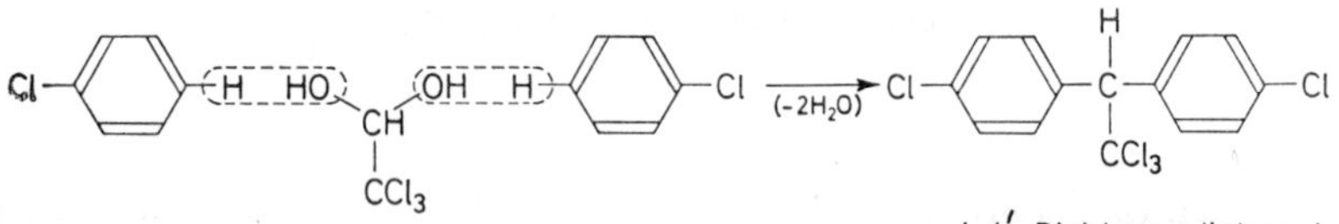

4,4'-Dichloro-diphenyl-trichloroethane

3. *Solvents*

Chlorinated hydrocarbons are very popular solvents, particularly in the realms of 'dry cleaning'. The common examples are

(a) *carbon tetrachloride*, made by the action of chlorine on carbon disulphide at high temperature

$$CS_2 + 3Cl_2 \rightarrow CCl_4 + S_2Cl_2$$

(b) '*Westrosol*', obtained by passing 1,1,2,2-tetrachloroethane over heated barium chloride

$$HC{\equiv}CH \xrightarrow{\ Cl_2\ } ClCH{=}CHCl \xrightarrow{\ Cl_2\ } Cl_2CH{-}CHCl_2 \xrightarrow[\text{(--HCl)}]{\ BaCl_2\ } ClCH{=}CCl_2 \quad \text{'Westrosol'}$$

4. *Refrigerants*

Freons, fluorochloro-derivatives of methane and ethane, are valuable as refrigerants because of their easy liquefaction and low reactivity. They are prepared by reactions such as that between hydrogen fluoride and carbon tetrachloride, in the presence of a catalyst and under pressure

$$2CCl_4 + 3HF \rightarrow CFCl_3 + CF_2Cl_2 + 3HCl$$

ANALYSIS

Halides, particularly if the halogen content is high, are not very inflammable—carbon tetrachloride is even used as a fire extinguisher—and this property is sometimes helpful in their analysis.

Beilstein's test can be used as an indication of the presence of a halogen. In this test, copper wire is heated to give a superficial layer of copper(II) oxide and then reheated with the compound. The appearance of a green flame indicates a halogen forming the comparatively volatile copper(II) halide.

The chlorine, bromine or iodine in alkyl halides is recognized by treatment with alcoholic silver nitrate solution, when a precipitate of the appropriate silver halide is obtained. The reaction with aryl halides is too slow to be worthwhile, but the halogen can be identified by carefully fusing the substance with a little sodium and treating the aqueous extract containing sodium halides with excess dilute nitric acid and then silver nitrate solution; a characteristic precipitate indicates that halogen is present.

The same basic chemistry is used in the quantitative estimation. A known weight of the substance is heated in a sealed tube with excess of concentrated nitric acid and silver nitrate until reaction is complete. The silver halide formed is washed, dried and weighed. From the results, the percentage by weight of halogen in the substance can be found and, provided that the percentage weights of the other elements are known, the simplest formula can be calculated. A knowledge of the molecular weight, e.g. from vapour density measurements, then permits the molecular formula to be evaluated.

Example—1·42 g of an alkyl halide when treated as described above, yielded 2·35 g of silver iodide. The vapour density was 71. Calculate the molecular formula.

$$C_nH_{2n+1}I \quad \rightarrow \quad AgI$$

$$(14n + 1) + 127 \qquad\qquad\qquad 235$$

Weight of iodine in 2·35 g of silver iodide $= 127/235 \times 2\cdot35$ g $= 1\cdot27$ g i.e. 1·42 g of iodide contained 1·27 g of iodine.

Assume 1 mole of I is contained in 1 mole of the alkyl iodide, then 127 g of iodine are contained in 142 g of the alkyl iodide; therefore

$$\text{molecular weight} = 142$$

This is confirmed from the vapour density, $(71 \times 2 = 142)$

$$\text{i.e. } (14n + 1) + 127 = 142$$

$$\text{or} \qquad\qquad n = (142 - 128)/14 = 1$$

hence

$$\text{molecular formula: } C_1H_3I_1 = CH_3I$$

SUMMARY

Alkyl monohalides undergo substitution of the halogen fairly readily. Polyhalides, where the halogen atoms are attached to the same carbon atom, are less reactive, as are the halides with halogen attached directly to an unsaturated carbon atom (including the benzene nucleus); this is at least partly due to interpenetration of the *p* orbitals

Some common reactions of bromoethane are shown in *Figure 20.1*.

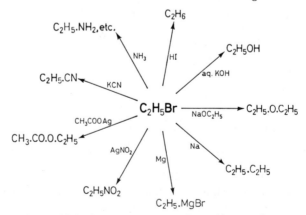

Figure 20.1. Some common reactions of bromoethane

QUESTIONS

1. Name the following:

$$CH_3-\underset{\underset{CH_3}{|}}{\overset{\overset{I}{|}}{C}}-CH=CH_2$$

$$\text{(benzene ring)}-Cl \quad -CH_2CHCl_2$$

$$CH_3C=CH\underset{\underset{I}{|}}{\overset{\overset{C_2H_5}{|}}{C}}H.CH_2\underset{Br}{|}$$

2. A compound has the following composition by weight: C = 66·4 per cent, H = 5·5 per cent, Cl = 28·1 per cent. Its vapour density is 63·25. Write down the structural formulae of substances which accord with this information. Suggest ways of identifying them.

3. Suggest methods for the following conversions:
(a) propanone to methanoic acid; (b) methanol to ethane; (c) propene to propyne; (d) benzene to diphenyl; (e) iodoethane to ethoxyethane; (f) 1-bromopropane to butanoic acid.

4. An alcohol, A, with sodium carbonate and iodine solution gives a yellow solid, B, and the sodium salt of an acid, C, which on heating with soda lime gives methane. Identify the substances and explain the reactions taking place.

5. 0·86 g of an unknown compound on combustion yielded 0·44 g of carbon dioxide and 0·225 g of water. When treated with silver nitrate it yielded 1·175 g of silver iodide. Calculate the empirical formula and suggest a method for determining the molecular formula.

6. Explain why the halogen derivatives of hydrocarbons are insoluble in water.

7. It has been stated that the ease with which alcohols react with hydrogen chloride is in the order tertiary⟩secondary⟩primary. Explain why this is so.

8. Comment on the following relative dipole moments: carbon tetrachloride, 0·00; chloromethane, 1·83; chlorobenzene, 1·55; 1,4-dichlorobenzene, 0·00.

9. What products are likely to be formed when alcoholic potassium hydroxide solution is refluxed with (a) 2-iodopropane and (b) 2-chloro-2-methylbutane?

10. Comment on the following data:

Compound	Percentage of alkene formed by dehydrobromination
CH_3CH_2Br	2
$CH_3CHBrCH_3$	80
$(CH_3)_3CBr$	100

11. Suggest why the removal of HBr takes place more readily from bromofumaric than from bromomaleic acid.

12. Discuss the possible nature of Grignard reagents.

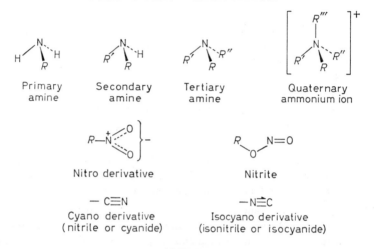

CHAPTER 21

NITROGEN DERIVATIVES

Primary amine	Secondary amine	Tertiary amine	Quaternary ammonium ion

Nitro derivative	Nitrite

$-C{\equiv}N$	$-N{\equiv}C$
Cyano derivative (nitrile or cyanide)	Isocyano derivative (isonitrile or isocyanide)

AMINES

Preparation

Progressive replacement of the hydrogen in ammonia by alkyl or aryl groups gives rise to *amines, primary* for the replacement of *one hydrogen, secondary* for *two,* and *tertiary* when all *three* have been removed:

$$NH_3 \rightarrow RNH_2 \rightarrow R_2NH \rightarrow R_3N$$

As the nitrogen atom possesses a lone pair of electrons, it can be protonated by acids to give the following types of ions:

$$R{-}\overset{+}{N}H_3 \qquad \overset{R'}{\underset{R}{\diagdown}}\overset{+}{N}H_2 \qquad \overset{R'}{\underset{R}{\diagdown}}R''{-}\overset{+}{N}H$$

The hydrogen atom in the last of these can itself be replaced by an alkyl or aryl group, with the formation of a *quaternary ammonium ion*:

$$\left[\begin{array}{c} R' \\ | \\ R{-}N{-}R'' \\ | \\ R''' \end{array}\right]^{+}$$

410

All the above types are obtained if ammonia and an alkyl halide are heated in a sealed tube, e.g.

$$RBr \xrightarrow{NH_3} \begin{cases} RNH_2 + HBr \rightarrow RNH_3^+Br^- \\ R_2NH + HBr \rightarrow R_2NH_2^+Br^- \\ R_3N\ + HBr \rightarrow R_3NH^+Br^- \\ R_3N\ + RBr \rightarrow R_4N^+Br^- \end{cases}$$

Separation of these products can prove tiresome, and this method is seldom employed for the preparation of pure amines.

Aryl halides react only with difficulty, but ammonolysis of chlorobenzene does take place at a pressure of 300×10^5 Nm^{-2} and a temperature of $300°C$ in the presence of copper(I) oxide, which removes the chloride ion as copper(I) chloride:

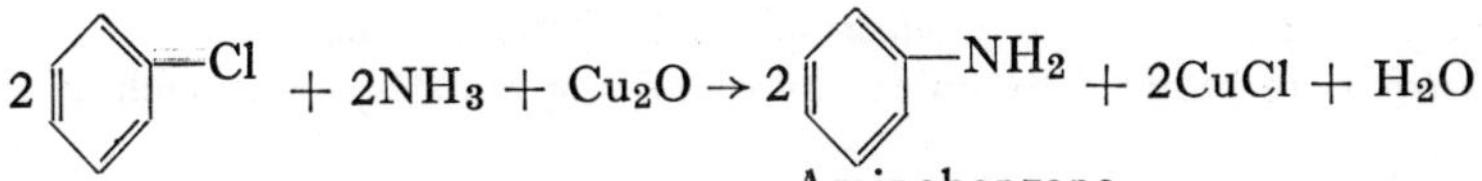

Aminobenzene
(aniline)

A modification of the ammonolysis reaction, which yields primary amines uncontaminated by further substituted derivatives, is that in which the ammonia is replaced by *phthalimide*; the sequence of reactions is indicated by the equations

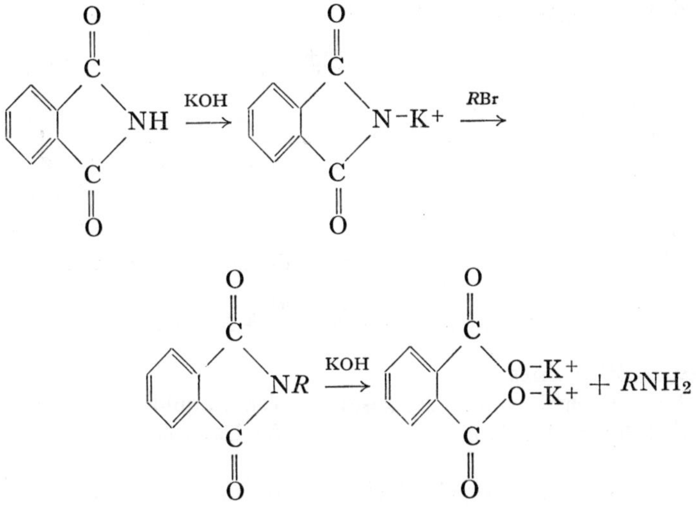

A method of preparation usually preferred to ammonolysis is the *Hofmann*

degradation of amides. Bromination of an amide followed by hydrolysis with hot, concentrated alkali results in decarboxylation via a rearrangement reaction:

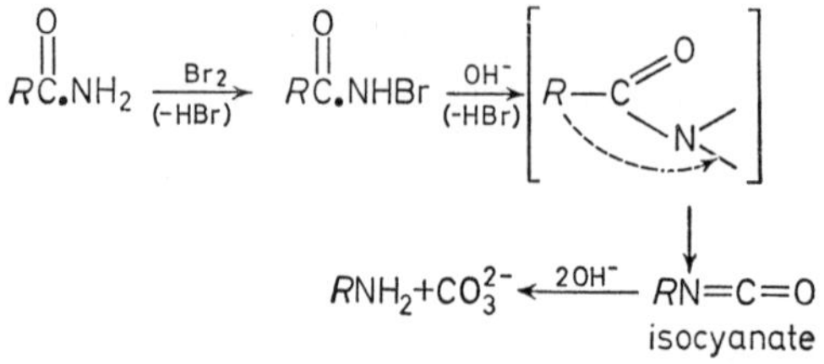

$$R\overset{\overset{O}{\|}}{C}.NH_2 \xrightarrow[(-HBr)]{Br_2} R\overset{\overset{O}{\|}}{C}.NHBr \xrightarrow[(-HBr)]{OH^-} \left[R-C\overset{O}{\underset{N}{\diagup}} \right]$$

$$RNH_2 + CO_3^{2-} \xleftarrow{2OH^-} RN=C=O$$
$$\text{isocyanate}$$

The term 'degradation' is applied to this reaction because the product possesses one carbon atom less than the reactant; it is therefore the key step in the descent of an homologous series (see p. 509).

Amines represent the reduction products of many nitrogenous compounds:

(*a*) *Nitriles* are reduced to amines by dissolving metal systems such as zinc in dilute acid:

$$RC\equiv N \xrightarrow{2H} RCH=NH \xrightarrow{2H} RCH_2.NH_2$$
$$\text{Imine}$$

(*b*) *Nitro compounds* are reduced fairly easily, but for reduction to proceed as far as the amine requires the dissolving metal to be in acid media. This method is particularly important in aromatic chemistry as the nitro-compound is readily obtained from the parent hydrocarbon:

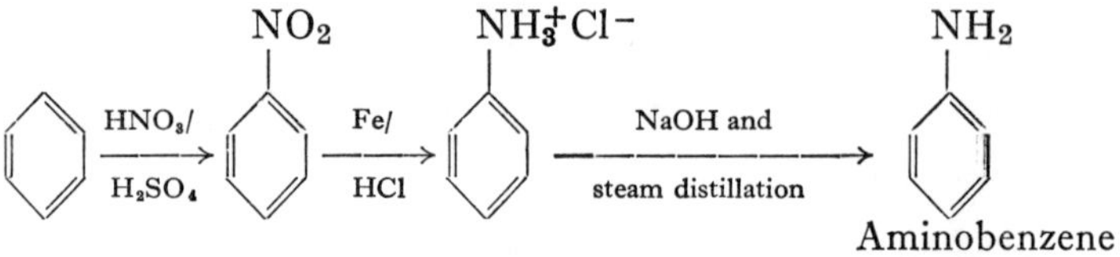

Aminobenzene
(aniline)

(*c*) Sodium and ethanol bring about the reduction of *oximes* to amines:

$$RCH=NOH \xrightarrow{Na/C_2H_5OH} RCH_2.NH_2$$

Properties of Amines

The lower aliphatic amines are gases or volatile liquids, very soluble in water, and with fishy smells. Aromatic amines are far less volatile and are insoluble in water.

The presence of a lone pair of electrons on the nitrogen atom ensures basic properties. Alkyl groups exert an electron-releasing inductive effect, and so, when attached to nitrogen, facilitate the acceptance of protons, with the result that aliphatic amines are more basic than ammonia and secondary amines are more basic than primary amines:

$$H-\overset{H}{\underset{H}{N}}: \qquad H-\overset{R}{\underset{H}{N}}: \qquad R'-\overset{R}{\underset{H}{N}}:$$

However, this explanation is only partly correct, as tertiary amines are less basic than both primary and secondary amines, a discrepancy which can probably be attributed to the steric hindrance afforded by three large alkyl groups surrounding the nitrogen atom.

The basicity of aromatic amines is much less than that of the aliphatic analogues or of ammonia because the lone pair of electrons on the nitrogen atom is conjugated with the aromatic system and is thus not so readily available for proton acceptance

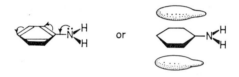

Table 21.1. Dissociation constants of amines

$$\overset{}{\underset{}{>}}N: + H_2O \rightarrow \overset{}{\underset{}{>}}NH^+ + OH^-$$

$$K_b = \frac{[\overset{}{\underset{}{>}}NH^+][OH^-]}{[\overset{}{\underset{}{>}}N:]}$$

Compound	K_b	pK_b	Compound	K_b	pK_b
Ammonia, NH_3	$1 \cdot 8 \times 10^{-5}$M	4·74	Aniline, $C_6H_5NH_2$	$4 \cdot 2 \times 10^{-10}$M	9·38
Methylamine, CH_3NH_2	$4 \cdot 4 \times 10^{-4}$M	3·36	N-methylaniline, $C_6H_5NHCH_3$	$7 \cdot 1 \times 10^{-10}$M	9·15
Dimethylamine, $(CH_3)_2NH$	$5 \cdot 2 \times 10^{-4}$M	3·28	N-dimethylaniline, $C_6H_5N(CH_3)_2$	$1 \cdot 2 \times 10^{-9}$M	8·92
Trimethylamine, $(CH_3)_3N$	$5 \cdot 5 \times 10^{-5}$M	4·26	2-Methylaniline	$2 \cdot 5 \times 10^{-10}$M	9·60
			4-Methylaniline	$11 \cdot 0 \times 10^{-10}$M	8·96

This basic character of amines means that they possess nucleophilic properties. Thus *primary and secondary amines are readily acetylated by treatment with an acid chloride;* benzoylation (the Schotten–Baumann reaction) and acetylation are particularly important, e.g.

$(CH_3)_2N{-}H$ + $Cl{-}CO.C_6H_5$ ⟶ $(CH_3)_2N.CO.C_6H_5$ + HCl

Dimethylamine   *N*, *N*-Dimethylbenzene carbonamide

$C_6H_5N{-}H$ + $Cl{-}COCH_3$ ⟶ $C_6H_5.NH.CO.CH_3$ + HCl

Aniline   *N*-Phenylethanamide (acetanilide)

The mechanism of the latter reaction is indicated by the sequence of reactions

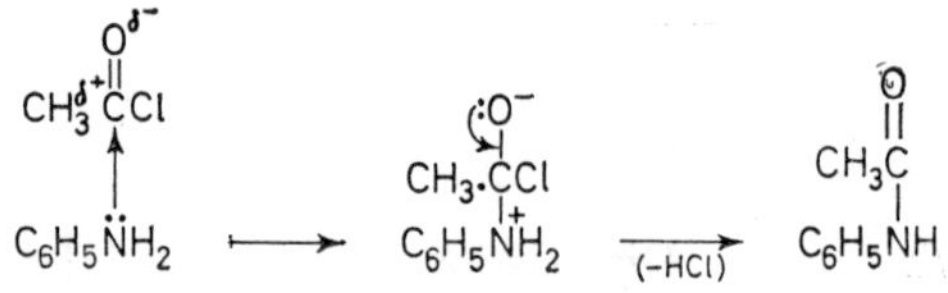

Diazotization

Primary and secondary amines also take part in nucleophilic attack on the nitrogen atom in compounds NOX, where X is an electron-withdrawing group. The classic example involves nitrous acid (produced *in situ* from cold, dilute hydrochloric acid and sodium nitrite solution). It has been shown that oxygen exchange takes place between $H_2^{18}O$ and nitrous acid, in keeping with the equilibrium:

$$2HNO_2 \rightleftharpoons H_2O + N_2O_3$$

The rate of reaction between amine and 'nitrous acid' has been found to be second order with respect to amine and N_2O_3

$$\text{i.e. rate} \propto [\text{amine}][N_2O_3]$$

Other substances, like nitrosyl chloride $NOCl$, can also bring about similar reactions and the following general mechanism has been proposed:

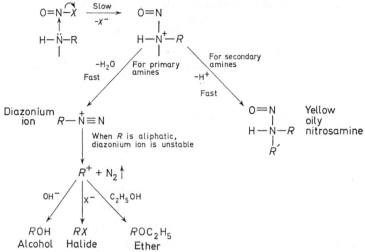

Provided the temperature is kept below 10°, the diazonium salts of aromatic amines can actually be isolated, although the resultant solids must be handled with care because of their explosive nature. However, as these compounds serve as very useful intermediates, there is no need to isolate them; instead, the solution obtained can be used directly

$$C_6H_5.NH_2 + H^+ + NO_2^- \rightarrow C_6H_5.\overset{+}{N}{\equiv}N + H_2O + OH^-$$

The reactions of the aromatic diazonium compounds can be conveniently classified into three types:

(a) *Nitrogen is evolved*

If the diazonium salt is refluxed with potassium iodide solution, the iodo deriva-tive is formed

$$Ar-\overset{+}{N}{\equiv}N + I^- \longrightarrow Ar-I + N_2\uparrow$$

Similar reactions do not take place with potassium bromide or chloride; to introduce either of these halogens into the aromatic nucleus, it is necessary to have copper or copper(I) halide present. Complex copper ions are believed to be the active agents, e.g.

$$Ar-\overset{+}{N}{\equiv}N \rightarrow ArCl + N_2\uparrow + CuCl_2^-$$

The diazo group can be replaced by a cyano group on treatment with copper(I) cyanide in the presence of potassium cyanide

$$ArN_2^+ + CuCN \xrightarrow{KCN} ArCN + Cu^+ + N_2 \uparrow$$

The above reactions are known collectively as *Sandmeyer reactions*.

Nitrogen is also evolved when a solution of a diazonium salt is boiled, the organic product being a phenol

$$ArN_2^+ + H_2O \longrightarrow ArOH + H^+ + N_2 \uparrow$$

The parent hydrocarbon is produced when a diazonium salt solution is boiled with hypophosphorous acid

$$ArN_2^+ + H_3PO_2 + H_2O \longrightarrow Ar.H + N_2 \uparrow + H_3PO_3 + H^+$$

(b) The multiple bond between the nitrogen atoms is reduced

Vigorous reducing agents, e.g. zinc in acid, produce the salt of the aromatic amine from which the diazonium salt was originally derived

$$ArN_2^+ + 3Zn + 7H^+ \longrightarrow ArNH_3^+ + NH_4^+ + 3Zn^{2+}$$

Milder reducing agents, such as sodium sulphite, reduce the multiple bond without cleaving it, to give the salt of an aryl-substituted hydrazine

$$ArN_2^+ + 4H^+ + 2Sn^{2+} \longrightarrow ArNH.NH_3^+ + 2Sn^{4+}$$

(c) The diazonium ion 'couples' with another organic molecule

The diazonium ion is an electrophilic reagent and therefore combines or 'couples' with compounds in which there are centres activated to such reagents, e.g. phenols and amines

The phenol can be further activated by dissolving in alkali to form the strongly nucleophilic ion

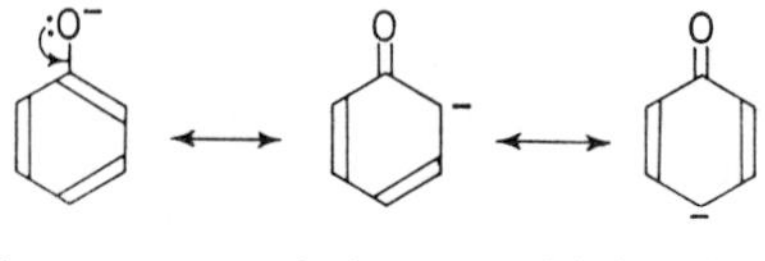

It also follows that the presence of electron-withdrawing groups in appropriate positions on the diazonium ion will further assist the progress of the reaction, as for example a sulphonic acid group in positions 2 or 4

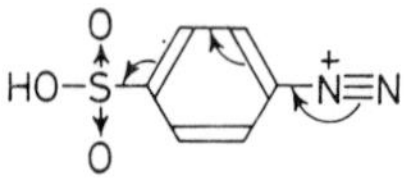

An important example of the use of this coupling reaction is the preparation of the indicator, *methyl orange*.

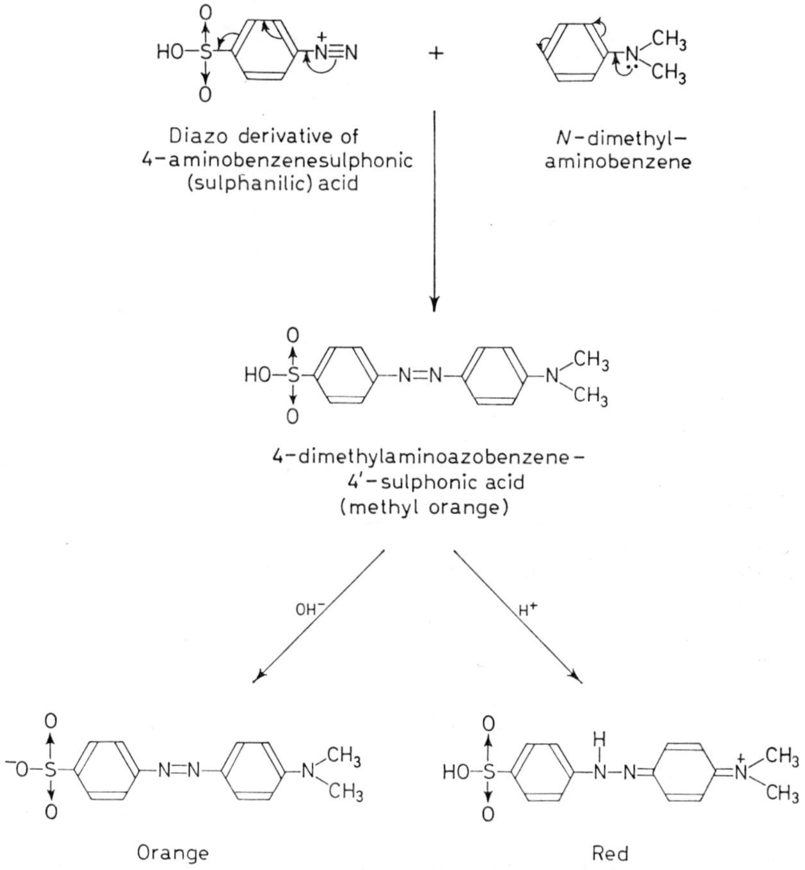
HO—S—N≡N
+
N—CH3 CH3
Diazo derivative of
4-aminobenzenesulphonic
(sulphanilic) acid
N-dimethyl-
aminobenzene
HO—S—N=N—N CH3 CH3
4-dimethylaminoazobenzene-
4′-sulphonic acid
(methyl orange)
OH−
H+
−O—S—N=N—N CH3 CH3
HO—S—N—N=N CH3 CH3
H
Orange
Red

The reactions of aromatic diazonium ions are summarized in *Table 21.2*.

Identification of Amines

The presence of nitrogen is best confirmed by fusion with sodium, followed by extraction with water. The nitrogen will then be present as sodium cyanide. This can be identified by making alkaline and adding iron(II) sulphate solution. The resulting mixture is boiled to convert some of the iron(II) ions to iron(III). Acidification then causes the iron(II) ions to react with the cyanide ions to give the complex hexacyanoferrate(II) ion which, in turn, reacts with the sodium and

Table 21.2. Reactions of the diazonium ion

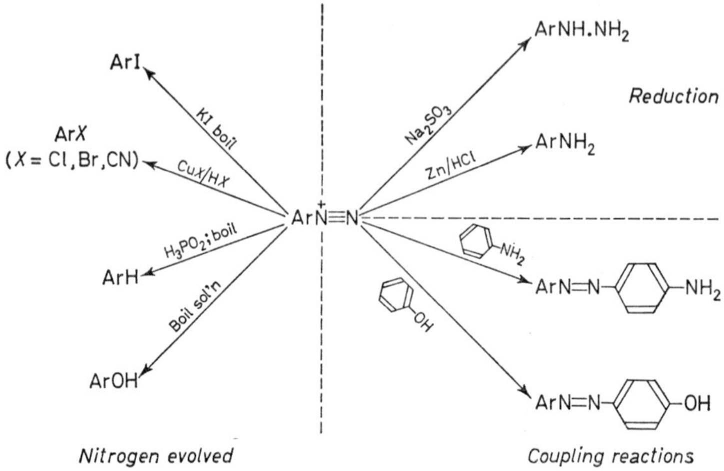

iron(III) ions to give the characteristic 'Prussian blue' colour of sodium iron(III) hexacyanoferrate(II)

$$Fe^{2+} + 6CN^- \longrightarrow [Fe(CN)_6]^{4-}$$

$$Na^+ + Fe^{3+} + [Fe(CN)_6]^{4-} \longrightarrow Na^+[Fe^{II}Fe^{III}(CN)_6]^- \downarrow$$

All primary amines react with chloroform in alcoholic potassium hydroxide to give evil-smelling isonitriles (the carbylamine test):

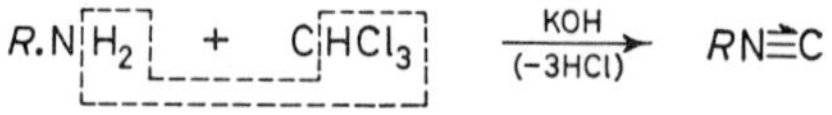

$$R.N H_2 + C HCl_3 \xrightarrow[(-3HCl)]{KOH} RN{\equiv}C$$

Nitrogen is evolved if a primary amine is treated with nitrous acid solution (p. 414). The formation of a yellow oil without the evolution of nitrogen is indicative of a secondary amine (p. 415), whilst tertiary amines give no visible reaction with nitrous acid.

The activating effect of the amino group when attached to the benzene nucleus means that aromatic amines are easily substituted in the 2-, 4- and 6-positions, and several of these reactions can be used for identification; for example, a white precipitate of 2,4,6-tribromoaniline is instantly formed when aniline is treated with bromine water

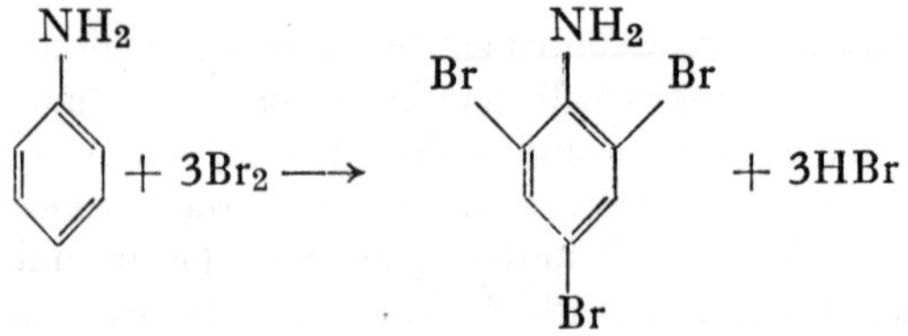

The more usual derivative for amines, however, is the benzamide, prepared from benzoyl chloride (the Schotten–Baumann reaction)

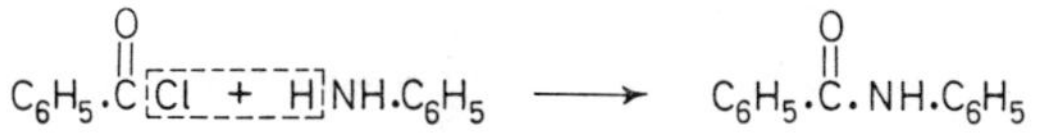

$$C_6H_5 . \overset{\overset{\textstyle O}{\|}}{C} \boxed{Cl \; + \; H} NH.C_6H_5 \quad \longrightarrow \quad C_6H_5 . \overset{\overset{\textstyle O}{\|}}{C} . NH.C_6H_5$$

N–Phenylbenzenecarbonamide

The ability of the lone pair on the nitrogen atom to undergo protonation is the basis for the purification of amines. If an impure ethereal solution of an amine is shaken with an acid, the amine forms a salt and dissolves in the aqueous phase. Separation of this phase, followed by treatment with alkali, leads to regeneration of the amine:

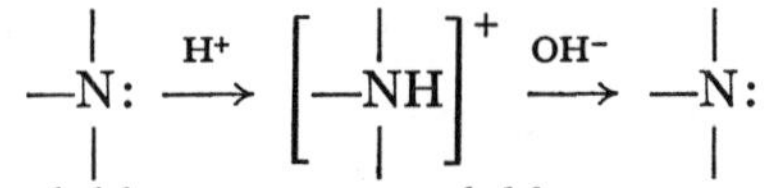

$$\overset{|}{\underset{|}{-N}}: \quad \overset{H^+}{\longrightarrow} \quad \left[\overset{|}{\underset{|}{-NH}} \right]^+ \quad \overset{OH^-}{\longrightarrow} \quad \overset{|}{\underset{|}{-N}}:$$

ether-soluble   water-soluble

NITRO COMPOUNDS

Nitro compounds can be prepared by direct nitration of a hydrocarbon using nitric acid, but the ease with which the reaction takes place depends upon whether there are any π electrons present in the hydrocarbon and upon the susceptibility to oxidation. Aromatic compounds, for example, possess the electron clouds attractive to the nitryl ion and so reaction takes place readily, particularly in the presence of substances such as concentrated sulphuric acid which favour the formation of this ion (p. 389)

$$HNO_3 + 2H_2SO_4 \rightarrow H_3O^+ + 2HSO_4^- + NO_2^+$$

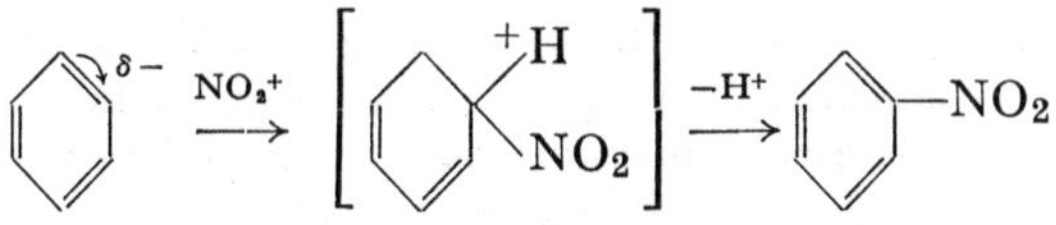

On the other hand, an alkane is nitrated only in the gaseous phase at a temperature of about 400°C (p. 391), e.g.

$$CH_4 + HNO_3 \rightarrow CH_3NO_2 + H_2O$$

All nitro compounds are fairly easily reduced, the extent of reduction depending on the conditions employed. Thus, nitrobenzene is reduced to phenylhydroxyl-amine in neutral solution with zinc and ammonium chloride, to azoxybenzene with

419

sodium methoxide, to azobenzene with zinc in caustic alkali and to aminobenzene with metals in acid solution

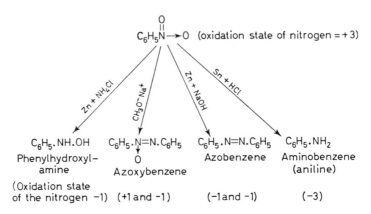

Primary and secondary nitro compounds, by virtue of the hydrogen atom on carbon-1, exhibit dynamic isomerism or *tautomerism*, in which the two forms differ only in the position of the hydrogen atom

$$-\underset{\underset{H}{|}}{C}-\underset{\underset{O}{\|}}{N}\rightarrow O \quad \rightleftharpoons \quad -\underset{|}{C}=\underset{\underset{OH}{|}}{N}\rightarrow O$$

As a result of this equilibrium, these compounds exhibit weakly acidic properties. Tertiary (including aromatic) nitro compounds are without the requisite hydrogen atom and are accordingly neutral.

NITRITES

Nitrites, $RO.N{=}O$, are isomeric with nitro compounds; the difference between them is that, whereas carbon is linked with nitrogen in the nitro compounds, in the nitrites carbon is attached to oxygen. Nitrites can, in fact, be regarded as esters of nitrous acid and can be prepared by reaction between nitrous acid (i.e. sodium nitrite and hydrochloric acid) and the alcohol

$$ROH + HO.N{=}O \rightarrow RO.N{=}O + H_2O$$

As would be expected from the structure, reduction of nitrites is unambiguous and always produces the corresponding alcohol

$$RO.N{=}O \xrightarrow{\text{[H]}} ROH$$

Consideration of the structures also suggests that nitro compounds are more polar than the nitrites; in accordance with this, there is more molecular interaction and nitro compounds are found to be less volatile than the nitrites.

Nitro compound	b.p. °C	Nitrite	b.p. °C
$CH_3.N \rightarrow O$ $\parallel$ O	101	$CH_3.O.N{=}O$	-12
$CH_3.CH_2.N \rightarrow O$ $\parallel$ O	113	$CH_3.CH_2.O.N{=}O$	16

NITRILES, $RC{\equiv}N$, AND ISONITRILES, $RN{\equiv}C$

Nitriles can be obtained in the following ways:
(*a*) dehydration of an amide with phosphorus(V) oxide

$$RC.NH_2 \overset{P_4O_{10}}{\longrightarrow} RC{\equiv}N + H_2O$$

(with O double bonded to the carbon of $RC.NH_2$)

(*b*) the action of cyanide on alkyl halides

$$RX + CN^- \longrightarrow RC{\equiv}N + X^-$$

Some isonitrile is also formed in this reaction; in fact, by using the covalent silver cyanide instead of the electrovalent potassium cyanide, isonitrile predominates

$$AgCN + RX \rightarrow AgX + RN{\equiv}C$$

Isonitriles are the evil-smelling substances referred to previously in the reaction between primary amines, chloroform and alcoholic potassium hydroxide

$$RNH_2 + CHCl_3 + 3OH^- \rightarrow RN{\equiv}C + 3Cl^- + 3H_2O$$

The bond between nitrogen and carbon in both compounds is multiple and the difference in electronegativity between the two elements results in some polarization, so that the molecule can undergo *nucleophilic attack at the carbon* atom and *electrophilic attack at nitrogen*. Thus, both nitriles and isonitriles can be hydrolysed by acid, the former giving rise to carboxylic acids and the latter to amines. In the case of nitriles, the initial attack is by a proton on the nitrogen

$$R-C{\equiv}N \quad \longrightarrow \quad R-\overset{+}{C}{=}NH$$
$$\uparrow$$
$$H^+$$

421

The overall reactions can be summarized as follows

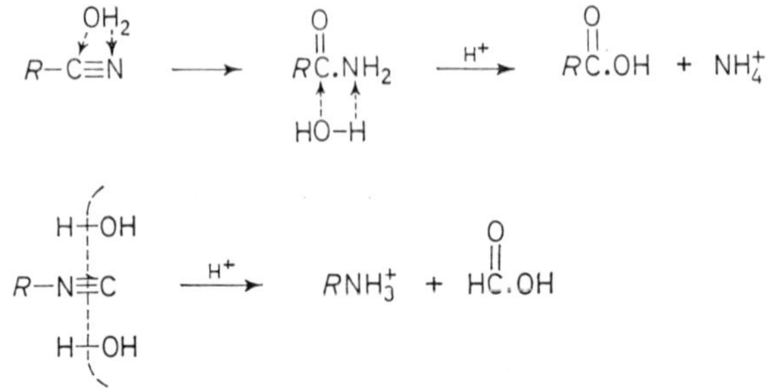

Reduction of nitriles gives primary amines, but isonitriles are converted into secondary amines. This is a natural consequence of the fact that, in the case of the latter, nitrogen is already attached to two carbon atoms

$$RC\equiv N + 4[H] \rightarrow RCH_2NH_2$$

$$RN{\equiv}C + 4[H] \rightarrow RNHCH_3$$

HETEROCYCLIC COMPOUNDS

Many naturally occurring compounds of great importance contain nitrogen as one or more members of a closed ring system, i.e. as part of a *heterocyclic* system. For example, they play a fundamental part in processes of biological inheritance and in the transfer of energy in the cell, as well as comprising the *alkaloids*, a group of physiologically active, basic substances derived from plants. Any book attempting a broad survey of the field of chemistry would be incomplete without some reference to such compounds.

Pyridine,

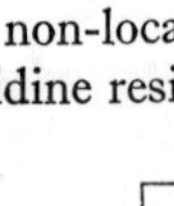

Pyridine is an unpleasant-smelling liquid present in coal tar. It is basic by virtue of the lone pair on the nitrogen but it is weaker than ammonia and comparable with tertiary amines. The p orbitals of nitrogen allow it to play a full part in the non-localized system of the hexagon, so that pyridine is truly aromatic.

The pyridine residue is present in the alkaloid nicotine and in vitamin B_6

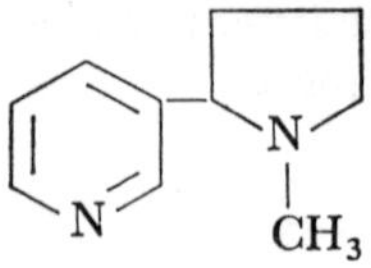

Nicotine 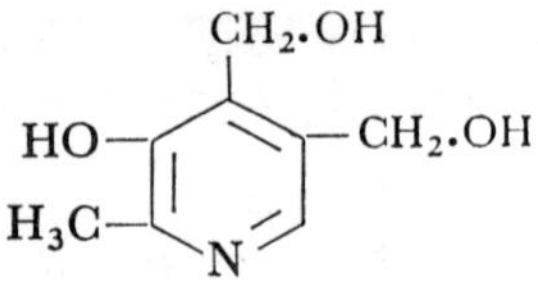   Vitamin B_6

Compounds containing two quaternary ammonium groups present as pyridine residues are found to interfere with the photosynthetic process in grasses and are finding application as non-emergent weed killers, e.g. as seed dressings such as

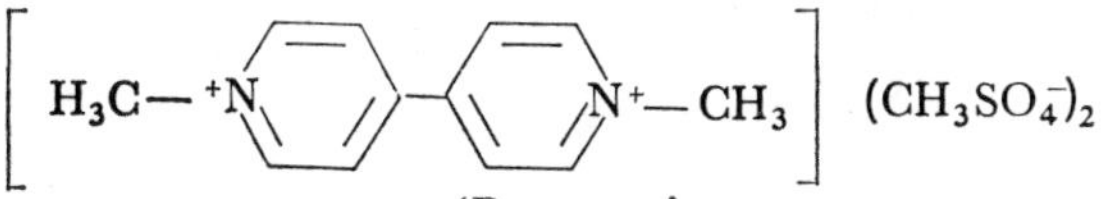

$$\left[H_3C-{}^+N \underset{}{\bigcirc}-\bigcirc{}N^+-CH_3 \right] (CH_3SO_4^-)_2$$

'Paraquat'

Pyrrole,

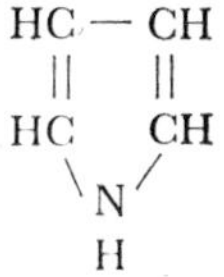

$$\begin{array}{cc} HC-CH \\ \| \quad \| \\ HC \quad CH \\ \diagdown N \diagup \\ H \end{array}$$

This compound has a five-membered ring with the lone pair on the nitrogen contributing to the resonance of the system; it is therefore not available for protonation, so that pyrrole is not basic: it is, in fact, slightly acidic.

Four pyrrole rings joined together by carbon 'bridges' comprise the interesting *porphyrin* framework; *haem* contains the iron (II) ion at the centre of the structure and *chlorophyll*, magnesium. Both haem and chlorophyll are *prosthetic* groups, i.e. they are attached to protein in the normal, active condition. In the case of haem, the complete substance, known as *haemoglobin*, plays a vital part in respiration, transporting oxygen in the form of *oxyhaemoglobin* from the lungs to the cells. Chlorophyll is of prime importance as a catalyst in photosynthesis (p. 475):

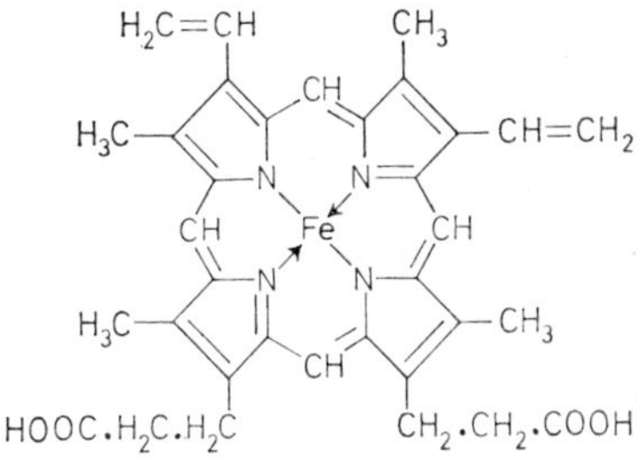

Haem

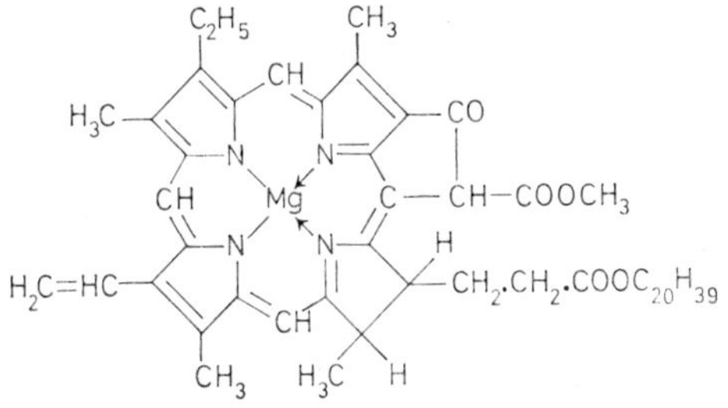

Chlorophyll-*a*

Both *quinoline* and *isoquinoline* are comparable to naphthalene, with one atom of nitrogen replacing one of carbon. Both residues are the basis of certain alkaloids

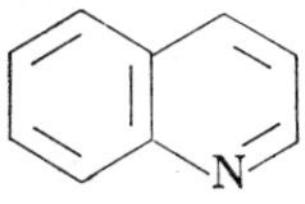

Quinoline 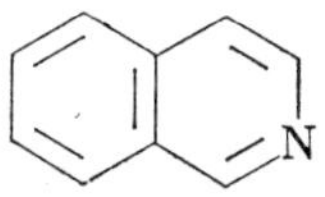   Isoquinoline 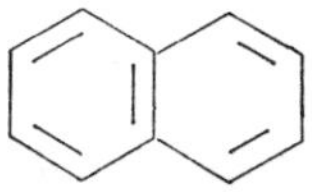   Naphthalene

423

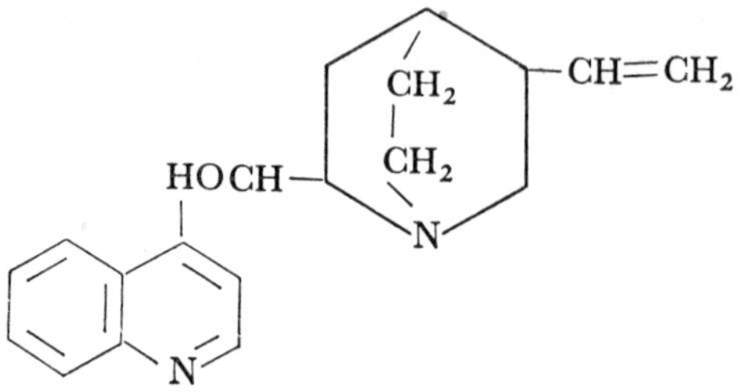

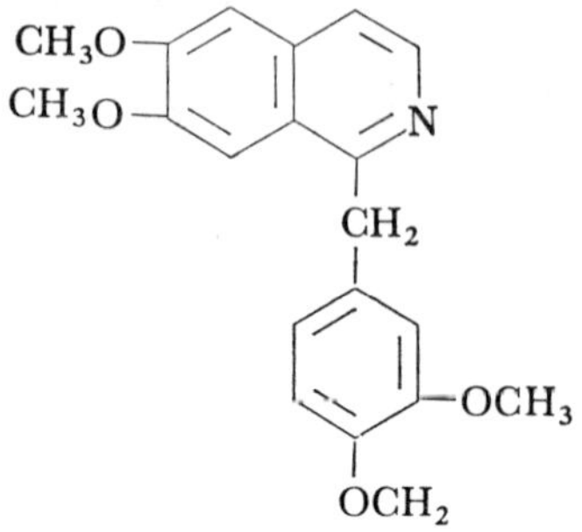

Quinine (from cinchona)   Papaverine (from opium poppy)

Pyrimidine and **purine** bases are of especial importance nowadays, with the present emphasis on research into the chemistry of genetics. Pyrimidines contain two nitrogen atoms replacing carbon in the benzene hexagon, whilst purines are pyrimidines attached to a 5-membered closed system containing two nitrogen atoms

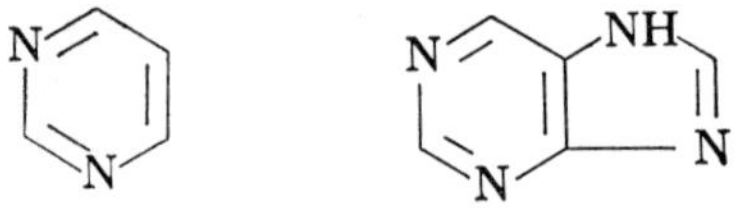

Pyrimidine   Purine

The genetic material DNA (deoxyribonucleic acid) contains the two pyrimidines *thymine* and *cytosine*, and the two purines *adenine* and *guanine*.

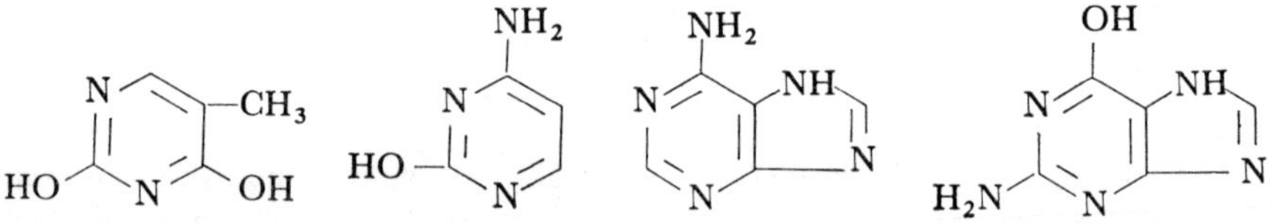

Thymine   Cytosine   Adenine   Guanine

These bases are linked to deoxyribose, which is also linked to a phosphoric acid residue, the whole unit being called a *nucleotide*.

424

An example of a nucleotide is adenylic acid

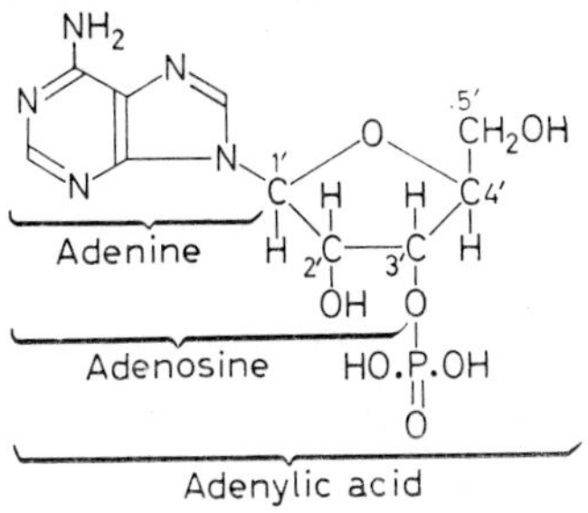

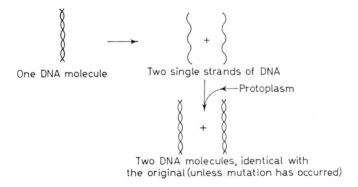

Condensation of this particular nucleotide with one molecule of phosphoric acid gives ADP and, with two molecules, ATP (p. 473).

Repeating of units like the adenylic acid shown above through position 5′ of the sugar and through a hydroxyl group of the phosphoric acid residue gives one-half of a DNA molecule, which in some respects can be compared to a spiral staircase. The other half is constructed in a similar way and the 'rungs' are established by hydrogen bonding between adenine and thymine and between guanine and cytosine (*Figure 21.1*). Only if one base is a purine and one a pyrimidine is there the 'right' geometric fit for the 'rungs' to be formed.

The affinity between the corresponding pairs of bases means that when a molecule splits lengthwise during cell division, two identical molecules, one for each new cell, can be constructed from materials available in the protoplasm; that is, a mechanism for replication is established:

One DNA molecule   Two single strands of DNA

←—Protoplasm

+

Two DNA molecules, identical with
the original (unless mutation has occurred)

The structure of the DNA molecule determines the protein produced, i.e. it provides the code for protein synthesis, but the intermediary, RNA (ribonucleic acid), is required to transport the information from the cell nucleus to the site of

425

the protein synthesis. This 'transfer' RNA differs from DNA in that the sugar unit is ribose instead of deoxyribose and uracil replaces thymine

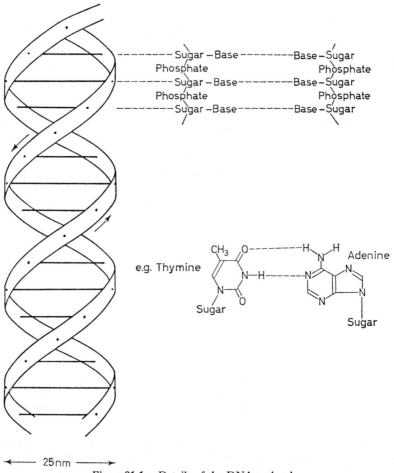

Ribose unit   Deoxyribose unit   Uracil

Uracil has an affinity for adenine and so it can serve a purpose similar to thymine.

25 nm

Figure 21.1. Details of the DNA molecule

426

This means that when a DNA molecule divides, a single strand, instead of making a new molecule of DNA, can synthesize an RNA molecule of complementary structure. Every 'transfer' RNA molecule is specific for a particular amino acid, and so the particular DNA, by controlling the nature of the 'transfer' RNA molecules made from it, also controls the nature of the subsequent protein that is synthesized. It can be said to carry the genetic information in terms of a four-symbol code, according to the way in which the four bases are repeated and arranged along the length of the molecule.

The production of 'transfer' RNA may be shown as

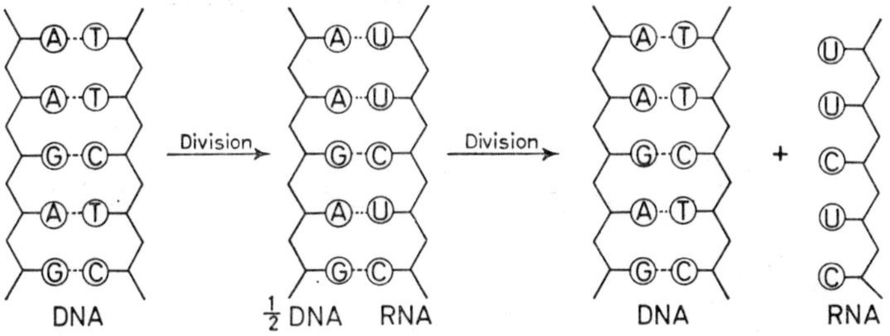

(For details of the role played by RNA in protein synthesis, see p. 512).

Pteridines resemble purines in that they too contain the pyrimidine residue attached to a ring containing two nitrogen atoms, but this latter ring is 6- instead of 5-membered

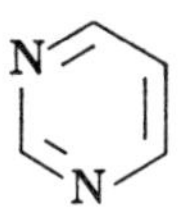

Pyrimidine	Purine	Pteridine

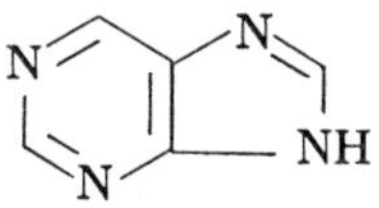

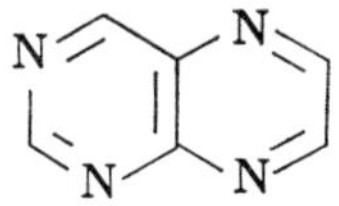

Pteridines (Greek *pteron* = wing) are present on the wings of insects, e.g. white *leucopterin* in cabbage white butterflies and yellow *xanthopterin* in brimstones

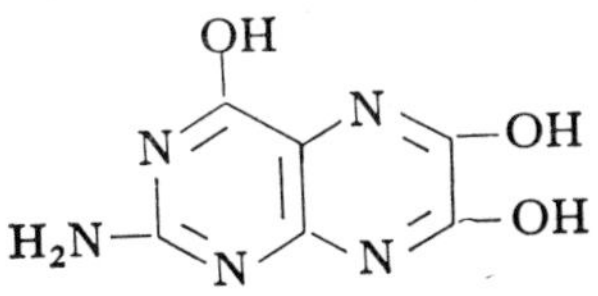

Leucopterin	Xanthopterin

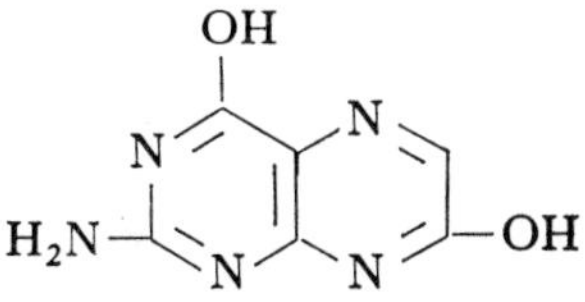

Apart from adornment, these substances appear to have growth-controlling

427

properties and indeed bear a close resemblance to two vitamins, *riboflavin* (Vitamin B_2) and folic acid

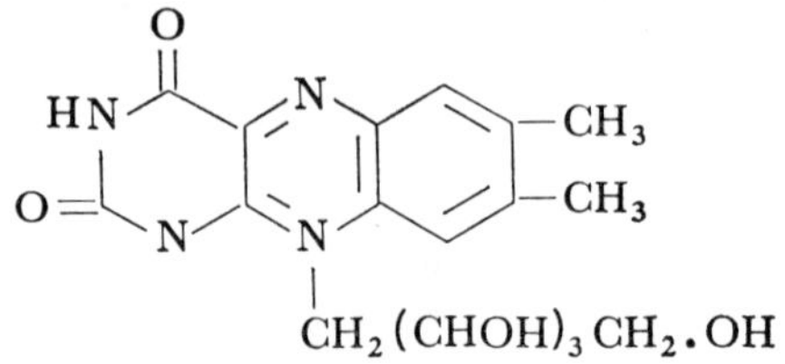

Riboflavin

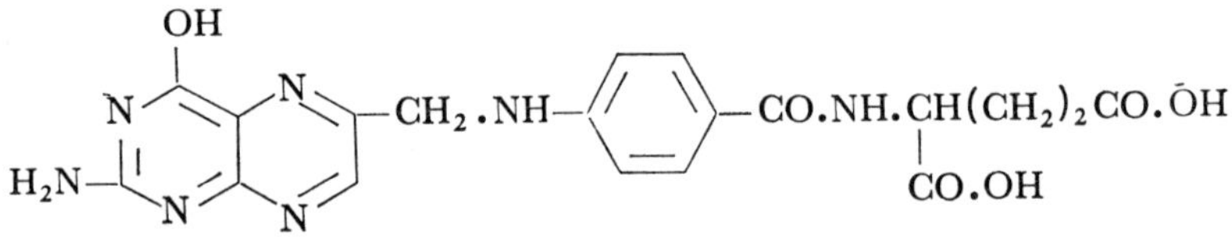

Folic acid

SUMMARY

Nitrogenous compounds in which a lone pair resides on the nitrogen atom are potentially basic. Amines and nitriles, for example, can be protonated in acid solution:

$$R_3N: + H^+ \rightarrow R_3NH^+ \qquad\qquad RC{\equiv}N + H^+ \rightarrow R\overset{+}{C}{=}NH$$

Amines can also act as nucleophilic reagents in reactions with acid chlorides and nitrous acid.

If the nitrogen atom is attached directly to an aromatic nucleus, then the lone

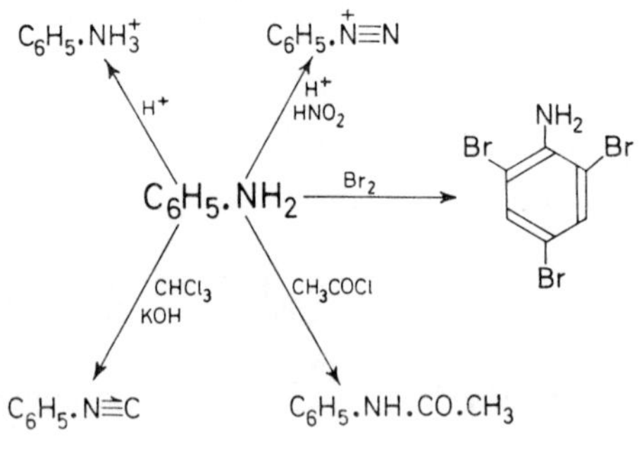

Figure 21.2. Some reactions of aminobenzene

428

pair forms part of the non-localized system. By activating the nucleus, it also encourages 2,4-substitution:

QUESTIONS

1. Starting from ethene, how could (*a*) *N*-methylaminoethane, (*b*) butane-1,4-dioic acid, (*c*) butane-1,4-diamine, (*d*) propanoic acid, (*e*) 2-hydroxy-1-(*N,N*-dimethyl)aminoethane be obtained?

2. Explain mechanistically the difference in the action of potassium cyanide and silver cyanide on alkyl halides.

3. Give a qualitative explanation of the differences in the dissociation constants of the amines listed in *Table 21.1*.

4. In the method referred to as Lassaigne's sodium test, if an organic substance containing nitrogen, a halogen, sulphur and phosphorus is fused with sodium, sodium cyanide, halide, sulphide and phosphide can be extracted by hot water. Devise an analysis scheme to identify the presence of the anions.

5. Account for the reactivity of the hydrogen atom attached to nitrogen in the phthalimide molecule.

6. A compound (*A*) of molecular formula $C_7H_7NO_2$, when treated with iron filings and dilute hydrochloric acid, yielded a compound (*B*), $C_7H_{10}ClN$. When dissolved in dilute hydrochloric acid and treated with sodium nitrite, this reacted to give (*C*), $C_7H_7ClN_2$. Half of the resultant solution was boiled and made alkaline; the remainder was then added to this and a bright-coloured precipitate, (*D*), was produced. Elucidate the above reactions and suggest possible structures for *A*, *B*, *C* and *D*.

7. What properties would you expect to be exhibited by $H_2N.CH_2.CH=CHCl$ and by

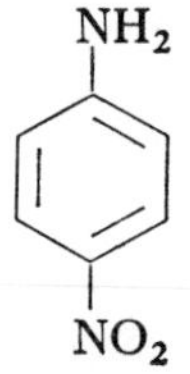

8. Explain the characteristic properties of nitrogen when present in:

(a) NH_3, (b) $RNHR'$, (c) $C_6H_5.NH_2$, (d) $RC{\equiv}N$.

9. Taylor (1928) found the rate of diazotization of an amine was given by the equation

$$\text{rate} \propto [\text{amine}][HNO_2]^2$$

Hammett subsequently (1933) modified this equation to

$$\text{rate} \propto [\text{amine}][N_2O_3]$$

How is it possible to reconcile these two rate equations?

What information can a rate equation give about the nature of a chemical reaction? What, for example, is it reasonable to conclude from the above data?

10. Write an essay on heterocyclic compounds.

11. Why should the acid concentration not be too high during diazotization, especially when the amine concerned is aliphatic?

12. Replication of a DNA molecule has been dealt with in this chapter from the point of view of the human being. How do the basic constituents of the DNA (or similar genetic material) vary throughout the animal world? How reasonable do you think is the view that these constituents emerged by chance in the geological past and since then have controlled the course of organic development?

CHAPTER 22

HYDROXY COMPOUNDS: ALCOHOLS AND PHENOLS

Organic compounds containing the hydroxy group (—OH) are normally classified as alcohols or phenols. The latter are those compounds in which the hydroxy group is directly linked to an aromatic ring. It has been seen in Chapter 20 how the aromatic ring can modify the properties of a halogen atom, and its effect on an amino group has been discussed in Chapter 21. In similar fashion it can considerably alter the properties of a hydroxy group.

The number of hydroxy groups in a compound is denoted by use of the terms monohydric, dihydric, etc. Monohydric alcohols can be further divided into three types, depending on the number of hydrogen atoms attached to the carbon atom carrying the hydroxy group

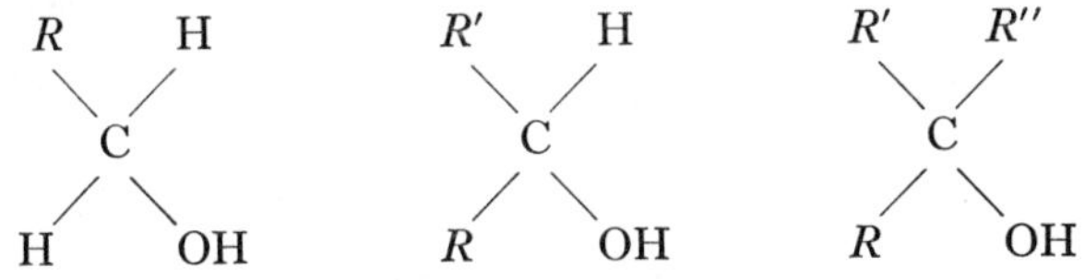

$$
\begin{array}{ccc}
R\diagdown C \diagup H & R'\diagdown C \diagup H & R'\diagdown C \diagup R'' \\
H \quad OH & R \quad OH & R \quad OH
\end{array}
$$

or

$$RCH_2.OH \qquad R'RCH.OH \qquad R'R''RC.OH$$
primary   secondary   tertiary

Alcohols are named by replacing the final -e of the corresponding hydrocarbon by -ol (monohydric), or adding -diol (dihydric), etc., together with the relevant number of carbon atoms where necessary. Phenols are named as hydroxy derivatives of benzene, etc.

Some typical hydroxy compounds are shown in *Table 22.1*.

Table 22.1. Representative hydroxy compounds

ALCOHOLS

Monohydric

$$CH_3.CH_2.OH \qquad \langle\!\!\bigcirc\!\!\rangle\!-CH_2.OH \qquad \langle\!\!\bigcirc\!\!\rangle\!\!\stackrel{H}{_{OH}} \qquad CH_3.\!\!\underset{CH_3}{\overset{CH_3}{\mid\,\mid}}\!\!C.OH$$

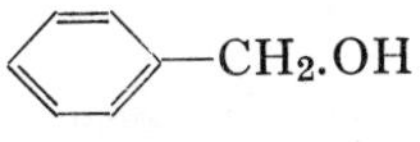

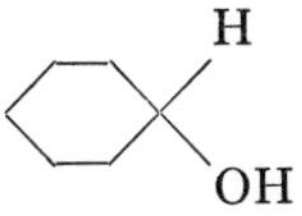

Ethanol (primary) Phenylmethanol (primary) Cyclohexanol 2-Methylpropan-2-ol
                   (benzyl alcohol)   (secondary)   (tertiary) •

continued

431

CUA—Q

Table 22.1—Continued

Dihydric	Trihydric	Polyhydric
CH$_2$.OH | CH$_2$.OH	CH$_2$.OH | CH.OH | CH$_2$.OH	CH$_2$.OH | (CH.OH)$_4$ | CH$_2$.OH
Ethanediol (primary) (ethylene glycol)	Propanetriol (primary and (glycerol) secondary)	Hexanehexol (mannitol)

PHENOLS

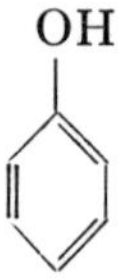

OH ⬡	⬡⬡—OH	OH ⬡ OH	OH ⬡—OH —OH
Hydroxybenzene (phenol)	2-Hydroxy- naphthalene (β-naphthol)	1,4-Dihydroxy- benzene (quinol)	1,2,3-Trihydroxy- benzene (pyrogallol)

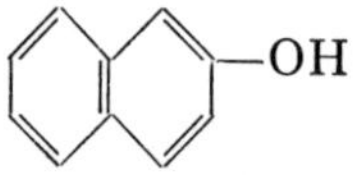

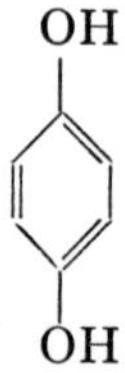

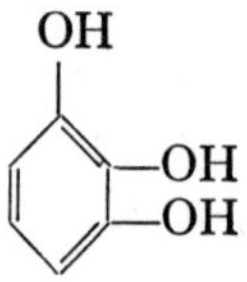

OCCURRENCE

Many alcohols occur as terpene derivatives in plants, e.g. *phytol*, C$_{20}$H$_{40}$O, containing four isoprene units, is produced in the hydrolysis of chlorophyll (p. 423):

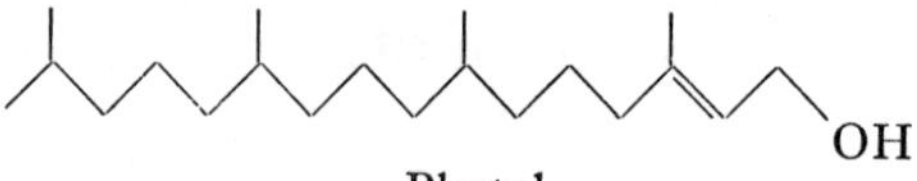

Phytol

and *menthol* is a reduced hydroxy derivative of limonene

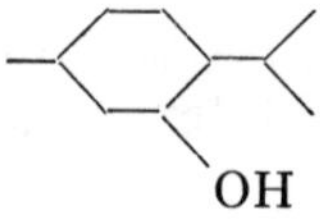

432

The formation of *cholesterol* (a precursor of Vitamin D) from the skeleton of squalene (p. 369) can be seen by reorientating the structure shown earlier

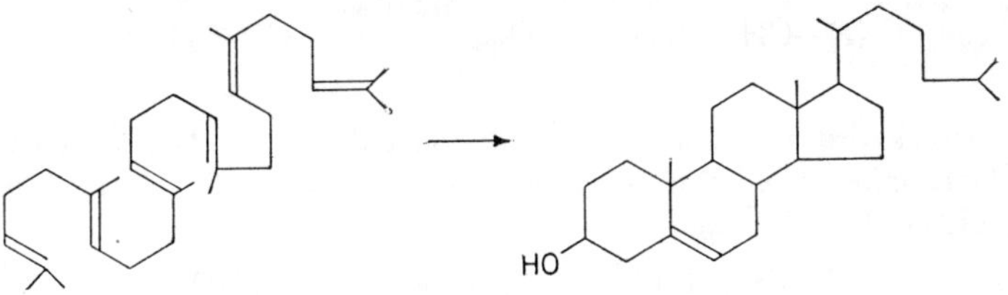

Squalene   Cholesterol

Vitamins D_1-D_4 are closely related to this compound, e.g. Vitamin D_2

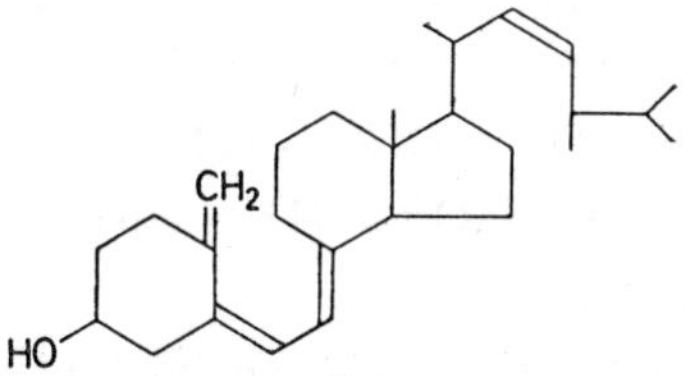

In some biochemical processes the benzene ring is readily hydroxylated, giving rise to many naturally occurring phenols such as the tannins, some phenolic ethers and Vitamin E

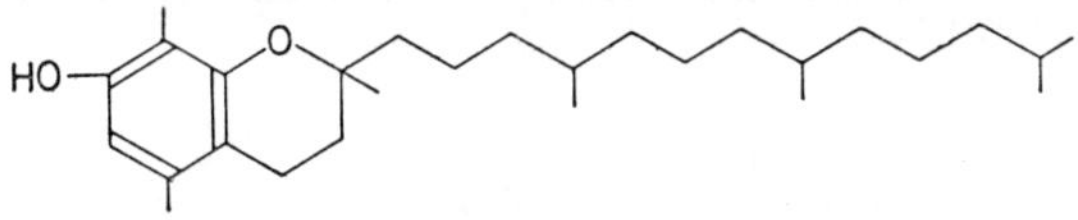

Vitamin E

PREPARATION

Preparation from the Hydrocarbon

(*a*) If an alkene is treated with concentrated sulphuric acid, an alkyl hydrogen sulphate results

$$RCH{=}CH_2 + HO{-}\underset{\underset{\displaystyle OH}{|}}{\overset{\overset{\displaystyle O}{\|}}{S}}{=}O \rightarrow RCH.CH_3 \atop \hspace{1cm} O.SO_2.OH$$

Addition of water converts the alkyl hydrogen sulphate into the alcohol, which can be removed by distillation

$$\begin{matrix} RCH.CH_3 \\ | \\ O.SO_2.OH \end{matrix} + H_2O \rightarrow \begin{matrix} RCH.CH_3 \\ | \\ OH \end{matrix} + H_2SO_4$$

Hydration of the alkene can also be achieved in the vapour phase under suitable conditions:

$$R\text{—}CH{=}CH_2 + H_2O_{(g)} \xrightarrow[\substack{pumice \\ 300°C}]{H_3PO_4 \text{ on}} RCH_2CH_2OH$$

(*b*) Aromatic hydrocarbons are slowly sulphonated by hot, concentrated sulphuric acid. As the reaction is believed to involve sulphur trioxide, the velocity is increased by using 'oleum' (p. 388), e.g.

$$C_6H_6 + H_2SO_4 \rightarrow C_6H_5.SO_2.OH + H_2O$$
$$\text{Benzenesulphonic acid}$$

Fusion of the sodium salt of benzene sulphonic acid with sodium hydroxide, followed by acidification, results in the formation of phenol. As the demand for phenol is considerable, this method is of economic importance:

$$C_6H_5.SO_2.O^- + OH^- \rightarrow C_6H_5.O^- + HSO_3^-$$
$$C_6H_5.O^- + H^+ \rightarrow C_6H_5.OH$$

Preparation from the Amine

Primary amines react with nitrous acid (i.e. cold dilute hydrochloric acid and cold dilute sodium nitrite solution) to give diazonium compounds (p. 414)

$$RNH_2 + HNO_2 \rightarrow RN_2^+ + H_2O + OH^-$$

Aliphatic diazonium compounds decompose immediately, so that the overall reaction is often simply represented by the equation

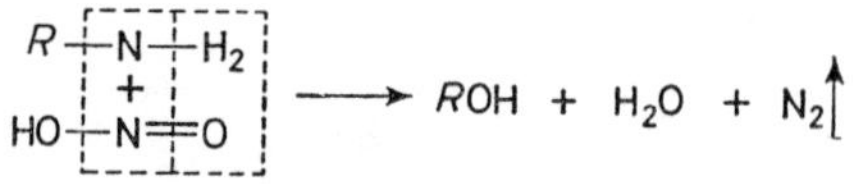

Aromatic diazonium compounds are, however, stable at low temperatures ($<10°$) and are converted into phenols on warming, e.g.

$$C_6H_5.\overset{+}{N}{\equiv}N \rightarrow C_6H_5^+ + N_2 \uparrow \qquad C_6H_5^+ + H_2O \rightarrow C_6H_5.OH + H^+$$

Preparation from the Halides

Halogen derivatives can be hydrolysed by refluxing with alkali. Alkyl halides are easily hydrolysed, e.g.

$$CH_3I + OH^- \rightarrow CH_3.OH + I^-$$
$$\text{Iodomethane} \qquad \text{Methanol}$$

whereas aryl halides must be subjected to a pressure of $200 \times 10^5 \text{ N m}^{-2}$ and a temperature of about $300°C$ (p. 401). Tertiary halides tend to give alkenes under these conditions, so that in these cases other methods are preferred.

Preparation from Carbonyl Compounds, etc.

Esters, ketones and aldehydes can be reduced to alcohols and phenols by a variety of methods, including treatment with a Grignard reagent or with lithium tetrahydridoaluminate, e.g.

$$R'C\overset{\displaystyle OR''}{\underset{\displaystyle R}{|}}=O + RMgX \rightarrow R'C\overset{\displaystyle OR''}{\underset{\displaystyle R}{|}}.O.MgX \xrightarrow{\;H_2O\;} R'C\overset{\displaystyle }{\underset{\displaystyle R}{|}}=O + MgX(OH) + R''OH$$

Ester   Grignard   Ketone
              reagent

$$R'C\overset{\displaystyle }{\underset{\displaystyle R}{|}}=O + R''MgX \rightarrow R'C\overset{\displaystyle R''}{\underset{\displaystyle R}{|}}.O.MgX \xrightarrow{\;H_2O\;} R'C\overset{\displaystyle R''}{\underset{\displaystyle R}{|}}.OH + MgX(OH)$$

    Ketone   Alcohol

$$RCH\underset{\displaystyle O}{\|} + H^- \xrightarrow{\;(from\;} RCH\overset{\displaystyle H}{\underset{\displaystyle O^-}{|}} \xrightarrow{\;H^+\;} RCH_2.OH$$
$$ \quad LiAlH_4)$$

Aldehyde   Alcohol

Preparation by Hydrolysis of Esters

Esters can be hydrolysed to acids, together with alcohols or phenols, by acid or base catalysis, e.g.

$$RC\overset{\displaystyle O}{\|}.OR' \longrightarrow RC\overset{\displaystyle O}{\|}.OH + R'OH$$

The base-catalysed hydrolysis of fats (see saponification, p. 504) is important as a means of manufacturing soaps and glycerol

$$\begin{matrix} CH_2.O.CO.C_{17}H_{35} \\ | \\ CH.O.CO.C_{17}H_{35} \\ | \\ CH_2.O.CO.C_{17}H_{35} \end{matrix} + 3NaOH \longrightarrow \begin{matrix} CH_2.OH \\ | \\ CH.OH \\ | \\ CH_2.OH \end{matrix} + 3C_{17}H_{35}.CO.O^-Na^+$$

'Tristearin'   Propanetriol   Sodium octadecanoate
                                   (glycerol)   (stearate)

Biochemical Processes

The anaerobic respiration of micro-organisms living in suitable media can produce a variety of compounds, including several alcohols (see 'fermentation', p. 478).

PROPERTIES

The high electronegativity of oxygen in the —OH group results in the hydroxy compounds being polarized (cf. water) and allows the lower homologues of the alcohols to be completely miscible with water, i.e. —OH is a hydrophilic group. As the hydrocarbon chain increases in size, the solubility of the alcohols decreases. An increase in the number of hydroxy groups also results in an increase in hydrophilic character, so that glycerol is hygroscopic and completely miscible with water but immiscible with ether. (The equilibrium between phenol and water is discussed on p. 97).

The polarization of the hydroxy compounds is sometimes so marked that hydrogen bonding is present, giving rise to aggregates of molecules of relatively low volatility (*Table 22.2*):

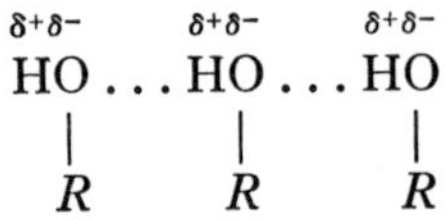

$$\overset{\delta+\delta-}{HO}\ldots\overset{\delta+\delta-}{HO}\ldots\overset{\delta+\delta-}{HO}$$
$$\underset{R}{\mid}\qquad\underset{R}{\mid}\qquad\underset{R}{\mid}$$

Table 22.2. Boiling points of hydrocarbons and hydroxy compounds of comparable molecular weight

Hydrocarbons	Molecular weight	b.p. °C	Hydroxy compounds	Molecular weight	b.p. °C
Ethane	30	−88	Methanol	32	65
Propane	44	−42	Ethanol	46	78
Butane	58	0	Propanol	60	98
Methylbenzene	92	110	Hydroxybenzene	94	182
1,4-Dimethylbenzene	106	138	1,4-Dihydroxybenzene	110	286

In the case of phenols, polarization is so pronounced that there is a measurable amount of ionization: they are particularly non-volatile and even acidic

$$C_6H_5.OH \rightleftharpoons C_6H_5.O^- + H^+$$

The forward reaction is favoured by the fact that lone pairs of electrons on the oxygen atom can form part of the delocalized π bonding of the benzene ring

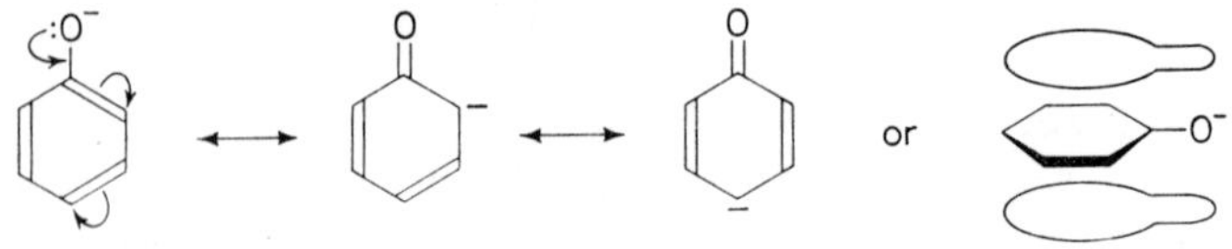

436

Because of their acidity, phenols, unlike alcohols, are soluble in sodium hydroxide solution, but generally their acidity is not great enough for them to dissolve in the low concentration of hydroxide ions provided by sodium carbonate solution. Their acidity is enhanced by the presence of electron-attracting groups in the 2- and 4-positions. For example, 4-nitrophenol exhibits stronger acid properties than does phenol itself

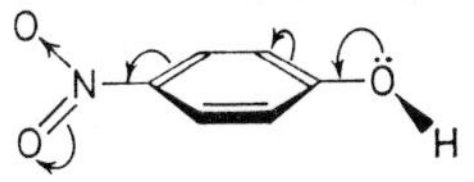

and 2,4,6-trinitrophenol (picric acid) will even decompose carbonates.

Although alcohols are not acidic, they will react with metallic sodium to form the corresponding *alkoxide*:

$$ROH + Na \rightarrow RO^-Na^+ + \tfrac{1}{2}H_2 \uparrow$$

The oxygen atom of alcohols, by virtue of its lone pairs of electrons, *can be protonated*

Protonation of this type is often the precursor to reactions given by alcohols and is responsible for the several differences between the chemistry of alcohols and phenols.

In the case of phenols, the lone pairs on the oxygen are less available, as they form part of the delocalized electron system. However, they do activate the ring towards electrophilic attack, especially at positions 2, 4 and 6. Thus, bromine water gives an immediate precipitate of 2,4,6-tribromophenol with phenol

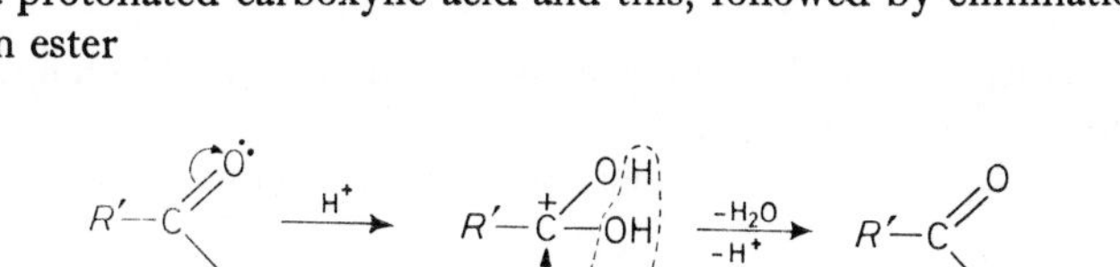

Ester formation (esterification)—The lone pair on the alcohol is able to attack the carbon of a protonated carboxylic acid and this, followed by elimination of water, produces an ester

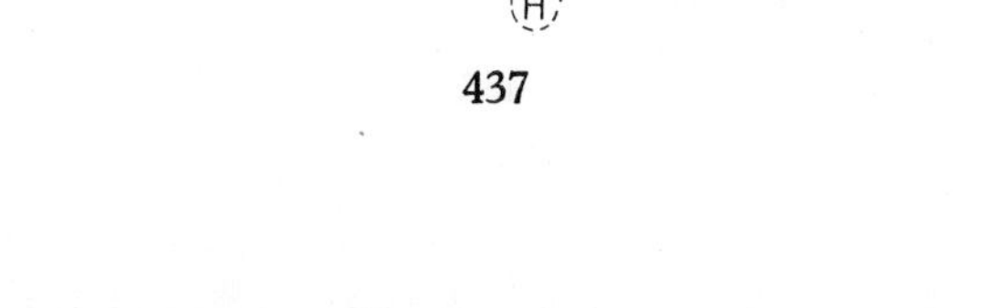

437

that is

$$\overset{O}{\overset{\|}{R'C.OH}} + ROH \rightarrow \overset{O}{\overset{\|}{R'C.OR}} + H_2O$$

That it is an oxygen of the carboxylic acid which is eliminated was shown to be the case by using an ^{18}O-enriched alcohol; after esterification, no enrichment of the oxygen in the water produced was found, so that the reaction is

$$RCO.\overline{|OH} + H\overline{|OR'}} \longrightarrow RCO.OR' + H_2O$$

in accordance with nucleophilic attack by the alcohol on protonated acid, and *not*

$$RCO.O\overline{|H} + HO\overline{|R'}} \longrightarrow RCO.OR' + H_2O$$

The reaction is reversible and the equilibrium can be favourably displaced for ester formation by using concentrated sulphuric acid, anhydrous zinc chloride or dry hydrogen chloride under reflux.

Many alcohols, particularly the secondary and tertiary ones, which readily form carbonium ions, easily eliminate the elements of water (p. 375) to produce alkenes at moderate temperatures in the presence of concentrated sulphuric acid or anhydrous zinc chloride, e.g.

$$\underset{CH_3}{\overset{CH_3}{CH_3\text{-}\underset{H}{\overset{+}{C}.OH}}} \xrightarrow{-H_2O} \underset{H_2CH}{\overset{CH_3}{CH_3\text{-}\overset{+}{C}}} \xrightarrow{-H^+} \underset{H_2C}{\overset{CH_3}{CH_3\text{-}C}}$$

so that dry hydrogen chloride must be used as catalyst.

Because of their acidic nature, phenols do not readily form esters using the methods just described for alcohols. It is generally necessary to reflux the phenol with an acid halide, although benzoylation of phenols can be effected by merely shaking with benzene carbonyl (benzoyl) chloride

$$C_6H_5.OH + C_6H_5.COCl \rightarrow \overset{O}{\overset{\|}{C_6H_5.C.O.C_6H_5}} \qquad + HCl$$

Phenyl benzene carboxylate
(phenyl benzoate)

The product is less stable than an ester produced from an alcohol and is more readily hydrolysed into its components.

Esters can also be formed with mineral acids such as sulphuric and hydrochloric acids, e.g.

$$ROH + H_2SO_4 \rightarrow RO.SO_2.OH + H_2O$$

$$ROH + HCl \quad \rightarrow RCl + H_2O$$

The readiness with which halides are formed from alcohols and hydrogen halides (p. 398) forms a method for distinguishing between primary, secondary and tertiary alcohols. The order of reactivity can be explained in terms of the electron-repelling effect exerted by alkyl groups; in primary alcohols, there is only one such effect tending to weaken the C—O bond; in a secondary alcohol there are two such effects, and in a tertiary three:

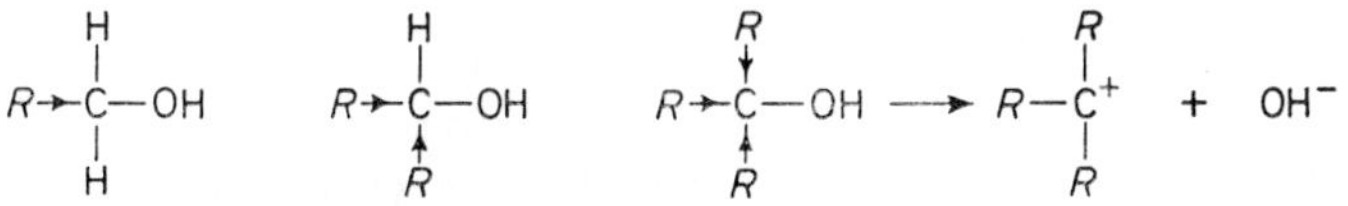

The mechanism of halide formation must involve reaction between the hydroxy group of the alcohol and the hydrogen of the hydrogen halide, i.e. the C—O bond of the alcohol is broken: a mechanism different from that in which a carboxylic acid is esterified.

Ether formation—Alcohols are protonated in the presence of cold concentrated sulphuric acid, and if there is insufficient acid to give complete protonation, then the following reaction can take place:

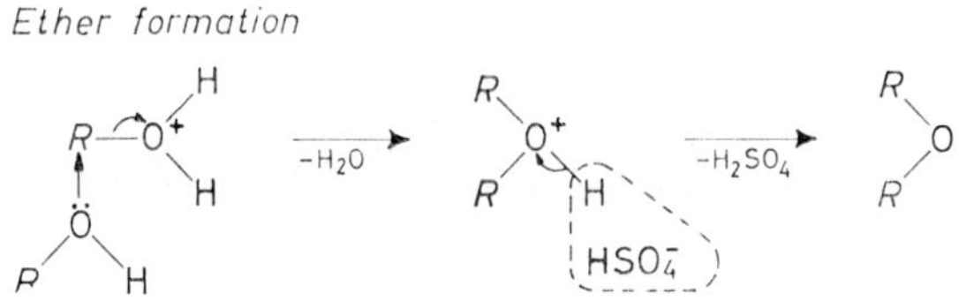

Ether formation

Alkene formation—High temperature and excess concentrated sulphuric acid result in an elimination reaction of the protonated alcohol (p. 375)

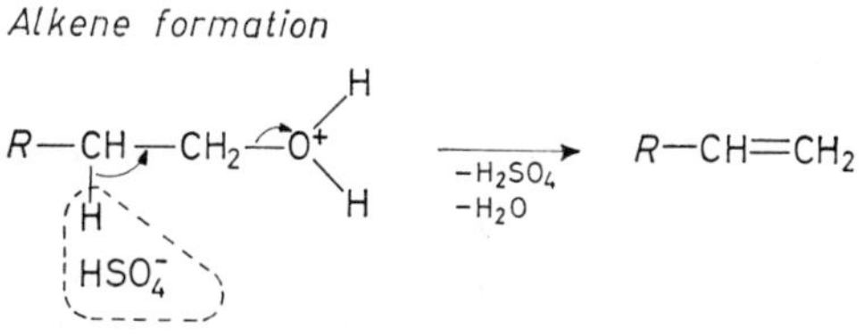

Alkene formation

Reduction and oxidation—Alcohols are very resistant to reduction, but if phenol vapour mixed with hydrogen is passed over a heated nickel catalyst, it is reduced to cyclohexanol

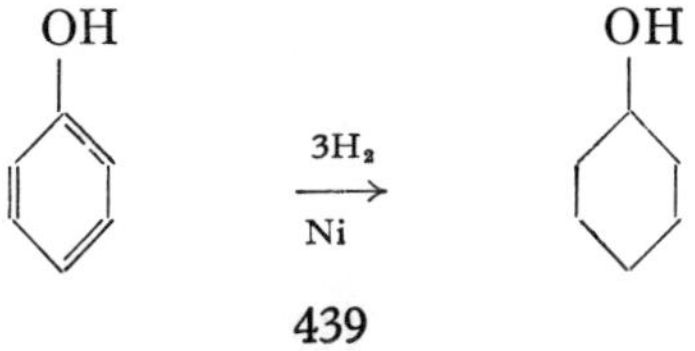

439

Phenols are oxidized only with great difficulty if rupture of the aromatic ring is involved. However, monohydric phenols are oxidized with persulphates in alkaline solution to dihydric phenols, e.g.

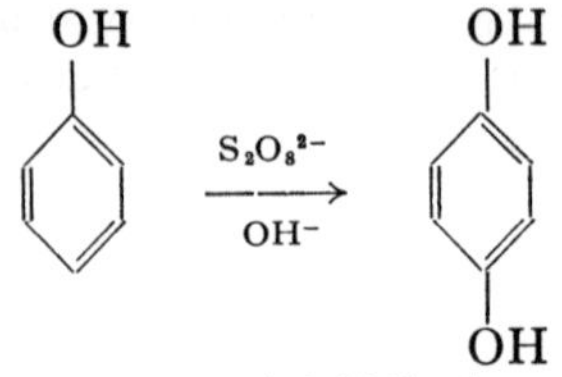

1,4-Dihydroxybenzene
(quinol)

Quinol and the related *N*-methyl-4-aminophenol (metol) are readily oxidized to benzo-1,4-quinone (quinone) and so are strong reducing agents, used, for example, in photographic developers:

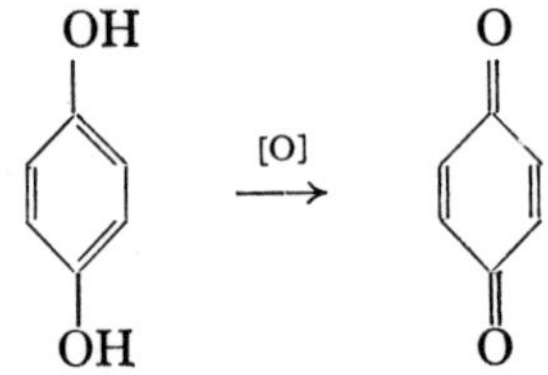

Primary and secondary alcohols are easily oxidized using, for example, sodium dichromate and sulphuric acid in the liquid phase or catalytic dehydrogenation in the vapour phase, e.g.

$$CH_3.CH_2.OH \xrightarrow[+H_2SO_4]{Na_2Cr_2O_7} CH_3.CHO \xrightarrow[+H_2SO_4]{Na_2Cr_2O_7} CH_3.CO.OH$$

Ethanol   Ethanal   Ethanoic (acetic)
          (acetaldehyde)   acid

The oxidation products afford a means of distinguishing between the three types of alcohols:

$$RCH_2.OH \xrightarrow{[O]} RCHO \xrightarrow{[O]} RCO.OH$$

Primary alcohol   Aldehyde   Carboxylic acid

$$\begin{matrix} R \\ R' \end{matrix} CH.OH \xrightarrow{[O]} RCR' \left[\xrightarrow[\text{difficult}]{[O]} \text{a mixture of carboxylic acids with fewer carbon atoms} \right]$$

Secondary alcohol $\overset{\parallel}{O}$
              Ketone

$$\begin{matrix} R' \\ | \\ R''.C.OH \\ | \\ R \end{matrix} \left[\xrightarrow[\text{difficult}]{[O]} \text{a mixture of carboxylic acids with fewer carbon atoms} \right]$$

Tertiary alcohol

440

INDUSTRIAL ASPECTS

Monohydric Alcohols

(a) *Methanol* can be manufactured from the products of petroleum cracking

$$3CH_4 + CO_2 + 2H_2O \xrightarrow[800°C]{Ni} 4CO + 8H_2 \xrightarrow[300 \times 10^5 \, N \, m^{-2}]{ZnO/Cr_2O_3} 4CH_3OH$$

(see Fischer-Tropsch process, p. 377)

It is used as a solvent and for the manufacture of methanal (formaldehyde).

(b) *Ethanol* can also be manufactured from cracked petroleum

$$CH_2 = CH_2 + H_2O \xrightarrow[\substack{pumice \\ 300°C}]{H_3PO_4 \text{ on}} CH_3CH_2OH$$

As ethanol is the by-product of the anaerobic respiration of yeast with carbohydrates as substrate (p. 478), it has been known to man for centuries. The 'fermentation' process can be summarized as

$$C_6H_{12}O_6 \xrightarrow[\text{(in yeast cells)}]{enzymes} 2C_2H_5.OH + 2CO_2 \uparrow$$

At a concentration approaching 15 per cent, development of yeast cells is inhibited and the fermentation ceases. Increase of the alcoholic content above this concentration requires distillation, leading to the production of 'spirits'. Ethanol, like methanol, is used widely as a solvent.

Polyhydric Alcohols

(a) *Ethane-1,2-diol* (*ethylene glycol*) is in demand as an 'antifreeze' (a substance which depresses the freezing point of water, p. 107), as a solvent and, because of its difunctional nature, for making the polyester 'Terylene' (*ter*ephthalic acid and eth*ylene* glycol):

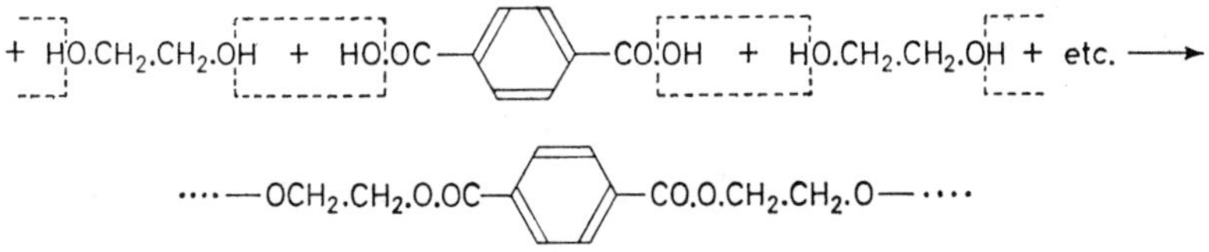

'Terylene'

Nearly all of the diol is at present manufactured from ethene (from petroleum) via epoxyethane (ethylene oxide); ethene is reacted directly with air or oxygen in the presence of a silver catalyst and the resultant epoxyethane is hydrolysed to ethanediol

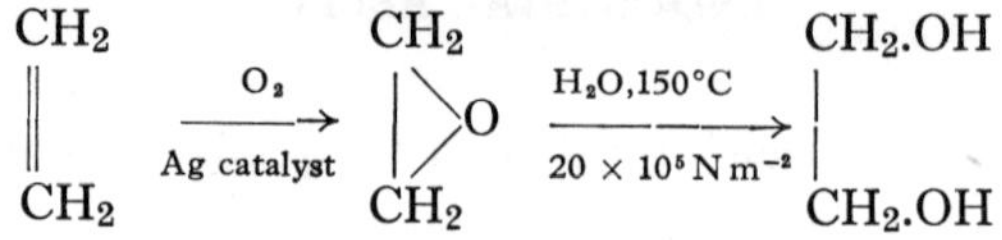

$$\begin{array}{c} CH_2 \\ \| \\ CH_2 \end{array} \xrightarrow[\text{Ag catalyst}]{O_2} \quad \begin{array}{c} CH_2 \\ | \diagdown \\ | \diagup O \\ CH_2 \end{array} \xrightarrow[20 \times 10^5 \, N\,m^{-2}]{H_2O, 150°C} \quad \begin{array}{c} CH_2.OH \\ | \\ CH_2.OH \end{array}$$

Important derivatives of ethanediol are the 'cellosolves' (pp. 452, 454).

(b) *Propane-1,2,3-triol* (*glycerol*)—Glycerol is used widely for the preparation of the explosive glyceryl trinitrate (commonly, and incorrectly, called nitroglycerine):

$$\begin{array}{c} CH_2.OH \\ | \\ CH.OH \\ | \\ CH_2.OH \end{array} + 3HONO_2 \xrightarrow{H_2SO_4} \begin{array}{c} CH_2.O.NO_2 \\ | \\ CH.O.NO_2 \\ | \\ CH_2.O.NO_2 \end{array} + 3H_2O$$

This substance is too dangerous to handle on its own and so is either adsorbed on to sawdust (which, with ammonium nitrate, is *dynamite*), mixed with 'gun cotton' (*gelignite*) or with gun cotton and vaseline (*cordite*).

Table 22.3. Routes to glycerol

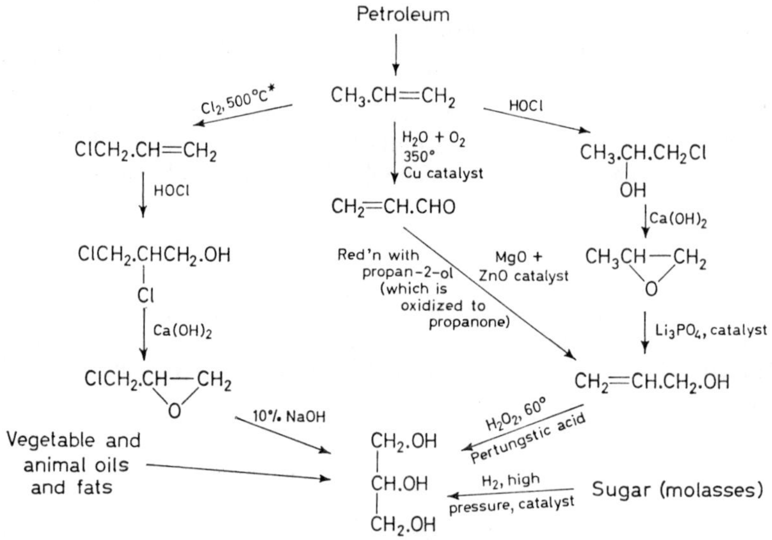

*At lower temperatures the 'expected' addition of chlorine to propene occurs

Glycerol is also a constituent of the *alkyd resins* (p. 501).

The esters of glycerol with the fatty acids are the main constituents of vegetable oils (p. 503) and animal fats, and so are one important source of glycerol as a by-product in the manufacture of soap. Other sources are available chiefly from propene, a product of petroleum cracking (p. 378). The reactions involved in the manufacture of glycerol are summarized in *Table 22.3*.

(c) *Tetra(hydroxymethyl)methane* (*pentaerythritol*)—This polyhydric alcohol is prepared by the reaction of methanal (formaldehyde) with ethanal (acetaldehyde)

$$4HCHO + CH_3.CHO + H_2O \xrightarrow[50°]{Ca(OH)_2} HO.CH_2-\overset{\overset{\displaystyle CH_2.OH}{|}}{\underset{\underset{\displaystyle CH_2.OH}{|}}{C}}-CH_2.OH + HCO.OH$$

The alcohol is an important constituent of alkyd resins, because of the number of functional groups present, and its tetranitrate is a powerful explosive.

Phenol

Phenol is in such demand industrially that it is essential to obtain it from a wide variety of sources. Some of the methods used are shown in *Table 22.4*.

Table 22.4. Routes to phenol

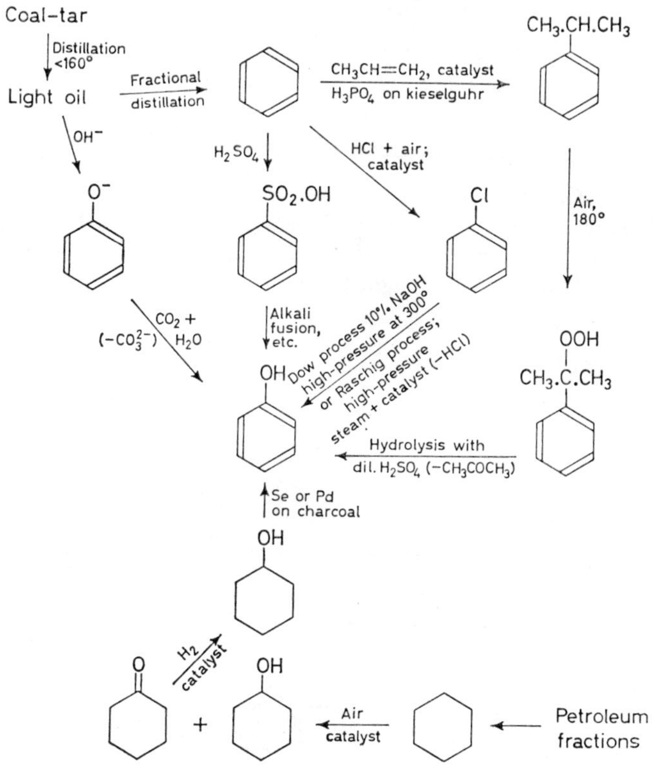

The activating effect of the hydroxy group for positions 2, 4 and 6 of the benzene ring allows many useful derivatives of phenol to be made. Among them are:

(*i*) *2,4,6-trichlorophenol* (*T.C.P.*) which is prepared from chlorine water and phenol (cf. tribromophenol) and is used as a germicide;

(*ii*) *2-acetoxybenzene carboxylic acid* (*acetyl salicylic acid—Aspirin*).

Chloroform is reacted with phenol in the presence of alkali; after acidification, when 2-hydroxybenzene carbonal is formed, the product is oxidized and acetylated.

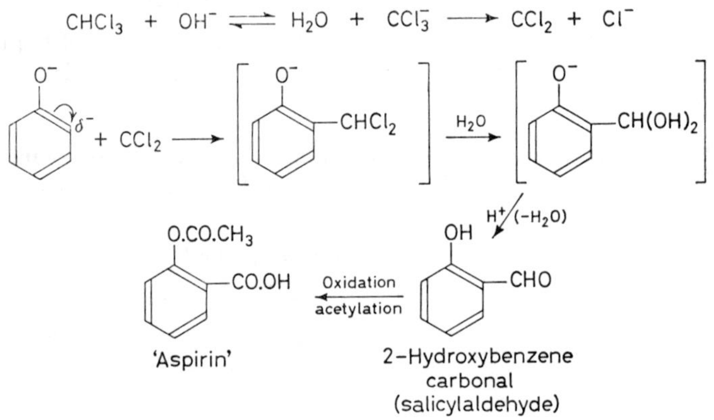

$$CHCl_3 + OH^- \rightleftharpoons H_2O + CCl_3^- \longrightarrow CCl_2 + Cl^-$$

'Aspirin'

2-Hydroxybenzene carbonal (salicylaldehyde)

(*iii*) Condensation Products

Thermosetting plastics (i.e. those which cannot be remoulded after the original condensation reaction is complete) of the *Bakelite* type are formed with methanal. The sequence of reactions is thought to be

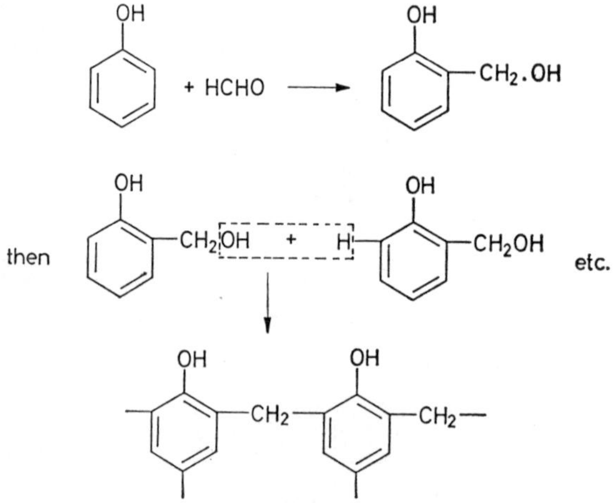

444

(*iv*) *Phenolphthalein* results when phenol is heated with benzene-1,2-dicarboxylic (phthalic) anhydride in the presence of concentrated sulphuric acid

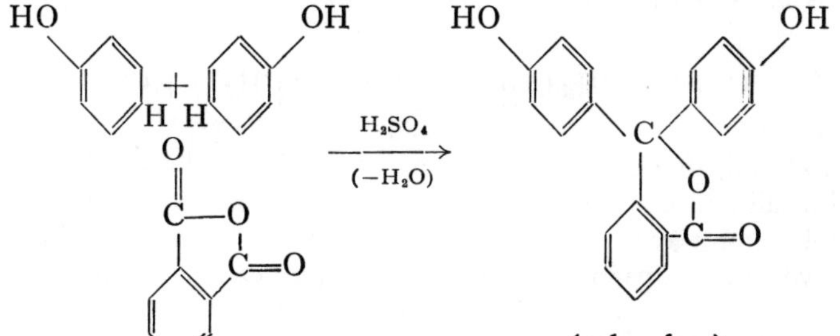

(colourless)

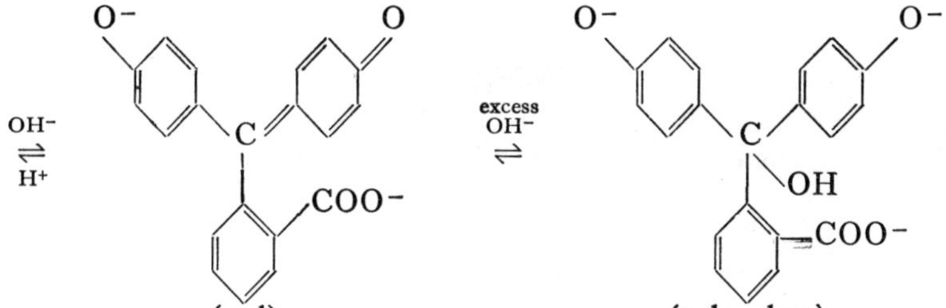

(red)   (colourless)

ANALYSIS

The amount of oxygen present in an organic compound is determined by heating the substance in a stream of nitrogen and passing the oxides of carbon so formed over heated carbon. This reduces any carbon dioxide to the monoxide, which is then reacted with iodine pentoxide

$$5CO + I_2O_5 \rightarrow 5CO_2 + I_2$$

The iodine liberated is estimated by titration against a solution of sodium thio-sulphate (p. 236).

The alcoholic hydroxy group can be recognized by the reaction it gives with sodium to produce hydrogen. Both alcohols and phenols react exothermically with phosphorus(V) chloride giving hydrogen chloride fumes. Many phenols give characteristic colours with neutral iron(III) chloride solution. A particular alcohol or phenol can be characterized by the preparation of the 3,5-dinitrobenzoate ester, e.g,

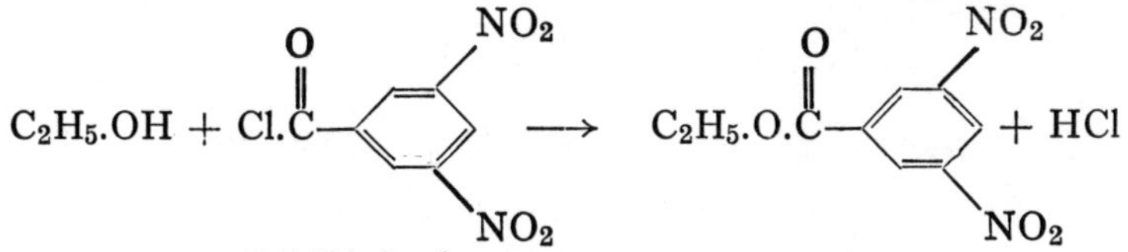

3,5-Dinitrobenzene
carbonyl chloride

The number of hydroxy groups in a polyhydric alcohol or phenol can be determined by completely acetylating the compound, using acetic anhydride or acetyl chloride:

$$—\overset{\displaystyle |}{\underset{\displaystyle |}{C}}.OH + CH_3.COCl \rightarrow —\overset{\displaystyle |}{\underset{\displaystyle |}{C}}.O.CO.CH_3 + HCl$$

The ester produced, after careful washing, may be hydrolysed with a known amount of standard alkali and the quantity of unused alkali determined by titration with standard acid. From these values the number of acetyl groups (and, therefore, the number of hydroxy groups in the original compound) can be found

$$—\overset{\displaystyle |}{\underset{\displaystyle |}{C}}.O.CO.CH_3 + OH^- \rightarrow —\overset{\displaystyle |}{\underset{\displaystyle |}{C}}.OH + CH_3.CO.O^-$$

SUMMARY

Many of the reactions of alcohols can be attributed to the availability of a lone pair on the oxygen atom. The fact that this lone pair is, because of interaction with

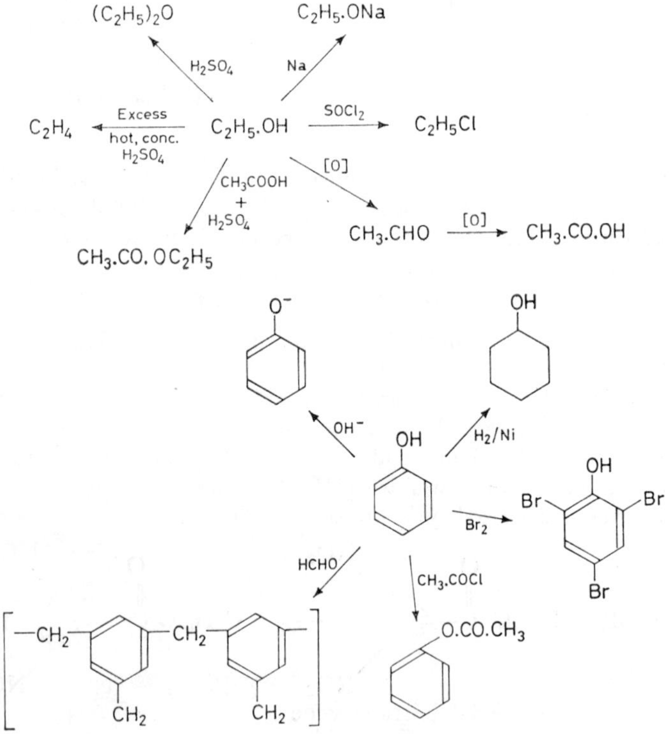

Figure 22.1. Chief reactions of ethanol and hydroxybenzene

446

the benzene nucleus, far less available in phenols than in alcohols leads to marked differences between them. The main reactions of ethanol and hydroxybenzene are summarized in *Figure 22.1*.

QUESTIONS

1. Give the five possible oxidation products of ethanediol in which the two carbon atoms remain attached.

2. Draw the various structures of the alcohols with the molecular formula $C_4H_{10}O$.

3. Suggest reasons for the following:
(*a*) phenol is soluble in sodium hydroxide solution but insoluble in sodium carbonate solution;
(*b*) of the three mononitrophenols, the strongest acidity is shown by 4-nitro-phenol;
(*c*) the melting point of 2-nitrophenol is only 43°, whereas 3- and 4-nitrophenol melt at the much higher temperatures of 97° and 114°, respectively;
(*d*) trinitrophenol is a stronger acid than phenol, whilst ethanol is neutral.

4. An organic compound has a molecular weight of 92. 0·46 g of the substance was completely acetylated and the product treated with 200 cm³ of M/10 sodium hydroxide solution. The resultant solution required 25 cm³ of M/5 hydrochloric acid for neutralization. Calculate the number of hydroxy groups in a molecule of the original compound and suggest a structural formula.

5. A substance A, of molecular formula C_3H_8O, with excess of concentrated sulphuric acid yields B, C_3H_6, and with a little concentrated sulphuric acid C, $C_6H_{14}O$. B reacts with bromine to give D, $C_3H_6Br_2$, which with aqueous potassium hydroxide produces E, $C_3H_8O_2$, and with alcoholic potash, F, C_3H_4. Identify compounds A–F and illustrate the reactions by suitable equations.

6. A compound of molecular formula $C_{10}H_{15}NO$ gives benzoic acid on oxidation; when treated with nitrous acid, a nitroso compound is formed. The substance forms a dibenzoyl derivative with benzoyl chloride and also, on heating with hydrochloric acid, aminomethane and 1-phenylpropan-2-one are formed. Deduce structures which would account for these properties.

7. Compare the properties of ethanol and phenol.

8. The ease of esterification of an alcohol with a carboxylic acid is in the order: primary > secondary > tertiary, whereas the reverse order holds for the esterification using hydrogen halides. Suggest reasons for these observations.

9. Many phenols and alcohols are nowadays in great commercial demand and their production illustrates the changing balance of industrial supply. Review the major sources of organic materials, with particular reference to hydroxy compounds.

10. A compound of formula C_3H_7Br is converted into C_3H_9N by reaction with alcoholic ammonia. This compound, in turn, is changed into C_3H_8O by treating with nitrous acid solution. Knowing that this latter compound gives a yellow precipitate with sodium carbonate solution and iodine, explain the above reactions.

CHAPTER 23

ETHERS

Ethers are derived from alcohols, the hydrogen of the hydroxyl group being replaced by alkyl or aryl. Alcohols and ethers both possess saturated oxygen atoms, unlike carbonyl and carboxyl compounds where the p electrons of the oxygen atoms are involved in 'double' bonds:

$$
\begin{array}{ccccc}
 & & & R' & OR' \\
 & & & | & | \\
ROH & ROR' & R-C{=}O & RC{=}O \\
\text{Alcohol} & \text{Ether} & \text{Carbonyl} & \text{Carboxyl} \\
 & & \text{compound} & \text{compound}
\end{array}
$$

Preparation of Ethers

It has already been stated (p. 437) that alcohols are protonated by strong acids

$$ ROH \xrightarrow{\text{H}^+} \overset{+}{ROH_2} $$

In the presence of excess concentrated sulphuric acid and at elevated temperatures, a proton is removed from the β-carbon atom, and simultaneous elimination of water gives rise to an alkene. At lower temperatures and in the presence of excess alcohol, so that not all the alcohol is protonated, a different reaction predominates by which an ether is formed (see p. 439).

This latter method is suitable for the preparation of the lower homologues only, because the decrease in volatility resulting from increased molecular weight means that the temperature of distillation is so high that alkene formation predominates.

Diols give, by the above reaction, cyclic ethers or *epoxy compounds*

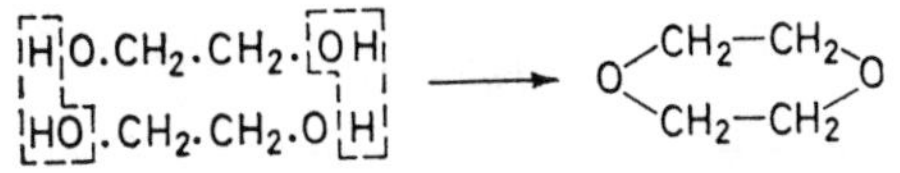

$$
\begin{array}{cc}
\boxed{H}O.CH_2.CH_2.\boxed{O\,H} & \\
& \longrightarrow \quad O\overset{\displaystyle CH_2-CH_2}{\underset{\displaystyle CH_2-CH_2}{\diagup\diagdown}}O \\
\boxed{HO}.CH_2.CH_2.O\boxed{H} &
\end{array}
$$

$$
\begin{array}{cc}
\text{Ethanediol} & \text{Dioxan}
\end{array}
$$

A general method of preparation of ethers involves the action of heat on a mixture of alkyl halide and sodium alkoxide or phenate (Williamson's synthesis):

$$ R'X + Na^+OR^- \rightarrow R'OR + Na^+X^- $$

The reaction is bimolecular, the alkoxide or phenate ion constituting a powerful nucleophile and attacking the carbon atom attached to the halogen. This carbon atom comes more and more under the influence of the nucleophile, and the halogen

449

atom steadily recedes, taking the electron pair of the original covalent bond with it; ultimately, a new bond is formed, as the old one is broken

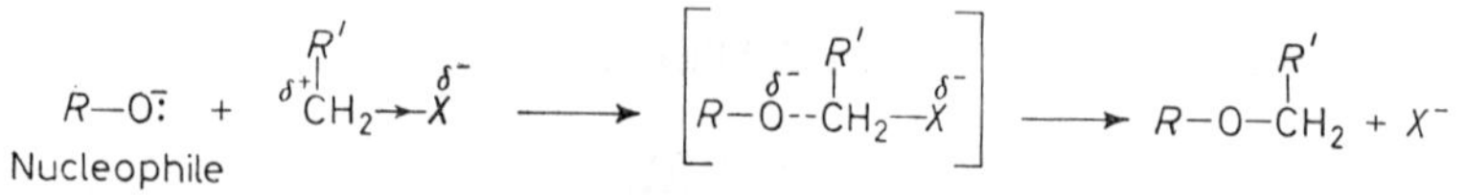

$$R-\overset{..}{O}{:} \quad + \quad \overset{\delta+}{C}\overset{R'}{\overset{|}{H_2}}\overset{\delta-}{\longrightarrow}X \quad \longrightarrow \quad \left[R-\overset{\delta-}{O}\cdots\overset{R'}{\underset{|}{CH_2}}\overset{\delta-}{\longrightarrow}X \right] \quad \longrightarrow \quad R-O-\overset{R'}{\underset{|}{CH_2}} + X^-$$

Nucleophile

Aryl halides cannot be used in this reaction; conjugation of the 'lone pair' on the halogen atom with the aromatic system leads to the latter being negatively rather than positively charged, and thus unattractive to the nucleophile.

Nomenclature

Ethers can be regarded as alkoxy derivatives of hydrocarbons, and this forms the basis for nomenclature. For example, diethyl ether is ethoxyethane and both anethole, a constituent of aniseed, and eugenol, found in oil of cloves, are methoxy derivatives of propenyl benzenes

$C_2H_5.O.C_2H_5$

Ethoxyethane

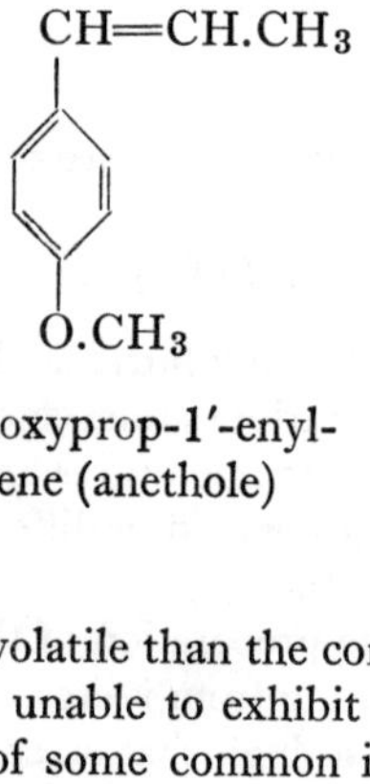

$CH{=}CH.CH_3$

$O.CH_3$

4-Methoxyprop-1'-enyl-benzene (anethole)

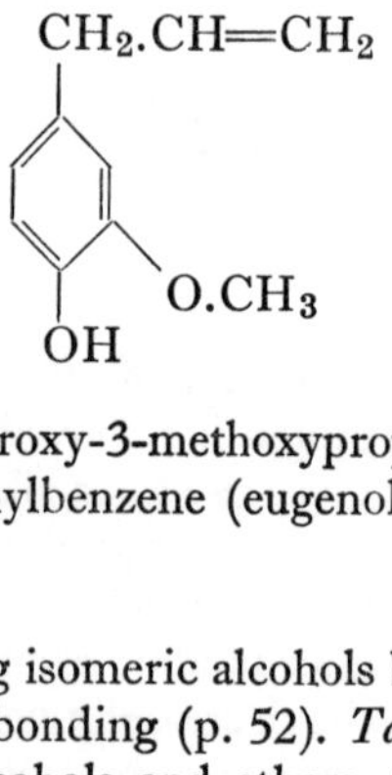

$CH_2.CH{=}CH_2$

$O.CH_3$

OH

4-Hydroxy-3-methoxyprop-2'-enylbenzene (eugenol)

Properties of Ethers

Ethers are much more volatile than the corresponding isomeric alcohols because, unlike the latter, they are unable to exhibit hydrogen bonding (p. 52). *Table 23.1* gives the boiling points of some common isomeric alcohols and ethers, together with hydrocarbons of comparable molecular weight.

Table 23.1

Molecular formula	Alcohol	b.p. °C	Ether	b.p. °C	Hydrocarbon	b.p. °C
C_2H_6O	ethanol, C_2H_5OH	78	methoxymethane, $CH_3.O.CH_3$	−24	propane, C_3H_8	−42
C_3H_8O	propan-1-ol, C_3H_7OH propan-2-ol, $(CH_3)_2.CHOH$	97 82	methoxyethane, $CH_3.O.C_2H_5$	8	butane, C_4H_{10}	0

It will be seen that, in terms of volatility, ethers are more like alkanes. Nor does the similarity end here; it extends to density, refractive index, immiscibility with water and inflammability. Ethers are more inflammable than the corresponding alcohols, but otherwise they are less reactive. Although the oxygen atom is capable of protonation by strong acids, the complexes, unlike those of alcohols, are incapable of eliminating water. Thus, although ethers dissolve in concentrated sulphuric acid, they are regenerated on dilution:

$$ROR' \xrightarrow{\text{H}_2\text{SO}_4} [ROR']^+ \text{HSO}_4{}^- \xrightarrow{\text{H}_2\text{O}} ROR' + \text{H}_2\text{SO}_4.\text{H}_2\text{O}$$
$$\overset{\displaystyle |}{\underset{\displaystyle H}{}}$$

Donation of the electron pair on the oxygen atom of an ether to other molecules is also common

$$R_2O \rightarrow BF_3 \qquad RMg \leftarrow O(C_2H_5)_2 \qquad (C_2H_5)_2O \rightarrow CrO_5$$
$$\overset{R}{\underset{\uparrow}{|}}$$
$$O(C_2H_5)_2$$

However, fission *does* result from attack by the powerful electrophile formed when acyl chlorides (p. 498) react with anhydrous zinc chloride; subsequent elimination of alkyl chloride from the intermediate gives an ester:

$$RC\overset{\frown}{=}O: \leftarrow\text{-ZnCl}_2 \longrightarrow R\overset{+}{C}.O.\bar{Z}nCl_2$$
$$\underset{Cl}{|} \qquad\qquad \underset{Cl}{|}$$

$$\underset{R\overset{+}{C}.O.\bar{Z}nCl_2}{\overset{Cl}{|}} \longrightarrow \underset{R.\overset{+}{C}.\bar{O}-ZnCl_2}{\overset{Cl}{|}} \longrightarrow RC=O\ (+RC=O) + R''Cl\ (+R'Cl) + ZnCl_2$$
$$\underset{\underset{R''}{|}}{:OR'} \qquad\qquad R'\overset{+}{-}OR'' \qquad\qquad\qquad \underset{OR'}{|} \qquad\quad \underset{OR''}{|}$$

Ethers are also attacked by constant-boiling-point hydriodic acid; this time, the strongly nucleophilic iodide ion is the active agent

$$R\underset{\substack{\uparrow \\ | \\ H^+}}{\overset{+}{O}}R' \longrightarrow R\overset{+}{O}R'; \quad R\overset{\frown}{\underset{I^-}{\overset{+}{O}}}{-}R' \longrightarrow RI + R'OH$$
$$\underset{H}{|} \quad \underset{H}{|}$$

The hydroxy compound formed, if aliphatic, can be converted by more hydrogen iodide into the alkyl iodide but phenols resist further attack. The alkyl iodide formed can be absorbed in alcoholic silver nitrate solution, and the resulting silver iodide can be weighed. This is the basis for the determination of methoxy groups by the Zeisel method

$$C_6H_5.O.CH_3 \xrightarrow{HI} C_6H_5.OH + CH_3I$$

$$\underset{142\ g}{CH_3I} + AgNO_3 \xrightarrow{H_2O} \underset{235\ g}{AgI \downarrow} + CH_3OH + HNO_3$$

CYCLIC ETHERS

Epoxyethane (ethylene oxide), $\underset{CH_2-CH_2}{\overset{O}{\diagup\ \diagdown}}$

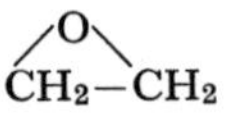

is prepared by treating 2-chloroethanol (ethylene chlorhydrin) with alkali

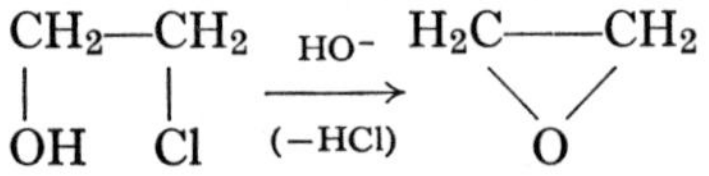

$$\underset{\overset{|}{OH}\quad \overset{|}{Cl}}{CH_2-CH_2} \xrightarrow[(-HCl)]{HO-} \underset{O}{H_2C\underline{\quad\quad}CH_2}$$

This compound can be compared with cyclopropane in that the strain imposed upon the bonds by excessive departure from the tetrahedral angle makes it unduly reactive. In this case, cleavage can readily take place in the presence of acid to give ethanediol or, in the presence of acid and alcohol, a 'cellosolve':

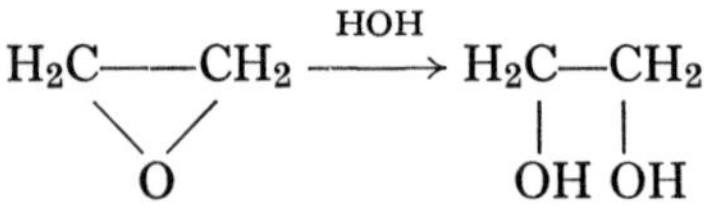

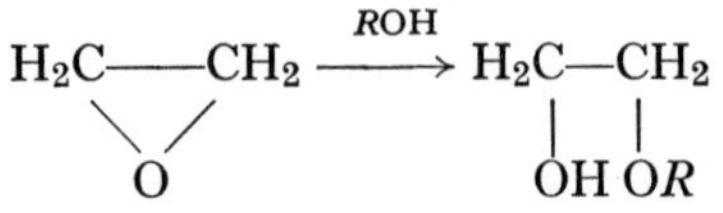

$$H_2C\underline{\quad}CH_2 \xrightarrow{HOH} \underset{\overset{|}{OH}\ \overset{|}{OH}}{H_2C-CH_2} \qquad H_2C\underline{\quad}CH_2 \xrightarrow{ROH} \underset{\overset{|}{OH}\ \overset{|}{OR}}{H_2C-CH_2}$$

A 'cellosolve'

Furan, $\underset{\diagdown O \diagup}{\overset{HC\underline{\quad}CH}{HC\quad\ CH}}$

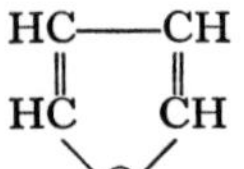

occurs in the distillates of most woody materials and is also obtained by heating mucic acid

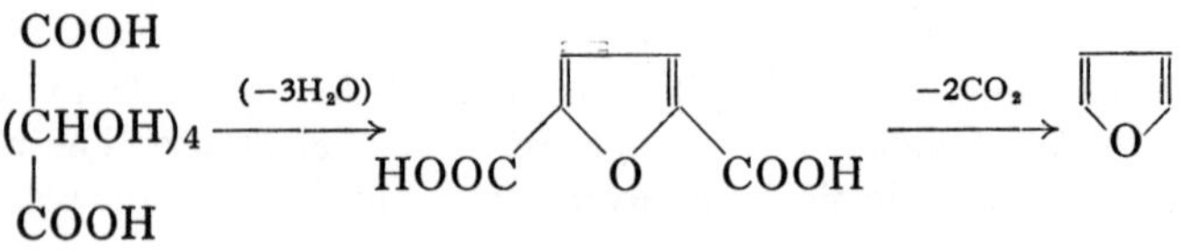

$$\underset{\overset{|}{COOH}}{\overset{COOH}{\underset{|}{(CHOH)_4}}} \xrightarrow{(-3H_2O)} \underset{HOOC\quad O\quad COOH}{\diagup\diagdown} \xrightarrow{-2CO_2} \overset{\diagup\diagdown}{\underset{O}{\ }}$$

452

It reveals its conjugated double-bond character by taking part in what is known as the Diels–Alder reaction with, for instance, maleic anhydride

$$
\begin{array}{ccc}
\text{HC=CH} & \text{HC—CO} & \text{HC—CH——CH—CO} \\
\mid \quad \diagdown & \parallel \quad \diagdown & \parallel \quad \diagdown \quad \mid \quad \diagdown \\
\mid \quad \quad \text{O} + & \quad \text{O} \longrightarrow & \quad \text{O} \quad \mid \quad \text{O} \\
\mid \quad \diagup & \parallel \quad \diagup & \parallel \quad \diagup \quad \mid \quad \diagup \\
\text{HC=CH} & \text{HC—CO} & \text{HC—CH——CH—CO}
\end{array}
$$

Pyran,
$$
\begin{array}{c}
\quad \text{CH}_2 \\
\text{HC} \quad \text{CH} \\
\parallel \quad \quad \parallel \\
\text{HC} \quad \text{CH} \\
\quad \text{O}
\end{array}
$$

is not known but the removal of an electron from the oxygen enables it to partici-pate in a non-localized π electron system and so form the pyrilium compounds that are so widespread as plant pigments, e.g.

4'-OH	pelargonidin (from pelargonium)
3',4'-OH	cyanidin (from cornflower)
3',4',5'-OH	delphinidin (from delphinium)

Tetrahydrofuran and tetrahydropyran provide the basis for the furanose and pyranose ring structures present in carbohydrates (pp. 471, 472).

Thiophen,
$$
\begin{array}{c}
\text{HC——CH} \\
\parallel \quad \quad \parallel \\
\text{HC} \quad \text{CH} \\
\quad \text{S}
\end{array}
$$

is similar to furan but contains an ether link through sulphur instead of oxygen. This —S— link is present in the penicillins, e.g. penicillin G

$$
\begin{array}{c}
\quad \quad \quad \quad \quad \quad \quad \quad \text{S} \quad \quad \text{CH}_3 \\
\quad \quad \quad \quad \quad \quad \quad \quad \diagup \quad \diagdown \diagup \\
\text{C}_6\text{H}_5.\text{CH}_2\text{—CONH—CH—CH} \quad \text{C—CH}_3 \\
\quad \quad \quad \quad \quad \quad \mid \quad \quad \mid \quad \quad \quad \mid \\
\quad \quad \quad \quad \quad \quad \text{CO—N——CH—COOH}
\end{array}
$$

These penicillins, produced by the mould *Penicillium*, are examples of *antibiotics*, i.e. substances produced by one organism and toxic to another.

INDUSTRIAL ASPECTS

The lower aliphatic homologues are of considerable importance as solvents (ethoxyethane is also used as an anaesthetic) and are manufactured on the large scale by treating alkenes (by-products from the cracking of petroleum) with cold, concentrated sulphuric acid. Addition of water in sufficient quantity converts some of the alkyl hydrogen sulphate formed by this reaction into alcohol; the ether results from interaction between this and more alkyl hydrogen sulphate, e.g.

$$CH_2{=}CH_2 \xrightarrow{H_2SO_4} CH_3.CH_2.O.SO_2.OH$$

$$CH_3.CH_2.O.SO_2.OH + H_2O \longrightarrow CH_3.CH_2.OH + H_2SO_4$$

$$CH_3.CH_2.O.SO_2.OH + CH_3.CH_2.OH \longrightarrow CH_3.CH_2.O.CH_2.CH_3 + H_2SO_4$$

'Cellosolves' are particularly good solvents because they contain both the hydrophilic alcohol group and the hydrophobic ethereal linkage, e.g.

$$HO.CH_2.CH_2.O.CH_3$$

Cyclic ethers can be polymerized to epoxy resins by treatment with a little water in the presence of acid or base as catalyst, e.g.

$$H_2C{-\!-\!-}CH_2 \rightarrow \ldots{-}O.CH_2.CH_2.O.CH_2{-}\ldots$$
$$\diagdown O \diagup$$

Copolymerization is effected by mixing the ether with a polyhydric alcohol: the use of glycerol, with its three hydroxy groups, establishes cross-linkages throughout the structure to give a three-dimensional resin. Epoxy resins are becoming increasingly important as cements and adhesives of high strength.

Compounds containing the ether link are also playing an expanding role in horticulture, e.g. as selective weedkillers such as 2,4-D.C.P.A. (see p. 490).

ANALYSIS

Advantage is taken of the presence of oxygen in ethers to distinguish them from the comparably inert alkanes; they dissolve, through protonation, in concentrated sulphuric acid but are regenerated on dilution.

Suitable derivatives of ethers are not easy to find, and so characterization is often effected by means of physical properties such as refractive index.

QUESTIONS

1. Give the structural formulae and the names of compounds of molecular formula $C_4H_{10}O$. Briefly say how you would distinguish between them chemically.

2. What products may be obtained from propan-1-ol and concentrated sulphuric acid, with and without treatment with water?

3. What ether may be prepared from 4-methoxybromobenzene and 4-methyl-phenol? What other ether would be produced by treating the product with potassium permanganate under reflux, followed by hydriodic acid?

4. How could you prepare $(CH_3)_2.CH.O.C_2H_5$, starting from ethanol and propan-1-ol? What is the name of this compound?

5. Suggest a synthesis of phenoxybenzene from benzene.

6. How would you separate a mixture of benzene, methoxybenzene and amino-benzene?

7. A pure substance was found to contain carbon, hydrogen and oxygen only. It dissolved in cold, concentrated sulphuric acid, but the addition of excess water served to regenerate it. 0·03 g of this compound displaced 13·2 cm³ of air in a Victor Meyer apparatus, measured dry at 27°C and 750 mmHg pressure. Deduce what you can from this information.

8. Write an essay on the role of ethers in either agriculture or adhesives.

CHAPTER 24

CARBONYL COMPOUNDS

The carbonyl group contains carbon attached to oxygen by a double bond:

$$>\!C\!=\!O \qquad \text{or} \qquad >\!C\overset{\sigma}{\underset{\pi}{}}O$$

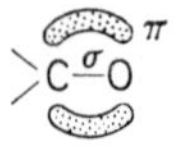

If the carbon atom of this unsaturated group is attached to two other carbon atoms, the compound is called a *ketone*; if to one carbon and one hydrogen, an *aldehyde*. Aliphatic aldehydes are named by replacing the final *-e* of the corresponding hydrocarbon by *-al*, aliphatic ketones by replacing it by *-one* and numbering the relevant carbon atom if necessary

$$\begin{array}{cc} CH_3\!-\!C\!=\!O & CH_3\!-\!C\!=\!O \\ | & | \\ H & CH_3 \\ \text{Ethanal} & \text{Propanone} \end{array}$$

PREPARATION

1) Aldehydes bear a close relationship to primary alcohols, and ketones to secondary alcohols

$$RCH_2.OH \longrightarrow \begin{array}{c} RC\!=\!O \\ | \\ H \end{array} \qquad RCH.OH \longrightarrow \begin{array}{c} R.C\!=\!O \\ | \\ R' \end{array}$$
$$\quad\;\; | \qquad\qquad\qquad\qquad\;\; R'$$

Consequently, an aldehyde is obtained by oxidation of a primary alcohol with acidified sodium dichromate. It should be distilled off immediately, without refluxing, to minimize further oxidation to the corresponding carboxylic acid, e.g.

$$CH_3.CH_2.CH_2.OH \xrightarrow{\text{[O]}} CH_3.CH_2.CHO$$
$$\text{Propan-1-ol} \qquad\qquad \text{Propanal}$$

Alternatively, dehydrogenation of the primary alcohol can be effected catalytically by passing the vapour over heated copper (in fact, the name 'aldehyde' is derived from '*al*cohol *dehyd*rogenatus'), e.g.

$$CH_3.CH_2.OH \longrightarrow CH_3.CHO$$
$$\text{Ethanol} \quad (-2H) \quad \text{Ethanal}$$
$$\text{(acetaldehyde)}$$

This latter method has the advantage that no aldehyde is lost by conversion to acid.

Ketones are produced from secondary alcohols in a similar manner, e.g.

$$(CH_3)_2CH.OH \longrightarrow (CH_3)_2CO$$

Propan-2-ol $(-2H)$ Propanone
(acetone)

2) Hydrolysis of compounds in which two halogen atoms are attached to the same carbon atom yields the corresponding carbonyl compound:

$$RCHCl_2 \xrightarrow[(-2HCl)]{2H_2O} [RCH(OH)_2] \longrightarrow RCHO + H_2O$$

This is of particular value in the case of benzene carbonal (benzaldehyde); methyl benzene (toluene) can be chlorinated to the required state and the product hydrolysed (p. 388).

3) Aromatic and arylalkyl ketones are readily prepared by the Friedel–Crafts reaction, i.e. by reacting aromatic hydrocarbons with an acid chloride in the presence of anhydrous aluminium chloride (p. 390)

$$C_6H_6 + RCOCl \xrightarrow[(-HCl)]{Al_2Cl_6} C_6H_5.COR$$

4) Acid chlorides can be selectively reduced to aldehydes by treatment with hydrogen in the presence of a palladium catalyst, partially deactivated with e.g. barium sulphate (Rosenmund reaction). The deactivation is essential to prevent further reduction of the aldehyde to alcohol

$$RC=O \xrightarrow[(-HCl)]{H_2} RC=O$$
$$\underset{Cl}{|} \qquad \underset{H}{|}$$

PROPERTIES

Because oxygen is more electronegative than carbon, there is a drift of electrons (the inductive effect) towards the oxygen in the carbonyl bond. That is, the bond is *polarized*

$$\overset{\delta+}{C}-\overset{\delta-}{O}$$

This condition may be represented as an increase in the probability of the valency electrons becoming associated with the oxygen atom, i.e. there is an increased electron density around the oxygen

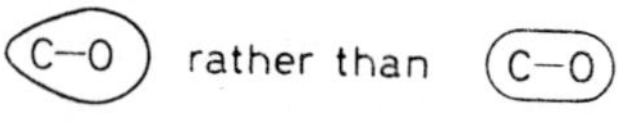

$$\boxed{C-O}$$ rather than $$\boxed{C-O}$$

The lower aliphatic homologues of the carbonyl compounds are freely soluble in water. However, as there is no prospect of hydrogen bonding with the solvent, aldehydes and ketones are not as a rule as soluble as the corresponding alcohols. For the same reason they are more volatile. There is, however, sufficient molecular interaction arising from polarization to ensure that the carbonyl compounds are considerably less volatile than the corresponding hydrocarbons.

The characteristic reaction of aldehydes and ketones involves nucleophilic attack by a base at the carbon atom of the carbonyl group

$$\text{>C=O} \longrightarrow \text{>C-O}^- \qquad (B = base)$$
$$\quad:B \qquad\qquad\qquad B^+$$

The reaction is sometimes reversible, and attempts to isolate the product may result in the reappearance of the original compound. In other cases, this preliminary addition is followed by the elimination of water or another small molecule; the combined reaction is then known as a *condensation*.

Generally, reactivity decreases as the molecular complexity increases, due at least partly to an increase in *steric hindrance*, i.e. to the carbonyl group being shielded from attack by large neighbouring groups.

Example of steric hindrance in an aromatic ketone

This effect will be more marked with ketones than with aldehydes where one adjacent group is always a small hydrogen atom. *It is the presence of this hydrogen atom that is responsible for some fundamental differences between aldehydes and ketones.* For example, aldehydes are readily oxidized to acids (with the same number of carbon atoms) and are therefore reducing agents; those soluble in aqueous media reduce Fehling's solution (which contains copper(II) ions, kept in alkaline solution as a tartrate complex) to copper(I) oxide or even, in the case of methanal, to metallic copper

$$2Cu^{2+} + 5OH^- + RCHO \rightarrow RCO.O^- + 3H_2O + Cu_2O \downarrow$$

Aliphatic aldehydes will also reduce a solution containing diammine silver(I) ions (ammoniacal silver hydroxide) to give a silver mirror

$$RCHO + 2Ag(NH_3)_2^+(OH)^- \rightarrow RCO.O^-NH_4^+ + 3NH_3 + H_2O + 2Ag \downarrow$$

Ketones, on the other hand, are oxidized with difficulty under more extreme conditions, to give acids with fewer carbon atoms than the original compound. They do not, therefore, give the reducing reactions of aldehydes, and this affords a means of distinguishing between the two classes of compounds.

Nucleophilic Attack by Nitrogen

Carbonyl compounds undergo many reactions with nitrogenous compounds. Initially, the lone pair on nitrogen is attracted to the carbon of the carbonyl group:

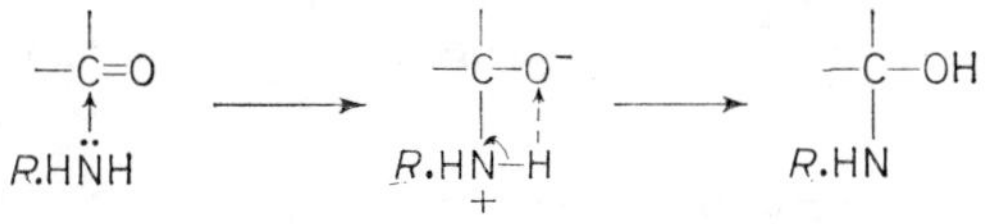

$$-\overset{|}{\underset{\underset{R.H\ddot{N}H}{\uparrow}}{C}}=O \longrightarrow -\overset{|}{\underset{\underset{+}{R.HN-H}}{C}}-O^- \longrightarrow -\overset{|}{\underset{R.HN}{C}}-OH$$

(a) *Ammonia*—When R = hydrogen, the attack is by ammonia and the initial product is a 1-amino-substituted alcohol (an 'aldehyde ammonia'), e.g.

$$CH_3.CHO + NH_3 \rightarrow CH_3.\underset{\underset{NH_2}{|}}{CH}.OH$$

This is seldom isolable, usually either reverting to the original substance or undergoing *condensation polymerization*. For example, methanal (formaldehyde) yields hexamethylenetetramine

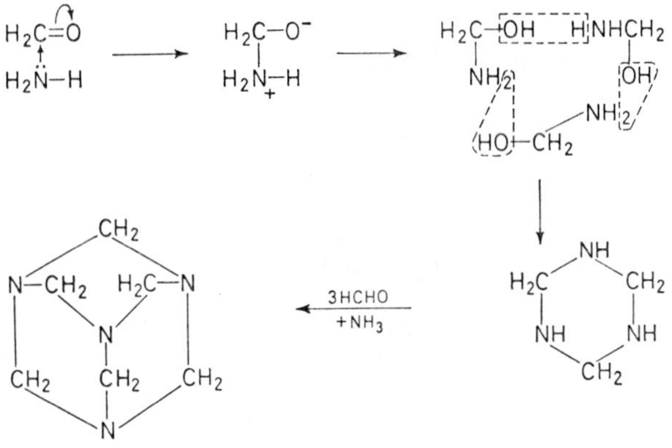

Hexamethylenetetramine

(b) *Amines*—A primary amine is involved if R = alkyl or aryl. Under acidic conditions, the product condenses to give an aldimine or Schiff's base, e.g.

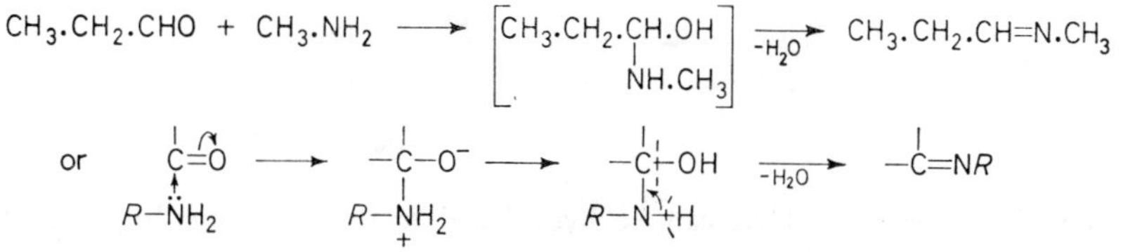

$$CH_3.CH_2.CHO + CH_3.NH_2 \longrightarrow \left[CH_3.CH_2.\underset{\underset{NH.CH_3}{|}}{CH}.OH \right] \xrightarrow{-H_2O} CH_3.CH_2.CH{=}N.CH_3$$

or

$$\underset{R-\ddot{N}H_2}{\overset{|}{C}{=}O} \longrightarrow \underset{R-NH_2}{\overset{|}{C}-O^-} \longrightarrow \underset{\underset{+}{R-N-H}}{-\overset{|}{C}-OH} \xrightarrow{-H_2O} -\overset{|}{C}{=}NR$$

459

(c) *Hydrazines*—In the case of hydrazine, i.e. when $R = NH_2$, condensation leads first to hydrazones and then to azines, e.g.

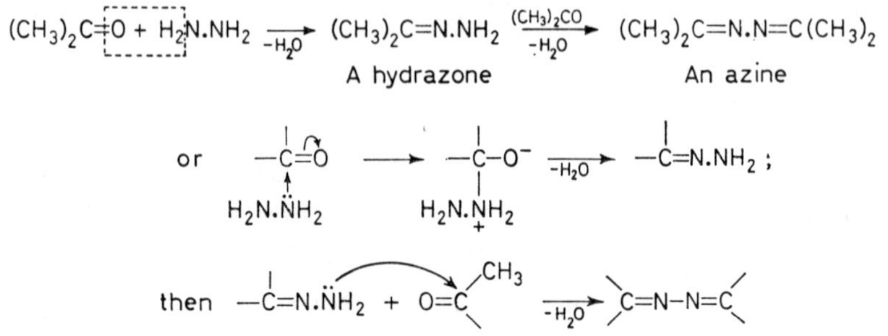

$$(CH_3)_2C{\doteq}O + H_2N.NH_2 \xrightarrow[-H_2O]{} (CH_3)_2C{=}N.NH_2 \xrightarrow[-H_2O]{(CH_3)_2CO} (CH_3)_2C{=}N.N{=}C(CH_3)_2$$

A hydrazone   An azine

$$\text{or} \quad \underset{H_2N.\overset{\cdot\cdot}{N}H_2}{-\overset{|}{\underset{|}{C}}{\doteq}O} \longrightarrow -\overset{|}{\underset{|}{C}}{-}O^- \xrightarrow[-H_2O]{} -\overset{|}{\underset{|}{C}}{=}N.NH_2 \ ;$$

$$\text{then} \quad -\overset{|}{\underset{|}{C}}{=}N.\overset{\cdot\cdot}{N}H_2 + O{=}C\overset{CH_3}{\underset{\diagdown}{\diagup}} \xrightarrow[-H_2O]{} \diagdown C{=}N{-}N{=}C\diagup$$

However, if phenylhydrazine is used, reaction stops at the hydrazone stage, and the product can be used to identify the carbonyl compound involved. Better crystalline products for melting-point determination are obtained if 2,4-dinitrophenylhydrazine is used, e.g.

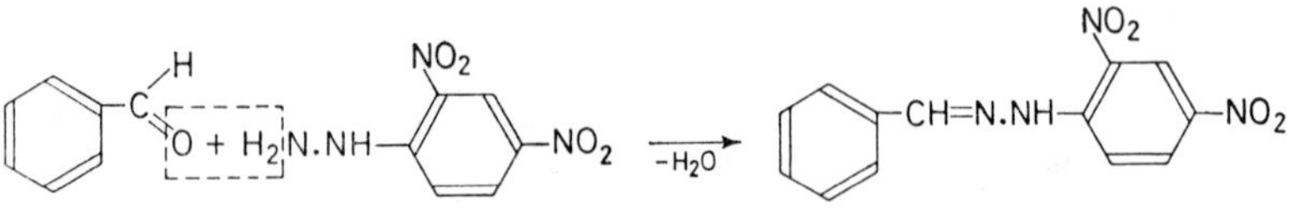

'Benzaldehyde
2,4-dinitrophenylhydrazone'

(d) *Hydroxylamine*—When $R = OH$, oximes are formed which can also be used for the characterization of aldehydes and ketones

$$>C{=}O + H_2N.OH \xrightarrow[-H_2O]{} >C{=}N.OH$$

e.g. $$(CH_3)_2C{=}O + H_2NOH \xrightarrow[-H_2O]{} (CH_3)_2C{=}N.OH$$

Nucleophilic Attack by Carbon

(a) *Cyanide*—An aqueous solution of cyanide ions, in the presence of dilute acid, converts carbonyl compounds into cyanohydrins. Typically, the initial attack is by the nucleophile, CN^-

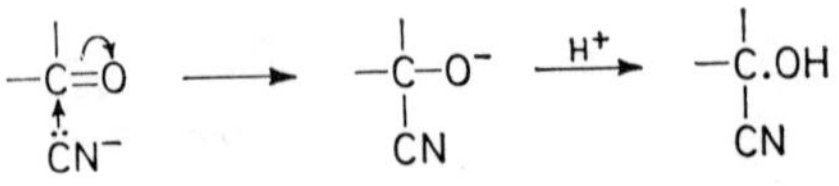

$$\underset{\overset{\cdot\cdot}{C}N^-}{-\overset{|}{\underset{|}{C}}{\doteq}O} \longrightarrow \underset{CN}{-\overset{|}{\underset{|}{C}}{-}O^-} \xrightarrow{H^+} \underset{CN}{-\overset{|}{\underset{|}{C}}.OH}$$

Cyanohydrin

As the nitrile group, —CN, can be hydrolysed to the corresponding carboxylic acid, the cyanohydrin synthesis is an important method of making hydroxy acids (p. 483).

In the case of aromatic aldehydes, the cyanohydrin can release a proton to form a *carbanion*, which then attacks a second molecule of aldehyde. In the case of benzene carbonal (benzaldehyde), ultimate loss of the elements of hydrogen cyanide gives benzoin, and this reaction is therefore known as the *benzoin condensation*

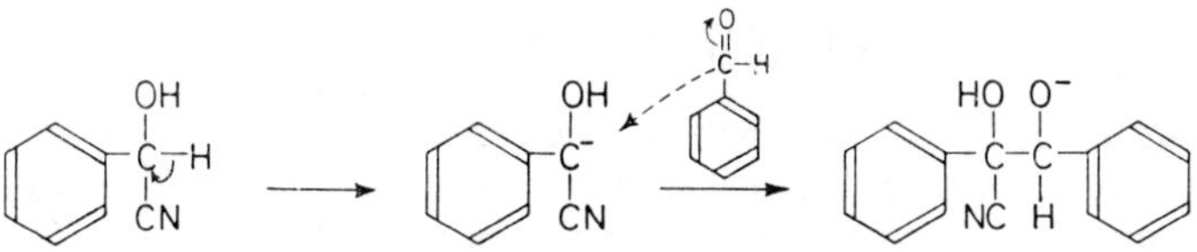

A carbanion

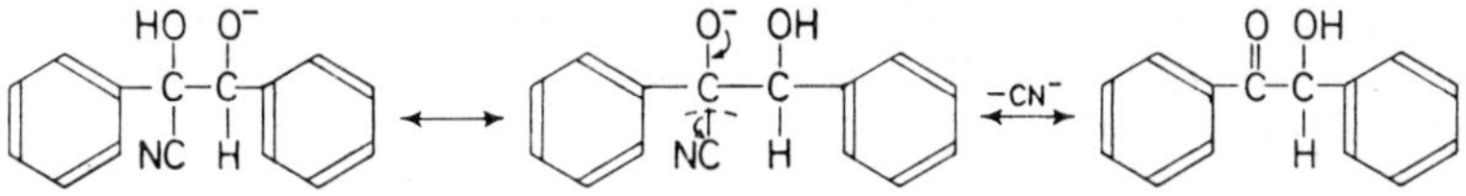

Benzoin

(*b*) *Aldol condensation*—Hydroxide ions are capable of removing a proton from the carbon atom adjacent to the carbonyl group. A carbanion is thereby formed and, being a nucleophile, can then attack a second carbonyl group, e.g. for ethanal

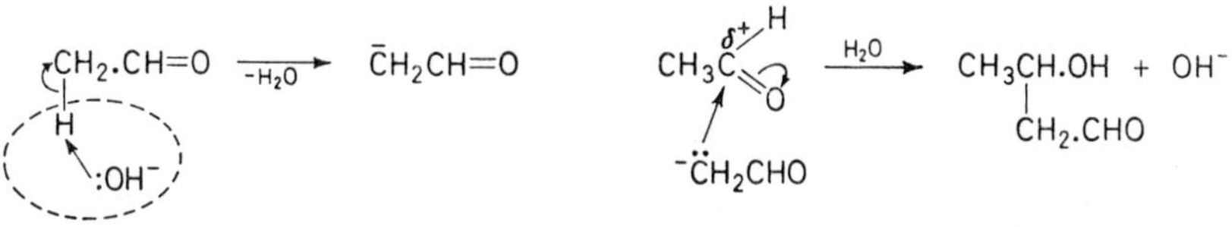

This addition reaction, paradoxically known as the aldol condensation, is a useful method of ascending an homologous series. Hydroxide ions are provided by an aqueous solution of potassium carbonate. In the presence of a large concentration of hydroxide ions, e.g. sodium hydroxide solution, a genuine condensation reaction (*the Claisen condensation*) develops with the formation of an unsaturated carbonyl compound

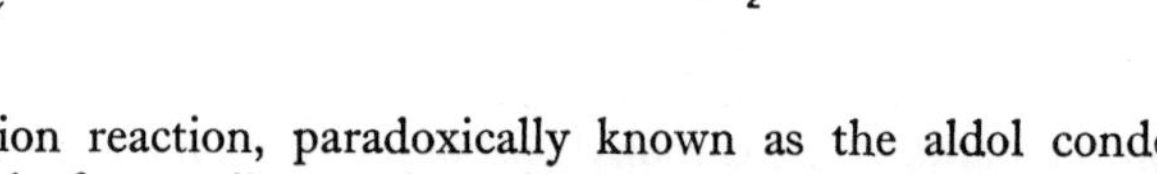

The carbonyl group in the product often perpetuates the process, so that resinous

461

polymers are formed when carbonyl compounds are boiled with sodium hydroxide solution. The possible reactions occurring with hydroxide ions are then, e.g.

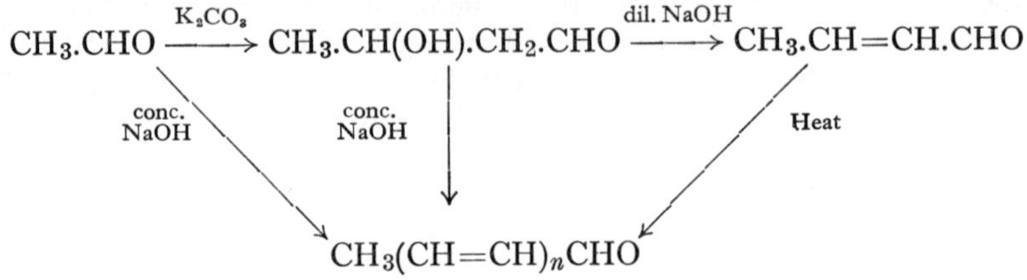

$$CH_3.CHO \xrightarrow{K_2CO_3} CH_3.CH(OH).CH_2.CHO \xrightarrow{dil. NaOH} CH_3.CH=CH.CHO$$

(with conc. NaOH, conc. NaOH, and Heat arrows converging to)

$$CH_3(CH=CH)_nCHO$$

(c) *Cannizzaro reaction*—Where there is no hydrogen attached to the carbon atom adjacent to the carbonyl group, e.g. in methanal and benzenecarbonal (benzaldehyde), a carbanion is not formed. Instead, the reaction in strong alkali takes a different course: there is preliminary attack by the hydroxide ion on the carbon of the carbonyl group, and the ion so formed attacks a second carbonyl group, the final products being a primary alcohol and the carboxylate ion, which mineral acid converts to carboxylic acid. This 'Cannizzaro reaction' is of wide application in aromatic chemistry, e.g.

$$2C_6H_5.CHO + OH^- \longrightarrow C_6H_5.CH_2.OH + C_6H_5.CO.O^-$$
Benzaldehyde · · · · · · · Benzyl alcohol · · · Benzoate

or, in general terms of reaction mechanisms

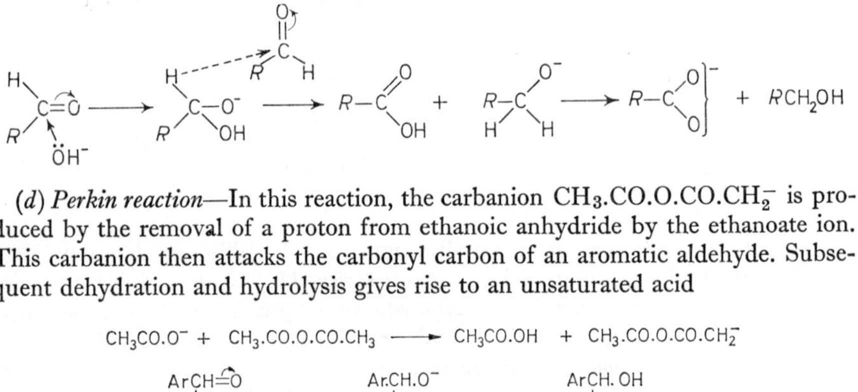

(d) *Perkin reaction*—In this reaction, the carbanion $CH_3.CO.O.CO.CH_2^-$ is produced by the removal of a proton from ethanoic anhydride by the ethanoate ion. This carbanion then attacks the carbonyl carbon of an aromatic aldehyde. Subsequent dehydration and hydrolysis gives rise to an unsaturated acid

$$CH_3CO.O^- + CH_3.CO.O.CO.CH_3 \longrightarrow CH_3CO.OH + CH_3.CO.O.CO.CH_2^-$$

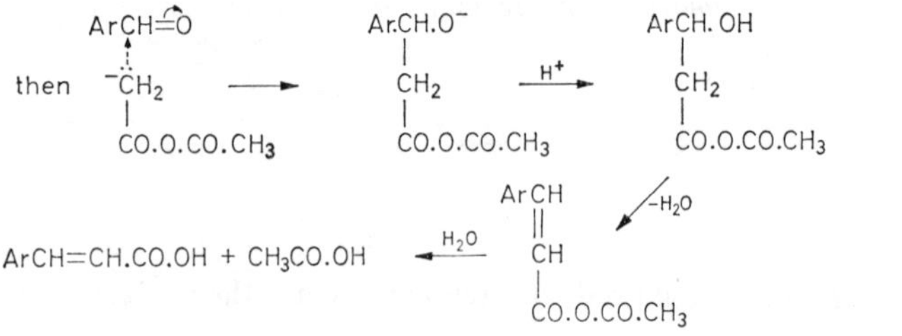

The overall reaction of *benzenecarbonal* (benzaldehyde) with ethanoic (acetic) anhydride and sodium ethanoate (acetate) is

$$C_6H_5.CH\boxed{O + H_2}CH.CO.O.CO.CH_3 \xrightarrow{CH_3.CO.ONa} C_6H_5.CH:CH.CO.OH + CH_3.CO.OH$$

Benzaldehyde   Acetic anhydride   Cinnamic acid   Acetic acid

Nucleophilic Attack by Oxygen

It is believed that aqueous solutions of the lower aliphatic aldehydes contain hydrates, the result of nucleophilic attack by the oxygen of the water molecule

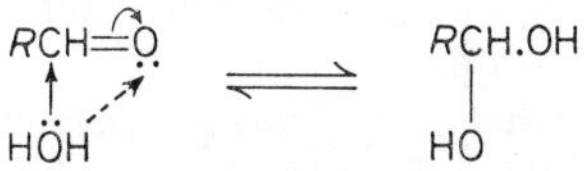

$$\begin{array}{ccc} RCH=\ddot{O} & \rightleftharpoons & RCH.OH \\ \uparrow & & | \\ H\ddot{O}H & & HO \end{array}$$

Attempts to isolate these hydrates result in their reconversion to the original aldehyde. The hydrate of trichloroethanal (chloral hydrate), however, is so stable that it is not even decomposed by moderate heat; the two hydroxyl groups in this compound are stabilized by the inductive effect of the chlorine atoms, and possibly by hydrogen bonding, so that the normal sequence of elimination of the two hydroxyl groups attached to the same carbon atom is prevented

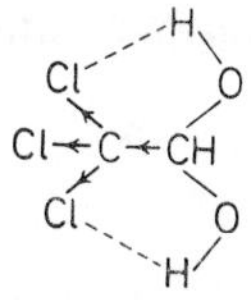

$$\begin{array}{c} Cl \cdots \cdots H \\ \quad \diagdown \quad O \\ Cl \leftarrow C \leftarrow CH \\ Cl \diagup \quad O \\ \quad \diagdown H \end{array}$$

If the nucleophile is an alcohol instead of water, 'acetals' are formed (with aldehydes)

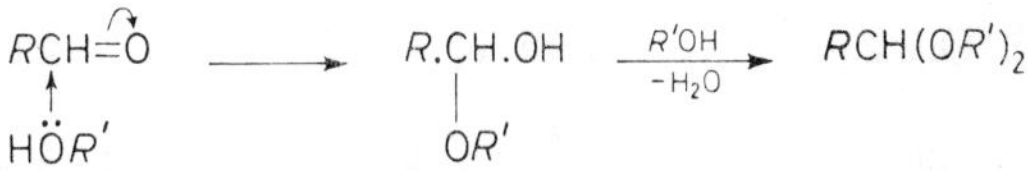

$$\begin{array}{ccccc} RCH=\ddot{O} & \longrightarrow & R.CH.OH & \xrightarrow[-H_2O]{R'OH} & RCH(OR')_2 \\ \uparrow & & | & & \\ H\ddot{O}R' & & OR' & & \\ & & \text{'Hemi-acetal'} & & \text{'Acetal'} \end{array}$$

Nucleophilic Attack by Sulphur

Thiols (mercaptans) react in a manner similar to alcohols

$$RCH=O \xrightarrow[(-H_2O)]{2R'SH} RCH(SR')_2$$

Carbonyl compounds react with a saturated solution of sodium hydrogen

463

sulphite to yield products that can often be readily crystallized. The nucleophile here is the hydrogen sulphite ion

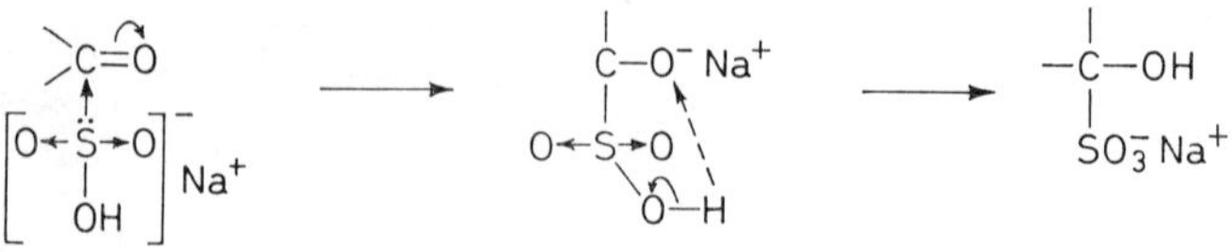

The original substance is easily liberated from the 'bisulphite' compound by treatment with acid or alkali. This reaction is therefore useful for the purification of aldehydes and ketones. An ethereal solution gives, on shaking with a saturated solution of sodium hydrogen sulphite, a solid derivative which can be removed, washed with ether and then hydrolysed to yield a pure product.

Aliphatic aldehydes and propanone (acetone) react so readily with sodium hydrogen sulphite solution that they *quickly* destroy Schiff's reagent (magenta bleached with sulphur dioxide) to give a red coloration.

Reaction with Halogens and Halogeno Compounds

Carbonyl compounds undergo the following reaction in alkaline solution

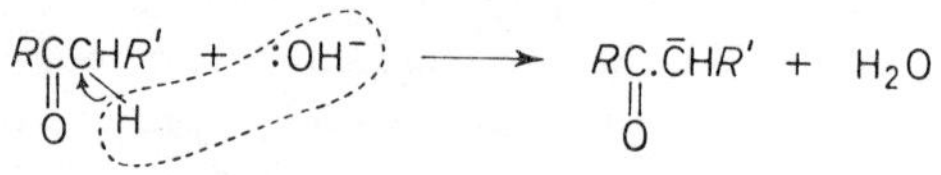

$$RCCHR' + :OH^- \longrightarrow RC.\bar{C}HR' + H_2O$$

The resultant anion undergoes reaction with halogens

$$RC.\bar{C}HR' + X-X \longrightarrow RC.CHR' + X^-$$

In the case of ketones containing the methyl group, i.e. when $R' = H$, the process can continue until all three hydrogen atoms are replaced by halogen. Hydrolysis then yields the haloform, e.g. iodoform (p. 399)

$$RCO.CH_3 \xrightarrow[I_2]{OH^-} RCO.CH_2I \xrightarrow[I_2]{OH^-} RCO.CHI_2 \xrightarrow[I_2]{OH^-} RCO.CI_3 \xrightarrow{OH^-} CHI_3 + RCO.O^-$$

Phosphorus pentachloride replaces the oxygen atom of the carbonyl group by chlorine, giving a dichloride

$$RCHO + PCl_5 \longrightarrow RCHCl_2 + POCl_3$$

Reduction of Aldehydes and Ketones

The carbonyl group can be reduced to different extents by various systems; most dissolving metal systems, e.g. sodium in ethanol, reduce the $>C=O$ to $>CHOH$, thus producing primary alcohols from aldehydes and secondary alcohols from ketones. Zinc amalgam and concentrated hydrochloric acid, however, can effect the complete reduction of >C=O to >CH_2 (the Clemmensen reduction).

The hydride ion, as provided by lithium tetrahydridoaluminate, selectively reduces the carbonyl group, leaving other unsaturated bonds intact, e.g.

$$\text{CH}_2\text{=CHCH}_2 \cdot \overset{\text{H}}{\underset{\uparrow}{\underset{\text{H}-}{\text{C}}}}\text{=O} \longrightarrow \left[\text{CH}_2\text{=CHCH}_2 \cdot \overset{\text{H}}{\underset{\text{H}}{\text{C}}}\text{-O}\right]^{-} \xrightarrow{\text{H}^+} \text{CH}_2\text{=CHCH}_2 \cdot \text{CH}_2\text{OH}$$

Complexed with Al

Reduction to alcohols is also effected by treatment with Grignard reagents, e.g.

$$\text{CH}_3.\underset{\underset{\text{O} \leftarrow\text{--} \text{MgBr}}{\|}}{\text{CH}} + \text{C}_2\text{H}_5 \longrightarrow \text{CH}_3.\underset{\underset{\text{O.MgBr}}{|}}{\text{CH}}.\text{C}_2\text{H}_5 \xrightarrow[-\text{Mg(OH)Br}]{\text{Hydrolysis}} \text{CH}_3.\underset{\underset{\text{OH}}{|}}{\text{CH}}.\text{C}_2\text{H}_5$$

Magnesium, activated by amalgamation, can initiate the reduction of ketones to 'pinacols'

$$\underset{\underset{\text{O}}{\|}}{\overset{\overset{R'}{|}}{R\text{-C}}} + \overset{..}{\text{Mg}} + \underset{\underset{\text{O}}{\|}}{\overset{\overset{R'}{|}}{\text{C-}R}} \longrightarrow \underset{\underset{\text{O-Mg-O}}{|}}{\overset{\overset{R'}{|}}{R\text{-C}}}\underset{\text{}}{\overset{\overset{R'}{|}}{\text{C-}R}} \xrightarrow{\text{Hydrolysis}} \underset{\underset{\text{OH}}{|}}{\overset{\overset{R'}{|}}{R\text{-C}}}\underset{\underset{\text{OH}}{|}}{\overset{\overset{R'}{|}}{\text{C-}R}}$$

The Directive Influence of the Carbonyl Group

The inductive effect of the oxygen atom in the carbonyl group, when attached directly to the benzene ring, results in the withdrawal of electrons from the benzene nucleus. This effect, as usual, is most apparent in positions 2, 4 and 6. Therefore, the benzene ring is deactivated, and substitution by the customary electrophiles takes place slowly, chiefly in positions 3 and 5: i.e. the carbonyl group is meta-directing to electrophilic reagents

CARBONYL COMPOUNDS IN INDUSTRY

It is clear from the foregoing that carbonyl compounds are reactive materials, and consequently they have many commercial uses.

Methanal (formaldehyde) is required in large quantities

(a) for polymerization with, for example, phenol and urea to give thermosetting resins which are useful for making moulded articles and adhesives (pp. 444, 513);

(b) for the manufacture of dyes;

(c) as a disinfectant and preservative (an aqueous solution used for this purpose is known as Formalin);

(d) for hardening gelatine.

Methanal is manufactured by the catalytic dehydrogenation or oxidation, in the vapour phase, of methanol, which is itself prepared from water gas (p. 441):

$$CO + H_2 \longrightarrow CH_3.OH \underset{-H_2}{\longrightarrow} HCHO$$

Ethanal (acetaldehyde), $CH_3.CHO$, is prepared commercially from the other plentiful two-carbon compounds, ethyne (acetylene) and ethene (ethylene).

The elements of water are added catalytically to ethyne by treating it with a dilute solution of mercury(II) sulphate acidified with sulphuric acid and kept at about 75°. The yield is about 95 per cent of the theoretical, but as it employs the highly endothermic ethyne, it is comparatively costly (p. 384)

$$CH{\equiv}CH + H_2O \longrightarrow CH_3.CHO$$

A cheaper route to ethanol involves direct oxidation of ethene in the Wacker process, the catalyst being a solution of palladium(II) chloride and copper(II) chloride. The reaction is very complex, but the net result can be summarized as

$$CH_2{=}CH_2 + O \longrightarrow CH_3.CHO$$

In more detail:

$$PdCl_2 + 2Cl^- \rightleftharpoons [PdCl_4]^{2-} \xrightarrow{\ C_2H_4\ } [PdCl_3C_2H_4]^-$$

$$\downarrow H_2O$$

$$[PdCl_2(OH)C_2H_4]^- \xleftarrow{\ -H^+\ } [PdCl_2(H_2O)C_2H_4]$$

$$-Cl^- \downarrow$$

$$Cl{-}Pd{-}CH_2CH_2OH \rightarrow Pd + HCl + CH_3CHO$$

Alternatively, ethene can be absorbed in cold, concentrated sulphuric acid and the intermediate hydrolysed to ethanol with water (p. 384); the ethanol, like methanol, can be dehydrogenated to the corresponding aldehyde, using a copper catalyst

$$\overset{\displaystyle CH_2}{\underset{\displaystyle CH_2}{\big\|}} \xrightarrow{\ H^+\ } \overset{\displaystyle CH_3}{\underset{\displaystyle +CH_2}{\big|}} \xrightarrow{\ OH^-\ } \overset{\displaystyle CH_3}{\underset{\displaystyle CH_2OH}{\big|}} \xrightarrow{\ -H_2\ } \overset{\displaystyle CH_3}{\underset{\displaystyle CHO}{\big|}}$$

Ethanal is used in the manufacture of ethanoic (acetic) acid, D.D.T. and various drugs.

Propanone (acetone) is now largely manufactured from propene, which results from the cracking of petroleum, by a series of reactions described above for ethanal

$$\underset{\displaystyle \overset{\big\|}{CH_2}}{CH_3.CH} \xrightarrow{\ H_2O\ } \underset{\displaystyle \overset{\big|}{CH_3}}{CH_3.CH.OH} \xrightarrow{\ -24\ } \underset{\displaystyle \overset{\big|}{CH_3}}{CH_3.C{=}O}$$

It is also obtained in considerable quantities as a by-product in the cumene process for phenol (p. 443).

Propanone is used as a solvent, e.g. for paints and ethyne, for the manufacture of chloroform (p. 339) and for making Perspex (p. 515).

Propenal (acrolein) manufacture is also a by-product of the petroleum industry; propene can now be oxidized directly to propenal

$$CH_2{=}CH.CH_3 + O_2 \xrightarrow{\text{Catalyst}} CH_2{=}CH.CHO + H_2O$$

Propenal, possessing a carbonyl group and the alkenic double bond, is very reactive. It is used in the manufacture of ring systems by the Diels–Alder reaction

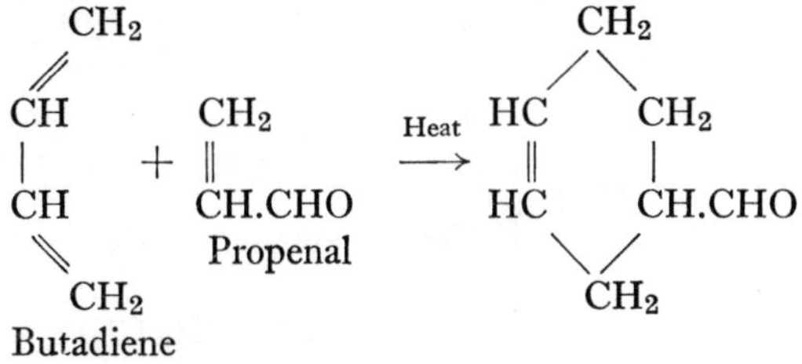

Reaction with aminobenzene (aniline) gives rise to pyridine derivatives

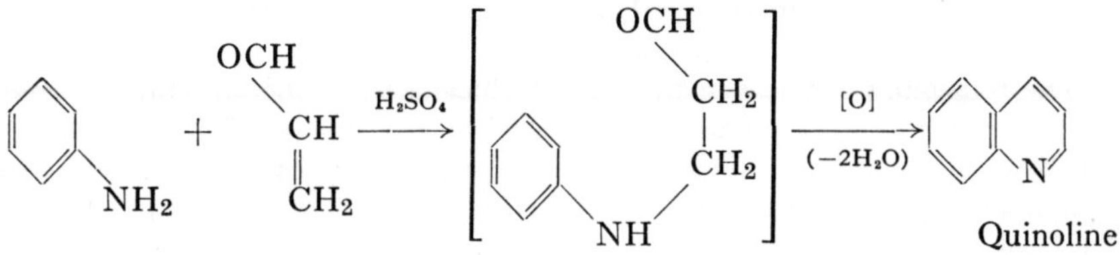

Other commercial applications include conversion to glycerol (p. 442), amino acids and plastics.

Benzenecarbonal (benzaldehyde) is in demand for perfumes, flavouring essences and dyes. The main source is coal tar. Methylbenzene, extracted from the benzol fraction, can be converted to benzenecarbonal in several ways

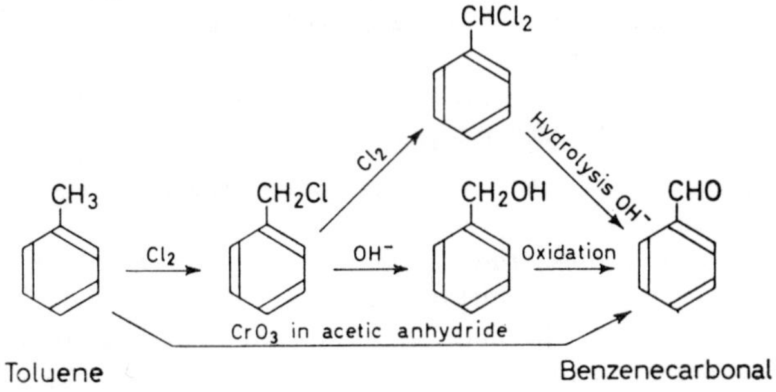

467

ANALYSIS OF CARBONYL COMPOUNDS

The presence of an unmodified carbonyl bond is indicated by the eventual formation of a solid bisulphite compound on shaking with a saturated solution of sodium hydrogen sulphite. Those aldehydes which are soluble in water quickly produce a colour in Schiff's reagent.

The derivative most often prepared is the condensation product with 2,4-dinitrophenylhydrazine.

The molecular formula can be determined by qualitative analysis, followed by quantitative analysis and calculation of the molecular weight, for instance from vapour density measurements. The presence of oxygen in the formula might suggest an alcohol or ether. However, the oxygen atom in a carbonyl group is equivalent to two hydrogen atoms, and so results in the molecule having two fewer hydrogen atoms than the corresponding saturated hydrocarbon, e.g.

$$C_3H_8O \begin{cases} \text{Propan-1-ol, } CH_3.CH_2.CH_2.OH \\ \text{Methoxyethane, } CH_3.CH_2.O.CH_3 \end{cases}$$

Propane, C_3H_8

$$C_3H_6O \begin{cases} \text{Propanal, } CH_3.CH_2.CHO \\ \text{Propanone, } CH_3.CO.CH_3 \end{cases}$$

(If the alkyl group is unsaturated, C_3H_6O also represents the molecular formula for $CH_2{=}CH.CH_2.OH$, prop-2-ene-1-ol.)

The remaining obstacle in the way of elucidation of the structural formula is the determination of the nature of the alkyl or aryl groups. This is usually overcome by degradation to simpler molecules capable of ready recognition and by final unambiguous synthesis of the substance.

CARBOHYDRATES

Carbohydrates make up the major part of the vegetable kingdom and contain potential carbonyl groups. They have the general formula $C_xH_{2y}O_y$—it was the ratio of hydrogen to oxygen that led the French to give them the unfortunate name 'hydrates de carbone'—and may be classified as

(a) *monosaccharides* or simple sugars, the most common of which are called hexoses and have the molecular formula $C_6H_{12}O_6$, e.g. glucose. They are soluble in water and have a sweet taste;

(b) *disaccharides*, which are usually the condensation products of two hexose units

$$2C_6H_{12}O_6 - H_2O \longrightarrow C_{12}H_{22}O_{11}$$

Sucrose (cane sugar) is one such example. They resemble the monosaccharides in sweetness and solubility;

(c) *polysaccharides*, representing the condensation of hundreds and possibly thousands of hexose units. Not surprisingly, in view of the size of their molecules, they are either completely insoluble in water, e.g. cellulose, or present in the colloidal form only, e.g. starch.

CARBOHYDRATES

It is important to realize that carbohydrates have a common origin (see Photosynthesis, p. 475) and are interconvertible to a large extent, simple hexoses being transformed to complex molecules by the process of condensation, and polysaccharides converted to simple sugars by hydrolysis, e.g.

$$C_{12}H_{22}O_{11} + H_2O \xrightarrow[\text{or enzymes}]{H^+} 2C_6H_{12}O_6$$

Structures of Carbohydrates

Hexoses—Hydroxyl groups are attached to five of the six carbon atoms, and the sixth forms a carbonyl group; if this latter is a terminal atom, then the carbonyl group is part of the aldehyde function, and the sugar is known as an *aldose*

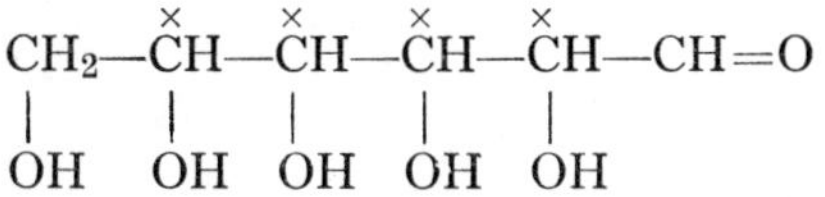

$$\overset{}{CH_2}-\overset{\times}{CH}-\overset{\times}{CH}-\overset{\times}{CH}-\overset{\times}{CH}-CH=O$$
$$\;\;\;|\quad\;\;|\quad\;\;|\quad\;\;|\quad\;\;|$$
$$\;\;OH\;\;\;OH\;\;\;OH\;\;\;OH\;\;\;OH$$

There are four asymmetric carbon atoms present (marked $\times$), and therefore 2^4 or 16 enantiomorphs exist. Two of these are (+)- and (−)-glucose:

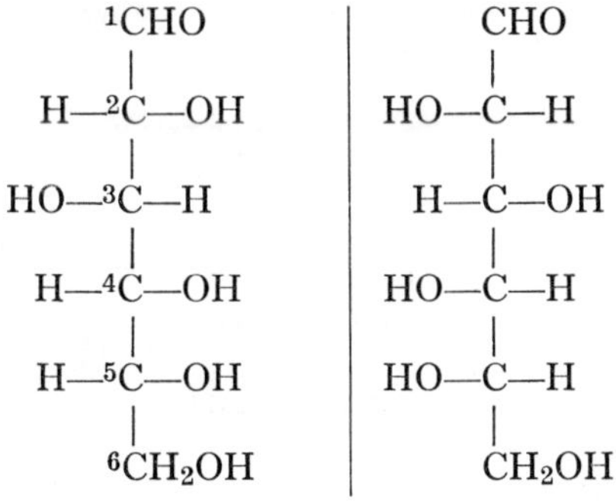

(+) (or D)-Glucose | (−)(or L-)-Glucose

It is now known that most of the hexose unit is in the form of a ring, usually 6- but sometimes 5-membered, brought about by addition to the double bond of the carbonyl group

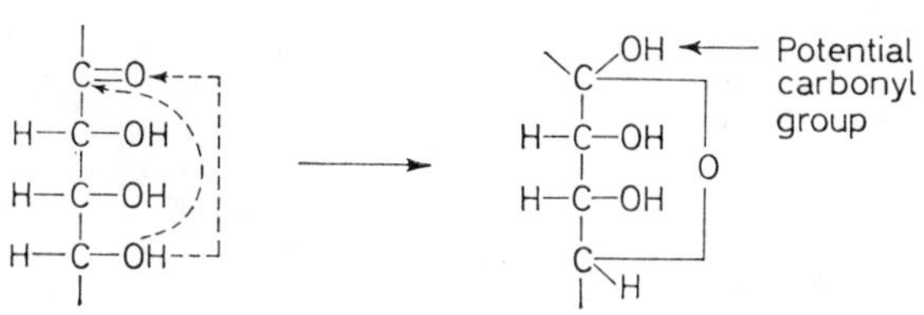

469

A further asymmetric carbon atom is therefore introduced and the number of optical isomers doubled. In the case of glucose, the possible forms are $\alpha(+)$, $\beta(+)$, $\alpha(-)$ and $\beta(-)$. The α and β forms are readily interconvertible, by the opening of the ring of one form and its reclosure into the other form. Some of the open form is always present in solution, and so hexoses have the properties of carbonyl compounds, although the possibility of modification by the hydroxyl groups must be borne in mind

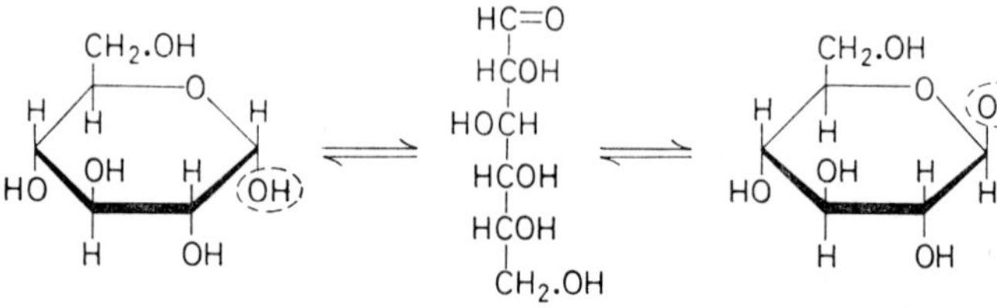

$\alpha(+)$-Glucose   (+)-Glucose $\beta(+)$-Glucose

Fructose is a reducing sugar, similar to glucose but containing a keto group instead of an aldehyde group. $(-)$-Fructose forms the same osazone (p. 474) as $(+)$-glucose, showing that the arrangement about the carbon atoms 3–6 is identical with that of $(+)$-glucose. The structure of fructose is

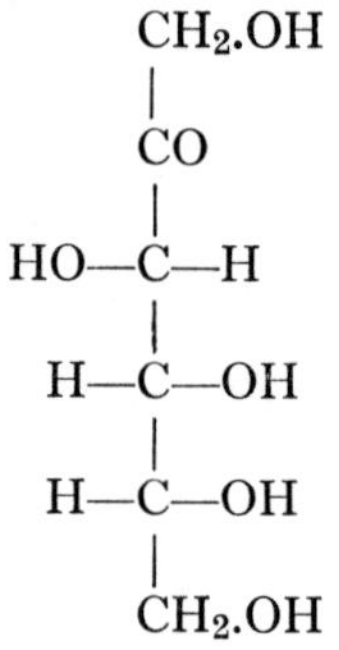

$$
\begin{array}{c}
CH_2.OH \\
| \\
CO \\
| \\
HO-C-H \\
| \\
H-C-OH \\
| \\
H-C-OH \\
| \\
CH_2.OH
\end{array}
$$

which in the free state exists also in the 6-membered ring structure

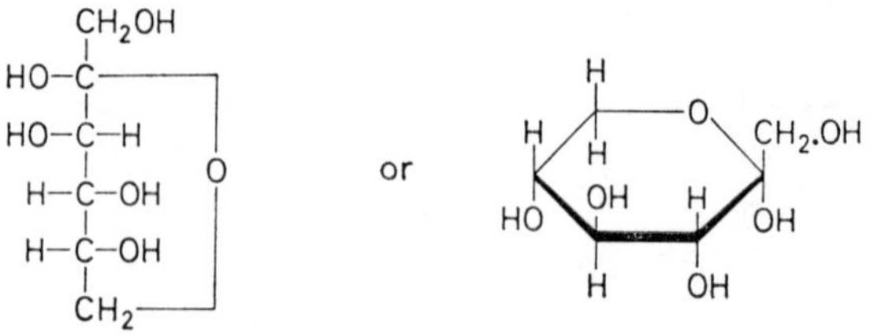

or

but in compounds, fructose often is in a 5-membered ring modification, e.g.

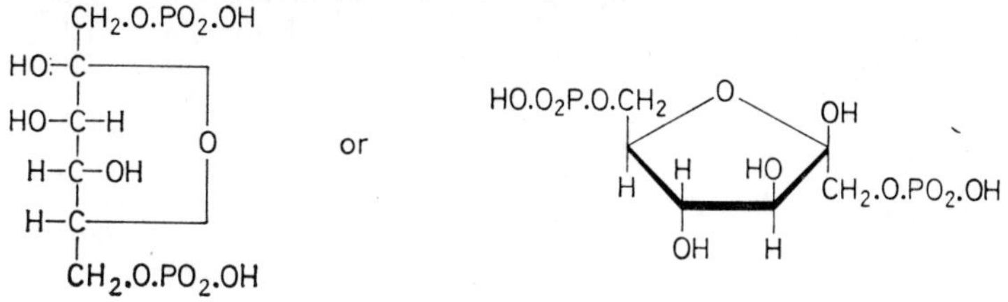

Fructose –1,6– diphosphate

Vitamin C (ascorbic acid), lack of which leads to nervous disorders and skin diseases, is derived from a hexose. The structure of the vitamin is

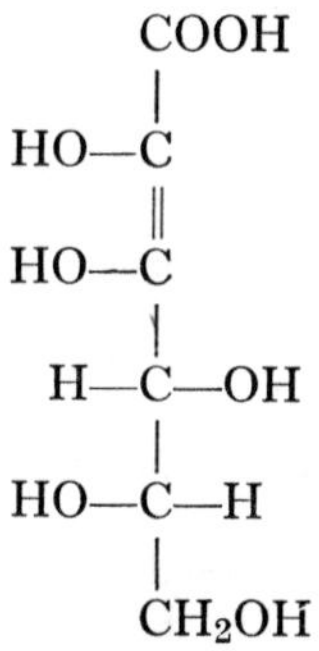

$$
\begin{array}{c}
\text{COOH} \\
|\\
\text{HO—C} \\
||\\
\text{HO—C} \\
|\\
\text{H—C—OH} \\
|\\
\text{HO—C—H} \\
|\\
\text{CH}_2\text{OH}
\end{array}
$$

and it normally exists as an internal ester (or lactone)

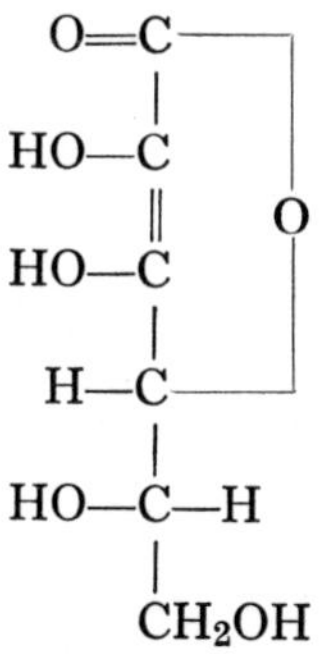

$$
\begin{array}{c}
\text{O=C} \\
|\\
\text{HO—C} \\
||\quad \text{O}\\
\text{HO—C} \\
|\\
\text{H—C} \\
|\\
\text{HO—C—H} \\
|\\
\text{CH}_2\text{OH}
\end{array}
$$

Disaccharides—One of the properties of the alcohol group is the ability to eliminate the elements of water to form an ether link

$$R-O+H \ + \ H-O+R' \longrightarrow R-O-R' + H_2O$$

The same sort of condensation results in monosaccharides giving rise to disaccharides. Provided that both carbonyl groups are not involved in the linkage, the product still retains some properties of an aldehyde or ketone, e.g. maltose

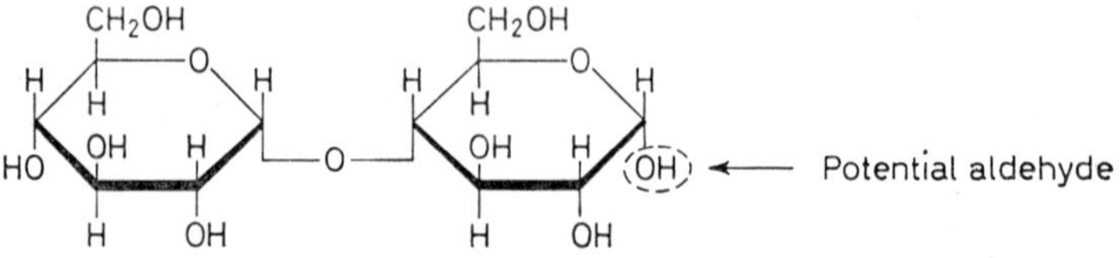

Potential aldehyde

$\alpha(+)$-Glucose $\qquad$ $\alpha(+)$-Glucose

If, on the other hand, the linkage involves both potential carbonyl groups, then the sugar is non-reducing, e.g. sucrose

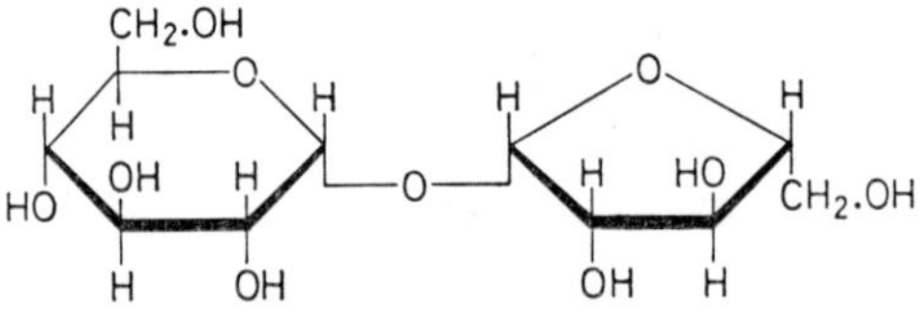

$\alpha(+)$-Glucose $\qquad$ $\beta(+)$-Fructose

Polysaccharides—Cellulose, so called because it is the chief component of the walls of the plant cell, is made up solely of $\beta(+)$-glucose units, the number involved being perhaps as many as **4 000**

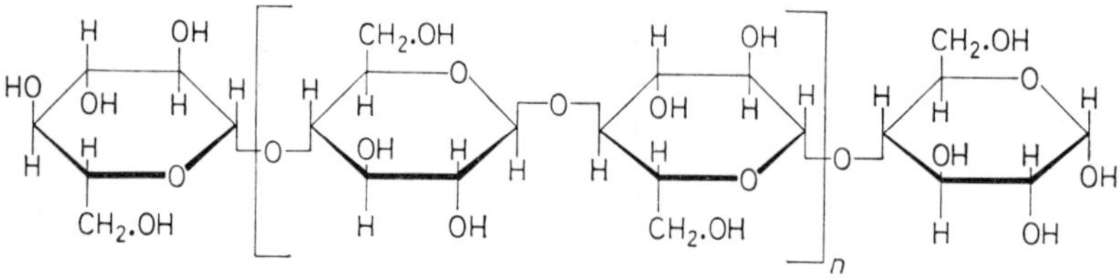

Starch, on the other hand, contains $\alpha(+)$-glucose units, partly at least in the form of a linear polymer

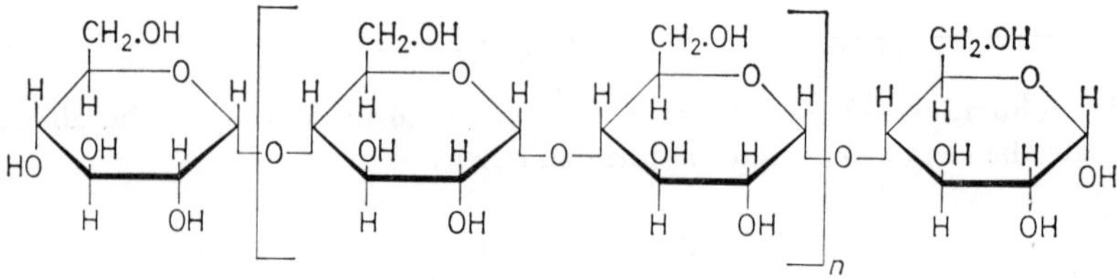

Starch represents the main energy reservoir in plants and animals; prior to its utilization in the respiratory processes of the cells (p. 477), it is hydrolysed first to maltose and then to glucose. Cellulose, however, cannot be digested in those animals which are not hosts to the intestinal bacteria that provide the cellulase enzymes; the cow is thus able to extract far more than the human being from grass. This point illustrates the specificity of enzyme action, when it is realized that the main difference between starch and cellulose is merely the configuration of one of the six carbon atoms of the hexose unit.

Glycosides, which occur in plants, are another important source of carbohydrates. On hydrolysis they split into saccharides and other substances (called aglycones): if the saccharide is glucose then the glycoside is called a glucoside, e.g.

$$\text{simple tannins} \xrightarrow{\text{Hydrolysis}} \text{glucose units} + \text{2,3,4-trihydroxybenzene carboxylic acid}$$
$$\text{(gallic acid)}$$

Properties of Carbohydrates

Esterification—Carbohydrates possess the characteristic alcoholic capacity for esterification. Acetic anhydride produces acetyl derivatives which are useful as a means of locating the positions of hydroxyl groups, and hence the carbon atoms, involved in the ether-type link of the ring system. Esters with phosphoric acid are fundamental in the processes of respiration and photosynthesis (pp. 476, 477). For instance, adenosine represents the condensation products of $(+)$-ribose (a pentose) with the base adenine; further condensation of the sugar with phosphoric acid gives rise to the substances adenosine di- and tri-phosphate (ADP and ATP) which play such a vital part in the energy transformations in the living cell:

Oxidation—Strong oxidation converts the terminal groups into carboxylic acids; e.g. glucose gives saccharic acid

$$\begin{array}{ccc} \text{CHO} & & \text{CO.OH} \\ | & \xrightarrow{\text{HNO}_3} & | \\ \text{(CH.OH)}_4 & & \text{(CH.OH)}_4 \\ | & & | \\ \text{CH}_2\text{.OH} & & \text{CO.OH} \end{array}$$

473

whilst mild oxidation affects the carbonyl groups only; e.g. glucose gives gluconic acid

$$
\begin{array}{ccc}
\text{CHO} & & \text{CO.OH} \\
| & & | \\
(\text{CH.OH})_4 & \xrightarrow{\text{Bromine water}} & (\text{CH.OH})_4 \\
| & & | \\
\text{CH}_2.\text{OH} & & \text{CH}_2.\text{OH}
\end{array}
$$

Carbohydrates which possess latent carbonyl groups and which are soluble in aqueous media, give a silver mirror with diammine silver(I) hydroxide solution

$$R\text{CHO} + [\text{Ag(NH}_3)_2]^+\text{OH}^- \rightarrow R\text{CO.OH} + \text{Ag} \downarrow$$

Reduction—Powerful reducing agents convert the carbonyl group to an alcohol

$$
\begin{array}{ccc}
\text{CHO} & & \text{CH}_2.\text{OH} \\
| & & | \\
\text{HO—C—H} & & \text{HO—C—H} \\
| & & | \\
\text{HO—C—H} & \xrightarrow{\text{P+HI}} & \text{HO—C—H} \\
| & & | \\
\text{H—C—OH} & & \text{H—C—OH} \\
| & & | \\
\text{H—C—OH} & & \text{H—C—OH} \\
| & & | \\
\text{CH}_2.\text{OH} & & \text{CH}_2.\text{OH} \\
\text{Mannose} & & \text{Mannitol}
\end{array}
$$

Identification—The presence of the carbonyl group is revealed by formation of oximes (p. 460) as well as by the reactions mentioned above.

Phenylhydrazine forms hydrazones by condensation, but these are not the ultimate products; an adjacent alcohol group is oxidized to carbonyl, which then condenses with more phenylhydrazine to give a double hydrazone or *osazone*. These derivatives are often of characteristic crystalline shape and can be useful in identifying a particular sugar:

$$
\begin{array}{ccc}
\text{HC=O} & & \text{CH=N.NH.C}_6\text{H}_5 \\
| & & | \\
\text{CH.OH} & & \text{C=N.NH.C}_6\text{H}_5 \\
| & & | \\
& & \uparrow \quad \text{Osazone} \\
\text{H}_2\text{N.NH.C}_6\text{H}_5 \Big| (-\text{H}_2\text{O}) & & \\
\downarrow & & \\
\text{CH=N.NH.C}_6\text{H}_5 & \xrightarrow[(-2\text{H})]{\text{H}_2\text{N.NH.C}_6\text{H}_5} & \text{CH=N.NH.C}_6\text{H}_5 \\
| & & | \\
\text{CH.OH} & & \text{C=O} \\
| & & | \\
\text{Hydrazone} & & (+\text{C}_6\text{H}_5.\text{NH}_2 + \text{NH}_3)
\end{array}
$$

PHOTOSYNTHESIS

Carbohydrates are produced naturally via photosynthesis. Basically, this process consists of the reduction, involving photochemical activation with chlorophyll as catalyst, of carbon dioxide by water. In the simplest terms it can be represented as

$$6CO_2 + 6H_2O \xrightarrow[\text{Energy}]{\text{Light}} C_6H_{12}O_6 + 6O_2 \qquad \Delta H = 2\,810 \text{ kJ}$$
$$\text{Efficiency} \sim 1 \text{ per cent}$$

This equation is misleading in its simplicity, since all of the oxygen evolved comes from the water and not from the carbon dioxide. In recent years, however, the use of radioactive tracers and other new techniques has permitted a much more detailed picture of the mechanism of photosynthesis to be drawn.

It is a significant fact that photosynthesis involves a relatively minute part of the electromagnetic spectrum. Chlorophyll appears green because it absorbs from white light the radiation from the red end of the spectrum. It is now believed that the energy of this radiation activates an electron from the chlorophyll, so that it can combine with a proton from water to give a hydrogen atom. The hydroxide ion remaining from the water molecule loses a low-energy electron to the electron-deficient chlorophyll molecule and then decomposes to water and oxygen. This part of the light-activated process can be summarized as

$$H_2O \longrightarrow H^+ + OH^-$$

$$H^+ + e \longrightarrow H\cdot \qquad OH^- - e \longrightarrow OH$$

$$4OH \longrightarrow 2H_2O + O_2$$

The hydrogen atom, in combination with enzyme systems, reduces 3-phospho-glyceric acid to 3-phosphoglyceraldehyde, which is converted to fructose-1,6-diphosphate. This either leaves the cycle as the precursor of carbohydrates and other essential substances or combines with phosphoglyceraldehyde to give, by various routes, a 5-carbon compound (ribulose-5-phosphate) which reacts with carbon dioxide to produce 3-phosphoglyceric acid and so complete the cycle (*Figure 24.1*). Clearly one molecule of fructose-1,6-diphosphate is formed for six molecules of carbon dioxide entering the cycle. The role of ATP and enzymes as catalysts in these processes must not be overlooked; for example, ATP activates molecules by phosphorylating them.

RESPIRATION

Photosynthesis is essentially the means of manufacturing foods, and appears important only when viewed against that process whereby the foods are consumed and energy (originally sunlight) is liberated. This process is known as respiration

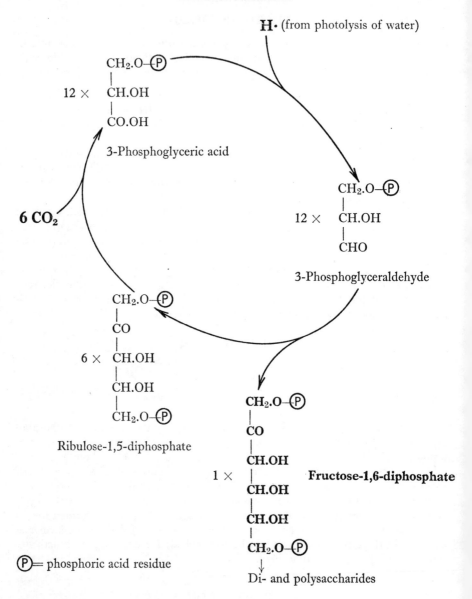

Figure 24.1. Photosynthesis (much simplified)

i.e. 1 molecule of fructose-1,6-diphosphate is manufactured from 6 'revolutions' of the cycle.

476

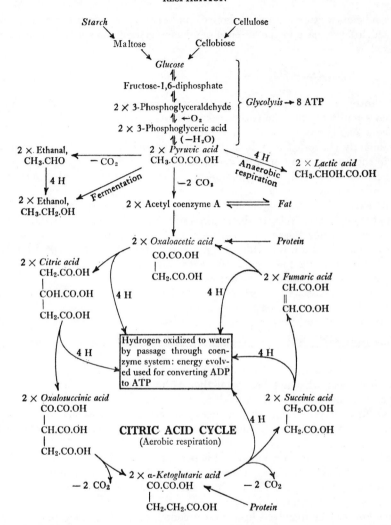

1 molecule of glucose is oxidized aerobically to carbon dioxide and water by two 'revolutions' of the citric acid cycle: oxidation of hydrogen in the course of this results in the manufacture of $(5 \times 4)/2 \times 3 = 30$ molecules of ATP. As 8 molecules are synthesized by glycolysis, the total number of molecules of ATP produced is 38, i.e.

$$C_6H_{12}O_6 + 6 O_2 \longrightarrow 6 CO_2 + 6 H_2O + energy \text{ (38 ATP)}$$

For the conversion of 1 mole ATP into ADP (adenosine diphosphate) 50 kJ energy are evolved.

∴ energy obtained from glucose $= 38 \times 50 \text{ kJ mol}^{-1} \approx 10 \cdot 5 \text{ kJg}^{-1}$

Figure 24.2. Respiration (simplified)

and can be of two types: *aerobic*, where oxygen is absorbed and utilized, and *anaerobic*, which takes place in the absence of oxygen. Aerobic respiration can be summarized as

$$C_6H_{12}O_6 + 6O_2 \longrightarrow 6CO_2 + 6H_2O \qquad \Delta H = -2810 \text{ kJ}$$

In this form, aerobic respiration is seen to be the reverse of photosynthesis, although this does not imply that the entire mechanism is reversed.

Figure 24.2 shows that starch and cellulose are hydrolysed to glucose, which is subsequently converted into 2-oxopropanoic (pyruvic) acid. In anaerobic respiration in the human body, pyruvic acid is changed into 2-hydroxypropanoic (lactic) acid; in the case of the micro-organism, yeast, it is transformed into ethanol, a process known as *fermentation*. In aerobic respiration, the pyruvic acid loses carbon dioxide and interacts with coenzyme A to give 'acetyl coenzyme A' which then enters the citric acid cycle. In this, carbon dioxide is eliminated, together with hydrogen, which after passage through a series of enzymes is oxidized to water. The energy released is used for synthesizing high-energy ATP from ADP. Protein can also be used in respiration, the amino acids resulting from hydrolysis being eventually absorbed into the citric acid cycle.

It should be realized that, because many of the processes mentioned are reversible, they can be part of both respiratory and photosynthetic processes. For example, fructose-1,6-diphosphate can appear as the result of photosynthesis from 3-phosphoglyceraldehyde or from the respiratory degradation of starch.

Commercial applications of carbohydrates
It is clear from what has been said that carbohydrates are an essential part of animal diet and that they are provided, in the first place, by plants. Starch is the reserve food of plants and is particularly concentrated in tubers, from some of which, e.g. potatoes, it is extracted. Sucrose, the most common sugar, is extracted from sugar cane and beet.

Cellulose is the main component of paper and for this purpose is obtained from wood by removal of the lignin, a complex polysaccharide, with solutions containing the hydrogen sulphite ion. The resultant slurry is converted into paper by the controlled removal of water.

Cellulose can be dissolved in solutions containing tetramminecopper (II) ions. If the solution is then forced through a narrow jet into a bath of acid, the tetramminecopper (II) complex is decomposed and the cellulose regenerated. This process therefore results in the production of cellulose in the form of filaments, suitable for use in the textile industry, where it is known as *cuprammonium rayon*. *Viscose rayon* is the product obtained by treating cellulose first with concentrated alkali and then with carbon disulphide when 'cellulose xanthates', $RO.CS.SNa$, are formed; these too are reconverted into cellulose by subsequent treatment.

Modified, as well as regenerated, cellulose is of wide application. Nitration to varying extents gives products ranging from *pyroxylin* (with up to two nitrate groups per hexose unit) to *gun cotton* (up to three nitrate groups per hexose unit) The former is used as the base of cellulose paints and the latter as an explosive

$$-\overset{|}{\underset{|}{C}}.OH + HO.NO_2 \xrightarrow[-H_2O]{} -\overset{|}{\underset{|_\bullet}{C}}.O.NO_2$$

Acetates, of greater stability than the nitrates, are manufactured by esterifying cellulose with acetic acid and are used in the manufacture of photographic films.

Fermentation processes can yield a wide variety of products depending on the conditions, e.g. pH and temperature, and the carbohydrate and micro-organism employed. The conversion of starch to ethanol has been mentioned already under the heading 'Respiration'; in the Weizmann process, *Clostridium acetobutylicum* converts starch into butan-1-ol and propanone.

Perhaps the chief potential of fermentation processes lies in the fact that the fermentation is roughly contemporaneous with the growth of the plant producing the carbohydrate, in sharp contrast with the coal and petroleum industries, where the interval might be 200 million years: the long-term implications are clear.

SUMMARY

Many reactions of the carbonyl group involve the addition of a nucleophile at the carbon atom; addition is often followed by elimination and the overall process is then known as condensation.

The more common reactions of ethanal are summarized in *Figure 24.3.*

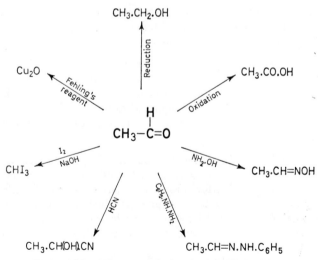

Figure 24.3. The more common reactions of ethanal

479

QUESTIONS

1. Give the equations and state the conditions for each of the processes shown in the flow chart.

$$(CH_3)_2C{=}N.NH_2 \qquad\qquad\qquad CH_3.CCl_2.CH_3$$

$$CH_3.CO.CH_3$$

$$(CH_3)_2CH.OH \qquad\qquad\qquad CHI_3$$

$$(CH_3)_2C(OH).CH_2.CO.CH_3$$

2. How might propanone be converted into: (a) 3-methyl-2-hydroxybutanoic acid; (b) propene; (c) 2-methyl-1-aminopropane?

3. Give the structures of the isomeric aldehydes and ketones of molecular formula C_4H_8O and suggest chemical methods for distinguishing between them.

4. Glycerol, on heating with anhydrous magnesium sulphate, yields an acrid-smelling distillate of molecular formula C_3H_4O, which immediately decolorizes bromine water and also restores the colour to Schiff's reagent. Indicate a possible structure for the compound and suggest its mode of formation from glycerol.

5. When propanone is distilled with concentrated sulphuric acid, a hydrocarbon of molecular formula C_9H_{12} is formed, which on oxidation gives $C_9H_6O_6$, an acid, of which 0·85 g required 12·2 cm³ of sodium hydroxide solution of concentration 4·0 g dm⁻³ for neutralization. Suggest a possible structure for, and mode of formation of, the hydrocarbon.

6. Ethanal, on treatment with concentrated sulphuric acid, gives $(CH_3.CHO)_3$, a trimer, and $(CH_3.CHO)_4$, a tetramer. Methanal also gives a trimer on standing. Suggest cyclic structures for these polymers.

7. When the vapour of an acid A, of empirical formula CH_2O, was passed over heated aluminium oxide, compound B, of empirical formula C_3H_6O, was produced. This had a vapour density of 29; it reacted with 2,4-dinitrophenylhydrazine to give an orange precipitate of C. Compound B did not reduce diammine silver(I) ions to metallic silver. Deduce the nature of A, B, and C and write equations to illustrate the above reactions.

8. The compound A, C_7H_9N, when reacted with chlorine, yielded B, $C_7H_7Cl_2N$; this was readily hydrolysed to C, C_7H_7NO. C oxidized easily on standing to D, $C_7H_7NO_2$, which exhibited a few properties characteristic of a carboxylic acid. C also reacted with nitrous acid to give a product which, after boiling with ethanol, gave a compound E, C_7H_6O. In the presence of traces of the cyanide ion this reacted to produce F, having the same empirical formula as E and a vapour density of 106.

Elucidate the reactions and name the compounds A–F.

9. A naturally occurring compound has the molecular formula $C_{14}H_{18}O_8$. On hydrolysis, it yields a hexose and a phenol of formula $C_8H_8O_3$. Treatment of the latter with HI and subsequent oxidation gives 3,4-dihydroxybenzenecarboxylic acid. Suggest a possible structure for the original compound.

10. Trace the chief energy changes in the conversion of sunlight to muscular contraction. How far is it true to say that the concentration of carbon dioxide in the atmosphere remains constant?

11. Trace the steps involved in the following syntheses.

(a) Benzene with ethanoyl chloride in the presence of anhydrous aluminium chloride, produces substance **A**. **A**, treated with methyl magnesium iodide gives **B**, which on hydrolysis forms **C**. **C**, on heating, gives **D**, which on reduction gives **E**.

(b) Ethene produces substance **F** on treatment with hypochlorous acid. **F**, with potassium cyanide forms **G**, which with alkali gives **H**.

(c) Cyclohexanol treated with concentrated sulphuric acid gives **J**, which with bromine forms **K** (molecular formula $C_6H_{10}Br_2$). **K** reacts with an excess of dimethylamine to form **L** (molecular formula $C_8H_{15}N$). **L** reacts with iodomethane to give a salt **M**, which reacts with silver 'hydroxide' to form **N** (molecular formula $C_9H_{19}NO$). **N** on heating forms **P**, a cyclohexadiene. By repeating the total process show how benzene may ultimately be prepared.

12. Discuss some of the evidence available for the mechanisms of respiration and photosynthesis suggested in this Chapter.

13. Explain why acetals are produced in the presence of dry hydrogen chloride but decomposed by dilute hydrochloric acid.

14. Write an essay on nucleophilic substitution.

CHAPTER 25

CARBOXYLIC ACIDS

Carboxylic acids are difunctional compounds, containing both the *carb*onyl and hydr*oxyl* groups, hence the name. Furthermore, both of these groups are attached to the same carbon atom; the inductive effect of the two oxygen atoms is sufficiently powerful for the hydrogen of the hydroxyl group to be released as a proton, so that these substances are acidic, although very weak in comparison with the common mineral acids:

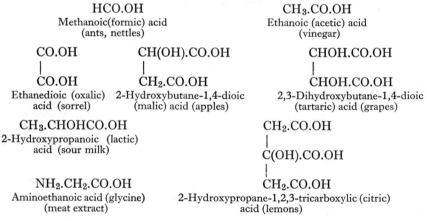

Other functional groups are often also present, for example, —NH_2 in amino acids and a second hydroxy group in hydroxy acids.

Carboxylic acids are named by replacing the final *-e* of the hydrocarbon by *-oic acid*.

Occurrence

Carboxylic acids are of widespread occurrence. Thus methanoic (formic) acid is a common irritant in both the plant and animal kingdoms (Lat. *formica* = ant); ethanoic (acetic) acid represents the end product of one type of fermentation of carbohydrates (when it is known as vinegar); ethanedioic (oxalic) acid is present in the Oxalis group of plants, e.g. sorrel; hydroxy acids, e.g. citric and tartaric acids, are the characteristic acids of citrus fruits, and amino acids are the building units of proteins:

HCO.OH
Methanoic(formic) acid
(ants, nettles)

CH_3.CO.OH
Ethanoic (acetic) acid
(vinegar)

CO.OH
|
CO.OH
Ethanedioic (oxalic)
acid (sorrel)

CH(OH).CO.OH
|
CH_2.CO.OH
2-Hydroxybutane-1,4-dioic
(malic) acid (apples)

CHOH.CO.OH
|
CHOH.CO.OH
2,3-Dihydroxybutane-1,4-dioic
(tartaric) acid (grapes)

CH_3.CHOHCO.OH
2-Hydroxypropanoic (lactic)
acid (sour milk)

CH_2.CO.OH
|
C(OH).CO.OH
|
CH_2.CO.OH
2-Hydroxypropane-1,2,3-tricarboxylic (citric)
acid (lemons)

NH_2.CH_2.CO.OH
Aminoethanoic acid (glycine)
(meat extract)

482

Methods of Preparation

1) As carboxylic acids are the products of oxidation of aldehydes and ketones, they can be obtained by refluxing these substances with acidified sodium dichromate and then distilling to separate the constituents of the reaction mixture, e.g.

$$RCH \xrightarrow{[O]} RC.OH$$
$$\parallel \qquad\quad \parallel$$
$$O \qquad\quad\ O$$

In view of the fact that aldehydes and ketones are themselves obtained by oxidation of primary and secondary alcohols, respectively, it is more usual to derive the acids directly from alcohols (or even, in some cases, from hydrocarbons)

$$CH_3.CH_2.CH_2.OH \xrightarrow{[O]} CH_3.CH_2.CO.OH$$
Propan-1-ol $\qquad\qquad\qquad$ Propanoic acid

Oxidation of aromatic hydrocarbons results in side chains being converted into carboxylic acid groups attached directly to the aromatic ring, as for example

$$C_6H_5.C_2H_5 \xrightarrow{[O]} C_6H_5.CO.OH$$
Ethylbenzene $\qquad$ Benzene carboxylic (benzoic) acid

Compound	$RCH_3 \to RCH_2.OH$	$\xrightarrow{-2[H]}$	$RCHO$	$\xrightarrow{[O]}$	$RCO.OH$
Oxidation state of $R.C-$	-3	-1	$+1$		$+3$

2) Cyano derivatives (nitriles) are converted into acids by hydrolysis in the presence of acid or alkali

$$CH_3.CN \xrightarrow{H_2O} CH_3.CO.OH$$
Cyanomethane $\quad$ Ethanoic acid

or, in more detail

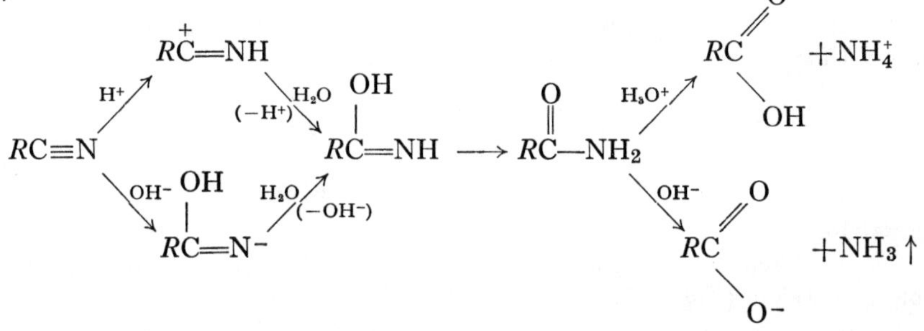

Carbonyl compounds react with the cyanide ion to give cyanohydrins (p. 460); hydrolysis of these compounds results in the formation of hydroxy-acids, e.g.

$$CH_3.CH \xrightarrow{HCN} CH_3.CH.CN \xrightarrow{H_2O} CH_3.CH.CO.OH + NH_3 \uparrow$$
$$\parallel \qquad\qquad\qquad | \qquad\qquad\qquad\qquad |$$
$$O \qquad\qquad\qquad OH \qquad\qquad\qquad\qquad OH$$
Ethanal $\qquad\qquad\qquad\qquad\qquad$ 2-Hydroxypropanoic (lactic) acid

483

3) Acids are often one of the products of the hydrolysis of natural products. When butter goes rancid, butanoic acid is one of the acids produced, whilst the turning sour of milk is caused by the appearance of lactic acid.

Fats and oils are usually compounds of higher aliphatic 'fatty' acids and glycerol. Boiling of these with sodium hydroxide solution converts them into the sodium salts of the acids and glycerol (p. 442).

Proteins, on prolonged refluxing with acid or alkali, yield amino acids, hydrolysis occurring at the 'polypeptide' link (p. 510). The nature of proteins ensures that the resultant hydrolysates are complex mixtures, but separation can usually be effected by paper chromatography (p. 518)

$$...R—CONH—R'—CONH... \longrightarrow NH_2.R.CO.OH + NH_2.R'.CO.OH + ...$$
$$\quad\quad\; HOH \quad\quad\quad HOH$$

4) Acids, especially of fairly complex structure, can be synthesized by using the esters of malonic or acetoacetic acids (p. 506).

5) The oxides of carbon are sometimes used as a source of acids. For instance, carbon monoxide reacts with sodium alkoxide at elevated temperatures and high pressure

$$RO^-Na^+ + CO \rightarrow RCO.O^-Na^+$$

and the free acid can be obtained from the salt by subsequent treatment with mineral acid.

Carbon monoxide and steam react with an alkene at elevated temperature and pressure in the presence of phosphoric acid as catalyst to produce a carboxylic acid, e.g.

$$RCH{=}CH_2 + CO + H_2O \rightarrow RCH_2.CH_2.CO.OH$$

Carbon dioxide, preferably in the solid state, can be reduced to acid by treatment with an ethereal solution of a Grignard reagent

$$CO_2 + RMgX \rightarrow RC{=}O \xrightarrow{H_2O} RC{=}O$$
$$\qquad\qquad\qquad\; | \qquad\qquad |$$
$$\qquad\qquad\quad O.MgX \qquad OH$$

Properties

Carboxylic acids are considerably less volatile than hydrocarbons of comparable molecular weight (*Table 25.1*).

Table 25.1

Hydrocarbon	Mol. wt.	b.p.°C	Acid	Mol. wt.	b.p.°C
Propane	44	−44	Methanoic	46	102
n-Butane	58	−1	Ethanoic	60	118
n-Pentane	72	36	Propanoic	74	141

Calculation of the van't Hoff factor (p. 111) from measurements of, for example, freezing-point determinations of solutions in non-ionizing solvents indicates that hydrogen bonding, arising from the marked inductive effect of oxygen, results in association into double molecules, with relatively low volatility

$$
\begin{array}{ccc}
 & O..H—O & \\
R—C & & C—R \\
 & O—H..O &
\end{array}
$$

Interaction with a polar solvent, e.g. water, causes the association to be replaced, at least partly, by ionization. Their aqueous solutions accordingly behave as acids

$$RCO.OH + H_2O \rightleftharpoons RCO.O^- + H_3O^+$$

This reaction does not occur to a very great extent and they are therefore very weak acids in aqueous solution, particularly the higher homologues which are in any case only slightly soluble in water. It follows that aqueous solutions of carboxylic acid salts are extensively hydrolysed

$$RCO.O^- + H_2O \rightleftharpoons RCO.OH + OH^-$$

For instance, if an aqueous solution of iron(III) ethanoate is boiled, a precipitate of basic salt is produced

$$(CH_3.CO.O)_3Fe + 2H_2O \rightarrow (CH_3.CO.O)Fe(OH)_2 \downarrow + 2CH_3.CO.OH$$

As the acidic strength depends on the extent of electron withdrawal from the relevant hydrogen atom, the dissociation constant will be increased by the introduction of electronegative groups into the molecule and decreased if hydrogen is replaced by a more electropositive group. Furthermore, the effect is most marked if the replacement is on the carbon atom next to the carboxyl group. For instance, methanoic acid is a stronger acid than ethanoic, where hydrogen has been replaced by an alkyl group of greater electron availability. Thereafter, as the series is ascended, the dissociation constant varies only slightly (see *Table 25.2*).

The acidity of the substituted carboxylic acids increases with the electronegativity of the element concerned. Fluorine has a greater effect than chlorine, chlorine than bromine, and bromine than iodine; the influence exerted increases with the number of halogen atoms present. Trichloroethanoic acid, for instance, has a dissociation constant comparable with those of mineral acids (see *Table 25.2*)

485

Table 25.2. Dissociation constants of acids

Acid	K_a	pK_a	Remarks
HCO.OH	$2 \cdot 0 \times 10^{-4}$M	3·7	acid weakened by electron-repelling effect of —CH₃; more distant alkyl groups have little effect
CH₃.CO.OH	$1 \cdot 6 \times 10^{-5}$M	4·8	
CH₃.CH₂.CO.OH	$1 \cdot 3 \times 10^{-5}$M	4·9	
(CH₃)₂CH.CO.OH	$1 \cdot 3 \times 10^{-5}$M	4·9	
FCH₂.CO.OH	$2 \cdot 0 \times 10^{-3}$M	2·7	halogen, with electron-attracting power, increases acidity, especially as the number of halogen atoms is increased
ClCH₂.CO.OH	$1 \cdot 6 \times 10^{-3}$M	2·8	
BrCH₂.CO.OH	$1 \cdot 4 \times 10^{-3}$M	2·9	
ICH₂.CO.OH	$7 \cdot 3 \times 10^{-4}$M	3·1	
Cl₂CH.CO.OH	$5 \cdot 0 \times 10^{-2}$M	1·3	
Cl₃C.CO.OH	$1 \cdot 2$M	−0·1	
benzoic acid (C₆H₅.CO.OH)	$6 \cdot 3 \times 10^{-5}$M	4·2	
2-nitrobenzoic acid (CO.OH with ortho NO₂)	$6 \cdot 5 \times 10^{-3}$M	2·2	electron-withdrawing effect of —NO₂ increases acidity, especially in position 2
3-nitrobenzoic acid (CO.OH with meta NO₂)	$3 \cdot 5 \times 10^{-4}$M	3·5	
4-nitrobenzoic acid (CO.OH with para NO₂)	$3 \cdot 7 \times 10^{-4}$M	3·4	
CO.OH–CO.OH (oxalic)	$6 \cdot 3 \times 10^{-2}$M	1·2	adjacence of a second carboxyl group increases acidity; the effect declines with increasing distance between the —COOH groups
H₂C(CO.OH)(CO.OH) (malonic)	$1 \cdot 6 \times 10^{-3}$M	2·8	
ortho-C₆H₄(CO.OH)₂ (phthalic)	$1 \cdot 0 \times 10^{-3}$M	3·0	
meta-C₆H₄(CO.OH)₂ (isophthalic)	$3 \cdot 2 \times 10^{-4}$M	3·5	

The benzene nucleus has a slight electron-donating effect relative to hydrogen, so that benzoic acid is weaker than methanoic acid. The introduction of electron-withdrawing groups into the benzene ring increases the acidity, however, especially when the substituents are in positions 2 and 4 (*Table 25.2*).

Dicarboxylic acids, by virtue of the increase in oxygen in the molecule, are stronger than the corresponding monobasic acids. Once again, the effect is less the further apart the acid groups (*Table 25.2*).

The acidity of all carboxylic acids can be increased by using a solvent capable of accepting a proton (a protophilic solvent), the extent depending on the particular solvent employed. For example, the acid will dissociate more completely in a base like pyridine than in water (p. 159)

$$RCO.OH + H_2O \rightleftharpoons RCO.O^- + H_3O^+$$

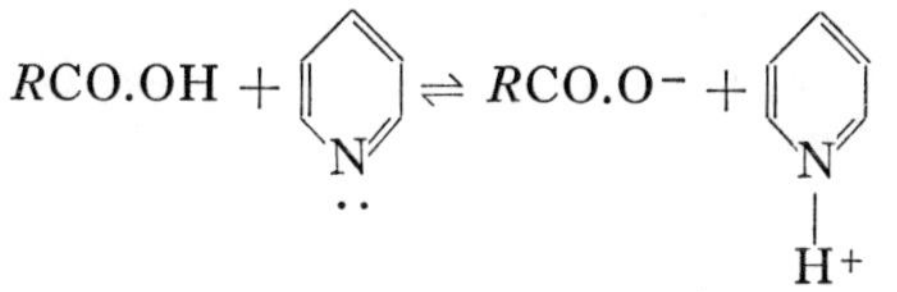

Reactions

The structure of the carboxyl group is normally represented as

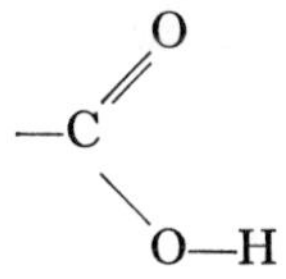

but the acids and the derived anions show very few properties of the individual groups because of hydrogen bonding in the former and resonance in the latter

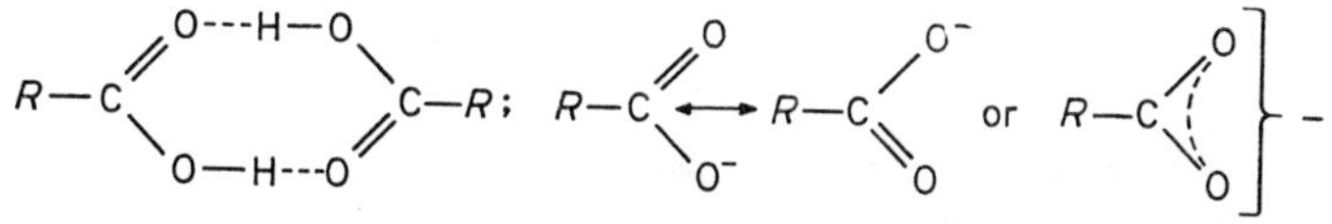

However, the carbonyl group can be reduced with lithium tetrahydridoaluminate, the final product being a primary alcohol

$$RCO.OH \xrightarrow{\text{LiAlH}_4} RCH_2.OH$$

The carbon of the carbonyl group also undergoes nucleophilic attack by the oxygen atom of alcohols, with the formation of esters, the reaction being catalysed by mineral acid (p. 437)

$$RCO.OH + HOR' \rightleftharpoons RCO.OR' + H_2O$$

Chlorine progressively substitutes hydrogen in the aliphatic chain in a manner reminiscent of carbonyl compounds (p. 464), e.g.

$$CH_3.CO.OH \xrightarrow[(-HCl)]{Cl_2} CH_2Cl.CO.OH \xrightarrow[(-HCl)]{Cl_2} CHCl_2.CO.OH \xrightarrow[(-HCl)]{Cl_2} CCl_3.CO.OH$$

Thionyl and phosphorus chlorides attack acids in a similar way to alcohols (p. 398), giving acid chlorides

$$RCO.OH \xrightarrow{SOCl_2} RCO.Cl$$

It follows from what has been said earlier that ionization of the carboxylic group will be favoured by treatment with alkali (i.e. salts will be formed)

$$RCO.OH + OH^- \longrightarrow RCO.O^- + H_2O$$

The resultant carboxylate ion possesses neither the carbonyl nor the hydroxyl group (see earlier) but has its own distinctive properties. For instance, it can be decarboxylated by treatment with alkali, e.g.

$$C_6H_5{\cdot}\underset{\underset{OH}{|}}{C}{=}O \longrightarrow C_6H_5{\cdot}C\underset{O}{\overset{O}{\diagup}} \Big\} - \xrightarrow{-CO_2} C_6H_6$$

Stereoismerism of Carboxylic Acids

Optical Isomerism—If an α-hydrogen atom of propanoic acid is replaced by a hydroxyl group, then that carbon atom is attached to four different groups, i.e. it is asymmetric (p. 362) and the structure can exist in two mirror-image (enantiomorphic), optically active forms

Lactic acid

and in a third inactive, or racemic, form which is an equimolecular mixture of the two active forms. This latter form of lactic acid is the variety that is present in sour milk. (+)-Lactic acid, rotating the plane of polarized light in a clockwise direction, occurs in muscle tissue and (−)-lactic acid, which rotates it in an anticlockwise direction, is made by the action of *Bacillus acidi laevolactiti* on sucrose solutions.

A similar situation exists if the α-hydrogen of propanoic acid is replaced by the amino group. Thus 2-aminopropanoic acid (alanine) is optically active. The enantiomorph derived by hydrolysis of protein, like all the other amino acids containing asymmetric carbon and used in protein synthesis, is the (−)-form. As the enzymes used in the manufacture of natural products are themselves protein, and therefore optically active, it is to be expected that the products will be optically active, but why they should all be of the (−)-form is not clear.

Geometric Isomerism—Unsaturated acids exhibit that form of stereoisomerism known as geometric isomerism (p. 362). The simplest examples are *cis*- and *trans*-but-2-ene-1,4-dioic (maleic and fumaric) acids, respectively

$$
\begin{array}{cc}
\text{HC.CO.OH} & \text{HO.OC.CH} \\
\| & \| \\
\text{HC.CO.OH} & \text{HC.CO.OH} \\
cis & trans
\end{array}
$$

Unlike optical isomers, geometric isomers show several differences in chemical as well as physical properties. Indeed, some of these permit separation and identification of the isomers. For instance, maleic acid is readily converted into the anhydride on heating; fumaric acid requires much stronger heating before an anhydride is formed, and then it gives maleic anhydride. This fact, together with the knowledge that the dipole moment of fumaric acid is virtually zero, is compelling evidence for the belief that maleic acid represents the *cis*-form:

$$
\begin{array}{ccc}
\text{HC.CO.OH} & \text{HC.C}{=}\text{O} & \text{HC.CO.OH} \\
\| \quad \xrightarrow{\text{easy}} & \| \quad {>}\text{O} \quad \xleftarrow{\text{difficult}} & \| \\
\text{HC.CO.OH} & \text{HC.C}{=}\text{O} & \text{HO.OC.CH} \\
\text{Maleic acid} & \text{Maleic anhydride} & \text{Fumaric acid}
\end{array}
$$

Some of the differences between the two acids are summarized in *Table 25.3*.

Table 25.3

Property	Maleic acid	Fumaric acid
Melting point, °C	130°	287°
Solubility (g/100 g water)	79	0·7
Acid dissociation constants:		
pK_a'	1·9208	3·02
pK_a''	6·2218	4·3768
Heat of combustion (kJ mol⁻¹)	1 370	1 340

INDUSTRIALLY IMPORTANT ACIDS

Methanoic (formic) acid, H.CO.OH

The first member of an homologous series often has atypical properties. Methanoic acid is no exception. It contains the aldehyde group and is consequently a strong reducing agent, reducing Fehling's solution to copper

$$Cu^{2+} + 2OH^- + HCO.OH \rightarrow Cu \downarrow + CO_2 \uparrow + 2H_2O$$

It is also easily oxidized to carbon dioxide by potassium permanganate

$$HCO.OH + [O] \rightarrow H_2O + CO_2 \uparrow$$

Concentrated sulphuric acid dehydrates formic acid to carbon monoxide

$$HCO.OH - H_2O \rightarrow CO \uparrow$$

It is a stronger acid than its homologues and is used in industry, e.g. dyestuffs, where a cheap and fairly strong acid is required; its fairly high volatility means that any residuum, e.g. on textiles, evaporates quickly. It is also used as a coagulant of rubber latex.

Methanoic acid is made on the large scale by passing carbon monoxide into a concentrated solution of sodium hydroxide at about 15×10^5 N m^{-2} pressure and $\sim 200°$ temperature. Methanoic acid is liberated from this solution by adding dilute sulphuric acid (concentrated sulphuric acid would dehydrate it—see above). Subsequent distillation gives an azeotrope rich in methanoic acid

$$CO + OH^- \rightarrow HCO.O^- \xrightarrow{H^+} HCO.OH$$

Ethanoic (acetic) acid, $CH_3CO.OH$

Ethanoic acid has been known for a long time as 'sour wine', the result of bacterial activity on ethanol, itself the result of the fermentation of carbohydrates. The process by which ethanol is converted into ethanoic acid is one of oxidation and takes place through the intermediate ethanal. It can be effected by oxidation of the vapour of the alcohol:

$$CH_3.CH_2.OH \xrightarrow{[O]} CH_3.CHO \xrightarrow{[O]} CH_3.CO.OH$$
$$\text{Ethanol} \quad _{-2\,[H]} \quad \text{Ethanal} \qquad \text{Ethanoic acid}$$

It is also economically feasible to proceed from the intermediate, ethanal, because of the relatively low cost of this chemical. Recently, ethanoic acid, together with methanoic and propanoic acids, have been obtained directly from the light-oil fraction of crude petroleum by oxidation of the liquid at elevated temperatures; separation of the products is effected by fractional distillation.

Ethanoic acid is required commercially for conversion to esters (p. 502), the manufacture of ethanoic anhydride (p. 501), as a solvent and in the drug, dye and foodstuffs industries.

Several aryloxy derivatives of ethanoic acid are plant hormones, e.g.

M.C.P.A. (2-methyl-4-chloro-
phenoxyacetic acid)

2,4-D.C.P.A. (2,4-dichloro-
phenoxyacetic acid)

Some of these are fairly specific in their action, so that application of the hormone in sufficient quantities results in excessive, irregular growth and subsequent death of certain plants only. Selective weed-killing is now, within certain limits, a practical possibility.

Hexane-1,6-dioic (adipic) acid has in recent years come into prominence as an intermediate in the manufacture of Nylon (p. 512). It is manufactured by oxidation of cyclohexanol, which is itself obtained from phenol, e.g.

Ethanedioic (oxalic) acid is the parent member of the series of dicarboxylic acids, i.e. it is in a comparable position to methanoic acid. The two acids have several properties in common. Ethanedioic acid is dehydrated to carbon monoxide (and carbon dioxide) by concentrated sulphuric acid

$$\begin{array}{c} CO.OH \\ | \\ CO.OH \end{array} \quad - H_2O \rightarrow CO \uparrow + CO_2 \uparrow$$

It is also oxidized readily by acidified potassium permanganate (particularly above 60°)

$$5\begin{array}{c} CO.OH \\ | \\ CO.OH \end{array} + 2MnO_4^- + 6H^+ \rightarrow 8H_2O + 2Mn^{2+} + 10\ CO_2 \uparrow$$

Furthermore, the two acids can be readily interconverted. Decarboxylation of ethanedioic acid, by heating its glyceryl monoester, gives rise to methanoic acid

$$\begin{array}{c} CH_2.O.CO.CO.OH \\ | \\ CH.OH \\ | \\ CH_2.OH \end{array} + H_2O \longrightarrow \begin{array}{c} CH_2.OH \\ | \\ CH.OH \\ | \\ CH_2.OH \end{array} + HCO.OH \uparrow + CO_2 \uparrow$$

Sodium methanoate on heating evolves hydrogen to give sodium ethanedioate. This is the basis of the large-scale production of the acid. The salt solution is treated with calcium chloride solution to cause precipitation of the calcium salt, from which the acid is liberated by addition of excess dilute sulphuric acid

$$\begin{array}{c} HCO.O^- \\ + \\ HCO.O^- \end{array} \xrightarrow{Heat} \begin{array}{c} CO.O^- \\ | \\ CO.O^- \end{array} + H_2 \uparrow$$

$$\begin{array}{c} CO.O^- \\ | \\ CO.O^- \end{array} \xrightarrow{Ca^{2+}} \begin{array}{c} CO.O^- \\ | \quad Ca^{2+} \downarrow \\ CO.O^- \end{array} \xrightarrow{H_2SO_4} \begin{array}{c} CO.OH \\ | \\ CO.OH \end{array} + CaSO_4 \downarrow$$

491

The acid is used in the dyestuffs industry and in the manufacture of inks. Some of its salts are used as photographic developers.

Benzenedicarboxylic (phthalic) acids

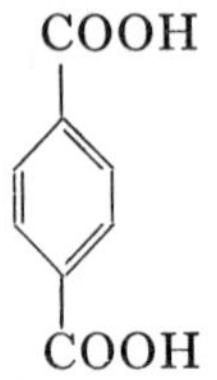

COOH

COOH

COOH
COOH

COOH

COOH

Benzene-1,4-dicarb-
oxylic (terephthalic) acid

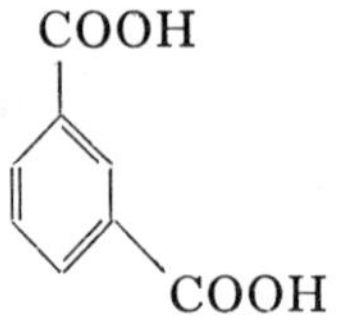

Benzene-1,3-dicarb-
oxylic (isophthalic) acid

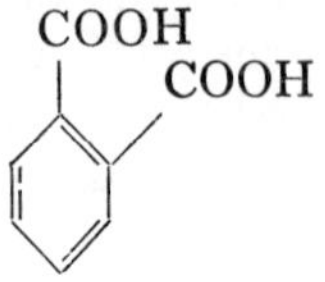

Benzene-1,2-dicarb-
oxylic (phthalic)
acid

Terephthalic acid is manufactured by the oxidation of 1,4-dimethylbenzene (*p*-xylene) with nitric acid

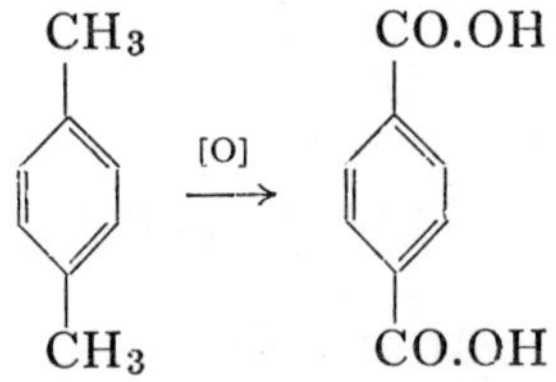

CH_3

$CO.OH$

[O]
$\longrightarrow$

CH_3

$CO.OH$

It is indirectly esterified with ethanediol to form 'Terylene' (p. 507).

Phthalic acid is manufactured via the anhydride. Passing naphthalene vapour and air over a hot vanadium pentoxide catalyst causes the reaction

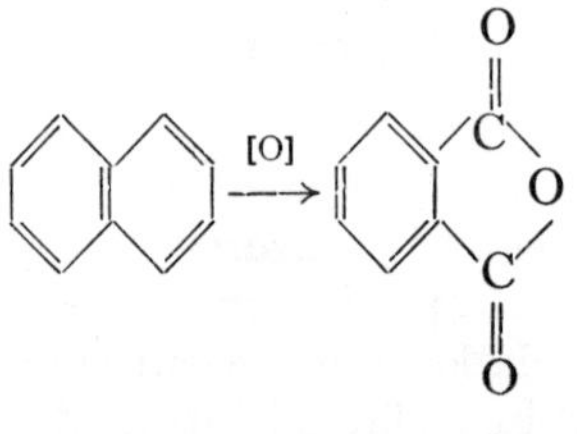

O

[O]
$\longrightarrow$

C
O
C
O

The anhydride is then boiled with sodium hydroxide solution to give the sodium salt, from which the free acid is liberated by the addition of excess mineral acid

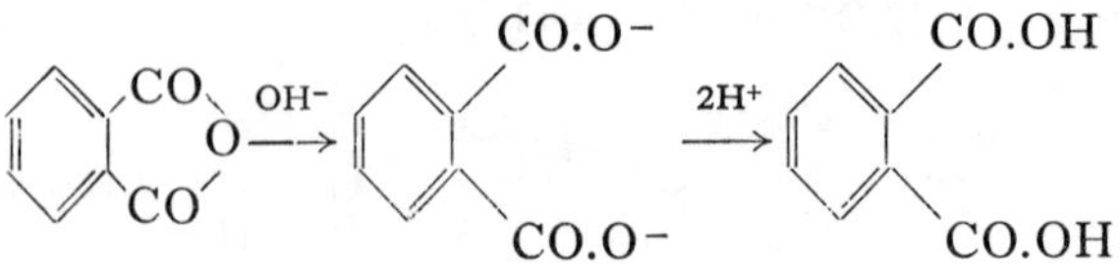

CO
O
CO

OH^-
$\longrightarrow$

$CO.O^-$

$CO.O^-$

$2H^+$
$\longrightarrow$

$CO.OH$

$CO.OH$

This acid is used for the manufacture of various esters, those with monohydric alcohols being used as plasticizers, e.g. dibutyl phthalate, and those with polyhydric alcohols, e.g. glycerol, as glyptal or alkyd resins (p. 501).

Analysis of Carboxylic Acids

The carboxylic acid group is readily identified by the fact that it reacts, with effervescence, with sodium carbonate solution. The anilide is often prepared as a derivative, first by converting the acid to the acid chloride and then reacting this with aniline (p. 428).

The elements present are estimated quantitatively in the usual way. Determination of the *true* molecular weight—association and dissociation often lead to erroneous results—permits evaluation of the molecular formula.

By adding excess silver nitrate solution to a neutral solution of an organic acid (prepared by boiling an excess of ammonium hydroxide with the carboxylic acid until free ammonia is no longer present), the silver salt is precipitated. The precipitate is washed, dried and weighed. By igniting the dry silver salt, a residue of silver is obtained; this, too, is weighed and the molecular weight of the acid calculated, providing the basicity has been separately determined. For a monobasic acid

$$RCO.OH \rightarrow RCO.O^-NH_4^+ \longrightarrow RCO.OAg \longrightarrow Ag \downarrow$$

$$\text{M g} \qquad\qquad\qquad (\text{M}-1) + 108 \text{ g} \quad 108 \text{ g}$$

Example—0·830 g of a silver salt of a dibasic acid, on ignition, gave 0·540 g of silver. Hence, 2×108 g of silver would result from $(2 \times 108 \times 0·830)/0·540$ g of the silver salt, i.e. 332·0 g, which must be the molecular weight of the silver salt. The molecular weight of the acid is then $332—(2 \times 108) + 2 = 118$.

SUMMARY

Carboxylic acids are capable of both association and dissociation. Association into double molecules is the result of hydrogen bonding and reduces the tendency to acidity

$$R-C \overset{\displaystyle O \ldots H - O}{\underset{\displaystyle O - H \ldots O}{}} C-R$$

Dissociation into ions is encouraged by polar solvents such as water which are able to accept protons

Both association and dissociation suppress the typical properties of the carbonyl and hydroxyl groups, so that carboxylic acids behave to a large extent as mono-functional compounds.

Some characteristic reactions of ethanoic acid are given in *Figure 25.1*.

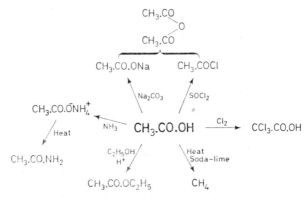

Figure 25.1. Characteristic reactions of ethanoic acid

QUESTIONS

1. How might the basicity of a carboxylic acid be determined?

2. How could you distinguish between the sodium salts of methanoic, ethanoic, ethanedioic, and benzenecarboxylic (benzoic) acids?

3. Give the structural formulae of isomers of molecular formula $C_4H_8O_2$. How would you attempt to distinguish between them?

4. Suggest methods of preparing the mononitrobenzoic acids. Account for the different acidities of the three acids.

5. What directive influence would you expect a carboxylic acid group to exert when attached to a benzene ring?

6. How might 2-benzoylbenzoic acid be prepared? The action of concentrated sulphuric acid on this compound produces a ring closure to give a quinone; suggest the structural changes involved and give the formula of the quinone produced.

7. A compound of molecular formula $C_9H_{10}O_3$ contains one carboxylic acid group, and although it does not react with bromine, it can be oxidized to a compound $C_9H_8O_3$, which imparts a coloration to Schiff's reagent. The original compound when strongly heated gives a substance of formula $C_9H_8O_2$, producing benzoic acid on oxidation. Suggest possible structures for the compound.

494

8. 4·59 g of the silver salt of a carboxylic acid yields, on complete ignition, 3·24 g of silver; 1·15 g of the acid just neutralizes 250 cm^3 of M/10 sodium hydroxide solution. Name the acid.

9. Suggest ways in which maleic and fumaric acids can be synthesized from ethene.

10. Write an essay on the optical activity of natural products.

11. Explain the following facts about methanoic, ethanoic and benzene carboxylic acids:
(*a*) methanoic acid is the strongest acid of the three,
(*b*) benzene carboxylic acid is the most readily decarboxylated,
(*c*) benzene carboxylic acid is the least soluble in water,
(*d*) methanoic acid is the only reducing agent of the three.

12. Write an account of the production of carboxylic acids by natural processes.

13. A 0·1 M solution of a monobasic organic acid in water had an osmotic pressure of 2·71 $\times$ 10^5 N m^{-2} at 27°C. A 0·1 M solution of the same acid in benzene had an osmotic pressure of only 1·23 $\times$ 10^5 N m^{-2} at the same temperature. What can you deduce from this information?

CHAPTER 26

ACID DERIVATIVES

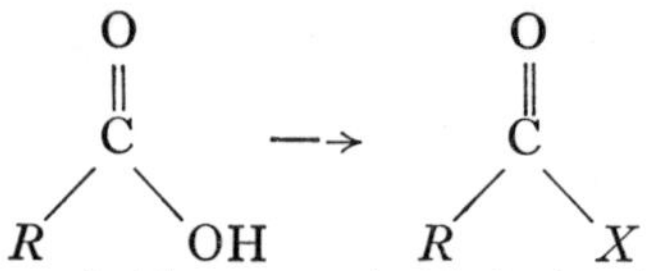

$$
\underset{\text{Acid}}{\overset{\displaystyle O}{\underset{R}{\overset{\|}{C}}}\text{—OH}} \longrightarrow \underset{\text{Acid derivative}}{\overset{\displaystyle O}{\underset{R}{\overset{\|}{C}}}\text{—X}}
$$

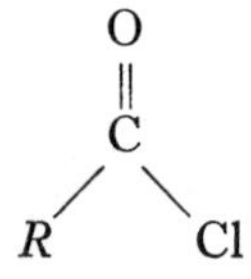

Acid chloride	Acid anhydride	Ester	Amide

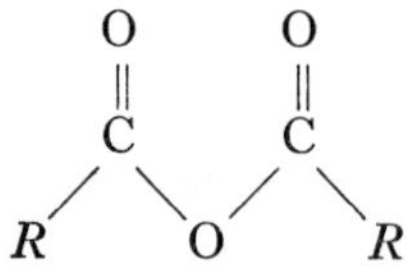

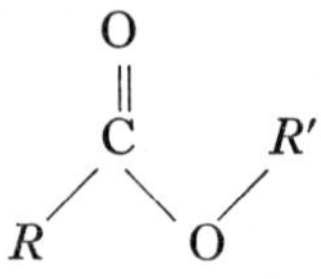

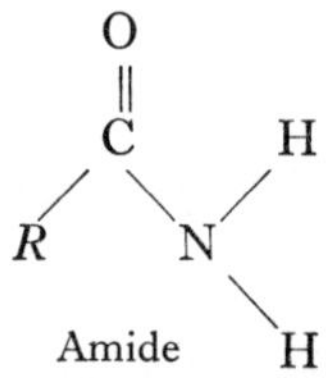

All carboxylic acid derivatives contain the carbonyl group, and it is interesting and helpful to consider how the characteristic properties of this group are modified by the substituent X. If X is hydroxyl, then the substance is a carboxylic acid and, as was seen in the previous Chapter, hydrogen bonding in the undissociated acid or resonance in the anion result in the virtual disappearance of carbonyl properties. Such is not the case with the acid derivatives; *nucleophilic addition at a carbon atom is still characteristic of these compounds*, although the readiness with which this reaction takes place can be affected by the electron-withdrawing capacity of the substituent X. In the case of acid chlorides $(X = \text{Cl})$, for instance, there is such a drift of electrons towards chlorine that the latter can even be ionized; e.g. it is claimed that ethanoyl (acetyl) chloride is ionized to the extent of 40 per cent. Consequently, the carbon atom is very polarized, and acid chlorides are easily the most reactive of the acid derivatives. On the other hand, the nitrogen atom of acid amides tends to share its lone pair with carbon, and this counteracts to some extent the polarization due to the carbonyl bond; amides are therefore the least reactive of acid derivatives:

$$
\overset{\displaystyle O}{\overset{\|}{R}C\text{Cl}} \longleftrightarrow \overset{\displaystyle O}{\overset{\|}{R}C^+} + \text{Cl}^-
$$

$$
\overset{\displaystyle O}{\overset{\|}{R}C\text{—}\ddot{N}H_2} \longleftrightarrow \overset{\displaystyle O^-}{\overset{|}{R}C\text{=}\overset{+}{N}H_2}
$$

496

It does not follow that addition of a nucleophile to the carbon will result in any significant and complete change taking place. Much depends on the relative 'leaving capacity' of the groups attached to carbon in the intermediate formed:

$$
\begin{array}{ccc}
\overset{\displaystyle O\searrow}{\underset{\underset{B}{\uparrow}}{\overset{\|}{R\overset{\,}{C}X}}} & \longrightarrow & \overset{\displaystyle O^-}{\underset{\underset{B^+}{|}}{\overset{|}{R\overset{\,}{C}X}}}
\end{array}
$$

B = base (nucleophile)

Consider, for example, the reaction with the nucleophile ROH. Aldehydes react in the presence of hydrogen chloride as catalyst to give acetals, which represent the end product of the reaction and from which the original aldehyde is readily regenerated (p. 463)

$$
2ROH + R'\overset{\overset{\displaystyle O}{\|}}{CH} \rightleftharpoons R'\overset{\overset{\displaystyle OR}{|}}{CH}.OR + H_2O
$$

(Ketones, with the carbon of the carbonyl group less polarized, do not even react in this way.) On the other hand, acid chlorides react vigorously with alcohols to give esters, chlorine leaving the 'acetal-type' intermediate

$$
R'\overset{\overset{\displaystyle O}{\|}}{C}Cl + ROH \rightarrow
\begin{bmatrix}
\overset{\overset{\displaystyle Cl}{|}}{R'\underset{\underset{ROH}{\overset{|}{|+}}}{C}.O^-}
\end{bmatrix}
\xrightarrow[-H^+]{-Cl^-}
R'\overset{\overset{\displaystyle O}{\|}}{C}.OR
$$

A similar reaction takes place with anhydrides, usually in the presence of a catalyst:

$$
R'\overset{\overset{\displaystyle O.CO.R''}{|}}{\underset{\underset{ROH}{\uparrow}}{C}{\rightleftharpoons}O}
\longrightarrow
\begin{bmatrix}
R'\overset{\overset{\displaystyle O.CO.R''}{\zeta}}{\underset{\underset{ROH}{\overset{|}{|+}}}{C}-O^-}
\end{bmatrix}
\xrightarrow{-R''COOH}
R'C\overset{\nearrow O}{\searrow_{OR}}
$$

Nucleophilic attack by an alcohol even takes place upon an ester, as is shown by the fact that the alkyl group can be replaced by another, if the ester is left in contact with an alcohol:

$$
R'\overset{\overset{\displaystyle OR''}{|}}{\underset{\underset{R\ddot{O}H}{\uparrow}}{C}{\rightleftharpoons}O}
\longrightarrow
\begin{bmatrix}
R'\overset{\overset{\displaystyle OR''}{\zeta}}{\underset{\underset{ROH}{\overset{|}{|+}}}{C}-O^-}
\end{bmatrix}
\xrightarrow{-R''OH}
R'C\overset{\nearrow O}{\searrow_{OR}}
$$

On the other hand, attack by alcohol upon an acid amide is negligible.

Volatility

Both esters and acid chlorides are more volatile than acids of comparable molecular weight, because there is no prospect of hydrogen bonding and the substances exist as discrete molecules. Hydrogen bonding exists, though, with acid amides, and these are less volatile than the corresponding acids

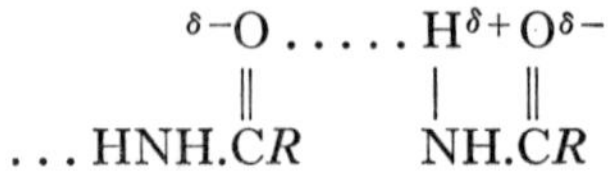

$$\delta^-O \ldots\ldots H^{\delta+}O^{\delta-}$$

$$\ldots HNH.CR \qquad NH.CR$$

The boiling or melting points of some common acids and their derivatives are shown in *Table 26.1*.

Table 26.1

		b.p. °C	m.p. °C		b.p. °C	m.p. °C
Acids	$CH_3.CO.OH$	118		$C_2H_5.CO.OH$	141	
Esters	$CH_3.CO.OCH_3$	56		$C_2H_5.CO.OCH_3$	79	
Chlorides	$CH_3.COCl$	52		$C_2H_5.COCl$	80	
Anhydrides	$(CH_3.CO)_2O$	140		$(C_2H_5.CO)_2O$	166	
Amides	$CH_3.CO.NH_2$		82	$C_2H_5.CO.NH_2$		77

ACID CHLORIDES, $RCOCl$

The systematic names of acid chlorides are formed from those of the parent acids by using the suffix *-yl* in place of the *-ic* of the acid, e.g.

$$C_2H_5.CO.OH \qquad C_2H_5.COCl$$
Propanoic acid   Propanoyl chloride

Because of their extreme reactivity, acid chlorides cannot be prepared from other acid derivatives; rather are they used for preparing the others, e.g.

$$RCOCl + R'OH \rightarrow RCO.OR' + HCl$$

They can be obtained, however, by distilling the carboxylic acid with thionyl or phosphorus chlorides; the former reagent is normally preferred because of the volatility of the by-products (and thus reduced contamination of the product itself), e.g.

$$CH_3.CO.OH + SOCl_2 \rightarrow CH_3.COCl + SO_2 \uparrow + HCl \uparrow$$

$$C_6H_5.CO.OH + PCl_3 \rightarrow C_6H_5.COCl + H_3PO_3$$

498

Acid chlorides are colourless, pungent liquids, fuming in moist air. Their great reactivity makes them important in synthetic work, e.g. ethanoyl (acetyl) chloride

$$CH_3.\overset{\displaystyle O}{\overset{\|}{C}}Cl$$

$$HO{-}H \longrightarrow CH_3.CO.OH + HCl$$

$$RO{-}H \longrightarrow CH_3.CO.OR + HCl$$

$$H_2N{-}H \longrightarrow CH_3.CO.NH_2 + HCl$$

$$RHN{-}H \longrightarrow CH_3.CO.NHR + HCl$$

$$RCOO^- \; Na^+ \longrightarrow CH_3.CO.O.OCR + Na^+Cl^-$$

The mechanisms of these reactions have been indicated above.

The reaction between an acid chloride and alcohols or phenols is used as a means of characterizing the latter. The chloride normally used is 3,5-dinitrobenzoyl chloride, and the reaction is carried out in the presence of sodium hydroxide solution (Schotten–Baumann reaction), e.g.

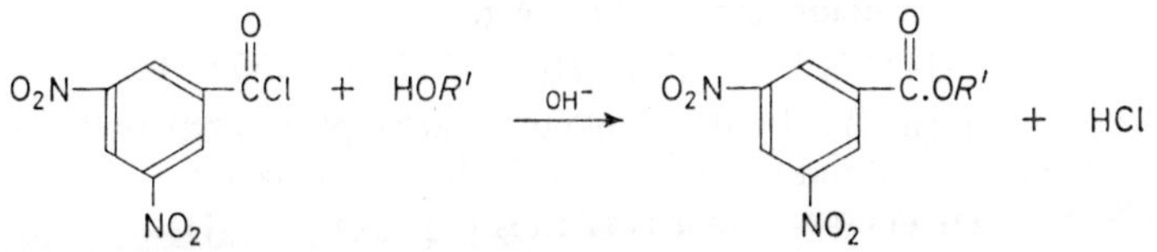

Amines can be converted into suitable derivatives in a similar way by treatment with benzoyl chloride in the presence of dilute sodium hydroxide solution

$$C_6H_5.COCl + H_2NR \rightarrow C_6H_5.CO.NHR + HCl$$

Anhydrous aluminium chloride encourages more complete ionization of acid chlorides by forming the complex $[AlCl_4]^-$ ion; the other ion formed, $[RCO]^+$, is a powerful electrophile and is capable of reacting with the benzene ring

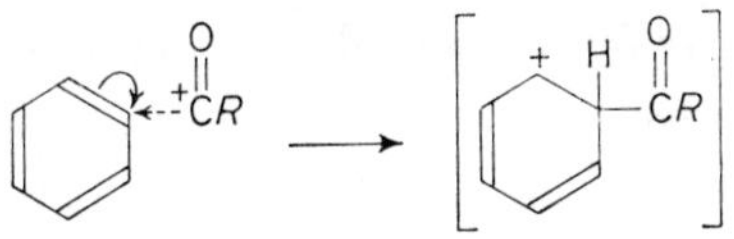

Expulsion of a proton from this intermediate results in the formation of a ketone (Friedel–Craft reaction, see also p. 390)

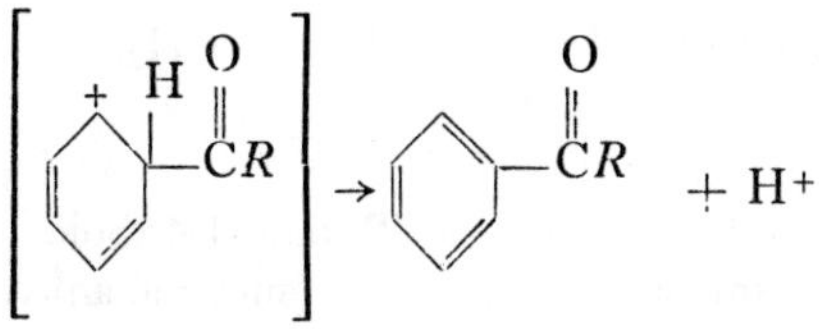

Acid chlorides are reduced by hydrogen in the presence of a palladium catalyst, first to aldehydes and then to alcohols (in the Rosenmund reaction (p. 457), the palladium catalyst is 'poisoned' with barium sulphate, so that the reaction stops at the aldehyde stage)

$$\underset{\text{RCCl}}{\overset{\text{O}}{\|}} \xrightarrow{\text{[2H]}} \underset{\text{RCH}}{\overset{\text{O}}{\|}} + \text{HCl}$$

$$\underset{\text{RCH}}{\overset{\text{O}}{\|}} \xrightarrow{\text{[2H]}} RCH_2.OH$$

Analysis

The chlorine in acid chlorides is so labile that hydrolysis rapidly takes place in the presence of aqueous silver nitrate; the resultant chloride ion then instantly reacts with the silver ion to give a white precipitate of silver chloride

$$RCOCl + H_2O \rightarrow RCO.OH + H^+ + Cl^-$$
$$Ag^+ + Cl^- \rightarrow AgCl \downarrow$$

Anilides provide suitable derivatives for characterization, and these are easily made by the Schotten–Baumann reaction, e.g.

$$CH_3.COCl + C_6H_5.NH_2 \rightarrow C_6H_5.NH.CO.CH_3$$

N.B. Methanoyl (formyl) chloride is unknown; attempts to prepare it always result in the production of carbon monoxide and hydrogen chloride:

$$HCO.OH + SOCl_2 \rightarrow SO_2 \uparrow + CO \uparrow + 2HCl$$

ACID ANHYDRIDES, $\underset{RC.O.CR'}{\overset{\text{O} \quad \text{O}}{\| \quad \|}}$

Acid anhydrides represent the removal of one molecule of water from two molecules of a monobasic acid; they are named from the parent acid by replacing 'acid' by 'anhydride'; e.g. $CH_3.CO.OH$, ethanoic (acetic) acid, gives rise to $CH_3.CO.O.CO.CH_3$, ethanoic (acetic) anhydride.

Dibasic acids often give *internal* anhydrides simply on distillation of the acid in an inert solvent, e.g.

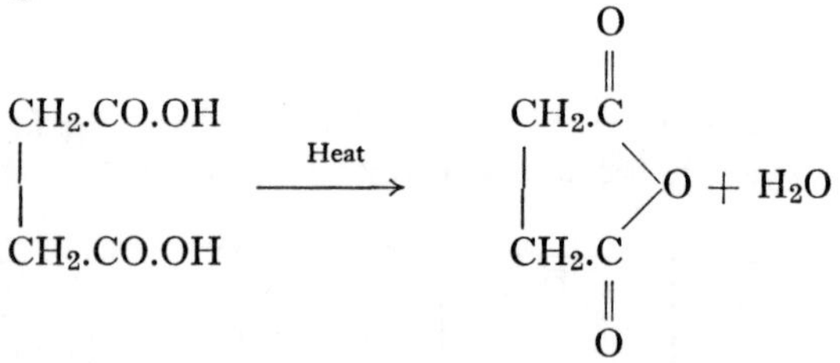

| Butane-1,4-dioic acid | Butane-1,4-dioic anhydride |
| (succinic acid) | (succinic anhydride) |

Anhydrides of monobasic acids, however, usually require a more devious method of preparation, namely the distillation of an acid chloride with the anhydrous sodium salt of the acid. Here the active nucleophile is the carboxylate ion, e.g.

$$\underset{\underset{CH_3COO^-}{\uparrow}}{\overset{\overset{O}{\|}}{CH_3.\overset{}{C}Cl}} \longrightarrow \overset{\overset{O\quad O}{\|\quad\|}}{CH_3.C.O.C.CH_3} + Cl^-$$

Acid anhydrides are pungent, colourless liquids. They are not so violently reactive as acid chlorides and are sometimes preferred to them on this account for making amides and esters, e.g.

$$\overset{\overset{O\quad O}{\|\quad\|}}{CH_3.C.O.C.CH_3} + NH_3 \longrightarrow \overset{\overset{O}{\|}}{CH_3.C.NH_2} + CH_3.CO.OH$$

$$\overset{\overset{O\quad O}{\|\quad\|}}{CH_3.C.O.C.CH_3} + C_2H_5.OH \longrightarrow \overset{\overset{O}{\|}}{CH_3.C.O.C_2H_5} + CH_3.CO.OH$$

Two very important anhydrides on the industrial front are butene-1,4-dioic (maleic) anhydride and benzene-1,2-dicarboxylic acid (phthalic) anhydride. The potential diacid functions present in these substances raise the possibility of condensation polymerization with polyfunctional alcohols to give polyesters; in fact, these are readily obtained. Known as 'alkyd resins' (*alco*hol + ac*id*), they have many desirable commercial properties

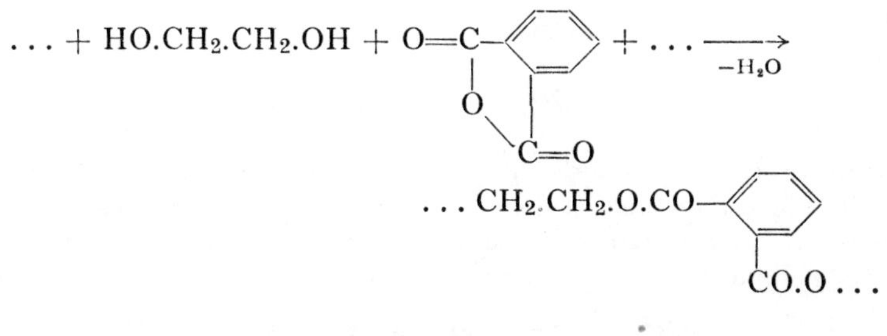

$$\ldots + HO.CH_2.CH_2.OH + O{=}C{-}\underset{\underset{C=O}{\overset{|}{O}}}{\bigcirc} + \ldots \xrightarrow[-H_2O]{}$$

$$\ldots CH_2.CH_2.O.CO{-}\underset{\underset{CO.O\ldots}{|}}{\bigcirc}$$

Phthalic anhydride is also esterified with butanol to give dibutyl phthalate, a widely-used plasticizer in the rubber and allied industries.

Both maleic and phthalic anhydrides are manufactured by vapour-phase oxidation of aromatic hydrocarbons over a vanadium(V) oxide catalyst, the

501

former from benzene, the latter from naphthalene or 1,2-dimethylbenzene (*o*-xylene):

$$
\text{(benzene)} \xrightarrow[\text{V}_2\text{O}_5]{\text{O}_2}
\begin{array}{c}
\text{CH.C} \\
\| \\
\text{CH.C} \\
\| \\
\text{O}
\end{array}
\begin{array}{c}
\text{O} \\
\| \\
\\
\text{O} \\
\end{array}
\qquad
\begin{array}{c}
\text{CH}_3 \\
\text{CH}_3
\end{array}
\xrightarrow[\text{V}_2\text{O}_5]{\text{O}_2}
\begin{array}{c}
\text{O} \\
\| \\
\text{C} \\
\text{O} \\
\text{C} \\
\| \\
\text{O}
\end{array}
$$

$$
\textbf{ESTERS, } R\overset{\text{O}}{\overset{\|}{\text{C}}}.OR'
$$

Esters are derived from acids by replacement of the acidic hydrogen by alkyl or aryl groups and are named accordingly, e.g. ethyl ethanoate, $CH_3.CO.OC_2H_5$, represents the replacement of the hydrogen atom in ethanoic acid, $CH_3.CO.OH$, by the ethyl radical.

The alkyl or aryl groups are usually provided by alcohols or phenols. Alcohols react with acids in a way which bears a superficial resemblance to the reaction of alkalis with acids but, on the other hand, the reaction velocity of *esterification* is fairly low and the reaction does not, in the absence of other substances, proceed to completion; in short, it is reversible and provides the classical example of chemical equilibrium (p. 131)

$$RCO.OH + R'OH \rightleftharpoons RCO.OR' + H_2O$$

Addition of concentrated sulphuric acid to the reaction mixture, however, not only catalyses the reaction (i.e. accelerates its rate) but, by removing the water (as hydrates) as it is formed, shifts the equilibrium in favour of ester formation.

Protons provided by the mineral acid protonate the oxygen of the carbonyl group

$$
\overset{\text{O:}}{\overset{\|}{R\text{C.OH}}} \xrightarrow{\text{H}^+} \overset{\text{OH}^+}{\overset{\|}{R\text{C.OH}}}
$$

The drift of electrons towards this oxygen atom is therefore increased, with the result that the carbon atom at the other end of the bond becomes more positively charged, i.e. more electrophilic, and attracts the oxygen atom of the alcohol. Eventually elimination of water gives the ester:

$$
\begin{array}{c}
\overset{\text{OH}^+}{\overset{\|}{R\text{C.OH}}} \\
\uparrow \\
R'\ddot{\text{O}}\text{H}
\end{array}
\rightarrow
\begin{array}{c}
\overset{\text{OH}}{\overset{|}{R\text{C.OH}}} \\
| \\
R'\text{OH}^+
\end{array}
\xrightarrow[-\text{H}^+]{-\text{H}_2\text{O}}
\overset{\text{O}}{\overset{\|}{R\text{C.OR}'}}
$$

502

Reaction of alcohols with acid chlorides and anhydrides is more pronounced than with the acid itself. Indeed, acid chlorides can be used to esterify phenols which, because of the modified properties of the hydroxy group when attached to an aromatic system (p. 436), do not react so readily with carboxylic acids, e.g.

$$R\overset{\delta+}{\underset{\underset{H}{Cl}}{C}}\overset{\overset{O}{\parallel}}{\leftarrow}:O.C_6H_5 \longrightarrow R\overset{+}{\underset{\underline{\underset{\cdot}{Cl} H}}{C}}.O.C_6H_5 \xrightarrow[-HCl]{} R\overset{\overset{O}{\parallel}}{C}.O.C_6H_5$$

Properties

The lower esters are colourless, fragrant liquids, mobile and insoluble in water. Many of them are found in nature as the 'essential oils' of fruit, but by far the largest quantity of ester occurring naturally is in the form of fats and oils. These are, respectively, solid and liquid condensation products derived from the trihydric alcohol, propanetriol (glycerol), and monobasic acids which almost always contain an even number of carbon atoms, in accordance with their synthesis from acetyl co-enzyme A units, $CH_3.CO—A$, formed from pyruvic acid during respiration (p. 477), e.g.

$$CH_2.O.CO.CH_2.CH_2.CH_3$$
$$|$$
$$CH.O.CO.CH_2.CH_2.CH_3$$
$$|$$
$$CH_2.O.CO.CH_2.CH_2.CH_3$$

Propanetriyl tributanoate
(which occurs in butter)

The melting point of a fat depends to a large extent upon the state of saturation of the acid involved; acids containing double bonds give esters of lower melting point than do saturated acids and therefore tend to be present in the liquid oils. For example, stearin, a glyceride of the saturated octadecanoic (or stearic) acid, $CH_3(CH_2)_{16}CO.OH$ (mol. wt. 284), melts at 65°C and is therefore solid at room temperature; olein, a glyceride of the unsaturated octadec-9-enoic (or oleic) acid, $CH_3(CH_2)_7CH{=}CH(CH_2)_7CO.OH$ (mol. wt. 282) melts at −6° and is therefore liquid at room temperature. Consequently, saturation by the addition of hydrogen across the double bonds provides a method for converting vegetable oils into fats. It is carried out by means of molecular hydrogen in the presence of a finely-divided nickel catalyst (Sabatier–Senderens reduction)

$$\underset{\substack{| \quad | \\ \text{—C}{=}\text{C—}}}{\overset{\substack{H \quad H \\ | \quad |}}{}} \xrightarrow[\text{Ni}]{H_2} \underset{\substack{| \quad | \\ H \quad H}}{\overset{\substack{H \quad H \\ | \quad | \\ \text{—C—C—} \\ | \quad |}}{}}$$

503

This reaction is of fundamental importance in the conversion of materials such as whale oil and groundnut oil into margarine.

Saturation can also be effected by the addition of atmospheric oxygen to the double bond. Linseed oil, containing glycerides of the unsaturated acids, octadec-9,12-dienoic (linoleic) and octadec-9,12,15-trienoic (linolenic), becomes saturated in this way when exposed to the air and accordingly finds application as a 'drying oil' in paints.

Esters can be converted into the corresponding acids and alcohols or phenols by hydrolysis, for example by refluxing the esters with strong alkali and then distilling. The distillate consists of the alcohol, whilst the acid remains as the non-volatile sodium or potassium salt in the residue from which it can be extracted by treatment with excess mineral acid. If the esters used are oils or fats, then sodium or potassium salts of long-chain acids result and, because these are soaps, the process is known as *saponification*:

$$
\begin{array}{ccc}
\text{CH}_2.\text{O}.\text{CO}R & \text{CH}_2.\text{OH} & R\text{CO}.\text{O}^- \\
| & | & + \\
\text{CH}.\text{O}.\text{CO}R' \;+\; 3\text{OH}^- \longrightarrow & \text{CH}.\text{OH} & + R'\text{CO}.\text{O}^- \\
| & | & + \\
\text{CH}_2.\text{O}.\text{CO}R'' & \text{CH}_2.\text{OH} & R''\text{CO}.\text{O}^- \\
& \text{Glycerol} &
\end{array}
$$

In the case of phenolic esters, the solution, after refluxing the ester with alkali, contains the alkali salts of both the acid and the phenol, and so no separation is effected by distillation. Acidification liberates the two acidic components

$$C_6H_5.\text{O}.\text{CO}R + 2\text{OH}^- \rightarrow C_6H_5.\text{O}^- + R\text{CO}.\text{O}^- + H_2\text{O}$$

$$C_6H_5.\text{O}^- + H^+ \rightarrow C_6H_5.\text{OH} \qquad R\text{CO}.\text{O}^- + H^+ \rightarrow R\text{CO}.\text{OH}$$

Advantage can now be taken of the marked difference in acidity between the two substances. If both are dissolved in ether and the ethereal phase is shaken with sodium carbonate, the carboxylic acid is the only component to react

$$2R\text{CO}.\text{OH} + \text{CO}_3^{2-} \rightarrow 2R\text{CO}.\text{O}^- + H_2\text{O} + \text{CO}_2 \uparrow$$

as it does, it forms the hydrophilic salt which dissolves in the aqueous phase, from which the acid is reliberated by treatment with excess mineral acid. Distillation of the ether phase leaves the phenol.

The mechanism of alkaline hydrolysis of an ester is based upon the addition of the nucleophilic hydroxide ion to the carbon of the carbonyl group and the subsequent departure of the alkoxy or phenoxy ion from the intermediate

$$
\begin{array}{ccccc}
\text{O} & \text{O}^- & \text{O} & & \\
\| & | & \| & & \\
R\text{C}.\text{O}R' \rightarrow & R\text{C}.\text{O}R' \rightarrow & R\text{C}.\text{OH} + {}^-\text{O}R' \rightarrow & R-\text{C} \begin{array}{c} \diagup \text{O} \\ \diagdown \text{O} \end{array}^- & + R'\text{OH} \\
\uparrow & | & & & \\
\overset{..}{\text{O}}\text{H}^- & \text{OH} & & &
\end{array}
$$

Nitrogen compounds can act in a similar way to hydroxyl groups. *Concentrated ammonia solution converts esters into amides* (particularly in the case of oxalates) and hydroxylamine brings about conversion into hydroxamic acids, e.g.

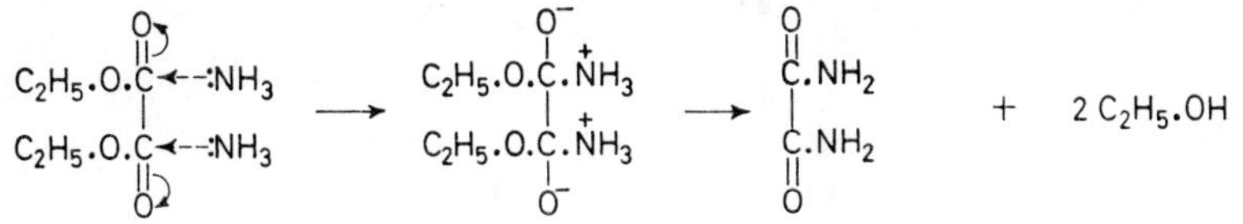

Diethyl oxalate　　Oxamide

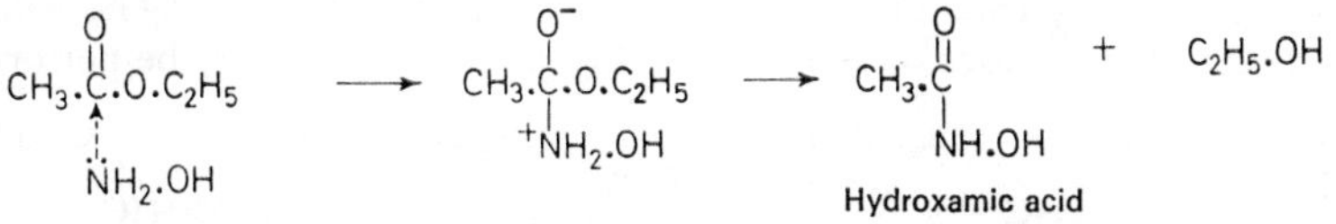

Hydroxamic acid

Hydroxamic acids give characteristic colours with iron(III) chloride, and the reaction therefore provides a useful test for esters, although acid chlorides also react in this manner.

The carbonyl group in an ester can be reduced to an alcohol by the use of lithium tetrahydridoaluminate:

$$R\text{CO.OR}' + 4[\text{H}] \rightarrow R\text{CH}_2\text{.OH} + R'\text{OH}$$

Grignard reagents react with esters to form tertiary alcohols:

$$R\overset{\overset{\displaystyle \text{O}}{\|}}{\text{C}}\text{.OR}' + R''\text{Mg}X \xrightarrow{\text{H}_2\text{O}} R\underset{\underset{\displaystyle R''}{|}}{\overset{\overset{\displaystyle \text{OMg}X}{|}}{\text{C}}}\text{—}OR' \rightarrow R\overset{\overset{\displaystyle \text{O}}{\|}}{\text{C}}R'' + \text{Mg(OH)}X + R'\text{OH}$$

$$R\overset{\overset{\displaystyle \text{O}}{\|}}{\text{C}}R'' + R''\text{Mg}X \rightarrow R''\underset{\underset{\displaystyle R}{|}}{\overset{\overset{\displaystyle \text{OMg}X}{|}}{\text{C}}}R'' \xrightarrow{\text{H}^+} R''\underset{\underset{\displaystyle R}{|}}{\overset{\overset{\displaystyle \text{OH}}{|}}{\text{C}}}R'' + \text{Mg(OH)}X$$

Malonic and Acetoacetic Esters

It has been pointed out that the presence of two oxygen atoms attached to the same carbon atom results in such a marked inductive effect that hydrogen attached to an oxygen is able to ionize

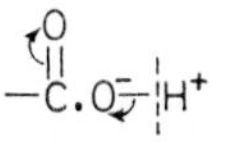

In the case of 1,3-diketones, electron drift towards the oxygen renders the hydrogen attached to carbon-2 so labile that the ketone is in equilibrium with the *enol* form. This type of dynamic isomerism is known as *tautomerism*:

$$
\begin{array}{ccc}
\overset{\displaystyle H}{\underset{1\ \ \ 2|\ \ \ 3}{-C-C-C-}} & \rightleftharpoons & \overset{\displaystyle H}{-C-C=C-}
\end{array}
$$

$$
\underset{\text{(keto)}}{\overset{\|\ \ |\ \ \|}{O\ \ H\ \ O}} \qquad \underset{\text{(enol)}}{\overset{\|\ \ \ \ \ |}{O\ \ \ \ OH}}
$$

Diethyl propane-1,3-dioate (*diethyl malonate*) is useful in synthesis because of this keto–enol isomerism. The hydrogen of the methylene group may be replaced by sodium, and then a series of reactions such as the following may be performed:

$$
\underset{\text{Keto-}}{H_2C\!\!\begin{array}{l}\diagup CO.OC_2H_5 \\ \diagdown \underset{\|}{\underset{O}{C.OC_2H_5}}\end{array}}
\rightleftharpoons
\underset{\text{Enol-}}{HC\!\!\begin{array}{l}\diagup CO.OC_2H_5 \\ \diagdown \underset{OH}{C.OC_2H_5}\end{array}}
\xrightarrow[-C_2H_5OH]{C_2H_5O^-Na^+}
HC\!\!\begin{array}{l}\diagup CO.OC_2H_5 \\ \diagdown \underset{O^-Na^+}{C.OC_2H_5}\end{array}
\rightleftharpoons
Na^+HC^-\!\!\begin{array}{l}\diagup CO.OC_2H_5 \\ \diagdown CO.\,OC_2H_5\end{array}
$$

$$
Na^+HC^-\!\!\begin{array}{l}\diagup CO.OC_2H_5 \\ \diagdown CO.OC_2H_5\end{array}
\xrightarrow[(-NaBr)]{RBr}
RHC\!\!\begin{array}{l}\diagup CO.OC_2H_5 \\ \diagdown CO.OC_2H_5\end{array}
\xrightarrow[(-2C_2H_5OH)]{2H_2O}
RHC\!\!\begin{array}{l}\diagup CO.OH \\ \diagdown CO.OH\end{array}
\xrightarrow[(-CO_2)]{Heat}
RCH_2.CO.OH
$$

Ethyl butane-3-oxo-1-carboxylate (*acetoacetic ester*) also exhibits tautomerism; in fact, the equilibrium mixture contains about 7 per cent of the enol form. Like diethyl malonate it is a very reactive compound and is used widely in synthetic work, in a manner similar to that of the former compound, e.g.

$$
H_2C\!\!\begin{array}{l}\diagup CO.OC_2H_5 \\ \diagdown \underset{\|}{\underset{O}{C.CH_3}}\end{array}
\rightleftharpoons
HC\!\!\begin{array}{l}\diagup CO.OC_2H_5 \\ \diagdown \underset{OH}{C.CH_3}\end{array}
\xrightarrow[-C_2H_5OH)]{C_2H_5O^-Na^+}
HC\!\!\begin{array}{l}\diagup CO.OC_2H_5 \\ \diagdown \underset{O^-Na^+}{C.CH_3}\end{array}
$$

$$\Updownarrow$$

$$
\overset{R}{\underset{H}{}}{\diagdown\!\!\!\diagup}C\!\!\begin{array}{l}\diagup CO.OC_2H_5 \\ \diagdown \underset{\|}{\underset{O}{C.CH_3}}\end{array}
\xleftarrow[(-NaX)]{RX}
Na^+HC^-\!\!\begin{array}{l}\diagup CO.OC_2H_5 \\ \diagdown \underset{\|}{\underset{O}{C.CH_3}}\end{array}
$$

The product can be hydrolysed in two ways:

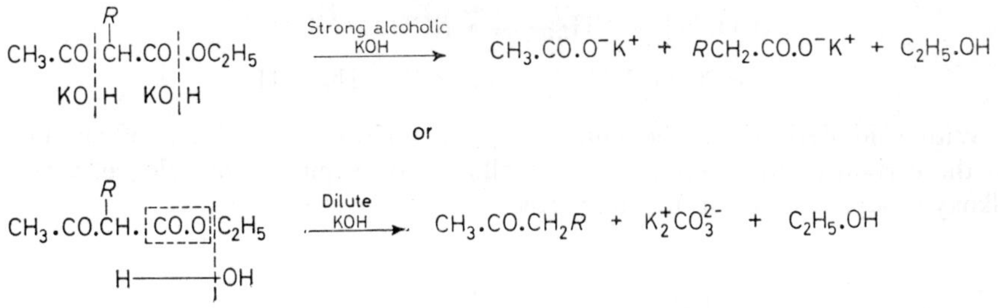

$$CH_3.CO \vert CH.CO \vert .OC_2H_5 \xrightarrow[\text{KOH}]{\text{Strong alcoholic}} CH_3.CO.O^-K^+ + RCH_2.CO.O^-K^+ + C_2H_5.OH$$

with R above the CH and $KO\vert H \quad KO\vert H$ below

or

$$CH_3.CO.CH.\vert CO.O\vert C_2H_5 \xrightarrow[\text{KOH}]{\text{Dilute}} CH_3.CO.CH_2R + K_2^+CO_3^{2-} + C_2H_5.OH$$

with R above and $H \quad\quad OH$ below

Analysis

The presence of an ester is suggested by the hydroxamic acid test. The ester should then be hydrolysed and the acid and alcohol or phenol separated and identified.

Industrial Applications

Esters are widely used as solvents, particularly for certain types of paints and varnishes. Since they provide the flavouring for many natural products, they can be used in synthetic essences.

Enormous quantities of oils and fats are hydrolysed and converted into soaps, sodium hydroxide producing hard, and potassium hydroxide soft, soaps.

In more recent years, considerable advances have been made in the manufacture of long-chain polyesters from difunctional acids and alcohols. For example, Terylene is obtained by polymerizing ethanediol with methyl benzene-1,4-dicarboxylate (methyl terephthalate)

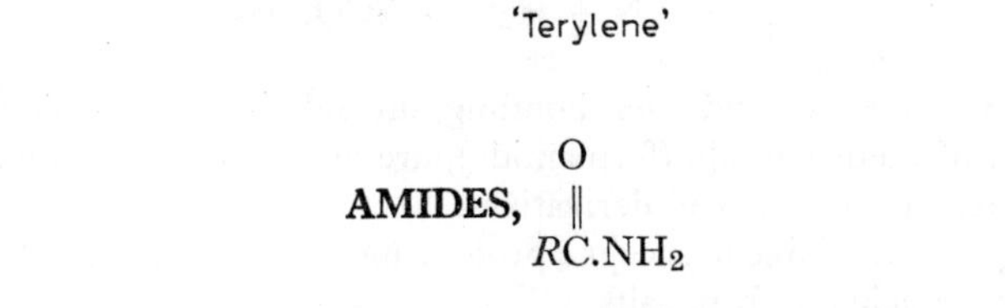

$$...+ H\vert O.CH_2.CH_2.O\vert H + CH_3.O\vert OC-\langle\text{benzene}\rangle-CO.\vert O.CH_3 +...$$

$$\xrightarrow{(-CH_3OH)} ..-O.CH_2.CH_2.O.OC-\langle\text{benzene}\rangle-CO-.....$$

'Terylene'

$$\textbf{AMIDES,} \quad RC\overset{O}{\underset{\|}{}}.NH_2$$

The formula of amides suggests that they can be prepared from **ammonia, and** this is indeed the case; acids or any of the derivatives mentioned earlier in this

507

Chapter can be used as the other reactant. With acids, the first stage of the reaction is salt formation, followed by elimination of water on distillation

$$RCO.OH + NH_3 \longrightarrow RCO.O^-NH_4^+$$

$$RCO.O^-NH_4^+ \longrightarrow RCO.NH_2 + H_2O$$

With acid derivatives, the initial stage is nucleophilic addition of ammonia to the carbon of the carbonyl group, followed by expulsion of halogen, acyl or alkoxy groups from chlorides, anhydrides or esters, respectively

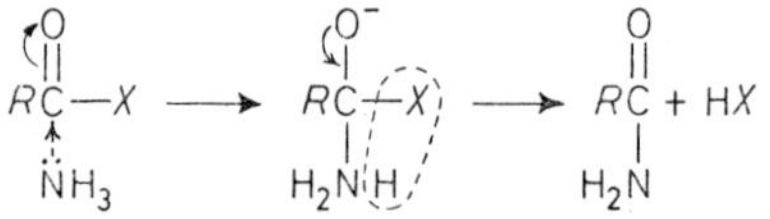

e.g. $CH_3.CO.OC_2H_5 + NH_3 \rightarrow CH_3.CO.NH_2 + C_2H_5.OH$
Ethyl ethanoate   Ethanamide
(ethyl acetate)   (acetamide)

$COCl_2 + 2NH_3 \rightarrow CO(NH_2)_2 + 2HCl$
Carbonyl   Carbamide
chloride   (urea)

Substituted amides, i.e. with alkyl or aryl groups attached to the nitrogen, can be similarly prepared by using amines instead of ammonia, e.g.

A different method of preparation involves the hydrolysis of nitriles with either acid or alkali. The preliminary step is the addition of the hydrogen ion to the carbon, or the hydroxide ion to the nitrogen, of the nitrile bond (p. 483)

$$R-C\equiv N + H_2O \rightarrow RCONH_2$$

Properties

Amides, on account of hydrogen bonding, are relatively non-volatile, and with the exception of methanamide (formamide), are all white crystalline solids. They are the least reactive of the acid derivatives.

The nitrogen atom is able to accept a proton, and so amides have basic tendencies. They dissolve in acids to form salts

$$RCO.NH_2 + H^+ \rightarrow RCO.NH_3^+$$

Amides liberate ammonia on boiling with alkali (unlike ammonium salts which liberate ammonia even with cold alkali)

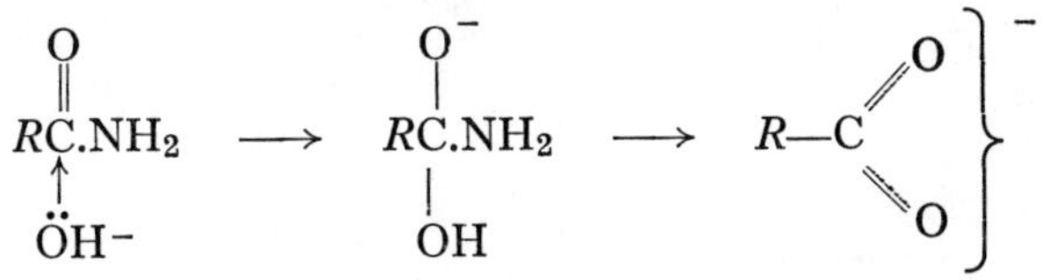

$$\underset{\underset{\ddot{O}H^-}{\uparrow}}{R\overset{\overset{\displaystyle O}{\parallel}}{C}.NH_2} \longrightarrow \underset{\underset{OH}{|}}{R\overset{\overset{\displaystyle O^-}{|}}{C}.NH_2} \longrightarrow \left. R-C\underset{\diagdown\diagdown O}{\diagup\diagup O} \right\}^-$$

Distillation of amides with phosphorus(V) oxide results in dehydration to the nitrile

$$RCO.NH_2 \rightarrow RC\equiv N + H_2O$$

If amides are treated with a cold, dilute mixture of sodium nitrite and hydrochloric acid, nitrogen is evolved and the carboxylic acid is formed

$$RCO.NH_2 + HONO \rightarrow RCO.OH + N_2 \uparrow + H_2O$$

This reaction should be compared with those of amines and nitrous acid (p. 414).

Another important reaction of amides is the *Hofmann degradation* (p. 412), so called because it provides a means of removing a carbon atom from an organic compound and, hence, of descending an homologous series, e.g.

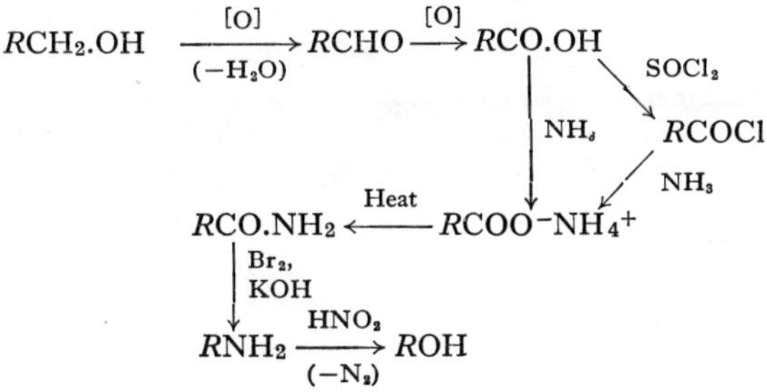

$$RCH_2.OH \xrightarrow[(-H_2O)]{[O]} RCHO \xrightarrow{[O]} RCO.OH$$

(with reaction scheme:)
$$RCO.OH \searrow SOCl_2$$
$$\downarrow NH_4 \qquad RCOCl$$
$$\xrightarrow{Heat} \qquad \nearrow NH_3$$
$$RCO.NH_2 \longleftarrow RCOO^-NH_4^+$$
$$\downarrow \begin{matrix}Br_2, \\ KOH\end{matrix}$$
$$RNH_2 \xrightarrow[(-N_2)]{HNO_2} ROH$$

Carbamide (urea), $O{=}C(NH_2)_2$
under similar treatment breaks down into carbonate and nitrogen

$$CO(NH_2)_2 + 4OH^- + Br_2 \rightarrow 2Br^- + CO_3^{2-} + 2H_2O + N_2 \uparrow$$

The nitrogen can be measured in a gas burette and affords a method of estimating urea quantitatively.

Sulphonamides

The amides so far described have been derived from carboxylic acids, but amides of sulphonic acids exist and are prepared by analogous methods, e.g.

$$C_6H_5.SO_2Cl \xrightarrow[(-HCl)]{NH_3} C_6H_5.SO_2.NH_2$$

Benzenesulphonyl chloride   Benzenesulphonamide

Many sulphonamide derivatives are valuable drugs; e.g. 'M & B 693' is derived from 4-aminobenzenesulphonic (sulphanilic) acid)

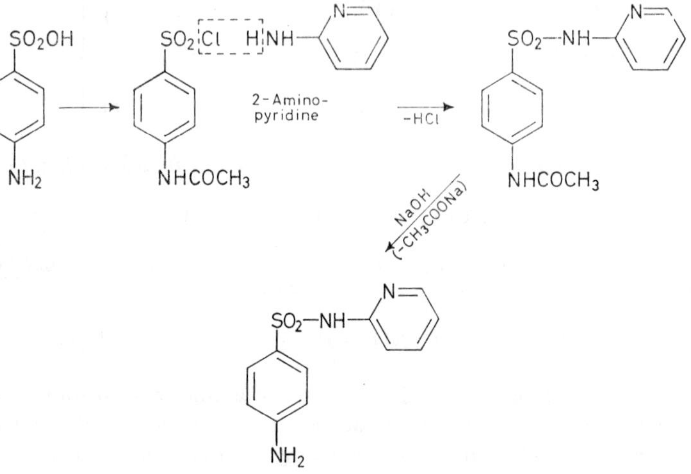

M & B 693

These drugs bear a resemblance to 4-aminobenzoic acid, and it is believed that the invading bacteria are unable to distinguish between the latter, vital to their well-being, and the drug (or at least a breakdown product of it). Metabolic processes consequently go awry, with disastrous results for the bacteria:

4-Aminobenzoic acid 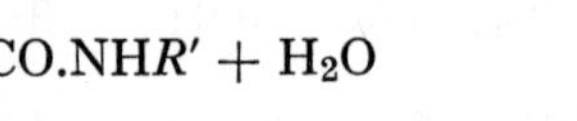   Sulphanilamide

Peptides and Proteins

Monobasic carboxylic acids and monoamines condense to form substituted amides

$$RCO.OH + R'NH_2 \rightarrow RCO.NHR' + H_2O$$

If amino acids are involved in condensation, then clearly the process can be perpetuated, and a long fibre results

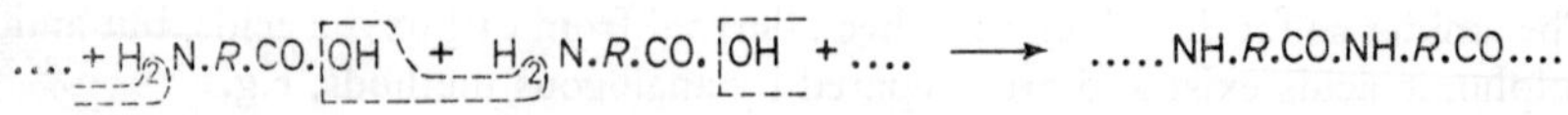

$$\dots + H_2N.R.CO.OH + H_2N.R.CO.OH + \dots \longrightarrow \dots NH.R.CO.NH.R.CO\dots$$

The result of this polymerization, if carried to sufficient lengths, is a *protein* molecule, and the repeating—CO.NH—linkage, representing the residues of the acid and amine functions, is called the *peptide* link.

Proteins are vital to living processes, and their production comprises a fundamental part of the activity of the living cell. An impressive feature of the metabolism of the cell is the specificity of protein synthesis, that is to say, the manner in which a cell makes the correct protein. For example, a muscle cell makes muscle and not hair tissue from the amino acids available in the cell.

This specificity of protein synthesis is a consequence of the precise and peculiar arrangement of the bases in the relevant RNA molecules (p. 427). As this arrangement is itself a consequence of the formulation of the base in the parent DNA molecule from which the RNA is generated, it follows that, unless something goes amiss and a mutation occurs, the types of proteins synthesized in a particular cell will be directed by the types of DNA present.

It is believed that every one of the twenty or so amino acids involved has its own particular 'transfer' RNA. The amino acid is first of all activated by reaction with an enzyme and a molecule of adenosine triphosphate (ATP). It is then transferred to its own RNA molecule by exchange of the adenine in the ATP residue for adenine in the RNA. The RNA molecule, with its amino acid attached, links with the appropriate and complementary site of the 'template' RNA through, it is thought, three base units. As the process continues with other 'transfer' RNA molecules and their corresponding amino acids, a situation is arrived at where, because of the complementary nature of the 'template' and the much smaller 'transfer' RNA molecules, amino acids are brought together in the correct sequence; condensation than takes place with the formation of a *polypeptide*, which subsequently leaves the RNA as a specific protein (*Figures 26.1* and *26.2*).

Proteins can be reconverted into amino acids by hydrolysis, during digestive processes; often the degradation is only partial and produces molecules intermediate between the protein and the individual amino acids, and known as *peptones*, which are capable of build-up into new proteins. The amino acids themselves can be further degraded and de-aminated by respiratory processes (p. 477)

$$RCH.CO.OH \xrightarrow[-2H]{} RC.CO.OH \xrightarrow{H_2O} RC.CO.OH + NH_3$$

$$\underset{NH_2}{|} \qquad \underset{NH}{||} \qquad \underset{O}{||} \searrow \qquad \searrow$$

$$\text{Pyruvic acid, etc.} \qquad \text{Urea, etc.}$$

A qualitative test for proteins is the *biuret* test. Compounds containing the peptide link,—CO.NH—, give a purple colour on treatment with sodium hydroxide and a drop of copper(II) sulphate solution. Biuret itself is obtained by the action of heat on urea, two molecules condensing together by the elimination of one molecule of ammonia. Because it contains the peptide link, biuret gives a positive result with the above test

$$H_2N.CO.NH|H + H_2N|CO.NH_2 \xrightarrow{-NH_3} H_2N.CO.NH.CO.NH_2$$
$$\text{Biuret}$$

511

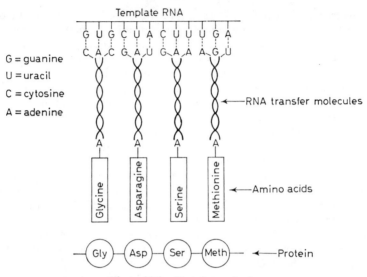

G = guanine
U = uracil
C = cytosine
A = adenine

Figure 26.1. Protein synthesis

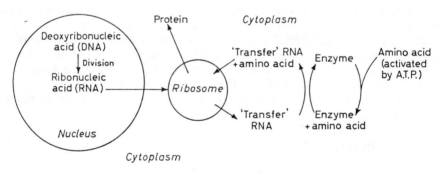

Figure 26.2. Cellular processes

Industrial Aspects

Nylon is undoubtedly the most common synthetic polyamide and is manufactured by the condensation polymerization of, for example, 1,6-diaminohexane with hexane-1,6-dioic acid (or the acid chloride):

$$...+ H_2N(CH_2)_6 NH_2 \; + \; HO.OC(CH_2)_4CO.OH +... \longrightarrow ...HN(CH_2)_6NH.CO(CH_2)_4CO...$$

6,6-Nylon

512

Both reactants are made from cyclohexane (which can itself be obtained from benzene or phenol):

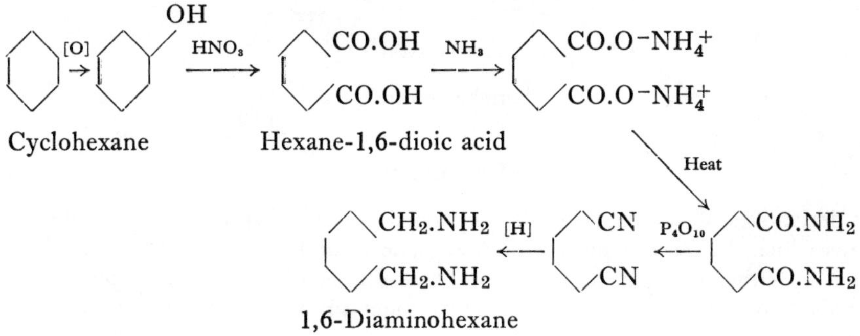

Cyclohexane   Hexane-1,6-dioic acid

1,6-Diaminohexane

Urea is utilized in the manufacture of resins. In the presence of alkali it reacts with methanal (formaldehyde);

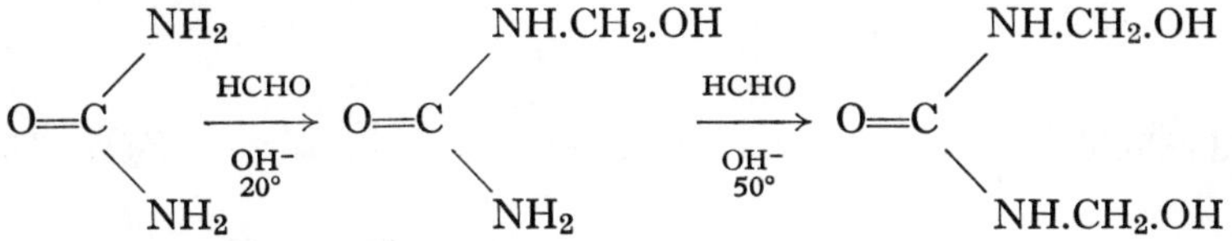

With a suitable catalyst, for example triethylphosphate, this product polymerizes to give two- and three-dimensional structures based on the unit

$$\ldots O.CH_2.NH.CO.NH.CH_2 \ldots$$

Urea also condenses with carboxylic acids; if dibasic acids are involved, then cyclic derivatives called *ureides* are formed, e.g.

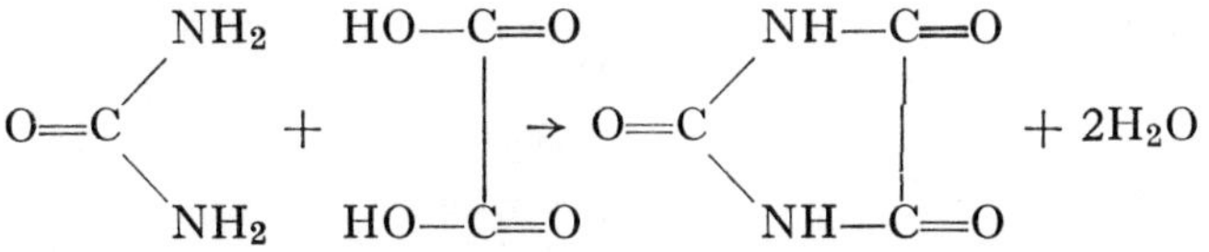

Malonic acid gives rise to *barbiturates,* useful as sedatives and hypnotics; for example, phenobarbitone is the ureide of ethylphenylmalonic acid:

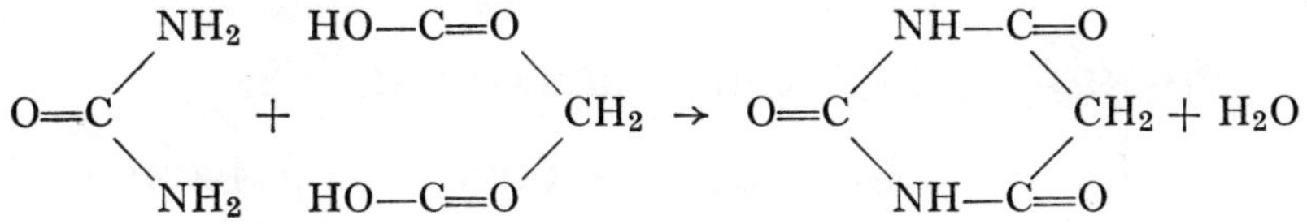

Barbituric acid

513

$$
\begin{array}{c}
\text{NH—C=O} \quad \text{C}_2\text{H}_5 \\
\text{O=C} \qquad \text{C} \\
\text{NH—C=O} \quad \text{C}_6\text{H}_5
\end{array}
$$

Phenobarbitone

SUMMARY

The derivatives of carboxylic acids mentioned in this Chapter possess carbonyl groups which, because of polarization, are able to undergo nucleophilic attack at carbon by either nitrogen or oxygen. The readiness with which reaction ensues depends upon the ease with which substituent groups leave the intermediate formed

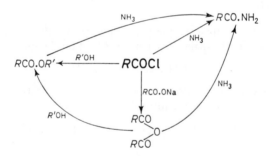

Acid chlorides are the most, and acid amides the least, reactive (*Figure 26.3*).

$$
\begin{array}{c}
\text{NH}_3 \longrightarrow R\text{CO.NH}_2 \\
\text{NH}_3 \\
R\text{CO.OR}' \xleftarrow{R'\text{OH}} R\text{COCl} \\
R\text{CO.ONa} \\
\text{NH}_3 \\
R'\text{OH} \\
\begin{array}{c} R\text{CO} \\ R\text{CO} \end{array}\!\!\Big\rangle\text{O}
\end{array}
$$

Figure 26.3

QUESTIONS

1. Name the compounds and give details of the reactions indicated in the flow chart

$$
\begin{array}{ccc}
\text{CH}_3.\text{NH}_2 & \text{CH}_3.\text{CO.OH} & \text{CH}_3.\text{CO.OC}_2\text{H}_5 \\
\uparrow & \uparrow & \uparrow \\
& \text{CH}_3\text{CO.NH}_2 \leftarrow \text{CH}_3.\text{COCl} \longrightarrow (\text{CH}_3.\text{CO})_2\text{O} \\
\downarrow & \downarrow & \downarrow \\
[(\text{CH}_3)_4\text{N}]^+\text{Br}^- & \text{CH}_3.\text{CN} & \text{CH}_3.\text{CHO}
\end{array}
$$

514

2. How could 2-methylprop-2-enoic acid (methacrylic acid) be prepared from propanone? Perspex is formed by polymerizing the methyl ester of this acid; give the structural unit of the polymer.

3. How might phthalimide, $\begin{array}{c}CO\\ \diagdown\\ NH,\\ \diagup\\ C\end{array}$ be prepared from naphthalene?

4. The following steps are involved in the synthesis of a compound, starting from methylbenzene: (a) treatment with chlorosulphuric acid and separation of the *ortho* derivative; (b) oxidation of this with potassium permanganate; (c) reaction with ammonia; and (d) heating of the product to eliminate water. Deduce the formula of the compound and give the experimental conditions for each step. What is the trivial name of the product?

5. A compound A, of molecular formula C_3H_5NO, on boiling with dilute hydrochloric acid produced an acid B, $C_3H_6O_3$, which on oxidation gave a dibasic acid, C. C formed an acid anhydride on heating alone. Identify A, B and C and give equations for the reactions involved.

6. Describe how bromoethane might be prepared from 1-bromopropane.

7. A compound A, C_3H_5N, when treated with dilute acid yielded B, $C_3H_6O_2$. This compound, when treated with ammonia and heated, gave C, C_3H_7ON. C when boiled with alkali liberated ammonia and a solid residue, D, remained. This was dissolved in water and acidified. Extraction with ether yielded a compound E which, upon treatment with lithium tetrahydridoaluminate, gave propan-1-ol. C when treated with bromine and alkali gave substance F which, with nitrous acid, yielded some ethanol. Deduce the nature of A–F and write equations for all the reactions.

8. Show, by means of structures, the possible reactions of linseed oil with oxygen.

9. Indicate how
(a) propan-1-ol can be converted into ethanol,
(b) propan-1-ol can be converted into butan-1-ol.

10. Write an account of the industrial utilization of esters.

11. Write an essay on proteins.

12. 4·0 g of commercial aspirin were heated on a water-bath for half an hour with 100 cm³ of м NaOH. After cooling, it was made up to 500 cm³ of solution and 25 cm³ were titrated against 0·1 м HCl, with phenol red as indicator: 31·2 cm³ of acid were required to neutralize the excess alkali. Say what you can about these reactions and about the condition of the original aspirin.

515

CHAPTER 27

MODERN EXPERIMENTAL METHODS

Introduction

The investigation of an unknown substance roughly comprises the following steps: isolation, purification and analysis. The ultimate aim is then the identification of the different atomic groups and their orientation in the molecule, so giving a complete *structural formula*. However, first the atomic ratios of the different atoms must be determined to provide the *empirical formula*, from which, knowing the molecular weight, the *molecular formula* can be deduced.

Example—A substance, of molecular weight 60, contains 40·0 per cent carbon, 6·7 per cent hydrogen and 53·3 per cent oxygen. Dividing by the relevant atomic weights gives the atomic ratios as 40·0/12 for carbon, 6·7/1 for hydrogen and 53·3/16 for oxygen. Changing these values to the smallest whole numbers gives the empirical formula as $C_1H_2O_1$. But as the molecular weight is 60, the molecular formula must be twice the empirical formula, i.e. $C_2H_4O_2$, and the possible structural formulae are

$$\underset{\text{CH}_3.\text{COH}}{\overset{\overset{\textstyle O}{\|}}{}} \quad \text{and} \quad \underset{\text{HCOCH}_3}{\overset{\overset{\textstyle O}{\|}}{}}$$

For complex molecules, particularly naturally occurring organic compounds, independent methods of synthesis are often explored to completely confirm their structures. Further elucidation involves a detailed examination of the physical properties of the compound and the mechanisms by which it takes part in chemical reactions, with a possible view to the commercial use of the substance.

Few reactions, particularly in organic chemistry, give simple, readily separable products, and this factor, together with the yield obtainable, limits the usefulness of many reactions. However, with the improvement in physical techniques, separation of many complex mixtures can now be made. For example, the separation of the amines resulting from the reaction between organic halides and ammonia under pressure (p. 401) now relies on the precise control which can be applied in fractional distillation, whereas the alternative has long been a lengthy chemical separation.

The modern use of *vacuum lines*, which consist of closed systems of reaction vessels suitably connected, so that the addition of reagents, their reaction together and the products formed can be maintained under controlled conditions, has allowed the separation of many new compounds which exist only under special circumstances, e.g. in inert atmospheres at low temperatures. The instability of many of these substances causes considerable difficulty in determining their molecular formulae, let alone their structures.

Similarly, *flash photolysis* and the use of *shock waves* result in the transient formation of new species. In the former method, high energies of suitable frequencies are supplied for a very short time (e.g. a few microseconds), whilst in the latter, the reaction vessel is divided by a partition which can be readily pierced so that, by building up a pressure on one side, a shock wave can pass along the rest of the tube when rupture of the membrane is made; the shock wave produces a high temperature by adiabatic compression. By these methods, species of short life can be observed spectroscopically (p. 523).

The increasing demand for high purity, required in, e.g. semi-conductor work, has meant that the classical methods of analysis can no longer be used. Instead, more sensitive methods, based on atomic and molecular properties (e.g. spectral analysis) have been developed. The use of these methods in turn requires high purity of the primary standards, and this may be achieved by, for example, chromatographic separation. In the majority of cases the criterion of purity is decided spectroscopically, reliance on the old-established methods of sharp melting and boiling points generally being impossible because of the instability of the substances at the high temperature necessary.

Methods of determining molecular weights have already been dealt with in Chapters 4 and 5. Besides these methods there are those based on mass spectroscopy (p. 521), measurements of diffusion in solutions and the rates of sedimentation under centrifugal forces, the latter being used exclusively for colloidal systems.

A certain fraction of the molecules of a compound can be 'labelled' by making one of the elements radioactive or by incorporating a different isotope into the molecule. Such methods find application in the investigation of reaction mechanisms, e.g. the sequence of reactions involving radioactive sulphur, $S\star$,

$$S\star + SO_3^{2-} \xrightarrow{\text{boil}} S\star SO_3^{2-} \xrightarrow{2H^+} S\star \downarrow + H_2O + SO_2 \uparrow$$

shows the non-equivalence of the sulphur atoms in the thiosulphate ion since, if they were in similar positions, the radioactivity of the sulphur precipitated would be halved, whilst in fact all the radioactivity is recovered in the sulphur liberated.

ISOLATION AND PURIFICATION

The methods available are:

(*i*) *Filtration*, which may be used when the particle size is too large for the substance to pass through the pores of a filter medium. It can also be carried out under reduced pressure as well as at any suitable temperature, particularly if use is made of the range of filter media of controlled porosity and inertness which are available.

(*ii*) *Fractional crystallization* may be used to separate solids of differing solubilities in a particular solvent; precise temperature control is necessary, and usually several recrystallizations are required to obtain satisfactory separation. Also, for this process to be suitable, conditions must be such that one of the substances will crystallize from the mother liquor before the solution becomes saturated with respect to the other solutes.

(*iii*) *Centrifuging* is an alternative to filtration, and by using high-speed (ultra-) centrifuges, values of up to 250 000 times the acceleration due to gravity may be obtained, allowing the separation of substances in the colloidal state.

(*iv*) *Distillation*—Steam and fractional distillations have been discussed already (pp. 29–103). Distillation under reduced pressure clearly allows a lowering of the normal boiling point and is therefore useful for substances which tend to decompose on heating. In *molecular distillation*, the mean free path (the average distance travelled by a gas molecule between successive collisions) of the vapour molecules produced by reducing the pressure to less than $0\cdot13$ N m^{-2} (10^{-3} mmHg), is increased to a few centimetres, allowing molecules from a liquid film to travel straight to an adjacent cool surface where condensation occurs (*Figure 27.1*); using molecular distillation, liquid mixtures may be fractionated at more than 100° below their normal boiling points.

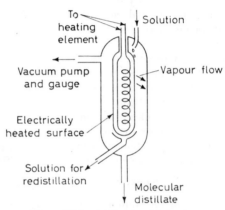

Figure 27.1. Molecular distillation

(*v*) *Solvent extraction*—The distribution of solutes between two immiscible solvents (see p. 97 for simple partition) has been made into a semi-continuous process—*countercurrent distribution*. After each extraction, the mixtures are allowed to separate, and the lighter fraction is transferred to another extraction tube containing the heavier solvent, whilst more of the lighter solvent is added to the residual heavy medium. The extraction is repeated in this manner many times, and as it proceeds from tube to tube, separation of the solutes occurs. Typical distribution curves of the solutes between the solvents are shown in *Figure 27.2*.

(*vi*) *Chromatographic separation* represents a continuous extraction process using one mobile solvent with either (*a*) a solid adsorbent replacing the second solvent (*adsorption chromatography*) or (*b*) the second solvent acting as a stationary phase because of its adsorption on to a solid. In *paper chromatography*, for instance, the moisture adsorbed on the cellulose fibres acts as the immobile second solvent. The mobile solvent may also be a gas (*gas chromatography*) *Figure 27.3*, and a further possibility is that, if the solutes carry any charge, then application of an electric potential across the immobile phase may aid separation (*continuous electrophoresis*). To a first approximation, the stationary phase takes up the substance of highest adsorbability or solubility to the exclusion of the other solutes; in any case, complete separation can be effected by drawing a suitable *eluting agent* through the column

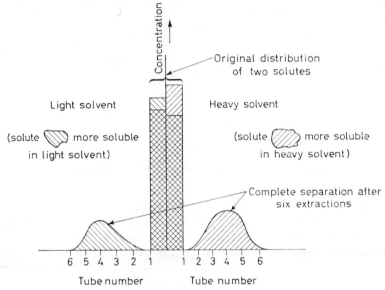

Figure 27.2. Solvent extraction

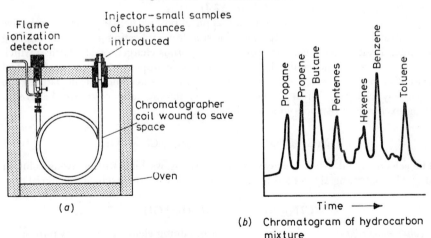

Figure 27.3. Gas chromatography: (*a*) The basic unit. (*b*) A chromatogram.

519

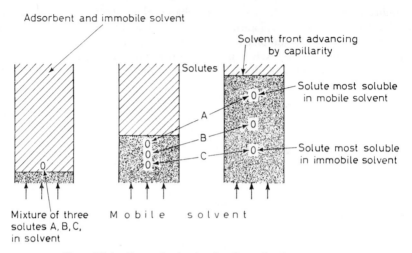

Figure 27.4. Stages in the development of a chromatogram

of the adsorbent. In gas chromatography, the vapour of the substance least adsorbed on the stationary phase is swept along by the carrier gas (e.g. argon or nitrogen), reaches the end of the adsorption column first and can be detected, for example, by observations based on the changes in the thermal conductivity of the gas mixtures. The extraction of the separated solutes on a *chromatogram* depends on firstly identifying their positions; for solutes which are not coloured, reliance is placed either on their response to ultra-violet light or to reagents added after drying the chromatogram (*Figure 27.4*).

(*vii*) *Ion exchange* finds widespread application, particularly for the selective concentration of certain ionic species. The simplest application is in the de-ionizing of mineral water, using columns packed with porous charged crystal lattices which maintain electrical neutrality by adsorbed ions of opposite charge. Zeolites and most clay minerals are cation exchange materials, since the silicate lattices are negatively charged, whilst artificial resins containing amine groups behave as anion exchange substances in acid solution, e.g.

$$—N(CH_3)_4^+OH^- \xrightarrow{Cl^-} —N(CH_3)_4^+Cl^- + OH^-$$

The selectively adsorbed ions can be displaced by washing with a strong solution of a salt containing the ions originally associated with the exchange material, e.g.

$$—N(CH_3)_4^+Cl^- \xrightarrow{OH^-} —N(CH_3)_4^+OH^- + Cl^-$$

(*viii*) *Zone refining* is now much used for obtaining elements in a very pure state It consists of extruding the impure substance slowly through a high-temperature

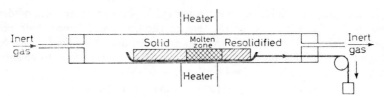

Figure 27.5. **Zone refining**

region. The impurities, provided their solubility increases with temperature, concentrate in the hotter part of the specimen, so that after several cycles most of the contaminants will have been swept along to one end of the sample (*Figure 27.5*).

ANALYSIS

Physical methods of analysis are now largely used in place of chemical methods, although the latter may still be required for primary standardization. The chemical methods of quantitative analysis depend upon the decomposition of the substance into simpler molecules which can be determined by reaction or absorption in suitable reagents (for example p. 392). The principles and applications of some of the physical techniques employed are:

1) *Mass spectroscopy* is a technique in which ions of the substance are produced by the use of electron bombardment at low pressures. The ions are focused by

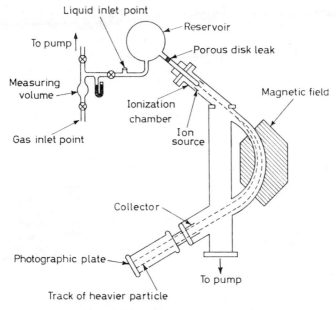

Figure 27.6. The mass spectrograph

521

electric and magnetic fields on to a photographic plate or other form of detector (*Figure 27.6*). The deflection of the ions in the fields, and hence their position on the plate, depends upon their mass and charge, and this affords a method of identifying the species present; for example, ethanol gives rise to twenty-four fragments, with the following ions predominating (*Figure 27.7*)

$$CH_3^+, C_2H_5^+, CH_2OH^+, C_2H_5O^+, C_2H_5OH_2^+$$

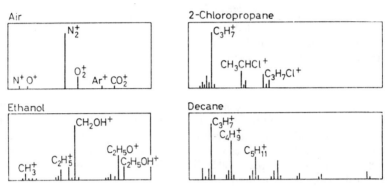

Figure 27.7. Mass Spectra

2) *Electromagnetic spectroscopy*—All substances are sensitive to some part of the electromagnetic spectrum (*Figure 27.8*), although comparatively few exhibit colour, i.e. respond to the visible part of the spectrum. Energy is absorbed in (*i*) *electronic transitions*, where electrons jump from one orbit to another, and involves larger energy changes than (*ii*) *vibrational transitions* which, in molecules, are represented by the stretching and contracting of covalent bonds, whilst (*iii*) *rotational transitions*, involving changes in the rotational energies of molecules,

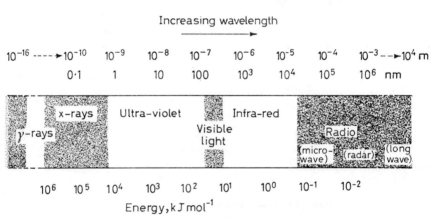

Figure 27.8. The electromagnetic spectrum

522

require only small energy changes. The energy diagram for a molecule can then be shown (*Figure 27.9*). Fortunately for the spectroscopist, changes cannot take place between all the energy levels shown, so that the spectral lines can in many cases be allotted to the correct energy levels.

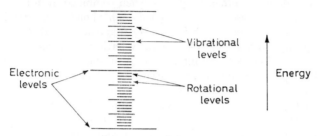

Figure 27.9. Molecular energy levels

Because of the relation $\Delta E = h\nu$ (where ΔE = the energy change, h = Planck's constant and ν = frequency of the spectral line) among the common spectral ranges, those of the infra-red involve the lowest frequencies, and the resulting spectra arise from the lowest energy transitions, i.e. from molecular vibration and rotation. For a diatomic group, the stretching vibration between the atoms is related to the bond strength, and spectral lines result from the vibrational frequency of the bond, as energy of the appropriate frequency is absorbed. Thus compounds containing this group produce a characteristic frequency, which is modified by the nature of the adjacent bonding. Rotational spectra are governed by the moments of inertia of the system, so that information regarding the mass and interatomic distances is obtained from these spectra. In general, infra-red spectra provide 'fingerprints' of the substances, permitting the identification of many groups (*Figure 27.10*).

The visible and ultra-violet spectra are more complicated, because electronic transitions are also possible. For substances called *fluorescers*, i.e. certain compounds containing loosely bound π electrons in delocalized systems, absorption of ultra-violet radiations results in the emission of visible light, of lower frequency

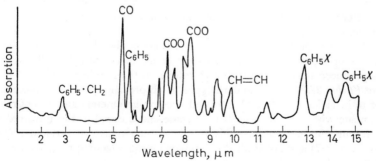

Figure 27.10. Infra-red spectrum of ethyl cinnamate

and energy, since the excited electrons do not return directly to their original energy states.

As indicated above, all these transitions are governed by quantum rules which explain the characteristic line and band structures in the spectra. The spectrum of a group may undergo a *shift* in frequency when combined in different molecules. This is particularly the case when conjugated double bonds, i.e. extended π clouds, are possible, and it results in a shift to lower frequencies so that, by introducing appropriate groups into a molecule originally having an ultra-violet spectrum, a spectrum in the visible range results: the new substance exhibits colour. Groups which produce this effect are called *chromophores* (*Table 27.1*).

The different types of spectroscopy are

(*a*) *emission spectroscopy*, in which energy is supplied to excite electrons to higher electronic states. When these electrons return to their lower orbits, characteristic radiations are emitted. The excitations may be accomplished by applying an electric discharge to the substance, resolving the spectrum using a prism or diffraction grating and finally recording the result photographically. In another form of apparatus (the *flame photometer*), a solution of the substances is drawn into a coal-gas flame under a constant air pressure and the radiations are passed through appropriate colour filters before measurement with a photoelectric cell. The flame

Table 27.1.   Chromophoric groups

Carbonyl	$>C{=}O$	Nitroso	$-N{=}O$
Nitro	$-N\!\!\nearrow^{O}_{\searrow O}$	Azo	$-N{=}N-$
Azoxy	$-N{=}N\!\!\nearrow_{\searrow O}$	Azoamino	$-N{=}N-N\!\!\nearrow_{\searrow H}$

photometer is used for the determination of sodium, potassium, calcium, strontium and barium ions in solution.

(*b*) *absorption spectroscopy*—Incident radiation is absorbed by molecules, provided it is of the correct frequency to satisfy the quantum rules. The absorption bands produced may give quantitative results on the number and types of bonds present (*Table 27.2*). The standard types of absorption spectroscopy include the use of ultra-violet, visible and infra-red spectrophotometers and also of measurements made with frequencies less than those encountered in the infra-red range. Such frequencies can produce only rotational transitions and inversions (e.g. see ammonia, p. 257) and are studied by employing radio techniques (*microwave spectroscopy*).

524

Spin resonance spectroscopy—As each extra-nuclear electron possesses spin (p. 22), so each nucleon also has spin. These spins are quantized and because both electrons and protons are charged, their spins give rise to magnetic moments just as do orbiting electrons. Under the influence of a suitable magnetic field, alignment of the spins either with or against the field occurs, producing two distinct energy levels. At suitable radio frequencies and high magnetic fields, adsorption of energy results, as nuclei jump from one energy level to another. This effect is called *nuclear magnetic resonance* (n.m.r.).

Nuclei capable of exhibiting n.m.r. are those which possess an odd number of protons, e.g. ^1_1H, $^{14}_7N$, and $^{19}_9F$. The method is generally used in identifying the structural units containing hydrogen nuclei, since the effective field at a nucleus depends not only on the applied field, but also on the shielding effect of the extra-nuclear electrons; for example, separate proton resonance lines are obtained for the hydrogen nuclei found in the $—CH_3$, $>CH_2$ and OH groups present in ethanol (*Figure 27.11*).

Table 27.2.

Spectral type	Wavelength range (nm)	Information obtained
n.m.r.	10^{11}—10^9	Environment of atom
e.s.r.	10^9—10^6	Electron distribution (and occupancy of orbitals)
Microwave	10^6—10^5	Shape of small molecules
Infra-red	10^5—10^3	Molecular geometry
Visible and Ultra-violet	10^4—10^2	Molecular energy levels

3) *Magnetic and electric measurements*—The pairing of electrons results in a substance possessing *diamagnetism* (i.e. it experiences a repulsion when placed in a magnetic field), whilst substances containing unpaired electrons exhibit *paramagnetism* (i.e. are attracted by a magnetic field) because of the increase in the number of lines of force passing through the specimen (*Figure 15.2*, p. 322). The latter phenomenon leads to the concept of *paramagnetic susceptibility*, the value of which is given by $\mu_M/3kT$, where μ_M = the magnetic moment, k = the Boltzmann constant and T = the absolute temperature (diamagnetic susceptibility is independent of the actual temperature). The importance of paramagnetism in indicating the number of unpaired electrons has been discussed in Chapter 15; determinations are made by finding the force resulting from the application of a powerful magnetic field to a long sample of the substance suspended in a glass tube from one arm of a balance.

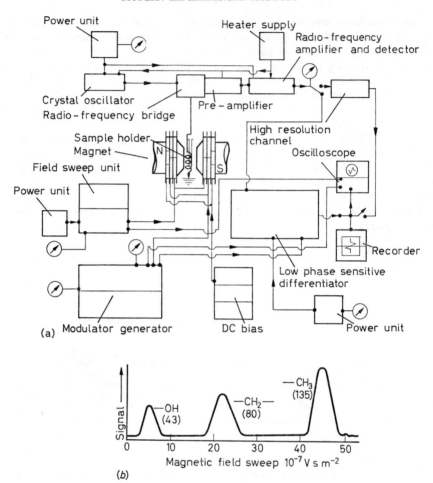

Figure 27.11. Nuclear magnetic resonance. (a) Layout of equipment. (b) n.m.r. spectrogram for ethanol

Dipole moments have been used for distinguishing between geometric isomers (e.g. p. 489); their values are found by measuring the capacities of a condenser when filled with the substance and when empty. The ratio of the two capacities gives the relative permittivity (*dielectric constant*) of the substance (*Figure 27.12*). The relation between the dielectric constant, ϵ, and the permanent *dipole moment*, μ, of a polar compound is given by

$$\frac{\epsilon - 1}{\epsilon + 2} \cdot \frac{M}{\rho} = \frac{4\pi}{3} L(\alpha_0 + \mu^2/3kT)$$

where M = molecular weight, ρ = density, L = Avogadro constant and a_0 = the distortion polarization arising from temporary displacement of the electrons and nuclei caused by the applied electrostatic field.

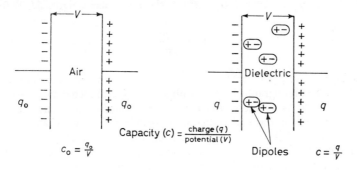

$$\epsilon = \frac{c}{c_0} = \frac{q}{q_0}$$

Figure 27.12. Dielectric constant, ϵ

4) *X-ray analysis*—Since the wavelengths of x-rays are comparable to the atomic diameters of most elements and therefore to the distance between the lattice planes in a crystal, reflections from these layers can produce interference patterns (*Figure 27.13*). In practice, monochromatic x-rays are used and, in order to produce a

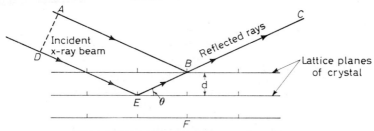

Figure 27.13. Diffraction of x-rays by a crystal
If the path difference between the reflected rays is equal to an integral number of wavelengths of the radiation, then a maximum occurs in the diffraction pattern.

sufficient number of diffraction spots on the photographic film, either a small crystal is rotated at the centre of a camera, moving laterally to and fro, or a stationary sample of powder (containing randomly orientated crystallites) is employed in conjunction with a stationary camera. In both cases, some lattice layers are bound to be in a position to produce diffraction patterns, but in the latter the powder photograph obtained is too complex for ready determination of the crystal structure, so that this method is employed only for comparing substances (cf. infra-red 'fingerprints').

CUA—T

X-ray diffraction is used initially to obtain the molecular weight from a determination of the cell dimensions of a crystal of the compound under investigation. The mass of substance in a unit cell of volume v is ρv, where $\rho = $ density, and by multiplying by the Avogadro constant, the molecular weight, M, is found

$$M = \frac{L\rho v}{n}$$

where n is the number of molecules in a unit cell. In practice, the information derived from the photographs is often sufficient also to predetermine the number n.

The final use of x-ray analysis in determining the absolute structure of a compound is time-consuming and difficult, particularly if no previous information, such as to the number and types of functional groups present, is available. The chief difficulty lies in determining the phase differences between the diffracted x-rays, so that the reflecting planes of the crystal can be correctly labelled. Sometimes a model is built, and the results expected from such a structure are compared with the actual diffraction pattern, giving further information from which the model can be refined, so that eventually the correct structure may be obtained.

5) *Polarography*—A certain potential is required to oxidize or reduce any particular ion, and when this value is reached, a surge of current results (*Figure 27.14*).

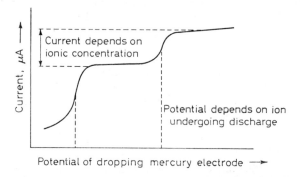

Figure 27.14. Polarography curve involving two dischargeable ons

For consistent results, the measuring electrode must not become contaminated with the products of the electrode reactions. This can be achieved by using a dropping mercury cathode (for redox reactions involving cations), the potential of which can be controlled, and a large pool of mercury as the anode (*Figure 27.15*). Polarographic techniques have revealed the existence of many ionic species previously unknown, besides furnishing standard methods of quantitative analysis at low concentrations (about 10^{-5} M).

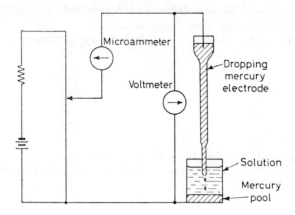

Figure 27.15. Basic circuit of the polarograph

QUESTIONS

1. Suggest how you could ascertain the molecular and structural formulae of sulphuric acid, phosphorus(V) chloride and ethanoic acid.

2. From *Figure 27.13* prove that the path difference on reflection from adjacent planes is equal to $2d \sin\theta$. (Hint: draw a perpendicular from B on to DE, produced to F). The relation then obtained for the constructive interference of the x-rays is known as Bragg's law.

3. Describe what ion exchange materials you would use in demineralizing tap water.

4. Suggest why the following compounds are coloured: butane-2,3-dione, prop-2-ene-1-al, 4-azoaminobenzene, cobalt(II) chloride and 'Prussian blue'.

5. Write an account of the various methods available to the chemist, for the purification of organic compounds.

6. Two sixth-form students obtained the following results for the gas chromatography of a sample of petrol.
One of the resistances of a Wheatstone bridge was placed in the stream of gas emerging from the adsorbent, its resistance being affected by its environment.

The bridge was adjusted for zero deflection before introduction of the sample of petrol: The following results were then obtained:

Deflection of galvanometer	Persistence of deflection (s)
1	31
2	40
2·5	152
6	30
2·5	30
22	170
1·5	60

(Rosser, W. E., *Sch. Sci. Rev.* 1967, **49**, 167, 180)

Say what you can about the composition of the petrol sample from these results.

7. Write an essay on spectroscopic methods in analysis.

CHAPTER 28

ASPECTS OF CHEMICAL INDUSTRY

Historical

We began this book by briefly tracing the development of chemical theory and from time to time we have mentioned industrial applications of certain chemical reactions. The impression might have been created that industrial processes are essentially developments of the last couple of centuries. If by 'industrial' we mean 'performed in a factory' this is no doubt true, but it would be wrong to suggest that scientific knowledge has only been applied for material benefit during this time. The Egyptians, for example, built up an impressive body of practical knowledge. Even the Romans, who are hardly remembered as scientific innovators, had well-laid down rules for performing technical tasks; for example, Vitruvius, writing about the making of concrete:

> 'Now where there is no quarry sand we must use washed river or sea sand. Sand which makes a noise when rubbed in the hand will be best; but that which is earthy will not have a like roughness . . . Also, if it is covered up in a white cloth and afterwards shaken up or beaten, and does not foul it and the earth does not settle therein, it will be suitable.'

Translated into modern English, these instructions would not disgrace a present day builders' manual. But this is merely a *recipe* and is no more scientific than cookery—it is based on shrewd observation and recording and little else. Clearly, if the questions 'How?' or 'Why?' were asked, they were not answered very successfully. A scientific revolution was indeed necessary and, as was shown in Chapter 1, it came with the investigations of Galileo and Kepler and the welding of apparently diverse information into Newton's theory of gravitation. This was the watershed that marked the separation of mystique and tradition from honest enquiry and open-mindedness. From this point on, scientific theory was more than a rationalization of fact—it embodied explanation in causal, deterministic terms, and provided a basis for speculation and further experiment.

On the face of it, this development would seem to have little connection with industrial life: it was more concerned with getting at the truth than with making things. But everything was changed by Newton. The entire Universe was now seen to be subject to natural laws and the invitation was there for man to discover them for himself. Societies were formed to discuss various scientific problems and to give lectures on specialized themes: one was the Lunar Society and James Watt was a member. By this time, steam was being used to drive engines for pumping water out of mines, but the whole process—even though using the knowledge and principles of the new scientific age—was very inefficient. Watt met Joseph

Black and talked over some of the problems facing the maker of steam engines; in the course of this he mentioned that a little steam was able to raise a much larger quantity of water to its boiling-point. Black then outlined his theory of latent heat to Watt. As Watt recounted later:

> 'I had not walked further than the Golf House, when the whole thing was arranged in my mind; the waste of heat could be avoided by keeping the cylinder at steam-heat and condensing the steam in a separate boiler.'

We see here the value of discussion between people of different background and interest and the contact of theory and practice (and by 'theory' is meant not a rationalization of practice, as was the case before the scientific revolution, but a device for explaining and unifying). Through the Lunar Society Watt was also to come into contact with Matthew Boulton (who was looking for a more efficient means of powering his Birmingham factory) and with the cannon-maker Wilkinson, who was able to provide the necessary expertise for boring cylinders. From this combination of talents emerged the Watt steam engine, which was to become the prime mover of the Industrial Revolution.

This development marks a shift from what has been called the *eotechnic* to the *palaeotechnic*. The former was a culture based on natural forces such as those of wind and water and natural materials such as wood and fur. It was pre-scientific in the sense outlined above and consisted of small societies with the members having fixed (ascribed) status and respecting what to them was a God-ordained hierarchy. The palaeotechnic culture, on the other hand, provided new forms of energy and transmuted natural materials into others more useful. Why was this change, though, begun in North-West Europe, rather than, for example, Italy where the Renaissance made its greatest mark and where Galileo, by attacking the dogma that had come down from the Greeks, had played a major part in over-throwing established scientific ideas?

When considering the development of industry there are other factors than those of pure science to take account of. Capital investment is needed, as well as people to take risks and to drive on others. The Protestant Ethic, arising out of the Reformation, is an important ingredient here—no longer was there nobility in poverty; rather was it regarded as a visitation of the sins of the fathers upon the children. 'Blest are the Poor' was reinterpreted as 'God helps those who help themselves'. Out of this new climate came the Middle Classes, the entrepreneurs and investors, working hard themselves, perhaps, but expecting a total effort from their employees, women and children included, who were clustered around the factories in the cheapest accommodation it was possible to provide. To some extent, the ideology of the palaeotechnic era was that of achievement-status, with the self-made man replacing the squire, and the factory-worker replacing the craftsmen. At least it was in theory—in reality the new order crystallized out into clear-cut divisions, the capitalist providing the money and the worker providing the labour, and from this was born Marxism and the Communist society where, theoretically the workers take over the means of production. Modern history tells us how the Western world was cleaved into two and how the Third World, that of the underdeveloped nations, makes up the economic trinity.

As societies become industrialized, they pass more and more from the *primary* industries of food and mining, to the *secondary* manufacturing industries and then to the *tertiary* service industries (*Table 28.1*). In Britain during the century 1814–1914, the percentage of the national labour force employed in agriculture and fishing fell from 34% to 8% and at the same time, manufacturing output increased 14-fold. At the end of the nineteenth century, Britain might have been excused for believing that she, more than any other nation, had achieved the industrial breakthrough and that a land of milk and honey was not far distant.

Table 28.1. Percentage of workers employed in primary, secondary and tertiary industries in Great Britain

Primary industry	9
Secondary industry	32
Tertiary industry	59

Even by this time, however, there were ominous signs that other nations were overhauling Britain. For example, whilst in the period 1876–1914 the industrial expansion in Britain was less than 2% per annum, in the United States it was 4·8% and in Germany 3·9%. Also in this period Britain's share of the export market fell from 38% to 27%. There were several reasons for this; for example, increasing competition as other nations became more industrialized, and also the fact that large amounts of capital were tied up in declining industries, such as cotton and coal. No matter what the complexion of the Government, this country has conspicuously failed to divert sufficient resources to the more promising new industries such as cars, electrical goods and chemicals. Inappropriate education has also meant in-efficiency at all levels of industry, whilst the industrial relations of the nineteenth century have left their mark in distrust between management and worker, in restrictive practices and in the insularity of various Trades Unions. The figures given for the chemical industry in *Table 28.2* are typical of industry generally and pinpoint the difficulties and problems that this country is facing.

Table 28.2. Number of British workers in the chemical industry required to do the work of one worker of country listed

U.S.	Canada	Italy	Sweden	Germany	France	Neths.	Spain
3·38	2·79	1·36	1·33	1·29	1·14	0·79	0·49

The Structure of Chemical Industry

Rationalization of industry, as economists are pleased to call it, to meet changing demand and increased competition in many cases requires the formation of larger from smaller organizations, and in recent years, take-over bids have been a common occurrence in this country. Provided they do not lead to the setting-up of monopolies and the elimination of home competition (and the Monopolies Commission exists to assess matters such as these) the Government looks kindly upon mergers and in some cases the Industrial Reorganization Corporation actively encourages amalgamation, so that firms generally are becoming larger and larger (*Table 28·3*). One

Table 28.3. The Twenty Largest Industrial Firms in the World in 1963

Company	Country	Annual Sales (10³ dollars)	Employees
General Motors	U.S.	16 494 818	640 073
Standard Oil	U.S.	10 264 343	147 000
Ford Motor	U.S.	8 742 506	316 568
Royal Dutch Shell	Neths/Britain	6 521 292	225 000
General Electric	U.S.	4 918 716	262 882
Socony Mobil Oil	U.S.	4 352 119	79 700
Unilever	Neths/Britain	4 297 384	290 000
U.S. Steel	U.S.	3 599 256	187 721
Chrysler	U.S.	3 505 275	120 447
Texaco	U.S.	3 415 746	55 040
Gulf Oil	U.S.	2 977 900	53 200
W. Electric	U.S.	2 832 988	147 210
Du Pont	U.S.	2 584 593	100 468
N.C.B.	Britain	2 520 000	550 900
Swift	U.S.	2 473 450	52 400
Standard Oil (Ind.)	U.S.	2 226 853	38 334
Standard Oil (California)	U.S.	2 202 512	43 764
British Petroleum	Britain	2 171 680	60 000
Shell Oil	U.S.	2 128 637	32 191
Westinghouse Electric	U.S.	2 127 307	115 170

NOTE 1. The dominance by the U.S. of large companies
     2. The importance of chemicals, especially oil
     3. The efficiency of oil firms in terms of output/employee

consequence of this is that the base of the pyramidal structure of the industry becomes broader and workers become more remote from decision-making machinery (see *Figure 28.1*). This can cause all sorts of social and psychological difficulties but a discussion of them is beyond the scope of this book. What concerns us more is how the scientist, and particularly the chemist, fits into the modern industrial organization.

One fairly obvious job of the chemist concerns the analysis of raw materials and final products, to ensure that the quality comes within the required limits, as well as the control of the chemical processes within the industry. For example, in a firm manufacturing plastics, it may be necessary for the chemist to carry out certain tests at appropriate times in order to discover how far the polymerization process has got. All these operations of the chemist would come within the sphere of '*Production*'.

If industry was completely static, the above tasks would presumably represent the sum total of the chemist's contribution to industry. A chemical firm must take account, however, of new relevant knowledge, of the customer's addiction to novelty

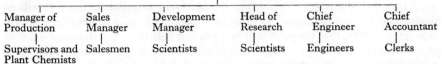

Figure 28.1a. The top of the industrial pyramid

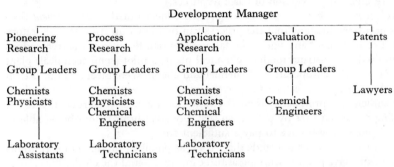

Figure 28.1b. A detail of part of the industrial organization

and to competition from other firms. It is therefore a necessity for there to be either improvements in existing processes or the discovery of new ones. These improvements are initially the responsibility of the *Research Department* and the chemists working there ideally bring both a large body of relevant chemical theory and a specialized knowledge of the industry to bear in the solution of its problems. Problems are not always readily identifiable, however, and there is the difficulty of trying to decide how much 'pure' research is justified in the hope that there will be some 'fall-out' of value to the firm. The entire problem of the financing of Research and Development is beset with difficulties, whether at Government or company level. Britain is, in fact, highly-placed in the league for the amount spent on research (*Table 28.4*) but, despite the fact that the expansion of the chemical industry is a little higher than that of industry as a whole, it can be argued that we get poor value for money.

Table 28.4. Amount spent on research and development (as percentage of GNP)

U.S.A.	3·2
U.K.	2·8
Sweden	1·6
Japan	1·5
Netherlands	1·4
W. Germany	1·3
France	1·2
Canada	0·7
India	0·2

535

Once the Research Department has made a breakthrough, however, the Board need have fewer qualms in providing the necessary financial backing. It will no doubt need some reassurance that there is a market for the product that will subsequently emerge, that the price will be right and that there is a sufficient quantity of raw material available for large-scale production in the foreseeable future. (It is partly because of the limits imposed on the chemical industry by the supply of raw material that so much research in industry is devoted to discovering new catalysts for the more-efficient conversion of basic resources.)

The project now becomes more and more the concern of the *Development* Section. It is necessary to show that the process can be blown up from the laboratory to the plant scale without any significant fall in potential. Because of the tremendous capital required to construct modern chemical plant, a pilot (semi-technical) plant is set up through the co-operation of chemist and chemical engineer and this is then *evaluated* thoroughly to see how far performance matches promise. If the indications are sufficiently promising, then the full-scale plant is constructed, but not before being protected first by patent, so that any rival firm, before being able to emulate the pioneer, will have to pay a sufficient sum.

Efficient use of resources demands that, as far as possible, production is continuous, so that expensive plant is working at full capacity around the clock. This aspect needs to be taken account of by those who design the chemical plant, particularly in view of the possibility of *automation*. Fluids are clearly to be preferred to solids with regard to continuous flow processes: the relative efficiency of the oil industry compared with coal, in terms of labour and output, is clear from *Table 28.3*. Now there is little point in installing expensive computer equipment if the data being processed is crude—working hand in hand with increased automation, therefore, is improved instrumentation. The information thus available passes to the control unit, which sends appropriate 'commands' to both the store and arithmetic unit: relevant data is supplied to the latter unit from the store and the subsequent results 'fed back' to the plant (*see Figure 28.2*). If this is fully automated, the continuous feedback results in appropriate action being taken mechanically.

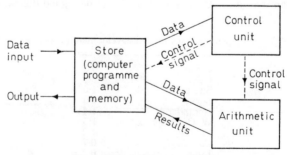

Figure 28.2 A computer circuit

Reference has already been made to the necessity for the chemical industry to exploit as fully as possible available natural resources. This has been shown earlier in the book with regard to the petroleum industry; as the market has changed in its demands (for example, from lamp-oil to petrol to aviation fuel) so the chemist has had to exercise his ingenuity to ensure that as far as possible all the products of the refinery find a commercial outlet and the processes such as 'cracking' and 'reforming' have led to the building up of a considerable fund of technical knowledge. Since industry is there to create wealth, waste is abhorrent and so the modern tendency is for complexes to develop where by-products can be isolated and utilized: Imperial Smelting Corporation's new £15 million complex at Avonmouth is a good example of this (*Figure 28.3*).

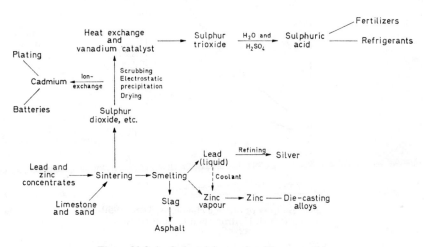

Figure 28.3 An industrial complex (Avonmouth)

Not only can the market change, however, so also can the supply of raw material. In recent years, for example the Gas Industry has increasingly taken advantage of the availability of petroleum products (*Figure 28.4*) and has supplemented coal-gas with naphtha. Because the latter is too 'rich' and gives too hot a flame, it was necessary to 'reform' it (that, is dilute it) by reacting it with steam in the presence of catalysts—the result was a larger volume of gas of lower calorific value, comparable to that of coal-gas, so that conventional burners could be used. The recent discovery of large quantities of natural gas under the North Sea has completely transformed the position, so that present policy is to use the 'rich' gas as such and convert appliances, so that they will burn it efficiently. The chemical industry is thus, like all industries, subject to new pressures, and how it adapts can have a vital bearing on its future. In this country, it is the most rapidly expanding part of

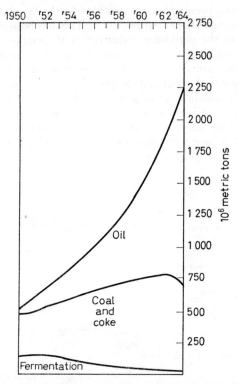

Figure 28.4. Sources of organic chemicals, U.K. 1950–1964
(from British Hydrocarbon Chemicals Ltd.)

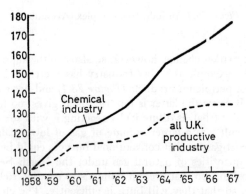

Figure 28.5 Chemicals production
Index of production. 1958 = 100. Seasonally adjusted

538

our economy (see *Figure 28.5*) and just how quickly our standard of living rises depends to a large extent on the skill of the chemist in developing new processes and exploiting changed conditions.

QUESTIONS

1. The terms 'eotechnic' and 'palacotechnic' have been explained in this chapter. What do you think 'neotechnic' means? Illustrate your answer by tracing the development of a particular industry or culture in terms of these sequences.

2. How far do you think it justified to call today the 'Age of Catalysis'?

3. List the basic raw materials for the Chemical Industry. Indicate which are fixed in terms of total availability and how they can be conserved. What substitutes for them may become available?

4. Britain has the highest expenditure of any European country on research and development, in terms of proportion of gross national product, and yet she has the lowest industrial growth. Explain this and suggest ways of improving the situation.

APPENDIX 1

CONSTANTS AND CONVERSION FACTORS

Atomic Weights

Atomic weights were originally based on a chemical standard—the natural mixture of the isotopes of hydrogen (i.e. $H = 1 \cdot 0000$). Since the introduction of accurate mass spectroscopy, physical standards have been adopted; until recently, the most abundant isotope of oxygen has been used as the standard (i.e. $^{16}O = 16 \cdot 0000$), but now the value $^{12}C = 12 \cdot 0000$ is employed.

The SI Units

SI (Système International) units have been used in this book where possible, but difficulty is met with in dealing with molar concentrations previously expressed in mol(e)/litre. Concentration, according to SI nomenclature, should be molal, i.e. mol kg^{-1}, although little difference will exist between mol dm^{-3} and mol kg^{-1} for dilute aqueous solutions and the symbol M for such solutions will be retained.

The SI fails to recognize g-molecule, g-equivalent, etc; instead mole is used for such quantities, where one mole is the amount of substance which contains as many elementary particles as there are atoms in $0 \cdot 012$ kilogramme of carbon-12. The elementary unit must be specified and may be an atom, molecule, ion, radical electron, photon, etc., or a specified group of such entities. Thus one mole of $\frac{1}{2}H_2SO_4$ is the same as one equivalent of H_2SO_4 and one mole of SO_4^{2-} is used in place of one g-ion of SO_4^{2-}.

Prefixes for SI units

Fraction	Prefix	Symbol	Multiple	Prefix	Symbol
10^{-1}	deci	d			
10^{-2}	centi	c			
10^{-3}	milli	m	10^3	kilo	k
10^{-6}	micro	μ	10^6	mega	M
10^{-9}	nano	n			
10^{-12}	pico	p			

Definitions of Units

Ampere. The ampere is that constant current which, if maintained in two straight parallel conductors of infinite length, of negligible circular cross section, and placed 1 metre apart in a vacuum, would produce between these conductors a force equal to 2×10^{-7} newton per metre of length. Symbol A

Newton. The newton is that force which, when applied to a body of mass 1 kilogramme, gives it an acceleration of one metre per second squared. Symbol $N = kg\ m\ s^{-2}$

Joule. The joule is the unit of energy and is the work done when the point of application of a force of 1 newton is displaced through a distance of 1 metre in the direction of the force. Symbol $J = kg\ m^2\ s^{-2}$

Coulomb. The coulomb is the unit of electric charge and is the quantity of electricity transported in 1 second by a current of 1 ampere. Symbol $C = A\ s$

Kelvin. The units of Kelvin and Celsius temperature interval are identical. A temperature expressed in degrees Celsius is equal to the temperature expressed in degrees Kelvin less 273·15, i.e. $K = °C + 273·15$

Ohm. The unit of electric resistance is the ohm and is the resistance between two points of a conductor when a constant difference of 1 volt, applied between these two points, produces in the conductor a current of one ampere, provided that the conductor itself is not a source of any electromotive force. Symbol $\Omega = V\ A^{-1}$

Volt. The volt is the unit of electric potential and is the difference of potential between two points across which a current of 1 ampere flows, when the power dissipated between these points is equal to 1 watt. Symbol $V = W\ A^{-1}$

Watt. The unit of power is the watt and is equal to 1 joule per second. Symbol $W = J\ s^{-1}$

Metre. The metre is the unit of length, equal to 1 650 763·73 wavelengths in vacuum of the radiation corresponding to the transition between the levels $2p_{10}$ and $5d_5$ of the krypton-86 atom. Symbol m

Kilogramme. The unit of mass is the kilogramme and is equal to the mass of the international prototype of the kilogramme made of platinum-iridium kept at Sèvres. Symbol kg

Second. The second is the duration of 9 192 631 770 periods of the radiation corresponding to the transition between the two hyperfine levels of the ground state of the caesium-133 atom. Symbol s

Conversion factors
Angstrom unit, $Å = 10^{-10}\ m = 10^{-1}\ nm = 10^2\ pm$
Micron unit, $\mu = 10^{-6}\ m = 1\ \mu m$
Calorie, $cal = 4·184\ J$
Litre, $l = 1\ dm^3$
Atmosphere, $atm = 760\ mmHg = 101\ 325\ N\ m^{-2} \approx 10^5\ N\ m^{-2}$
Logarithm, $ln\ (\log_e) = 2·3 \times \log_{10}$

Values of Constants

Speed of light in vacuum,	$c = 3 \times 10^8 \text{ m s}^{-1}$
Electronic charge,	$e = 1 \cdot 6 \times 10^{-19} \text{C}$
Avogadro constant,	$L = 6 \cdot 023 \times 10^{23} \text{ mol}^{-1}$
Boltzmann constant,	$k = 1 \cdot 38 \times 10^{-23} \text{ J K}^{-1}$
Universal gas constant,	$R = Lk = 8 \cdot 31 \text{ J K}^{-1} \text{ mol}^{-1}$
Faraday constant,	$F = Le = 9 \cdot 65 \times 10^4 \text{ C mol}^{-1}$

APPENDIX 2

TABLE OF ATOMIC WEIGHTS AND NUMBERS

(By courtesy of the International Union of Pure and Applied Chemistry)

	Symbol	Atomic Number	Atomic Weight		Symbol	Atomic Number	Atomic Weight
Actinium	Ac	89		Mendelevium	Md	101	
Aluminium	Al	13	26.98	Mercury	Hg	80	200.61
Americium	Am	95		Molybdenum	Mo	42	95.95
Antimony	Sb	51	121.76	Neodymium	Nd	60	144.27
Argon	Ar	18	39.944	Neon	Ne	10	20.183
Arsenic	As	33	74.91	Neptunium	Np	93	
Astatine	At	85		Nickel	Ni	28	58.71
Barium	Ba	56	137.36	Niobium	Nb	41	92.91
Berkelium	Bk	97		Nitrogen	N	7	14.008
Beryllium	Be	4	9.013	Nobelium	No	102	
Bismuth	Bi	83	209.00	Osmium	Os	76	190.2
Boron	B	5	10.82	Oxygen	O	8	16
Bromine	Br	35	79.916	Palladium	Pd	46	106.4
Cadmium	Cd	48	112.41	Phosphorus	P	15	30.975
Calcium	Ca	20	40.08	Platinum	Pt	78	195.09
Californium	Cf	98		Plutonium	Pu	94	
Carbon*	C	6	12.011	Polonium	Po	84	
Cerium	Ce	58	140.13	Potassium	K	19	39.100
Caesium	Cs	55	132.91	Praseodymium	Pr	59	140.92
Chlorine	Cl	17	35.457	Promethium	Pm	61	
Chromium	Cr	24	52.01	Protactinium	Pa	91	
Cobalt	Co	27	58.94	Radium	Ra	88	
Copper	Cu	29	63.54	Radon	Rn	86	
Curium	Cm	96		Rhenium	Re	75	186.22
Dysprosium	Dy	66	162.51	Rhodium	Rh	45	102.91
Einsteinium	Es	99		Rubidium	Rb	37	85.48
Erbium	Er	68	167.27	Ruthenium	Ru	44	101.1
Europium	Eu	63	152.0	Samarium	Sm	62	150.35
Fermium	Fm	00		Scandium	Sc	21	44.96
Fluorine	F	9	19.00	Selenium	Se	34	78.96
Francium	Fr	87		Silicon	Si	14	28.09
Gadolinium	Gd	64	157.26	Silver	Ag	47	107.880
Gallium	Ga	31	69.72	Sodium	Na	11	22.991
Germanium	Ge	32	72.60	Strontium	Sr	38	87.63
Gold	Au	79	197.0	Sulphur	S	16	32.066†
Hafnium	Hf	72	178.50	Tantalum	Ta	73	180.95
Helium	He	2	4.003	Technetium	Tc	43	
Holmium	Ho	67	164.94	Tellurium	Te	52	127.61
Hydrogen	H	1	1.0080	Terbium	Tb	65	158.93
Indium	In	49	114.82	Thallium	Tl	81	204.39
Iodine	I	53	126.91	Thorium	Th	90	232.05
Iridium	Ir	77	192.2	Thulium	Tm	69	168.94
Iron	Fe	26	55.85	Tin	Sn	50	118.70
Krypton	Kr	36	83.80	Titanium	Ti	22	47.90
Kurchatovium		104		Tungsten	W	74	183.86
Lanthanum	La	57	138.92	Uranium	U	92	238.07
Lawrencium	Lr	103		Vanadium	V	23	50.95
Lead	Pb	82	207.21	Xenon	Xe	54	131.30
Lithium	Li	3	6.940	Ytterbium	Yb	70	173.04
Lutetium	Lu	71	174.99	Yttrium	Y	39	88.92
Magnesium	Mg	12	24.32	Zinc	Zn	30	65.38
Manganese	Mn	25	54.94	Zirconium	Zr	40	91.22

* The basis for atomic weights is $^{12}C = 12$.
† Because of natural variations in the relative abundance of the isotopes of sulphur the atomic weight of this element has a range of ± 0.003.

APPENDIX 3

LOGARITHM AND ANTILOGARITHM TABLES

LOGARITHMS

	0	1	2	3	4	5	6	7	8	9	1	2	3	4	5	6	7	8	9
10	0000	0043	0086	0128	0170	0212	0253	0294	0334	0374	4	9	13	17	21	26	30	34	38
											4	8	12	16	20	24	28	32	37
11	0414	0453	0492	0531	0569	0607	0645	0682	0719	0755	4	8	12	15	19	23	27	31	35
											4	7	11	15	19	22	26	30	33
12	0792	0828	0864	0899	0934	0969	1004	1038	1072	1106	3	7	11	14	18	21	25	28	32
											3	7	10	14	17	20	24	27	31
13	1139	1173	1206	1239	1271	1303	1335	1367	1399	1430	3	7	10	13	16	20	23	26	30
											3	7	10	12	16	19	22	25	29
14	1461	1492	1523	1553	1584	1614	1644	1673	1703	1732	3	6	9	12	15	18	21	24	28
											3	6	9	12	15	17	20	23	26
15	1761	1790	1818	1847	1875	1903	1931	1959	1987	2014	3	6	9	11	14	17	20	23	26
											3	5	8	11	14	16	19	22	25
16	2041	2068	2095	2122	2148	2175	2201	2227	2253	2279	3	5	8	11	14	16	19	22	24
											3	5	8	10	13	15	18	21	23
17	2304	2330	2355	2380	2405	2430	2455	2480	2504	2529	3	5	8	10	13	15	18	20	23
											2	5	7	10	12	15	17	19	22
18	2553	2577	2601	2625	2648	2672	2695	2718	2742	2765	2	5	7	9	12	14	16	19	21
											2	5	7	9	11	14	16	18	21
19	2788	2810	2833	2856	2878	2900	2923	2945	2967	2989	2	4	7	9	11	13	16	18	20
											2	4	6	8	11	13	15	17	19
20	3010	3032	3054	3075	3096	3118	3139	3160	3181	3201	2	4	6	8	11	13	15	17	19
21	3222	3243	3263	3284	3304	3324	3345	3365	3385	3404	2	4	6	8	10	12	14	16	18
22	3424	3444	3464	3483	3502	3522	3541	3560	3579	3598	2	4	6	8	10	12	14	15	17
23	3617	3636	3655	3674	3692	3711	3729	3747	3766	3784	2	4	6	7	9	11	13	15	17
24	3802	3820	3838	3856	3874	3892	3909	3927	3945	3962	2	4	5	7	9	11	12	14	16
25	3979	3997	4014	4031	4048	4065	4082	4099	4116	4133	2	3	5	7	9	10	12	14	15
26	4150	4166	4183	4200	4216	4232	4249	4265	4281	4298	2	3	5	7	8	10	11	13	15
27	4314	4330	4346	4362	4378	4393	4409	4425	4440	4456	2	3	5	6	8	9	11	13	14
28	4472	4487	4502	4518	4533	4548	4564	4579	4594	4609	2	3	5	6	8	9	11	12	14
29	4624	4639	4654	4669	4683	4698	4713	4728	4742	4757	1	3	4	6	7	9	10	12	13
30	4771	4786	4800	4814	4829	4843	4857	4871	4886	4900	1	3	4	6	7	9	10	11	13
31	4914	4928	4942	4955	4969	4983	4997	5011	5024	5038	1	3	4	6	7	8	10	11	12
32	5051	5065	5079	5092	5105	5119	5132	5145	5159	5172	1	3	4	5	7	8	9	11	12
33	5185	5198	5211	5224	5237	5250	5263	5276	5289	5302	1	3	4	5	6	8	9	10	12
34	5315	5328	5340	5353	5366	5378	5391	5403	5416	5428	1	3	4	5	6	8	9	10	11
35	5441	5453	5465	5478	5490	5502	5514	5527	5539	5551	1	2	4	5	6	7	9	10	11
36	5563	5575	5587	5599	5611	5623	5635	5647	5658	5670	1	2	4	5	6	7	8	10	11
37	5682	5694	5705	5717	5729	5740	5752	5763	5775	5786	1	2	3	5	6	7	8	9	10
38	5798	5809	5821	5832	5843	5855	5866	5877	5888	5899	1	2	3	5	6	7	8	9	10
39	5911	5922	5933	5944	5955	5966	5977	5988	5999	6010	1	2	3	4	5	7	8	9	10
40	6021	6031	6042	6053	6064	6075	6085	6096	6107	6117	1	2	3	4	5	6	8	9	10
41	6128	6138	6149	6160	6170	6180	6191	6201	6212	6222	1	2	3	4	5	6	7	8	9
42	6232	6243	6253	6263	6274	6284	6294	6304	6314	6325	1	2	3	4	5	6	7	8	9
43	6335	6345	6355	6365	6375	6385	6395	6405	6415	6425	1	2	3	4	5	6	7	8	9
44	6435	6444	6454	6464	6474	6484	6493	6503	6513	6522	1	2	3	4	5	6	7	8	9
45	6532	6542	6551	6561	6571	6580	6590	6599	6609	6618	1	2	3	4	5	6	7	8	9
46	6628	6637	6646	6656	6665	6675	6684	6693	6702	6712	1	2	3	4	5	6	7	7	8
47	6721	6730	6739	6749	6758	6767	6776	6785	6794	6803	1	2	3	4	5	5	6	7	8
48	6812	6821	6830	6839	6848	6857	6866	6875	6884	6893	1	2	3	4	4	5	6	7	8
49	6902	6911	6920	6928	6937	6946	6955	6964	6972	6981	1	2	3	4	4	5	6	7	8

LOGARITHMS

	0	1	2	3	4	5	6	7	8	9	1	2	3	4	5	6	7	8	9
50	6990	6998	7007	7016	7024	7033	7042	7050	7059	7067	1	2	3	3	4	5	6	7	8
51	7076	7084	7093	7101	7110	7118	7126	7135	7143	7152	1	2	3	3	4	5	6	7	8
52	7160	7168	7177	7185	7193	7202	7210	7218	7226	7235	1	2	2	3	4	5	6	7	7
53	7243	7251	7259	7267	7275	7284	7292	7300	7308	7316	1	2	2	3	4	5	6	6	7
54	7324	7332	7340	7348	7356	7364	7372	7380	7388	7396	1	2	2	3	4	5	6	6	7
55	7404	7412	7419	7427	7435	7443	7451	7459	7466	7474	1	2	2	3	4	5	5	6	7
56	7482	7490	7497	7505	7513	7520	7528	7536	7543	7551	1	2	2	3	4	5	5	6	7
57	7559	7566	7574	7582	7589	7597	7604	7612	7619	7627	1	2	2	3	4	5	5	6	7
58	7634	7642	7649	7657	7664	7672	7679	7686	7694	7701	1	1	2	3	4	4	5	6	7
59	7709	7716	7723	7731	7738	7745	7752	7760	7767	7774	1	1	2	3	4	4	5	6	7
60	7782	7789	7796	7803	7810	7818	7825	7832	7839	7846	1	1	2	3	4	4	5	6	6
61	7853	7860	7868	7875	7882	7889	7896	7093	7910	7917	1	1	2	3	4	4	5	6	6
62	7924	7931	7938	7945	7952	7959	7966	7973	7980	7987	1	1	2	3	3	4	5	6	6
63	7993	8000	8007	8014	8021	8028	8035	8041	8048	8055	1	1	2	3	3	4	5	5	6
64	8062	8069	8075	8082	8089	8096	8102	8109	8116	8122	1	1	2	3	3	4	5	5	6
65	8129	8136	8142	8149	8156	8162	8169	8176	8182	8189	1	1	2	3	3	4	5	5	6
66	8195	8202	8209	8215	8222	8228	8235	8241	8248	8254	1	1	2	3	3	4	5	5	6
67	8261	8267	8274	8280	8287	8293	8299	8306	8312	8319	1	1	2	3	3	4	5	5	6
68	8325	8331	8338	8344	8351	8357	8363	8370	8376	8382	1	1	2	3	3	4	4	5	6
69	8388	8395	8401	8407	8414	8420	8426	8432	8439	8445	1	1	2	2	3	4	4	5	6
70	8451	8457	8463	8470	8476	8482	8488	8494	8500	8506	1	1	2	2	3	4	4	5	6
71	8513	8519	8525	8531	8537	8543	8549	8555	8561	8567	1	1	2	2	3	4	4	5	5
72	8573	8579	8585	8591	8597	8603	8609	8615	8621	8627	1	1	2	2	3	4	4	5	5
73	8633	8639	8645	8651	8657	8663	8669	8675	8681	8686	1	1	2	2	3	4	4	5	5
74	8692	8698	8704	8710	8716	8722	8727	8733	8739	8745	1	1	2	2	3	4	4	5	5
75	8751	8756	8762	8768	8774	8779	8785	8791	8797	8802	1	1	2	2	3	3	4	5	5
76	8808	8814	8820	8825	8831	8837	8842	8848	8854	8859	1	1	2	2	3	3	4	5	5
77	8865	8871	8876	8882	8887	8893	8899	8904	8910	8915	1	1	2	2	3	3	4	4	5
78	8921	8927	8932	8938	8943	8949	8954	8960	8965	8971	1	1	2	2	3	3	4	4	5
79	8976	8982	8987	8993	8998	9004	9009	9015	9020	9025	1	1	2	2	3	3	4	4	5
80	9031	9036	9042	9047	9053	9058	9063	9069	9074	9079	1	1	2	2	3	3	4	4	5
81	9085	9090	9096	9101	9106	9112	9117	9122	9128	9133	1	1	2	2	3	3	4	4	5
82	9138	9143	9149	9154	9159	9165	9170	9175	9180	9186	1	1	2	2	3	3	4	4	5
83	9191	9196	9201	9206	9212	9217	9222	9227	9232	9238	1	1	2	2	3	3	4	4	5
84	9243	9248	9253	9258	9263	9269	9274	9279	9284	9289	1	1	2	2	3	3	4	4	5
85	9294	9299	9304	9309	9315	9320	9325	9330	9335	9340	1	1	2	2	3	3	4	4	5
86	9345	9350	9355	9360	9365	9370	9375	9380	9385	9390	1	1	2	2	3	3	4	4	5
87	9395	9400	9405	9410	9415	9420	9425	9430	9435	9440	0	1	1	2	2	3	3	4	4
88	9445	9450	9455	9460	9465	9469	9474	9479	9484	9489	0	1	1	2	2	3	3	4	4
89	9494	9499	9504	9509	9513	9518	9523	9428	9533	9538	0	1	1	2	2	3	3	4	4
90	9542	9547	9552	9557	9562	9566	9571	9576	9581	9586	0	1	1	2	2	3	3	4	4
91	9590	9595	9600	9605	9609	9614	9619	9624	9628	9633	0	1	1	2	2	3	3	4	4
92	9638	9643	9647	9652	9657	9661	9666	9671	9675	9680	0	1	1	2	2	3	3	4	4
93	9685	9689	9694	9699	9703	9708	9713	9717	9722	9727	0	1	1	2	2	3	3	4	4
94	9731	9736	9741	9745	9750	9754	9759	9763	9768	9773	0	1	1	2	2	3	3	4	4
95	9777	9782	9786	9791	9795	9800	9805	9809	9814	9818	0	1	1	2	2	3	3	4	4
96	9823	9827	9832	9836	9841	9845	9850	9854	9859	9863	0	1	1	2	2	3	3	4	4
97	9868	9872	9877	9881	9886	9890	9894	9899	9903	9908	0	1	1	2	2	3	3	4	4
98	9912	9917	9921	9926	9930	9934	9939	9943	9948	9952	0	1	1	2	2	3	3	4	4
99	9956	9961	9965	9969	9974	9978	9983	9987	9991	9996	0	1	1	2	2	3	3	3	4

545

ANTILOGARITHMS

	0	1	2	3	4	5	6	7	8	9	1	2	3	4	5	6	7	8	9
·00	1000	1002	1005	1007	1009	1012	1014	1016	1019	1021	0	0	1	1	1	1	2	2	2
·01	1023	1026	1028	1030	1033	1035	1038	1040	1042	1045	0	0	1	1	1	1	2	2	2
·02	1047	1050	1052	1054	1057	1059	1062	1064	1067	1069	0	0	1	1	1	1	2	2	2
·03	1072	1074	1076	1079	1081	1084	1086	1089	1091	1094	0	0	1	1	1	1	2	2	2
·04	1096	1099	1102	1104	1107	1109	1112	1114	1117	1119	0	1	1	1	1	2	2	2	2
·05	1122	1125	1127	1130	1132	1135	1138	1140	1143	1146	0	1	1	1	1	2	2	2	2
·06	1148	1151	1153	1156	1159	1161	1164	1167	1169	1172	0	1	1	1	1	2	2	2	2
·07	1175	1178	1180	1183	1186	1189	1191	1194	1197	1199	0	1	1	1	1	2	2	2	2
·08	1202	1205	1208	1211	1213	1216	1219	1222	1225	1227	0	1	1	1	1	2	2	2	3
·09	1230	1233	1236	1239	1242	1245	1247	1250	1253	1256	0	1	1	1	1	2	2	2	3
·10	1259	1262	1265	1268	1271	1274	1276	1279	1282	1285	0	1	1	1	1	2	2	2	3
·11	1288	1291	1294	1297	1300	1303	1306	1309	1312	1315	0	1	1	1	2	2	2	2	3
·12	1318	1321	1324	1327	1330	1334	1337	1340	1343	1346	0	1	1	1	2	2	2	3	3
·13	1349	1352	1355	1358	1361	1365	1368	1371	1374	1377	0	1	1	1	2	2	2	3	3
·14	1380	1384	1387	1390	1393	1396	1400	1403	1406	1409	0	1	1	1	2	2	2	3	3
·15	1413	1416	1419	1422	1426	1429	1432	1435	1439	1442	0	1	1	1	2	2	2	3	3
·16	1445	1449	1452	1455	1459	1462	1466	1469	1472	1476	0	1	1	1	2	2	2	3	3
·17	1479	1483	1486	1489	1493	1496	1500	1503	1507	1510	0	1	1	1	2	2	2	3	3
·18	1514	1517	1521	1524	1528	1531	1535	1538	1542	1545	0	1	1	1	2	2	2	3	3
·19	1549	1552	1556	1560	1563	1567	1570	1574	1578	1581	0	1	1	1	2	2	3	3	3
·20	1585	1589	1592	1596	1600	1603	1607	1611	1614	1618	0	1	1	1	2	2	3	3	3
·21	1622	1626	1629	1633	1637	1641	1644	1648	1652	1656	0	1	1	2	2	2	3	3	3
·22	1660	1663	1667	1671	1675	1679	1683	1687	1690	1694	0	1	1	2	2	2	3	3	3
·23	1698	1702	1706	1710	1714	1718	1722	1726	1730	1734	0	1	1	2	2	2	3	3	4
·24	1738	1742	1746	1750	1754	1758	1762	1766	1770	1774	0	1	1	2	2	2	3	3	4
·25	1778	1782	1786	1791	1795	1799	1803	1807	1811	1816	0	1	1	2	2	2	3	3	4
·26	1820	1824	1828	1832	1837	1841	1845	1849	1854	1858	0	1	1	2	2	3	3	3	4
·27	1862	1866	1871	1875	1879	1884	1888	1892	1897	1901	0	1	1	2	2	3	3	4	4
·28	1905	1910	1914	1919	1923	1928	1932	1936	1941	1945	0	1	1	2	2	3	3	4	4
·29	1950	1954	1959	1963	1968	1972	1977	1982	1986	1991	0	1	1	2	2	3	3	4	4
·30	1995	2000	2004	2009	2014	2018	2023	2028	2032	2037	0	1	1	2	2	3	3	4	4
·31	2042	2046	2051	2056	2061	2065	2070	2075	2080	2084	0	1	1	2	2	3	3	4	4
·32	2089	2094	2099	2104	2109	2113	2118	2123	2128	2133	0	1	1	2	2	3	3	4	4
·33	2138	2143	2148	2153	2158	2163	2168	2173	2178	2183	0	1	1	2	2	3	3	4	4
·34	2188	2193	2198	2203	2208	2213	2218	2223	2228	2234	1	1	2	2	3	3	4	4	5
·35	2239	2244	2249	2254	2259	2265	2270	2275	2280	2286	1	1	2	2	3	3	4	4	5
·36	2291	2296	2301	2307	2312	2317	2323	2328	2333	2339	1	1	2	2	3	3	4	4	5
·37	2344	2350	2355	2360	2366	2371	2377	2382	2388	2393	1	1	2	2	3	3	4	4	5
·38	2399	2404	2410	2415	2421	2427	2432	2438	2443	2449	1	1	2	2	3	3	4	4	5
·39	2455	2460	2466	2472	2477	2483	2489	2495	2500	2506	1	1	2	2	3	3	4	5	5
·40	2512	2518	2523	2529	2535	2541	2547	2553	2559	2564	1	1	2	2	3	4	4	5	5
·41	2570	2576	2582	2588	2594	2600	2606	2612	2618	2624	1	1	2	2	3	4	4	5	5
·42	2630	2636	2642	2649	2655	2661	2667	2673	2679	2685	1	1	2	2	3	4	4	5	6
·43	2692	2698	2704	2710	2716	2723	2729	2735	2742	2748	1	1	2	3	3	4	4	5	6
·44	2754	2761	2767	2773	2780	2786	2793	2799	2805	2812	1	1	2	3	3	4	4	5	6
·45	2818	2815	2831	2838	2844	2851	2858	2864	2871	2877	1	1	2	3	3	4	5	5	6
·46	2884	2891	2897	2904	2911	2917	2924	2931	2938	2944	1	1	2	3	3	4	5	5	6
·47	2951	2958	2965	2972	2979	2985	2992	2999	3006	3013	1	1	2	3	3	4	5	5	6
·48	3020	3027	3034	3041	3048	3055	3062	3069	3076	3083	1	1	2	3	4	4	5	6	6
·49	3090	3097	3105	3112	3119	3126	3133	3141	3148	3155	1	1	2	3	4	4	5	6	6

ANTILOGARITHMS

	0	1	2	3	4	5	6	7	8	9	1	2	3	4	5	6	7	8	9
·50	3162	3170	3177	3184	3192	3199	3206	3214	3221	3228	1	1	2	3	4	4	5	6	7
·51	3236	3243	3251	3258	3266	3273	3281	3289	3296	3304	1	2	2	3	4	5	5	6	7
·52	3311	3319	3327	3334	3342	3350	3357	3365	3373	3381	1	2	2	3	4	5	5	6	7
·53	3388	3396	3404	3412	3420	3428	3436	3443	3451	3459	1	2	2	3	4	5	6	6	7
·54	3467	3475	3483	3491	3499	3508	3516	3524	3532	3540	1	2	2	3	4	5	6	6	7
·55	3548	3556	3565	3573	3581	3589	3597	3606	3614	3622	1	2	2	3	4	5	6	7	7
·56	3631	3639	3648	3656	3664	3673	3681	3690	3698	3707	1	2	3	3	4	5	6	7	8
·57	3715	3724	3733	3741	3750	3758	3767	3776	3784	3793	1	2	3	3	4	5	6	7	8
·58	3802	3811	3819	3828	3837	3846	3855	3864	3873	3882	1	2	3	4	4	5	6	7	8
·59	3890	3899	3908	3917	3926	3936	3945	3954	3963	3972	1	2	3	4	5	5	6	7	8
·60	3981	3990	3999	4009	4018	4027	4036	4046	4055	4064	1	2	3	4	5	6	6	7	8
·61	4074	4083	4093	4102	4111	4121	4130	4140	4150	4159	1	2	3	4	5	6	7	8	9
·62	4169	4178	4188	4198	4207	4217	4227	4236	4246	4256	1	2	3	4	5	6	7	8	9
·63	4266	4276	4285	4295	4305	4315	4325	4335	4345	4355	1	2	3	4	5	6	7	8	9
·64	4365	4375	4385	4395	4406	4416	4426	4436	4446	4457	1	2	3	4	5	6	7	8	9
·65	4467	4477	4487	4498	4508	4519	4529	4539	4550	4560	1	2	3	4	5	6	7	8	9
·66	4571	4581	4592	4603	4613	4624	4634	4645	4656	4667	1	2	3	4	5	6	7	9	10
·67	4677	4688	4699	4710	4721	4732	4742	4753	4764	4775	1	2	3	4	5	7	8	9	10
·68	4786	4797	4808	4819	4831	4842	4853	4864	4875	4887	1	2	3	4	6	7	8	9	10
·69	4898	4909	4920	4932	4943	4955	4966	4977	4989	5000	1	2	3	5	6	7	8	9	10
·70	5012	5023	5035	5047	5058	5070	5082	5093	5105	5117	1	2	3	4	6	7	8	9	11
·71	5129	5140	5152	5164	5176	5188	5200	5212	5224	5236	1	2	4	5	6	7	8	10	11
·72	5248	5260	5272	5284	5297	5309	5321	5333	5346	5358	1	2	4	5	6	7	9	10	11
·73	5370	5383	5395	5408	5420	5433	5445	5458	5470	5483	1	3	4	5	6	8	9	10	11
·74	5495	5508	5521	5534	5546	5559	5572	5585	5598	5610	1	3	4	5	6	8	9	10	12
·75	5623	5636	5649	5662	5675	5689	5702	5715	5728	5741	1	3	4	5	7	8	9	10	12
·76	5754	5768	5781	5794	5808	5821	5834	5848	5861	5875	1	3	4	5	7	8	9	11	12
·77	5888	5902	5916	5929	5943	5957	5970	5984	5998	6012	1	3	4	5	7	8	10	11	12
·78	6026	6039	6053	6067	6081	6095	6109	6124	6138	6152	1	3	4	6	7	8	10	11	13
·79	6166	6180	6194	6209	6223	6237	6252	6266	6281	6295	1	3	4	6	7	9	10	11	13
·80	6310	6324	6339	6353	6368	6383	6397	6412	6427	6442	1	3	4	6	7	9	10	12	13
·81	6457	6471	6486	6501	6516	6531	6546	6561	6577	6592	2	3	5	6	8	9	11	12	14
·82	6607	6622	6637	6653	6668	6683	6699	6714	6730	6745	2	3	5	6	8	9	11	12	14
·83	6761	6776	6792	6808	6823	6839	6855	6871	6887	6902	2	3	5	6	8	9	11	13	14
·84	6918	6934	6950	6966	6982	6998	7015	7031	7047	7063	2	3	5	6	8	10	11	13	15
·85	7079	7096	7112	7129	7145	7161	7178	7194	7211	7228	2	3	5	7	8	10	12	13	15
·86	7244	7261	7278	7295	7311	7328	7345	7362	7379	7396	2	3	5	7	8	10	12	13	15
·87	7413	7430	7447	7464	7482	7499	7516	7534	7551	7568	2	3	5	7	9	10	12	14	16
·88	7586	7603	7621	7638	7656	7674	7691	7709	7727	7745	2	4	5	7	9	11	12	14	16
·89	7762	7780	7798	7816	7834	7852	7870	7889	7907	7925	2	4	5	7	9	11	13	14	16
·90	7943	7962	7980	7998	8017	8035	8054	8072	8091	8110	2	4	6	7	9	11	13	15	17
·91	8128	8147	8166	8185	8204	8222	8241	8260	8279	8299	2	4	6	8	9	11	13	15	17
·92	8318	8337	8356	8375	8395	8414	8433	8453	8472	8792	2	4	6	8	10	12	14	15	17
·93	8511	8531	8551	8570	8590	8610	8630	8650	8670	8690	2	4	6	8	10	12	14	16	18
·94	8710	8730	8750	8770	8790	8810	8831	8851	8872	8892	2	4	6	8	10	12	14	16	18
·95	8913	8933	8954	8974	8995	9016	9036	9057	9078	9099	2	4	6	8	10	12	15	17	19
·96	9120	9141	9162	9183	9204	9226	9247	9268	9290	9311	2	4	6	8	11	13	15	17	19
·97	9333	9354	9376	9397	9419	9441	9462	9484	9506	9528	2	4	7	9	11	13	15	17	20
·98	9550	9572	9594	9616	9638	9661	9683	9705	9727	9750	2	4	7	9	11	13	16	18	20
·99	9772	9795	9817	9840	9863	9886	9908	9931	9954	9977	2	5	7	9	11	14	16	18	20

These logarithm and antilogarithm tables originally appeared in *Four Figure Tables and Constants for the Use of Students*, published by Her Majesty's Stationery Office. They are reproduced here by courtesy of the Controller of Her Majesty's Stationery Office.

The copyright of the part of the tables giving the logarithms of numbers from 1000 to 2000 is the property of Messrs. Macmillan & Co. Ltd., who have authorised their reprint for educational purposes.

APPENDIX 4

FOR FURTHER READING

The following books are recommended to those readers who would like to pursue certain topics in more detail than it has been possible to provide in a book of this size.

Addison, W. E. *Structural Principles in Inorganic Compounds*. London, Longmans Green, 1961

— *The Allotropy of the Elements*. London, Oldbourne, 1965

Basolo, F. and Johnson, R. *Coordination Chemistry*. New York, Benjamin, 1964

Denaro, A. R. *Elementary Electrochemistry*. London, Butterworths, 1965

Dewar, M. J. S. *An Introduction to Modern Chemistry*. London, Athlone Press, 1965

Ewing, G. W. *Instrumental Methods of Chemical Analysis*. New York and London, McGraw-Hill, 1960

Hargreaves, G. *Elementary Chemical Thermodynamics*. London, Butterworths, 1963

Moody, G. J. and Thomas, J. D. R. *Noble Gases and Their Compounds*. Oxford, Pergamon Press, 1964

Seaborg, G. T. *Man-made Transuranium Elements*. New York and London, Prentice-Hall, 1963

Sisler, H. H. *Chemistry in Non-Aqueous Solvents*. London, Chapman & Hall, 1961

Stevens, B. *Atomic Structure and Valency*. London, Chapman & Hall, 1964

— *Chemical Kinetics*. London, Chapman & Hall, 1965

Sykes, P. *A Guide Book to Mechanism in Organic Chemistry*. London, Longmans Green, 1961

APPENDIX 5

ANSWERS TO QUESTIONS

Chapter 1

3 48
4 238; 6
5 28·2

Chapter 2

1 54 per cent isotope 79
4 1 800 s
5 $19{\cdot}9 \times 10^8$ kJ mol^{-1}

Chapter 3

4 18 kJ mol^{-1}
7 + 6, − 1, + 7, + 5, + 4

Chapter 4

3 40
4 36·5
5 221

Chapter 5

2 1·5 g
6 71·2 per cent
7 (*a*) 30·8 g; (*b*) 34·6 g
8 67 per cent N_2, 33 per cent O_2
10 28·4 V

Chapter 6

1 + 122 kJ mol^{-1}
2 412 kJ
5 6·4; CH_3COOH = 3·2 g, C_2H_5OH = 8·6 g, $CH_3COOC_2H_5$ = 10·65 g, H_2O = 8·55 g
6 0·05; 5·6 per cent dissociation
7 $2{\cdot}5 \times 10^{-4}$ s^{-1}
11 10^2 kJ mol^{-1}
12 10^{17}

Chapter 7
  4 0.2; 1.25×10^{-3}
  8 122 kJ
  9 (*a*) 3.7 (*b*) 0.6
15 0.53

Chapter 13
  3 $K_4Fe(CN)_6$

Chapter 15
  9 (*i*) 1, 1, 2 (*ii*) 2, 2, 3

Chapter 19
  7 C_7H_{14}
  8 $CH_2:CH.CH:CH_2$

  9 Chief product

Chapter 20
  2 Benzyl chloride or a chlorotoluene
  4 $(CH_3)_2$ CHOH
  5 C_2H_5I

Chapter 21
  6 A = a nitrotoluene

Chapter 22
  4 3; glycerol
  5 A = propanol

  6 $\bigcirc$—CH_2—$\overset{\overset{\textstyle OH}{|}}{\underset{\underset{\textstyle CH_3}{|}}{C}}$—NH—$CH_3$

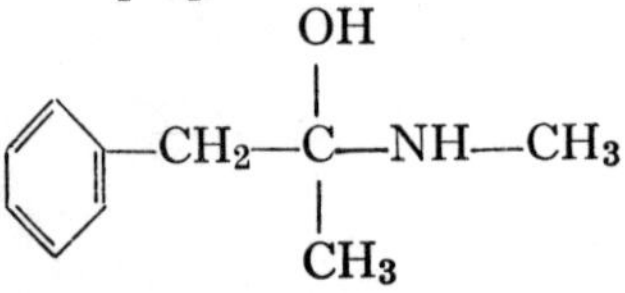

Chapter 23
  7 Mol. wt = 60

Chapter 24
  5 1, 3, 5-trimethylbenzene
  7 A = ethanoic (acetic) acid
  8 A = a methylaniline
  9 $C_6H_{11}O_5.OCH_2$—$\bigcirc$ with O—CH_2

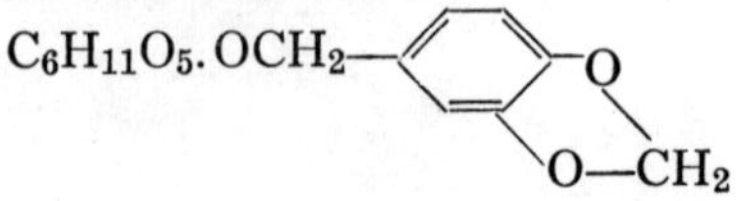

11 **E** = 2-methyl-1-phenylethane
  H = 3-hydroxypropanoic acid

Chapter 25

7

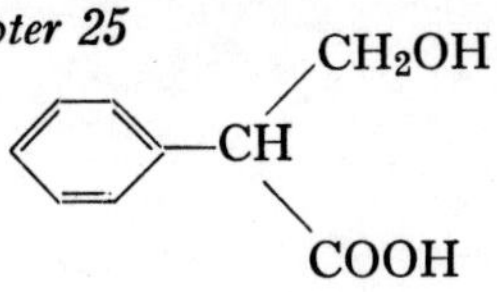

8 Methanoic (formic) acid

13 Acid 10 per cent dissociated in aqueous solution and fully associated in benzene

Chapter 26
  4 Saccharin
  5 *A* = 2-cyanoethanol
  7 *A* = cyanoethane

INDEX

INDEX